Kinetics and Mechanism of Ozone Reactions with Organic and Polymeric Compounds in Liquid Phase

KINETICS AND MECHANISM OF OZONE REACTIONS WITH ORGANIC AND POLYMERIC COMPOUNDS IN LIQUID PHASE

S. RAKOVSKY
AND
G. ZAIKOV

Nova Science Publishers, Inc.
Commack, New York

Editorial Production:	Susan Boriotti
Office Manager:	Annette Hellinger
Graphics:	Frank Grucci and John T'Lustachowski
Information Editor:	Tatiana Shohov
Book Production:	Donna Dennis, Patrick Davin, Christine Mathosian and Tammy Sauter
Circulation:	Maryanne Schmidt
Marketing/Sales:	Cathy DeGregory

Library of Congress Cataloging-in-Publication Data available upon request

ISBN 1-56072-642-3

6080 Jericho Turnpike, Suite 207
Commack, New York 11725
Tele. 516-499-3103 Fax 516-499-3146
e-mail: Novascience@earthlink.net
e-mail: Novascil@aol.com
Web Site: http://www.nexusworld.com/nova

Printed in the United States of America

Rakovsky devotes
this monograph
to his parents and family
with the greatest love and
deepest respect!

Contents

INTRODUCTION

The history of ozone chemistry as a research field began immediately after the discovery of ozone in 1840 by Schönbein [1]. Ozone is a poisonous and highly reactive gas - allotropic modification of ordinary atmospheric oxygen [2]. Upon studying the reaction of ozone with ethylene, which appears to be the first ozonolysis ever performed, it was shown that ozone attacks the double bonds breaking them. At the beginning of this century Harries showed that natural rubber reacts rapidly with ozone He used ozone as " a chemical scissors" to cut the rubber macromolecules and on the basis of the cut-off products structure and composition succeeded in evaluating the elastomer structure [4-7]. At present the classic mechanism of the reaction of ozone with isolated double bonds proposed by R.Criegee [8-11] is widely accepted. The development of novel instrumental methods for analysis opens great perspectives for intensive and expanded research on the reactions of ozone with various chemical compounds such as alkanes, arenes, oxygen-, nitrogen-, phosphor-, sulphur-containing compounds, fulerenes, elementorganic compounds, polymers, elastomers, metals, metal oxides, salts, halogens, carbon, nitrogen and sulphur oxides, actually with all anthropogenic contaminants, etc. [12-20]. As a result of the extensive investigations since 1958 to nowadays, a number of reviews and monographs describing and discussing the various aspects of ozone chemistry have been published [21-31. The intensive research on this topic is illustrated convincingly by the huge number of publications, more than 10 000 titles in patent and scientific literature, covering all aspects of ozone chemistry and physics, ozone preparation, application, storage, decomposition, etc. Many laboratories world-wide make investigations in the field of ozone and its behaviour. Fundamental contributions to the ozone reactions were made by various research groups in Russia, USA, Japan, France, Canada, Germany, Italy, Spain, China, India.

The study of the kinetics and mechanism of ozone reactions is an important field in the modern science closely related to the solution of the problem with "ozone holes', the development of physical-, organic-, inorganic-, polymer- and bio-chemistry with ozone participation, chemical kinetics, theory and utilisation of the reaction ability of chemical compounds towards ozone, development of new highly efficient technologies for chemical industry, electronics, fine organic synthesis, solution of ecological and medical problems by employing ozone, degradation and stabilisation of organic, polymer, elastomer and biological materials, etc. against its harmful action [32].

The oxidative, antibacteriological and antivirus properties of ozone make it very attractive in preparation of drinking water, treatment of natural, waste and process water, artificial pools, waste gases and contaminated soils, in human and veterinary medicine, sterilisation, etc.

The natural concentration of ozone in the lower atmosphere is 1 ppb, in some cases it may approach values up to 1 ppm. It is generated photochemically in the atmosphere and is believed to protect life on Earth from overexposure to UV and other high energy radiation. Ozone is also produced by photochemical reactions in "smog", phenomenon, during the decay of some seaweed, during thunder and lighting storms, near high voltage conduits, in the environment of radiation sources, UV-applying units, laser and radar units, electrolysis, galvanic and welding moulding apparatus. The control of all these sources of ozone generation is rather difficult and in many cases they cause ecological pollution of dwelling and working places with harmful ozone levels. Ozone is highly toxic in concentrations greater than 0.1 mg/m^3. The presence of ozone in the lower atmosphere leads to the appearance of undesirable phenomena related to Earth's flora and fauna , degradation of valuable organic, inorganic, polymer and biological objects, materials and articles. In this connection the harmful effect of ozone on human, rubber goods, polymer materials should be outlined [33-68].

The purposeful application of ozone promotes invention and development of novel and improvement of well known methods for its generation and analysis, means and methods for its more effective application, etc. [61-73]. A number of laboratory and industrial methods for its synthesis are proposed. The modern ozonation stations have a capacity of 10^2 -10^3 kg O^3/day with ozone concentrations not higher than 5-7 % vol. In some particular cases, solid or liquid ozone, or ozone deposited on inorganic supports like silica gel, or absorbed in freons, etc. [74-84] is used.

The investigations in the field of ozone chemistry in Bulgaria has been intensively developed since 1972. This research was initiated by the interest of Bulgaria towards development of new scientific fields, of great importance for solving problems of reactivity, physical, organic chemistry, fine organic synthesis, environment protection, climatology, utilisation of novel scientific methods and approaches, development of new highly effective technologies for the chemical industry and human health protection.

In this connection the Chemical Stabilisation Laboratory at the Institute of Catalysis of the Bulgarian Academy of Sciences was founded and its activity during that time is illustrated by : two PhD. theses were successfully defended, at present four more PhD theses are in preparation, one scientific specialisation was realised in Holland, three International projects were developed and at present investigations are carried out within a project with the Institute of Biochemical Physics of the Russian Academy of Sciences, many foreign scientists have visited our laboratory, an International Conference on Ozone Chemistry was held in 1984, we have reported our results in many countries in Europe, Asia and USA. The National Fund of Science is regularly providing finance support to our projects. Some of the results in this Monograph are obtained within the frame of these projects [85-88].

The results up to 1983 were summarised in a monograph: Ozone and its Reaction with Organic Compounds (in Russian) by Razumovskii, S.D., Rakovski, S.K., Shopov, D.M., Zaikov, G.E , published by Publ. House of Bulgarian Academy of Sciences, Sofia, 1983. We should also mention that more than 70 publications and reviews have been published in Bulgarian and International journals, and more than 120 participations in International and national conferences and symposia have been realised. Twenty six patents were registered for the time from 1973. The citations in literature, more than 80, reflect the importance and the interest of the International scientific community to our investigations and respectively our contributions to the reactions of ozone with double bonds and to the degradation and stabilisation of rubbers. In parallel to the fundamental research carried out during all these years, we have also paid serious attention to scientific-applied studies leading to the solution of two large-scale problems: the first one was related to the highly efficient stabilisation of pneumatic tyres and rubber-technical goods in the rubber plants in the town of Vidin and Pazardzchik. We have also participated in evaluating the projects for ozonation of drinking water in the town of Sofia in the ozonation stations at Pancharevo and Bistritza. We have contributed to the development of Bulgarian standards in the field of rubber industry and participated in projects for purification of waste waters from dyes pollutants in chemical plant in Kostenetz, amine contaminants in plant " Bulgarian Furrier Industry"-Sofia, and phenol admixtures in "Lakprom" plant, Sofia. Our two ozonators designed for ecological and medical purposes were awarded two golden medals at the exposition EXPO-91, Plovdiv.

The great importance of ozone chemistry and our results and contributions for more than two decades gave us ground to summates them in this monograph, as well as to outline some trends for the future research in this field.

REFERENCES

1. Schonbein, C.F., *Pogg. Ann.*, 49, 616 (1840); *Helb.Seances Acad. Sci.*, 10,.706 (1840).
2. Nebel, C. Ozone. *Kirk Othmer Encyclopedia of Chemical Technolog (third edition)*, 16, 650-713 (1981); Horwath, M., L. Vitzky, G. Huttner. *Ozone*. Budapest Academy Kiadoi, 1985.
3. Staudinger, H., *Chem.Ber.*, 58, 1088 (1925).
4. Harries, C.D. *Chem.Ber.*, 37, 2708, (1904); 38, 1195, (1905); 45, 936, (1912);
5. Harries, C.D., *Untersuchungen uber die natrurlichen und kunstlichen Kautschukarten*. Berlin, 1919.
6. Harries, C.D. *Chem.Ber.*, 40, 4905, (1907); 41, 1227 (1908).
7. Harries, C.D., V. Weiss. *Liebegs Ann.Chem.,*, 343, 369, (1905).
8. Criegee, R., *Ann.*, 564, 9 (1949).
9. Criegee, R., *Ann.*, 583, 1 (1953).
10. Criegee, R., *R. Rec. Chem. Prog.*, 18, 111 (1957).
11. Criegee, R., *Andew. Chem.*, 21, 765 (1975).
12. Hatakeyama, S., H Lai, S Gao, K Murano. *Chemistry Letters*, : 8, 1287-1290, (1993).
13. Schutt, WS, ME Sigman, YZ Li. *Analytica Chimica Acta*, 319: 3, 369-377 (1996).

14. Lee, MJ , H Arai, T Miyata. *Chemistry Letters* : 6, 1069-1070 (1994).
15. Hon, YS , JL Yan. *Tetrahedron Lett.*, 35: 11, 1743-1746 (1994).
16. Hon, YS , L Lu. *Tetrahedron, Lett.*, 34: 33, 5309-5312 (1993).
17. Atkinson, R, EC Tuazon, J Arey, SM Aschmann. *Atmospheric Environment*, 29, 3423-3432 (1995).
18. Rakitskaya, TL , AY Bandurko, AA Ennan, VV Litvinskaya. *Kinetics and Catalysis*, 35: 5, 705-707 (1994).
19. Mcelvany, SW , JH Callahan, MM Ross, LD Lamb, DR Huffman. *Science*, 260: 5114, 1632-1634 (1993).
20. Keshavaraj, R, RW Tock. Oxidative Effects of Ozone on the Aging of Structural Silicone Elastomers. *Advances in Polymer Technology*, 13: 2, 149-156 (1994).
21. Razumovskii, S.D., S.K. Rakovski, D.M. Shopov, G.E. Zaikov. *Ozone and Its Reactions with Organic Compounds (in Russian)*. Publ. House of Bulgarian Academy of Sciences, Sofia, 1983.
22. Bailey, P.S. "*Ozonattion in Organic Chemistry*", Academic Press, New York, v.I, 1978, v.II, 1982, *Adv. Chem. Ser. - A*, (Ed. HH Wasserman), 39 -I, 39 - II, 1978, 1982.
23. Razumovskii, S.D., G.E. Zaikov. *Ozone and Its Reactions with Organic Compounds*. Elseviar, Amsterdam, 1984.
24. Bailey, P.S., *Cem. Rev.*, 58, 925 (1958).
25. Ozone Chemistry and Technology. *Adv. Chem. Ser.*, 21, (1959); 112 (1972).
26. Ishmuratov, GY, RY Kharisov, VN Odinokov, GA Tolstikov. The ozonolysis of unsaturated compounds in the synthesis of insect pheromones and juvenoids. *Uspekhi Khimii*, 64: 6, 580-608 (1995).
27. Dhandapani, B., Oyama, S.T. Gas phase ozone decomposition catalysts. *Appl Catal B-Environ*, 11: 2, 129-166 (1997).
28. Rakovsky, S.K., Diene Rubbers, Conventional (Ozone Degradation and Stabilization)., *Polym. Mater. Encyclopedia (ed. P.Solomone)*, 1878 - 1884 (1996).
29. *Am. Chem. Soc., Div. Petr. Chem.*, Preprints, 16, (1972)
30. *Hendbook of Ozone Technology and Application*, v.II (ed. RG Rice and A. Netzer), Butterworth Publishers, Boston, 1983.
31. The Journal of the International Ozone Association: *Ozone Science&Engineering*, 1 Nos. 1-4 (1978) - 20 Nos. 1-4 (1998), Chief Editor Rip G. Rice, 1331 Patuxent Dr., Ashton, MD 20861, USA.
32. Chock, DP , G Yarwood, AM Dunker, RE Morris, AK Pollack, CH Schleyer. Sensitivity of urban airshed model results for test fuels to uncertainties in light-duty vehicle and biogenic emissions and alternative chemical mechanisms - Auto oil air quality improvement research program. *Atmospheric Environment*, 29: 21, 3067-3084 (1995); RD Bojkov, VE Fioletov. Estimating the global ozone characteristics during the last 30 years. *Journal of Geophysical Research - Atmospheres* 100: D8, 16537-16551 (1995).
33. Giovannoni, J.M. Muller, F. Clappier, A. Russell, A.G. An air quality moder that incorporates aqueous-phase chemistry for calculating ozone as well as nitrogen and sulfur-containing pollutants in Europe. *Tropospheric Modelling and Emission Estimation*, 7, 111-119 (1997).
34. George, S.M., Livingston, F.E. Dynamic ice surface in the polar stratosphere, *Surf Rev Letters*, 4: 4, 771-780 (1997).
35. Benjamin, M.T. Winer, A.M. Estimating the ozone-forming potential of urban trees and shrubs, *Atmos Environ*, 32: 1, 53-68 (1998).
36. Frakes, B. Yarnal, B. A procedure for blending manual and correlation-based synoptic classifications. *Int J Climatol*, 17: 13, 1381-1396 (1997).
37. Coy, L., Swinbank, R. Characteristics of stratospheric winds and temperatures produced by data assimilation, *J Geophys Res-Atmos*, 102: D22, 25763-25781 (1997).
38. Stockwell, W.R., Kirchner, F. Kuhn, M., Seefeld, S. A new mechanism for regional atmospheric chemistry modeling. *J Geophys Res-Atmos*, 102: D22, 25847-25879 (1997).
39. Ramacher, B. Rudolph, J., Koppmann, R. Hydrocarbon measurements in the spring arctic troposphere during the ARCTOC 95 campaign. *Tellus B-Chem Phys Meteorol*, 49: 5, 466-485 (1997).

40. Sander, R., Vogt, R., Harris, G.W., Crutzen, P.J. Modeling the chemistry ozone, halogen compounds, and hydrocarbons in the arctic troposphere during spring. *Tellus B-Chem Phys Meteorol*, 49: 5, 522-532 (1997).
41. Ariya, P.A., Catoire, V., Sander, R., Niki, H., Harris, G.W. Trichloroethene and tetrachloroethene: tropospheric probes for Cl- and Br-atom reactions during polar sunrise, *Tellus B-Chem Phys Meteorol*, 49: 5, 583-591 (1997).
42. Rudolph, J., Ramacher, B., Plassdulmer, C., Muller, K.P., Koppmann, R. The indirect determination of chlorine atom concentration in the troposphere from changes in the patterns of non-methane hydrocarbons. *Tellus B-Chem Phys Meteorol*, 49: 5, 592-601 (1997).
43. Ozone hole over South Pole expands. *Bull Amer Meteorol Soc*, 78: 11, 2681 (1997).
44. Newman, P.A., Gleason, J.F., Mcpeters, R.D., Stolarski, R.S. Anomalously low ozone over the Arctic. *Geophys Res Lett*, 24: 22, 2689-2692 (1997).
45. Pierce, R.B., Fairlie, T.D., Remsberg, E.E., Russell, J.M., Grose, W.L. HALOE observations of the Arctic vortex during the 1997 spring: Horizontal structure in the lower stratosphere. *Geophys Res Lett*, 24: 22, 2701-2704 (1997).
46. Mccormack, J.P., Hood, L.L., Nagatani, R., Miller, A.J., Planet, W.G., Mcpeters, R.D. Approximate separation of volcanic and 11-year signals in the SBUV-SBUV/2 total ozone record over the 1979-1995 period. *Geophys Res Lett*, 24: 22, 2729-2732 (1997).
47. Dubey, M.K., Smith, G.P., Hartley, W.S., Kinnison, D.E., Connell, P.S. Rate parameter uncertainty effects in assessing stratospheric ozone depletion by supersonic aviation. *Geophys Res Lett.*, 24: 22, 2737-2740 (1997).
48. Solberg, S., Krognes, T., Stordal, F., Hov, O., Beine, H.J., Jaffe, D.A., Clemitshaw, K.C. Penkett, S.A. Reactive nitrogen compounds at Spitsbergen in the Norwegian Arctic. *J Atmos Chem*, 28: 1-3, 209-225 (1997).
49. Schmitt, R., Volzthomas, A. Climatology of ozone, PAN, CO, and NMHC in the free troposphere over the southern north Atlantic. *J Atmos Chem*, 28: 1-3, 245-262 (1997).
50. Nevison, C., Holland, E. A reexamination of the impact of anthropogenically fixed nitrogen on atmospheric N_2O and the stratospheric O_3 layer. *J Geophys Res-Atmos*, 102: D21, 25519-25536 (1997).
51. *World Environmental Conferences* held in Rio de Janeiro, 1992 - "Firmly Growth"; Kyoto, 1998 - "Green House Effect".
52. Ko MKW, ND Sze, MJ Prather. Better Protection of the Ozone Layer. *Nature*, 367: 6463, 505-508 (1994).
53. Basiuk, VA. Organic reactions on silica surface: Synthetic applications. *Uspekhi Khimii*, 64: 11, 1073-1090 (1995).
54. Cataldo, V, O Ori. Ozone reaction with C-60 fullerene. A study on the antiozonant activity of C-60 fullerene in dienic rubber. *Polymer Degradation and Stability*, 48: 2, 291-296 (1995).
55. Murray, RW. Ozonolysis of Organic Compounds. *Techn. Meth. Org. Orgmetal. Chem. (Ed. D.B. Dennay)*, M. Decker, New York, 1, 1-32 (1969),
56. *Proceedings of the Second Intern.Symp.Ozone Technol.*, (Eds. R.Rice, P. Pichet, Marc-A Vincet), Montreal, Canada, May 11-14, 1975.
57. Ozone Reactions with Organic Compounds (Symposium Chairmen prof. R.S. Bailey). *Adv. Chem. Ser.*, 112, 1-125 (1972).
58. Dewar, MJS, JC Hawng, DR Kuhn. An AM1 Study of the Reactions of Ozone with Ethylene and 2-Butene. *J. Am. Chem. Soc.*, 113: 3, 735-741 (1991).
59. Belew, JS. Chapter 6:"Ozonization". *Oxidation*, v.1, (1969)
60. Kuczkowski, RL. *Chem. Soc. Rev.*, 21, 79 (1992).
61. Prousek, J., Klcova, A., Photocatalytic degradation of benzimidazole and pyridine in aqueous semiconductor dispersions of ZnO. *Chem Listy*, 91, 8, 575-579 (1997).
62. Zaror, C.A. Enhanced oxidation of toxic effluents using simultaneous ozonation and activated carbon treatment. *J Chem Technol Biotechnol*, 70, 1, 21-28 (1997).
63. Tarunin, B.I., Tarunina, V.N., Klimova, M.N., Ratkova, M.Y. Oxidation of certain organotin compounds by ozone in aqueous medium. *Zh Obshch Khim*, 67, 5, 806-810 (1997).

64. Otal, E., Mantzavinos, D., Delgado, M.V., Hellenbrand, R., Lebrato, J., Metcalfe, I.S., Livingston, A.G. Integrated wet air oxidation and biological treatment of polyethylene glycol-containing wastewaters. *J Chem Technol Biotechnol,* 70, 2, 147-156 (1997).
65. Benitez, F.J., Beltranheredia, J., Acero, J.L., Pinilla, M.L. Simultaneous photodegradation and ozonation plus UV radiation of phenolic acids - Major pollutants in agro-industrial wastewaters, *J Chem Technol Biotechnol,* 70, 3, 253-260 (1997).
66. Finlaysonpitts, B.J., Pitts, J.N. Tropospheric air pollution: Ozone, airborne toxics, polycyclic aromatic hydrocarbons, and particles. *Science,* 276, 5315, 1045-1052 (1997).
67. Roscoe, H.K., Clemitshaw, K.C. Measurement techniques in gas-phase tropospheric chemistry: A selective view of the past, present, and future. *Science,* 276, 5315, 1065-1072 (1997).
68. Tiefenbrunner, FH, HG Moll, A Grohmann, D Eihelsdorfer, K Seidel, G Golderer. Ozone treatment of Small - Size Swiming Pools and Whirlpools. *Ozone Sci.Eng*., 12, 393-401 (1990).
69. Ko, MKW, ND Sze, MJ Prather. Better Protection of the Ozone Layer. *Nature,* 367: 6463, 505-508 (1994).
70. Finlaysonpitts, B.J., Pitts, J.N. Tropospheric air pollution: Ozone, airborne toxics, polycyclic aromatic hydrocarbons, and particles. *Science,* 276, 5315, 1045-1052 (1997).
71. Hegeler, F., Akiyama, H. Ozone generation by positive and negative wire-to-plate streamer discharges. *Jpn J Appl Phys Pt 1,* 36, 8, 5335-5339 (1997).
72. Rakovski S., V. Podmasterev, S. Razumovskii, G. Zaikov. The Mechanism of Crack Formation in Rubbers Under the Effect of Ozone. *Int.J.Polym.Mater.*, 15, (2), 123 - 133 (1991).
73. Kerr, RA. Atmospheric chemistry - Ozone-destroying chlorine tops out. *Science,* 271: 5245, 32 (1996).
74. Ufimkin, DP, VV Lunin, LV Sabitova, LN Burenkova, VA Voblikova, SN Tkachenko. The effect of ozone on heterogeneous-catalytic oxidation of carbon monoxide in the presence of cement-containing catalyst. *Zhurnal Fizicheskoi Khimii,* 69: 11, 1964-1967 (1995).
75. Dhandapani, B , ST Oyama. Kinetics and mechanism of ozone decomposition on a manganese oxide catalyst. *Chemistry Letters,* : 6, 413-414 (1995).
76. Naydenov, A, R Stoyanova, D Mehandjiev Ozone decomposition and CO oxidation on CeO_2. *Journal of Molecular Catalysis A - Chemical,* 98: 1, 9-14 (1995).
77. Martynov, IV, VI Demidyuk, SN Tkachenko, MP Popovich Kinetics of Ozone Decomposition on Oxide Catalysts. *Zhurnal Fizicheskoi Khimii,* 68: 11, 1972-1976 (1994).
78. Merz, E, F Gaia. Comparison of Economics of Various Ozone Generation Systems. *Ozone Sci. Eng*., 12, 401-423 (1990).
79. Hegeler, F., Akiyama, H. Ozone generation by positive and negative wire-to-plate streamer discharges. *Jpn J Appl Phys Pt 1,* 36, 8, 5335-5339 (1997).
80. Kitayama, J., Kuzumoto, M. Theoretical and experimental study on ozone generation characteristics of an oxygen-fed ozone generator in silent discharge. *J Phys-D-Appl Phys,* 30, 17, 2453-2461 (1997).
81. Basiuk, VA. Organic reactions on silica surface: Synthetic applications. *Uspekhi Khimii,* 64: 11, 1073-1090 (1995).
82. Rimmer, S, JR Ebdon, MJ Shepherd. Polymer-supported reagents as aids in controlled degradation of polymers containing ozonides in the main chain to give telechelic oligomers. *Reactive & Functional Polymers,* 26: 1-3, 145-155 (1995).
83. Ruban, LV, SK Rakovsky, AA Popov Kinetics and Mechanism fo Cyclohexane Ozonoizatio. Izv. AN SSSR ser.khim., 40, (9), 1950 - 1956 (1976).
84. Ufimkin, DP, VV Lunin, LV Sabitova, LN Burenkova, VA Voblikova, SN Tkachenko. The effect of ozone on heterogeneous-catalytic oxidation of carbon monoxide in the presence of cement-containing catalyst. *Zhurnal Fizicheskoi Khimii,* 69: 11, 1964-1967 (1995).
85. Rakovsky SK. PhD Theses: "Kinetics and Mechanism of the Ozone Reactions with Paraffins in Liquid Phase. Moscow, Inst. Chem. Phys., 1975.
86. Anachkov MP. PhD Theses: "Kinetics and Mechanism of the Ozone Reaction with Diene Rubbers in Solution. Moscow, Inst. Neftekhimii, 1982.
87. I-st Intern. Micro Symposium: "Ozone Degradation of Polymers", Sofia, 14-18 May 1984.

Chapter 1

OZONOLYSIS OF HYDROCARBONS

The ozonolysis of hydrocarbons is one of the rare reactions of paraffins that takes place both at low and ambient temperatures and allows the transformations into new valuable oxygen-containing products [1-12].

The hydrocarbon ozonolysis has been investigated by a number of authors, but the results and the conclusions obtained in most cases differ from one another, perhaps because of differences and complicated contents of formed products [13-15]. Of special interest is the action of ozone as an initiator in oxidation processes [16,-18], modifying polymer agent [19-21], oxidizer at the preparation of alcohols, ketones, and some others [22-24]. Recently, the problems concerning the kinetics and mechanism of this reaction has been intensively discussed [25-46]. The main parts of our investigations are also dedicated to these problems [47-64].

1.1 REACTION MECHANISMS

The first studies on the reaction of ozone with paraffins have been carried out with methane, ethane, propane and butane in gas phase [2-8]. It has been established that the absorption of 1 mol ozone yields 1 mol products. The analysis of the experimental data reveals two mechanisms of ozone action : 1) ozone is decomposed to atomic oxygen, which initiates the oxidation process [2] (M.1.1) and 2) ozone interacts directly with the alkane [3-5] (M.1.2.):

$$O_3 = O_2 + O$$
$$RH + O \rightarrow R^{\bullet} + {}^{\bullet}OH \qquad (M.1.1)$$

$$RH + O_3 \rightarrow RO^{\bullet} + {}^{\bullet}O_2H \qquad (M.1.2)$$

Our opinion is that mechanism M.1.1.can be realized only in gas phase at temperatures >>50-60°C, because in this temperature range the value of the rate constant for the thermal decomposition of ozone is <<~10^{-4} M^{-1} .s^{-1} [28, 29] and can not ensure the occurrence of the oxidation process. Thus this reaction is of no importance in liquid phase whereby the reactions are usually carried out at • 50-60°C.

Mechanism M.1.2. supposes the formation of H_2O_2 in quantities proportional to the absorbed ozone which is however not observed experimentally since only traces of H_2O_2 has been found in the liquid phase. Really only a few amounts of hydroperoxide is formed [1]. It allowed us to suggest that the role of M.1.2. in ozonataion of alkanes is negligible.

The ozonolysis of alkanes in gas phase leads to the formation of excited species which determines the peculiar character of these reactions [65].

The first studies in liquid phase have been carried out by Azinger and Co., on ozonation of octadecane [16]. On the basis of the composition of the carboxylic acids and their yield they come to the conclusion that ozone attacks directly the CH_2 groups. Upon ozonolysis of decalin and adamantane the tert-CH- groups are attacked leading to the formation of the respective tert alcohols. In cyclohexane ozone reacts with the CH_2 -groups giving rise to cyclohexanol and cyclohexanone [16].

The kinetics of intermediate and end products formation has been studied and the mechanism of ozone reaction with tetradecane is suggested as [66]:

$$RH + O_3 \rightarrow RO_2^{\bullet} + {}^{\bullet}OH \qquad (M.1.3)$$

Although the basic reaction in M.1.3. is exothermal its occurrence in two steps appears to be more entropy favorable than in one as assumed by this mechanism..

Hamilton and co., [6, 67] consider as the most probable mechanism the ozone attack on the CH-bonds via a 1,3-transition state (TS):

$$RH + O_3 \Leftrightarrow TS \rightarrow \underset{\text{cage}}{[R^{\bullet} + HO_3^{\bullet} \Leftrightarrow R^{\bullet} + HO_3^{-}]} \rightarrow ROOOH\ (1) \qquad (M.1.4)$$

$$\downarrow \qquad\qquad\qquad\qquad \downarrow$$

$$ROH + O_2\,(s)\ (2); \qquad R^{\bullet}\uparrow + \uparrow^{\bullet}OH + O_2\,(t)\ (3)$$

$$\downarrow$$

$$RO_2^{\bullet} + O_2,$$

which is followed by H-atom abstraction and hydrotrioxide (1), alcohol, singlet oxygen (2), peroxy radical and triplet oxygen (3) outlet from the kinetic cage into the bulk liquid. They postulate that the products in the cage have a radical character, while according to Myurei they are more likely ionic pairs [68].

Nangiya and Benson [28] assume mechanism M.1.4. as more probable, but the interaction between ozone and CH-bonds is not synchronic one.

Denissov and co. [26] considered the reaction of ozone with C-H-bonds in the frames of parabolic model of the potential surface and also adhere to mechanism M.1.4.

Another concept of the mechanism of alkane ozonolysis is that R, HO -radicals are kinetically responsible for the proceeding of the reaction and ROH, $R^{\bullet}$, $HO^{\bullet}$, and O_2 leave the kinetic cage and pass into the solution volume [51]:

$$RH + O_3 \rightarrow [R^{\bullet} + HO^{\bullet} + O_2], \qquad (M.1.5)$$

However one can not exclude the possibility for the $RO_2^{\bullet}$-radicals formation from ROOOH as a precursor via the mechanism of ozone incorporation into the CH-bonds

$$RH + O_3 \rightarrow ROOOH \qquad (M.1.6)$$

ROOOH could act as a precursor in all mechanisms discussed in the literature [9, 67]. This reaction probably have to be entropic disadvantageous, because it goes through a cyclic TS and it is necessary to use the assumption five coordinated C-atom.

1.1.1 Thermochemical Analysis

We have determined the heats of ozone reaction with paraffins, according to the different mechanisms on the example of methane. The latter has been chosen since the energy of its CH-bond is the highest and thus the results from the calculations for it can be extended for other paraffins. For this purpose we have used the experimentally and calculated, marked by (*), values of the bond energies and heats of formation [69, 70]:

CH_3-H = 104; OO-O = 24; CH_3O-H = 102 , CH_3OO-H = 77?, CH_3OOO-H = 70?, CH_3OO-OH = 51? , CH_3O-OOH = 47?, CH_3-OH = 91, CH_3-OOH = 69?, CH_3-OOOH = 61?, H-$O^{\bullet}$ = 102; H-$OO^{\bullet}$ = 77; H-$OOO^{\bullet}$ = 62; C-$O^{\bullet}$ = 78; C-$OO^{\bullet}$ = 59; C-$OOO^{\bullet}$ = 48 kcal/mol,

After substituting with the corresponding thermochemical value the heats of the reactions (Q) are:

Mechanism:	M.1.1	M.1.2	M.1.3	M.1.4	M.1.5	M.1.6
Q, kcal/mol:	26	-27	-33	66	26	-5

The reactions of mechanisms M.1, M.1.3, M.1.6. are exothermal while those included in the other mechanisms are endothermal, and therefore the former reactions are thermodynamically more favorable.

For alkanes with secondary and tertiary C-H-bonds with lower bond energies, Q for each one reaction will be higher by 10-15 kcal.

1.2. Structural-Kinetic Investigations

As we have shown above there exist two approaches concerning the interaction of ozone with paraffins: ozone introduction along the CH-bonds and H-atom abstraction. We should note, however, that the effect of paraffins composition in this reaction has not been systematically studied which limits the possibilities for analysis and elaborations of the mechanism of these reactions [25, 59].

The variety of the proposed mechanisms and the insufficient kinetic data impose the performance of a systematic investigation of the reaction applying a wide spectrum of conditions and theoretical and experimental methods of study.

1.2.1. Linear and Isoparaffins

We have studied the kinetics of ozone reaction with a series of paraffins of different structure [47, 59]. The rates of the reactions (W) have been determined in a close and open system. By following the change of $[O_3]$ in time (t) in static conditions we have determined the rate constant of the pseudomonomolecular reaction $k' = k.[RH]$ and the bimolecular constant -k.

$$W = d\,[O_3]/dt = k\,[RH]_0.[O_3] = k'\,[O_3] \qquad \text{(Eq.1.1)}$$

The constant k' has been determined during the hafl-decomposition time:

$$k' = \ln 2/\tau_{1/2} \qquad \text{(Eq.1.2)}$$

or from the slope of the curve in coordinates $\ln([O_3]_o/[O_3]/t$.

The reaction in open system has been carried out at bubbling of ozone through the paraffin and following spectrophotometrically its concentration in the UV-region (254-300 nm). The course of the concentration change has the following pattern: first it drops off abruptly from $[O_3]_o$ - the concentration at the reactor inlet to $[O_3]_g$ - the concentration at the reactor outlet after which for a long time it remains parallel to the x axis, at a distance of $\Delta[O_3] = \{[O_3]_o - [O_3]_g\}$ The analysis of the kinetic curves has been done on the basis of its stationary part when the rate of the chemical reaction becomes equal to the rate of ozone consumption $W = W_{O3}$ or:

$$k.[RH].[O_3]_l = \omega.\Delta[O_3] \qquad \text{(Eq.1.3)},$$

from which:

$$k = \omega.\Delta[O_3]/([RH].[O_3]_l), \qquad \text{(Eq.1.4)}$$

where $\omega = v/V$ is the relative rate; v - the rate of the gas flow; V - the volume of the liquid phase; in one of the used models $[O_3]_l = \alpha.[O_3]_g$ - the ozone concentration in the liquid; α - the Henry's coefficient [31]. This model is valid in all cases when the rate of ozone absorption is bigger than the rate of chemical reactions.

Several criteria could by applied if the Henry's low is performed during the course of absorption and proceeding of irreversible first order chemical reaction (k_1):

$$k_1.\tau << 1,$$

where: τ - $1/\omega$

or

$$D_{O3}.k_1/k_L^2<<1$$

where: D_{O3}.- diffusion coefficient of ozone in solution; k_L= D_{O3}/δ - coefficient of mass transfer in liquid phase and δ - thickness of the bound layer in hydrodynamic model - renovation surface or $k_L=(D_{O3}.s)^{1/2}$, where s - time of renovation; at bubbling method with small bubbles with diameters to 2.5-3 mm - $k_L=0.31.(g\nu)^{1/3}.(D_{O3}/\nu)^{2/3}$, where $\nu=\eta/\rho$ - kinematics viscosity of the solvent, η - viscosity of the solvent, ρ - solvent density, g - earth acceleration. Usually k_L value is 0.1-0.05 cm/s.

In all other cases $[O_3]_l$ is a complicated function of $[O_3]_g$ è $[O_3]_0$. and it requires special investigations [54].

Conventionally, v = 0.1l/min, V = 10 ml and respectively ω = 0.167 s^{-1}, $[O_3]_o \approx 10^{-6}$- 10^{-3} M, solvent - CCl_4, for which depending on the temperature $\alpha \approx$ 1.8-3, [RH]$\approx$ 10^{-3}-10^1 M and for the paraffins $\alpha \approx$ 1.8 - 2.5. The bubbling reactors are glassy with grating G2, G3 or G4 built in its lower end, with cylindrical form with inner diameter ϕ = 1.7-3.7 cm and height h = 7-15 cm. The accuracy of thermostating was ±0.1°C.

By applying the kinetic methods pointed above we have studied the influence of concentrations, composition, solvent nature and temperature on the kinetics of a definite reaction.

The rates of the reaction studied were linearly dependent on $[O_3]$ and with some exceptions - linearly on [RH] (for example adamantane in the first case and 3-methylpentane for the second one Fig 1.1).

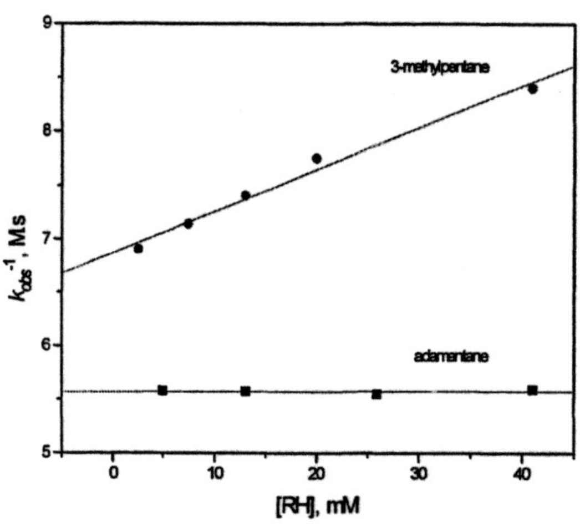

Fig.1.1. The dependence of $1/k_{obs}$ v [RH]

For 3-methylpentane a linear dependence of $1/k_{obs}$ from [RH] has been observed and for adamantane $1/k_{obs}$ does not depend on [RH]. The dependence is true only at [RH]<0.3 M and ratio of [RH]/ $[O_3]$<1.10^4.

Williamson and Cvetanovich [17] have also registered an increase of W at low [RH] and ratios [RH]/ $[O_3]$<1.10^4 at ozonolysis of n-octane, n-pentane, 3-methylpentane and iso-pentane. They explain this fact rather by the ozone interaction with microimpurities and by the initiation of the chain reactions at low paraffins concentration, than by the reaction mechanism.

We have tried to explain the dependence of $1/k_{obs}$ from [RH] assuming that the reaction takes place via a complex multistep mechanism.

For this purpose we calculated and analyzed several kinetic schemes including steps of complex formation between the paraffin and ozone. Scheme 1.1 provides a good agreement between the applied kinetic and chemical criteria and the experimental data:

Scheme 1.1

$$RH + O_3 \underset{k_{-1}}{\overset{k_1}{\rightleftharpoons}} [R...H...O_3] \xrightarrow{k_2} \text{products}$$

$$[R...H...O_3] + RH \xrightarrow{k_3} 2RH + O_3$$

which includes the formation of complex - $[R...H...O_3]$.

We have studied the case when the equilibrium concentration of the complex is not affected by the occurrence of reactions 2 and 3.

The material balance relative to the complex is:

$$[R...H...O_3] = [O_3]_0 - [O_3] \ ,$$

At quasi-stationary approximation in relation to $[R...H...O_3]$ and taking into consideration the material balance, we obtain for k_{obs} :

$$k_{obs} = k_1k_2/\{k_{-1} + k_2 + (k_1 + k_3)\,[RH]\}, \qquad \text{(Eq.1.5)}$$

and in reciprocal form:

$$1/k_{obs} = (1/k_1 + k_{-1}/k_1k_2) + \{(k_1 + k_3)/k_1k_2\}.[RH] \qquad \text{(Eq.1.6)}$$

It follows from Eq. 1.6 that $1/k_{obs}$ will increase linearly with the increase of [RH]. From the ordinate intercept one can determine ($1/k_1 + K_d/k_2$), where $K_d = k_{-1}/k_1$ - the dissociation constant of the complex and from the slope angle the ratio $(k_1+ k_3)/k_1k_2$. At our calculations we have assumed that $k_2 << k_{-1}$ according to Scheme 1.1. Hence the intercept gives K_d/k_2 and at $k_1 >> k_3$ $1/k_2$ can be found from the slope. The last assumption transforms Scheme 1.1 into a mechanism of Michaelis-Menten type.

Scheme 1.2

$$RH + O_3 \underset{k_{-1}}{\overset{k_1}{\rightleftharpoons}} [R...H...O_3] \xrightarrow{k_2} \text{products}$$

Then Eq.1.5 is transformed in:

$$k_{obs} = k_1k_2/(k_{-1} + k_2 + k_1[RH]), \qquad \text{(Eq.1.7)}$$

and Eq.1.6. in:

$$1/k_{obs} = (1/k_1 + k_{-1}/k_1.k_2) + (1/k_2)[RH] \qquad \text{(Eq.1.8)}$$

In the case when the equilibrium constant of $[R...H...O_3]$ varies as a result of the proceeding of reactions 2 1 and 3, $1/k_{obs}$ will be equal to:

$$1/k_{obs} = (1/k_1 + k_{-1}/k_1k_2) + k_3/k_1k_2.[RH] \qquad \text{(Eq.1.6')}$$

and $1/k_1$ will be determined from the intercept since $K_d << k_2$, and k_3/k_1k_2 from the slope, respectively.

In order to assess the validity of the kinetic schemes we have experimentally determined the rate constants of ozone interaction with 24 paraffins of various structure. The rate constants depending on [RH] are given in Table 1 whereby in column 3 are shown the constants calculated from the ordinate' intercepts and in column 4 are those obtained from the slope according to Eq. 1.6'/Eq.1.8.

Table 1.1

Rate constants of ozone reactions with paraffins in CCl_4 at 20°C [47]

No	Hydrocarbon	k ±10% [M^{-1}. s^{-1}]	$(k_3/k_2)/k_2$ [M^{-1}]	No	Hydrocarbon	k ±10% [M^{-1}. s^{-1}]	$(k_3/k_2)/k_2$ [M^{-1}]
1.	n-Pentane	0.015	0.001/15	13.	2-Me-hexane	0.13	5.2/0.025
2.	n-Hexane	0.019	0.0013/15	14.	3-Me-hexane	0.20	7.4/0.027
3.	n-Heptane	0.021	0.0012/17	15.	2,2-di-Me-hexane	0.015	<<0.001/<<15
4.	n-Octane	0.023	-	16.	2,4-di-Me-hexane	0.13	-
5.	n-Nonane	0.026	-	17.	2,2,5-tri-Me-hexane	0.19	9.0/0.021
6.	n-Decane	0.029	<0.001<29	18.	3-Me-heptane	0.20	9.0/0.022
7.	n-Tetradecane	0.036	-	19.	Hexa-Me-ethane	0.0002	-
8.	n-Octadecane	0.048	<0.001/<48	20.	Ethyl-n-propane	0.05	-
9.	3-Me-pentane	0.15	0.75/0.2	21.	Cyclohexane	0.01	<<0.001/<<14
10.	2,3-di-Me-pentane	0.29	11/0.026	22.	Norbornane	0.014	<<0.001/<<14
11.	2,4-di-Me-pentane	0.08	-	23.	Adamantane	0.22	<<0.1/<<2.2
12.	2,3,4-tri-Me-pentane	0.59	40/0/015	24.	Isooctane	0.14	-

The analysis of the data in Table 1.1. shows that the kinetic models, most likely, do not conform the experimental results although they formally describe some of the experimental observations in Fig.1.1 and Table 1.1. For some of the representatives among n-, iso- and cyclo-paraffins no relationship of $1/k_{obs}$ on [RH] has been established: Nos 4, 5, 7, 11, 16, 19, 20, 24 or this dependence is within the limits of the graphic presentation: Nos 1, 2, 3, 6, 8, 15, 21, 22, 23. A good correlation has been found for Nos 9, 10,12, 13, 14, 17, 18 since all these are iso-parrafins with 1, 2 or 3 CH_3- groups in the side chains. These isomers contain always hardly separable microimpurities which are more reactive than the parent paraffins and can affect the processes of di-, tri- merization of the hydrocarbon molecules.

It is seen that the values of k_1 depend on the structure and type of the CH-bonds in the alkane molecule as for the primary ones (I) it is 0.0002 No 19, for the secondary ones (II) - 0.048 No 8 and for the tertiary (III) - 0.20 No 18. These bonds differ basically in their energies which are in the ratio:1:0.98:0.96 [69, 70].

The strong effect of the bond energies supposes that the CH-bond in the transition state is highly distorted and the heat of the reaction may be zero or negative [75]. The enthalpy of the process experimentally determined is within the limits of 0-10kcal. The Hammond law states that the TS would be more similar to the reaction products than to the initial compounds, i.e. the reaction coordinate is relatively high. This justifies to a great extent that the dependence of $1/k_{obs}$ on [RH] is not so much related to the chemistry of the reaction but to the experiments conditions because it is observed only in separate cases and is concentration limited. In addition W has a constant value in solvents of various properties (CCl_4, $CHCl_3$, CH_3CN, decane). The experimentally determined activation energies for the ozonolysis of paraffins approaches values of 14-15 kcal/mol, while the reactions of complex formation have E_a<10 kcal/mol. The UV-spectrum of ozone in n-hexane solution is similar to that in the gas phase. The insignificant bathochromic shift of 6 nm does not lead to broadening of the absorption band. The hypsochromic effect leads to change in the molecular extinction (ε) by 10 % from 1820 to 2002 $M^{-1}.cm^{-1}$. The blue color of ozone is preserved in the n-hexane solution. The IR-spectra of the gaseous ozone and of its solution in n-hexane are practically identical and new bands are not identified. In the NMR-spectra of the h-hexane solutions of ozone at temperature -80°C and after a long stay no changes in the spectrum of n-hexane are observed. Only upon ozonolysis of olefins at -100°C in the IR-cell and the analysis of the IR spectra is presumed the formation of π-complexes [76, 77].

Thus the reaction of ozone with paraffins is more likely a bimolecular process, excluding preliminary complex formation, which leads directly to the formation of the reaction products:

$$RH + O_3 \rightarrow \text{products (k)}$$

and then the data extrapolated to $[RH]_{\rightarrow 0}$ in column 3 of Table 1.1. will correspond to the bimolecular constant *k*.

We have assumed that the reaction of ozone with the homologies series of n-paraffins is isoentropic and the difference in their reactivity is determined by the energies of C-H-bonds. In order to eliminate random factors we have applied the relative rate constants on example of n-hexane:

$$k^n/k^6 = \exp(-0.3\Delta D/RT)$$

(Eq.1.9),

where $\Delta D = D^n - D^6$ is the difference between the energies of C-H bonds in kcal/mol; n - index for the corresponding paraffin; 0.3 - empirical coefficient.

At 293K, R = 1.99 cal/(mol.K) and $k_6 = 0.019\ M^{-1}.s^{-1}$:

$$k^n = 0.019.\exp(-0.514.\Delta D)$$

(Eq.1.10),

The energies of C-H-bonds for primary (I), secondary)II) and tertiary (III) C-atom are calculated according to the following equations [70]

$$D_I = 93.6 + 8\Sigma 0.4^{f, g, I, j,...}$$

(Eq.1.11),

$$D_{II} = 85.6 + 8\Sigma 0.4^{f, g, I, j,...}$$

(Eq.1.12),

$$D_{III} = 77.6 + 8\Sigma 0.4^{f, g, I, j,...}$$

(Eq.1.13),

where: f, g, I, j,... are the number of C-atoms in each linear hydrocarbon chain bonded to C-atom, whereby the energy of the C-H-bonds is calculated.

The estimated by Eq.1.10. k values for paraffins comprising only primary C-H-bonds (for example methane, neopentane) are within the limits of 0.00078÷0.000013; with secondary C-H bonds (n-paraffins)- 0.48÷0.00078 and with tertiary CH-bonds (sec-butane, alkylsubstituted n-paraffins) - 2.93÷0.048. The ratio in relation to the maximum and minimum rates is 1 : 61 :3756 and 1 : 60 : 3692, respectively.

The calculated values of k and D of the C-H-bonds for some paraffins are presented in Table 1 and a comparison with the experimental data have been made.

Table 1.2.

Calculated and experimental kinetic and thermodynamic parameters of ozone reactions with some paraffins

No.	Paraffin	k_{cal}/k_{exp} $M^{-1}.s^{-1}$	D_{C-H} kcal	No.	Paraffin	k_{cal}/k_{exp} $M^{-1}.s^{-1}$	D_{C-H} kcal
1.	Methane	$1.2.10^{-5}/6.6.10^{-4}$	101.6	18.	2,3-di-Me-pentane	0.08/0.29	84.64
2.	Ethane	$1.52.10^{-4}/6.3.10^{-3}$	96.8	19.	2,4-di-Me-pentane	0.06/0.08	85.02
3.	n-Propane	$1.8.10^{-3}/3.4.10^{-3}$	92	20.	2,3,4-tri-Me-pentane	0.04/0.59	85.92
4.	n-Butane	$5.10^{-2}/5.8.10^{-3}$	90	21	2-Me-hexane	0.10/0.13	84.20
5.	n-Pentane	0.013/0.015	88.2	22.	3-Me-hexane	0.23/0.20	82.59
6.	n-Hexane	0.019/0.019	87.4	23.	2,2-di-Me-hexane	0.01/0.015	88.42
7.	n-Heptane	0.029/0/021	86.6	24.	2,4-di-Me-hexane	0.17/0.13	83.10
8.	n-Octane	0.033/0.023	86.3	25.	2,2,5-tri-Me-hexane	0.08/0.19	84.61
9.	n-Nonane	0.039/0.026	86	26.	3-Me-heptane	0.26/0.20	82.28
10.	n-Decane	0.041/0.029	85.9	27.	Ethylcyclopropane	0.15/0.05	88.6?
11.	n-Undecane	0.043/	85.8	28.	Neopentane	$1.12.10^{-4}/3.8.10^{-5}$	97.4
12.	n-Dodecane	0.046/	85.7	29.	Hexamethylethane	0.00016/0.00002	96.67
13.	n-Tridecane	0.046/	85.66	30.	Cyclohexane	0.01/0.01	88.6?
14.	n-Tetra-decane	0.047/0.036	85.66	31.	Norbornane	0.014/0.014	99?
15.	n-Octa-decane	0.048/0.048	85.607	32.	Adamantane	0.22/0.22	82.6?
16.	Polyethylene	0.048/	85.6	33.	2,2,4-tri-Me-pentane	0.05/0.14	85.54
17.	3-Me-pentane	0.15/0.15	83.36				

It is seen that the calculated values of k and D are in a very good fit with the experimental results of Table 1.1. and are in a complete conformity with the literature data [2-15] too. A slight difference is noticed for hydrocarbons containing primary C-H-bonds. Because of their lower reactivity insignificant amounts of microimpurities might increase the experimental values of k. Thus the difference between the calculated and experimental values of k could be used as a measure for checking the purity of paraffins.

The energies of C-H bonds, given in column 4 for linear and branched paraffins are calculated according to Eq. 1.11 -1.13 and for cycloparaffins Nos 27, 30 -32- from the rate constants. The latter can be regarded as a very good method for determination of thermodynamic parameters on the basis of kinetic data.

The good correlation between the calculated and experimental values of k testifies the H-atom abstraction mechanism is more adequate for describing the interaction of ozone with alkanes.

1.2.2. CYCLOPARAFFINES

We have studied the kinetics and mechanism of ozone reaction with cycloparaffins ranging from cyclopentane to cyclododecane. From thermodynamically point of view cycloparaffins contain only equivalent C-H-bonds and their energies calculated according to Eq. 1.11 - 1.13, using a model of infinite chain, are 88.9 - 88.5 kcal/mol and for cyclopropane and cyclobutane are 91 and 89.5 kcal/mol. The literature values for the former are within the limits of 94±3 kcal/mol and for the latter are 100.4±2 and 95±3 kcal/mol [28]. The deviations in the calculated and literature values could be attributed to the approximations of the model chosen which considers both the small and big cycles as infinite large. In both cases, however it could be stated that after cyclobutane, D does not actually depend on the cycle size.

From stereochemical point of view, however, cycloparaffins differ considerably with respect to number, stability and symmetry of the conformers in which they exist , to the number of equatorial, axial , pseudoequatorial and pseudoaxial, exo- and endo H-atoms, to the deviations from the equilibrium values of the bond lengths, valent and torsion angles and steric hindrance in the cycles. Upon ozonolysis the energy of the system would either decrease and the reaction will be accelerated or increase and the reaction will be retarded and the reactivity of cycloparaffins would be different irrespective of the similar energies of the C-H bonds in them. The rate will be also determined by the product of the numbers of symmetry:

$$\sigma = \prod_{i=1}^{j} C_V$$

where: C_v - the axis of symmetry; ν - the order of identity per one motion; j - the number of axes of symmetry (for example, cyclopropane has one axis of symmetry of third order or three-fold C_{3v} and then σ = 3, cyclobutane - three axis of symmetry of second order - C_{2v} and σ = 8, cyclohexane - one axis of third order C_{3v} and one of second C_{2v} and σ = 6) of the initial compounds and the activated complex (AC), which upon other other conditions the increase of symmetry will increase W.

Using *pcmod4* program based on molecular mechanics methods [78, 79] we have calculated the steric energies (SE) of cycloparaffins and their radical forms (c-C_nH_{2n} and c- $C^{\bullet}_nH_{2n-1}$). First of all, we have carried out calculations on the various stereoisomers and select the conformer with the lowest energy (for example envelope for C_5 and bath for C_6) and SE_{RH}. Further we simulate the H-atom abstraction at consecutive move of the reaction center at each nonequivalent C-H-bond in this conformer and thus we select the model with minimum SE_R. Five calculations have been made for cyclo-C_5, 2 for cyclo- C_6, 7 - for cyclo- C_7, etc.
Thus by the difference $\Delta SE = SE_{RH} - SE_R$ we have estimated whether the transition $sp^3 \rightarrow sp^2$ is favorable or not. The results from the calculations are given in Table 1.3.

Table 1.3

Conformations, numbers of symmetry (σ) calculated SE_{RH} and SE_R, ΔSE and SE_{RH} and the experimental SE_R of cycloparaffins

No.	c-C_n	Conformation	σ	SE_{RH}, kcal/mol	$SE_{R\cdot}$, kcal/mol	ΔSE, kcal/mol	SE_{RH}, (exp) kcal/mol
1.	Propane		3	26.2	39.9	-13.7	27.6
2.	Butane		4	24.0	21.7	2.3	26.4
3.	Pentane		1	5.5	3.4	2.1	6.5
4.	Hexane	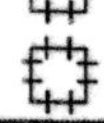	6	0.0	0.0	0.0	0
5.	Heptane	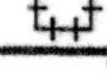	2	7.1	4.1	3.0	6.3
6.	Octane	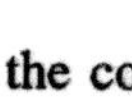	8	12.5	8.0	4.5	9.6
7.	Nonane	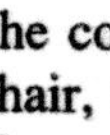	1	15.6	9.8	5.8	12.6
8.	Decane	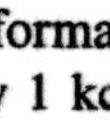	10	15.3	9.3	6.0	12.0
9.	Undecane	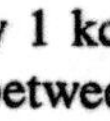	1	14.6	9.9	4.7	-
10.	Dodecane	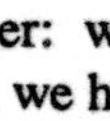	16	10.2	6.2	4.0	3.6

Note: the designations of the conformations of cycloparaffins from top to bottom are: triangle, plane square, envelope, chair, twist chair, twist crown, twist crown, twist crown, twist crown and square with 4 butane segments in chair conformation. For the cycles over C_7, the conformations shown are among thosediffering not more than by 1 kcal/mol but have the highest symmetry. In real conditions the intrinsic (real) number of symmetry is between 1 and its maximum magnitude and will be determined by the partial participation of each conformer: we have looked for a coincidence between the theoretical and the experimental values of SE [80]; we have calculated SE_{RH}.(exp) from the thermochemical values of the heats of combustion of cycloparaffins in gaseous phase according to the expression (Q_n/n-Q_6/6), where n is the number of methylene groups in the cycle, and Q_n and Q_6 are the heats of combustion of cycloparaffin and cyclohexane, respectively.

The reaction with cyclopropane and cyclohexane appears to be the slowest due to the increase in energy cycle upon $sp^3 \rightarrow sp^2$ transition. In the other cases the energy decreases and this leads to rise of the reaction rates.

In Table 1.4 are depicted the parameters of cycloalkane ozonolysis [48, 51] experimentally obtained.

Table 1.4

Kinetic parameters of ozone reaction with cycloparaffins (20°C, CCl_4)

Parameters	Pentane	Hexane	Heptane	Octane	Nonane	Decane	Dodecane
$k.10^3 \pm 10\%, M^{-1}.s^{-1}$	24.9	10.0	229.0	480.0	890.0	408.0	62.0
$A.10^{-7}, M^{-1}.s^{-1}$	2.7	38.2	2.2	1.7	24.3	170.0	168.0
$E_a \pm 0.5$, kcal/mol	12.2	14.3	10.8	10.2	11.4	13.0	14.1

For elimination of some factors general for the reactivity of cycloparaffins we have used the values of the relative constants:

$$k^n/k^6 = (A_n/A_6).\exp(-\Delta E/RT)$$

(Eq.1.14),

where: n - the number of carbon atoms in the cycle; $\Delta E = E_n - E_6$ -the difference between the activation energies of n-th cycloparaffin and that of cyclohexane.

Taking into consideration the experimental data of Table 1.1. the magnitudes of the parameters in the right term of Eq.1.14 will be:

	C5	C6	C7	C8	C9	C10	C12
ΔE, kcal/mol	-2.1	0	-3.5	-4.1	-2.9	-1.3	-0.2
exp(-ΔE/RT)	35.9	1	383.7	1064.2	138.4	9.1	1.4
A_n/A_6	0.071	1	0.058	0.045	0.636	4.45	4.34

The decrease of ΔE to the left and right from cyclohexane indicates the $sp^3 \rightarrow sp^2$ transition in cyclohexane as energetically most unfavourable. The variations in A_n/A_6 and ΔE show the occurrence of compensation effect (CEF) which in this particular case could be ascribed to the different SE of the cycles and their symmetry.

On the base of the expression for the rate constant in the theory of the activation complex including the entropy of the activation complex [81, 82]:

$$k = e^2.(RT/N_a h).\exp(\Delta S/R).\exp(-\Delta E/RT) \qquad \text{(Eq.1.15)}$$

where: e -the base of natural logarithm, N_a - Avogadro's number, h - Planck's constant.

the values of ΔS have been calculated:

	C5	C6	C7	C8	C9	C10	C12
ΔS, cal/(mol.K)	-19.46	-14.46	-19.87	-20.39	-15.06	-11.17	-11.20

The values of entropy increases for small - C_5 and medium - C_7 - C_9 - cycles, and are lower for the large C_{10} and C_{12} but even at the latter the rate constants are higher than that of cyclohexane.

The combined solution of Eq.1.14 and Eq.1.15 leads to:

$$k^n/k^6 = \exp(\Delta\Delta S/R).\exp(-\Delta E/RT) \qquad \text{(Eq.1.16),}$$

where: $\Delta\Delta S = \Delta S_n - \Delta S_6$

The change of $\Delta\Delta S$ and ΔE is an additional evidence for the occurrence of CEF.

We have found out an empiric relationship between the reactivity of cycloparaffins and their steric energy (Eq.1.17) [48]:

$$\lg(k^n/k^6) = \xi.(SE)^{1/2}$$

(Eq.1.17)

where: ξ is an empiric coefficient; SE - steric energy

Thus by applying Eq.1.17 (the data for the steric energies are available) it is possible to evaluate k for new compounds.

The experimental dependence of the reactivity of cycloalkanes on their steric energy is shown in Fig.1.2.

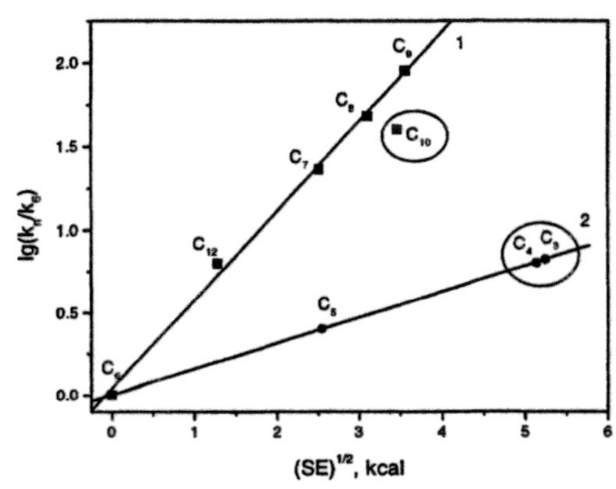

Fig.1.2. The dependence of the cycloalkanes reactivity on their steric energy.1 - for cycles > C_6; 2 - for cycles < C_6.

It follows from Fig.1.2. that a=0.54 and 0.16 for for curve 1 and curve 2, respectively. The experimental point for C_{10} which deviates from the straight line by 16% (the experimental value of lg (k_n/k_6) is 1.6 and the extrapolated one is 1.9 (designated by a circle on curve 1). The same designation is made on curve 2 for the extrapolated points for lg (k_n/k_6) of C_4 and C_3 with values of 0.8 and 0.83 respectively. The calculated rate constants for the ozone reaction with these two cycloalkanes will amount to 63 and 67 $M^{-1}.s^{-1}$, respectively. The difference in the slopes of the two curves is most likely associated with the more hindered $sp^3 \rightarrow sp^2$ transition in smaller cycles. This fact is also supported by the data for the values of SE in Table 1.3.

The theoretical and kinetic studies reveal that ozone reacts with cycloalkanes via hydrogen atom abstraction. The reactivity of cycloalkanes depend on the direction of the steric energy change during the $sp^3 \rightarrow sp^2$ transition in AC.

In Fig.3 is presented the dependence of the rate constant of cyclooctane ozonolysis on its concentration.

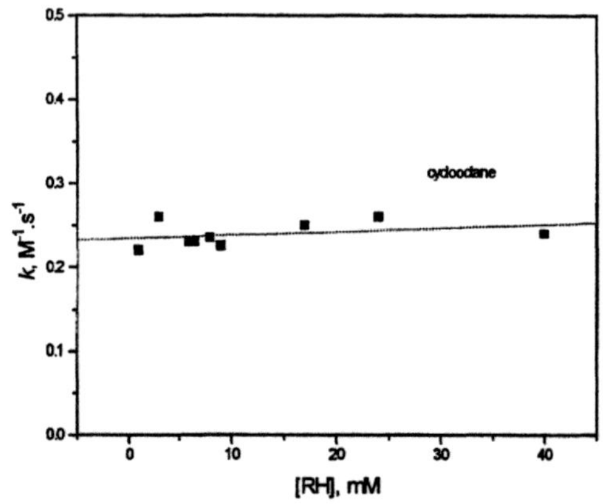

Fig.1.3. Dependence of the rate constant of ozone reaction with cyclooctane at 10°C.

The variation of [RH] and [O] up to 3 orders does not affect *k* and the rate of the reaction is described by the following equation:

$$W = k.[RH]^1.[O_3]^1$$

(Eq.1.18)

The semilogarithmic anamorphosis of the kinetic curves of ozone consumption at [RH]>>[O_3] upon ozonolysis of H_{12} and D_{12} cyclo-C_6 is shown in Fig. 1.4.

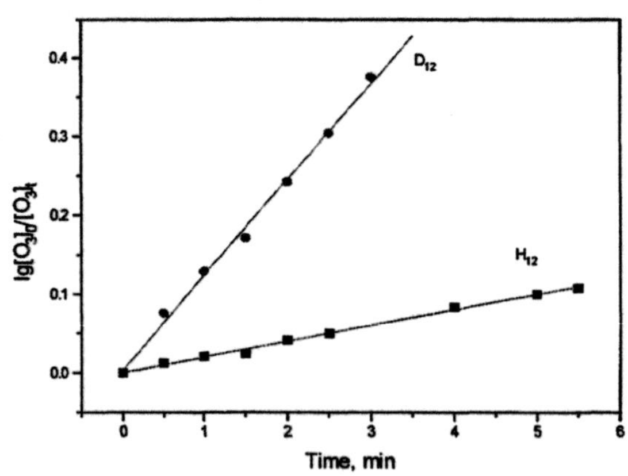

Fig. 1.4. Semilogarithmic anamorphosis of the kinetic curves of ozone consumption upon ozonolysis of H_{12} and D_{12} cyclo-C_6 ozone consumption at 28°C.

The reaction of ozone decomposition follows first order kinetics and the rate of the reaction with D_{12} is found to be higher. The ratio k_H/k_D=5.4 is indicative of the occurrence of primary isotope effect which testifies that the H-atom abstraction is the rate limiting step of the reaction.

1.2.3. Method for Estimation of Reaction Mechanisms (ERM)

The method for estimation of reaction mechanism (ERM), developed by us, is based on the comparison of calculated by means of activated complex theory and experimental values of preexponents. According to the concepts developed in the AC theory [81-83] the equation for the rate constant k of a bimolecular reaction is:

$$k = \gamma.\chi.[RT/(N_a h)].[F^{\#}/(F^a F^b)].F_{fr}\exp[-E_a/(RT)] \quad \text{(Eq.1.19)}$$

where: $F^{\#}$, F^a and F^b are the corresponding total statistical sums of the activation complex and the initial compounds A and B.

The total statistical sums are a product of the separate statistical sums of the state:

$$F = F_t.F_r.F_v.F_e.F_n \quad \text{(Eq.1.20)},$$

where: the indexes t, r, v, e and n denote the corresponding translational, rotational, vibration, electronic and nuclear sums of the state.

The translational sum of the state at 3 orders of freedom is described by the following equation:

$$F_t = h^{-3}.(2\pi m kT)^{3/2} \quad \text{(Eq.1.21)},$$

where: m is the mass of the molecule ; k - Boltzmann constant; T -absolute temperature.

The values of F_t are within the limits of 10^{24} - 10^{25} cm^3 per molecule.
The rotational sum of the state for nonlinear molecule is:

$$F_r = 8\pi^2(8\pi^3 I_A I_B I_C)^{1/2}(kT)^{3/2}h^{-3} \quad \text{(Eq.1.22)},$$

where: $I_A I_B I_C$ - are the principal moments of inertia of the molecule when the origin of the coordinate lies on the mass center of the molecule.

The values of these sums do not exceed 10^2 - 10^3 cm^3/molecule.
The equation for estimation of the statistical sum of the vibration movement is:

$$F_v = (1 - e^{-h\nu/kT})^{-1} \quad \text{(Eq.1.23}$$

where: ν - vibrational frequency.

The magnitude of F_v for one degree of freedom is of the order of 1-10 cm^3 per molecule, as the vibrations in the molecules are within the limit of 100-3000 cm^{-1} , i.e. ν is in the interval of 3.10^{12} - 9.10^{13} c^{-1} that is close to the value of $kT/h = 6.1.10^{12}$ c^{-1} at 100 K.
The statistical sums of the free rotator is:

$$F_{fr} = (8\pi^3 I' kT)^{1/2} h^{-1} \sigma^{-1} \quad \text{(Eq.1.24)},$$

where: I' is the reduced moment of inertia towards the axis of rotation.

The electronic sum of the state for adiabathic processes, including also ozone reactions, is demonstrated by:

$$F_e = 1 \quad \text{(Eq.1.25)},$$

The nuclear sum of the state for Frank-Condon processes, such as are most of the chemical transformations, is given by:

$$F_n = 1 \quad \text{(Eq.1.26)}$$

The transmitional coefficient (γ) takes account of the AC probability to be converted into reaction products. In case of adiabathic reactions it is equal to 1, and for multiplied nonadiabathic reactions it may be much smaller than 1.

The coefficient λ takes account of the tunnel effects associated with the reaction pathway via quantum way through percolation "below barrier" of the potential surface.

The probability for tunneling depends on:1) the mass of the particle m and the smaller is it , the greater is the probability for transition; 2) the barrier width δ - its decrease leads to increase of the tunneling probability; 3) its height E_a - with its decrease the probability for tunneling increases and 4) the steepness of the potential surface at the depression point - (d^2U/dx^2) - its rise results in increasing the tunneling probability.

According to Gamov [83] the probability for tunneling via triangle barrier is:

$$\chi = [2m(E_a\text{-}U)]^{1/2}.\exp[-8\pi\delta/(3h)] \quad \text{(Eq.1.27)},$$

where: U is the height of tunneling occurrence.

In case of reactions proceeding with redistribution of atoms at room and higher temperatures the tunnel effect is extremely small and λ is equal to 1. This is also true for the reactions studied by us.

For calculation of the rate constant according to Eq.1.19 it is necessary to estimate the potential surface of the reaction and the statistical sums of AC, RH and O_3.

The calculation of the potential surfaces is a difficult and labour-consuming problem and it has been solved only for small particles. This is a typical quantum-chemical task which has been the subject of intensive studies of various researchers and we will not focus our attention on the difficulties and successes in this respect [84-88]. Easier methods for calculation of potential surfaces are semi-empirical ones, like *mopac6* [87]. Its systhem using will be presented in Chapter 5.

The estimation of the statistical sums is much more simple problem but it provides an important information for the mechanism of chemical reactions. For performing the calculations it is necessary to know the geometry of the starting molecules and AC and the vibration frequency of the bond on the reaction coordinate. The geometry of the starting molecules can be usually determined with high degree of confidence, because of the great number of experimental and theoretical data concerning the interatomic interactions, the conformational states, bond lengths, the angles between atoms and their masses [70]. The question about the determination of the geometrical parameters of the activated complex is more complicated since the potential surface of the reaction is unknown. It is difficult, usually only by indirect data, to determine the reaction coordinate. In order to find an acceptable solution of this problem we have applied the Franck-Condon principle [90] and the Hammond empirical rule [75] which connects the thermal effect of the reaction with the reaction coordinate. These two principles allow the determination of the geometrical parameters of the activated complex with accuracy up to 10 %.

The preexponential in the Arrhenius equation in the framework of the AC theory can be described in the following way:

$$A = \gamma.\chi.[RT/(N_a h)].[F^{\#}/(F^a F^b)].F_{fr} \quad \text{(Eq.1.28)}$$

These considerations were used in developing a computer program for calculating A and the statistical sums of internal rotation [89]. In addition the program includes also calculation of A and the steric factor - p, according to the collision theory. It also takes into account the number of equivalent H-atoms and proceeding the reaction in liquid phase.

For calculation of A it is necessary to define the geometrical parameters of AC. Principally there exist two forms of the activated complex - linear and cyclic. In the first one an inner rotation is possible while in the second it is absent. The values of A calculated for the linear AC in each case will be higher than those calculated with the cyclic form.

The calculated values of A were compared with the experimental ones and on the bases of the coincidence degree one or another mechanism could be preferred. By this manner the method ERM gives additional possibilities to clarify the investigated reaction mechanism.

1.2.4. Application of ERM

Using ERM we have made the analysis of mechanisms M.1.1-1.6. The reactions according to them take place via two forms of the activated complex - linear (LC) and cyclic(CC):

Scheme 1.3

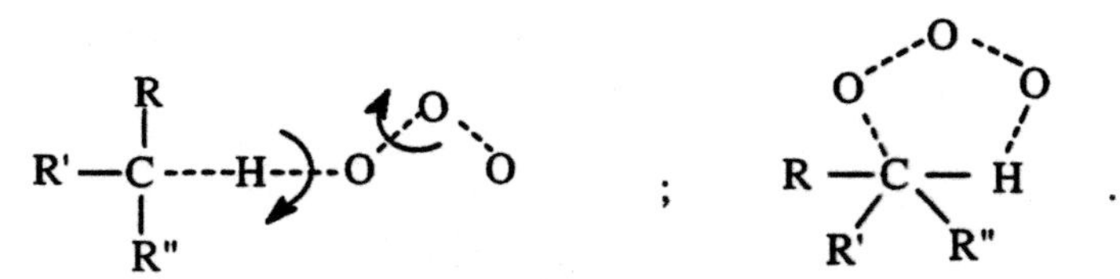

LC CC

LC is associated with H-atom cleavage mechanism and CC with ozone incorporation along the C-H bonds. LC is characterized by the ozone interaction with the C-H bond via one of its terminal atom and thus there is a possibility for free rotation around H-O and O-O while in the CC it forms five member cycle with participation of five coordinated carbon. In Table 1.5 are summarized the calculated values of A for the two AC forms.

Table 1.5

Preexponentials (A) of the ozone reaction with cycloalkanes at 300 K calculated by the method of the activated complex

Cycle	σ	$J.10^{-7}$, $M^{-1}.s^{-1}$	$I_AI_BI_C.10^{-6}$, au.• 2	$\sigma.F_{fr}$	$^{\#}I_AI_BI_C.10^{-8}$ au.• 2	$^{\#}F_r.10^{-6}$	$F_{fr}.10^{-3}$	$A_0.10^{-7}$	$A.10^{-7}$, $M^{-1}.s^{-1}$	$A_1.10^{-7}$, $M^{-1}.s^{-1}$
C_5	1	2.54	0.8	0.71	0.44	0.52	1.24	6.84	0.59	1.65
C_6	6	2.29	2.5	1.23	0.59	0.60	1.28	26.6	2.29	6.40
C_7	2	2.12	8.3	2.25	2.02	1.10	1.32	8.56	0.74	2.06
C_8	6	1.99	13.5	2.88	2.54	1.24	1.34	21.5	1.85	5.17
	1	1.99	13.5	-	2.54	1.24	1.34	3.6	0.31	0.86
C_9	1	1.88	34	4.53	5.40	1.81	1.37	3.2	0.28	0.77
C_{10}	2	1.81	97	7.68	12.2	2.72	1.39	5.6	0.48	1.34
C_{12}	16	1.69	278	13	32	4.40	1.41	40.3	3.55	9.95

where: $J = (kT/h) . (F_t^{\#}/F_t^{RH}F_t^{O3})$

Further we will focus our attention on estimation of the value of A calculated for each separate reaction [48, 52].

The ratio between the statistical sums of electronic and nuclear movement of the activated complex and the initial compounds is 1 since the change of the energy is smooth and the electron-nucleic states are not changed; γ and λ are equal to 1 under the experimental conditions.

The value of A can be presented by two multipliers [63]: A_o is a geometric factor which is determined by the AC geometry and A_1 - the contribution by the internal degrees of freedom. In this case we can write the following expressions:

$$A_0 = (kT/h) . (F_t^{\#}/F_t^{RH}F_t^{O3}).(F_r^{\#}/F_r^{RH}F_r^{O3}).F_{fr}^{\#} \quad \text{(Eq.1.29)}$$

$$A_1 = (F_v^{\#}/F_v^{C\text{-}H}).(F_v^{R\#}/F_v^{R}).(F_v^{O3\#}/F_v^{O3}).\varphi \quad \text{(Eq.1.30)}$$

where: F_v are the statistical sums of the vibrations of the reaction center , i.e C-H-bond in the initial alkane;

$\varphi = \varphi_o . \exp(-E_h/RT)$ - coefficient considering the highest energy that the rotator must overcome at its twist to 360^0.

The sums for 300 K were calculated by using the following equations:

$$F_t = 3.86.10^9[(m_{O3} + m_{RH})/m_{O3}.m_{RH}] \quad \text{(Eq.1.31)}$$
$$\sigma.F_r = 78\ (I_A I_B I_C)^{1/2} \quad \text{(Eq.1.32)}$$
$$\sigma.F_{fr} = 6.24\ I'^{\,1/2} \quad \text{(Eq.1.33)}$$
$$F_v = (1 - e^{-h\nu/kT})^{-1} \quad \text{(Eq.1.34)},$$

It should be noted that the values of A_o thus calcultated are relatively reliable. For ozone are used literature values [29, 64] for the length of O-O bond 1.278±0.003 A and for the OOO angle - $116^0\ 49' \pm 30'$; the main inertia moments $I_A I_B I_C$ are $7.875.10^{-40}$, 62.844. 10^{-40} and 70.888. 10^{-40}, respectively; $\sigma = 2$, F_e and $F_n = 1$. For cycloalkane the values of the length of the C-C and C-H bonds are 1.54 A and 1.09 A respectively and 109.5 for the CCC=CCH=HCH angles have been used. In Table 1.5 are tabulated the values of $I_A I_B I_C$ calculated according to well known methods [1, 90]. We have selected the most stable conformers of the cycles known [80] confirmed also by the calculations according to Alinjer method (Table 1.3). Thus the preferred conformation for cyclopentane is envelope (C_s), for cyclohexane - chair, for cycloheptane - twisted chair (C_{2v}), for cyclooctane - twisted crown with 3 axes C_{2v}, regular crown with 2 axes C_{2v} or knock crown without any elements of symmetry. The product $I_A I_B I_C$ is nearly the same for all conformations . However, bearing in mind that the different degrees of symmetry would affect the calculation of A we have used two conformations with a number of symmetry of 6 and 1.

The conformation state of cyclononane is a crown with a symmetry number of 1 but it is insufficiently studied and for this reason we have combined the calculations by the Alinjer method with the X-ray data for cyclononylamine bromohydrate [80]. The conformation of cyclododecylamine-1,6 dihydrohydrate ,which has only one axis of symmetry of second order, is taken as a base for cyclodecane. According to the X-ray data cyclododecane has a square shape conformation with butane segments as its sides. This conformation has one axis of fourth and two of second order and its number of symmetry is 16. Some of the most stable conformations of cycloalkanes are presented in Fig.1.5.

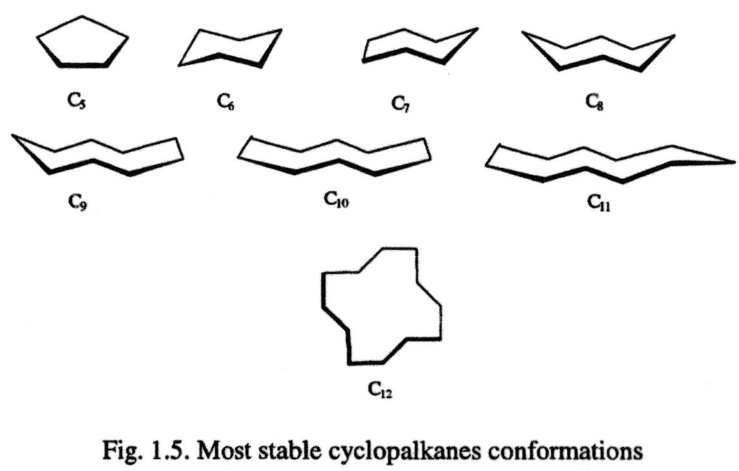

Fig. 1.5. Most stable cyclopalkanes conformations

The structure of LC was elaborated on the basis of the following geometrical parameters : the H...O length was estimated according to Pauling formulae [91]; $d_{H..O} = d_{H\text{-}O} - 0.6.\ln(r)$, where: r is the bond order, $d_{H...O}$ the bond length in the complex and $d_{H\text{-}O}$ - the length of ordinary H-O - bond equal to 0.95• . Considering that the coordinate of the maximum on the potential surface can be approximately determined the bond order was defined in a wide range from 0.2 to 0.8 with $d_{H...O}$ varying from 1.92 to 1.08• . The actual value of $d_{H...O}$ lies most likely in this interval and thus the value of 1.35• at r=0.513 was used for the calculations. The lengths of C-H and O-O bonds in LC are increased by 10 % as compared to their normal values, the H...O-O angle is accepted as 100^0 (in water HOH is 104.5^0 and in HOOH the HOO angle is 96.6^0).

The calculation of F_{fr} was carried out for two asymmetry rotators: one with rotation about axis x coinciding with the H...O bond and the second with rotation about y axis coinciding with the O-O bond:

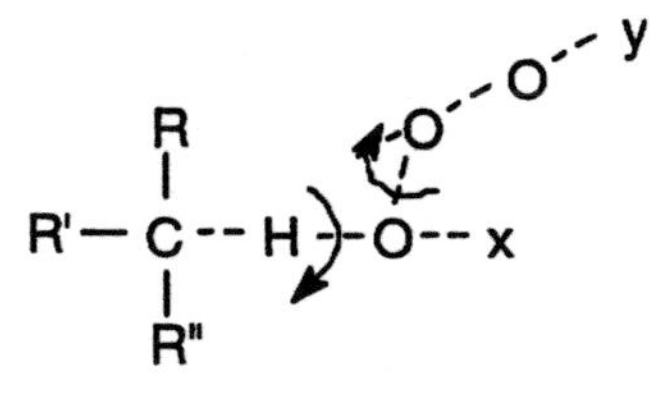

The reduced moments of inertia about y axis are found to be equal for all cycles and have the value of 21 au.• ². However those about the x axis increase with the rise of the cycloparaffins masses and their values are;

C5	C6	C7	C8	C9	C10	C12
50	53	56	58	60	62	64

In Table 1.5 are presented the values of A_o which depend on the LC geometry only and to a lesser extent on the potential surface. The values of A_1 could be calculated with less accuracy for they depend to a greater extent on the form of the potential surface.

For the accurate calculation of F_{fr} it is necessary to know the values of the rotation barrier E_{fr}. The latter were calculated by us using *mopac 6* and for the rotation around x and y axes they are 1.8 and 7.5 kcal, respectively. These values correlate well with the energy of NO rotation in the nitromethane molecule [91]. The correction coefficient φ was also determined using the Pitzer tables [92]. Although the latter are designed for symmetric rotators as Pitzer has shown, they can be successfully applied for the approximate estimation of the thermodynamical functions of asymmetric rotators.

The ratio of the vibration statistical sums $(F^{\#RH}_v.F^{\#O3}_v)/(F^{RH}_v.F^{O3}_v)$ which stands for the variations of the vibration frequencies is unit. At the same time for the ozone molecule which is characterized by two valent vibrational frequencies at 1106 and 1036 cm^{-1} and one deformational at 703 cm^{-1} one can expect a decrease of the frequency at 703 cm^{-1}, but not lower than the value within the range of 150-200 cm^{-1} to became kinetic active at 300K.

The contribution of the vibration sums in the thermodynamical functions is small and is accepted unit for calculating the ratio $F^{\#rc}_v / F^{C\text{-}H}_v$ at such moderate temperatures like 300K, where $F^{\#rc}_v$ is the vibration statistical sum of the reaction center. For higher temperatures, however, this ratio may be much higher [93, 94].

The accurate determination of the vibration frequency of the reaction center requires knowledge of the form of the potential surface near the maximum point. For that reason this value could be only approximately evaluated.

The frequency of the deformational vibrations of the reaction center C..H...C lies in the range of 100-400 cm^{-1} [94]. For the reaction center C-H...O this center is probably shifted to the higher frequency kinetically nonactive range. This is associated with the fact that $\nu_{O\text{-}H}$ lies in the range of 1000-1600 cm^{-1} and that of C-H in 400-1000 cm^{-1}. Hence the contribution of the vibrational sums should not differ from 1. However one can not rule out the probability of the vibration sum of the reaction center for bigger cycles to reach values by 5-10 % higher than those for the small cycles. Even in this case the trend in the calculations remains the same. Upon increasing the inertia moment of cycloalkanes as shown above, the vibration frequency should also increase since it is inverse to its square root. For example, if we accept that ν =200 cm^{-1} for C_5, then for C_{12} it will be 120 cm^{-1}. In this case the vibration term will be increased 3.5-fold.

It should be noted, however, that the calculations have been made for reactions in the gas phase, while for those in the liquid medium the values of A would be different. According to the collision [95] and activated complex [81-83, 96] theories the number of collisions in the liquid phase is by 2-3 times higher than those in the gas phase. The both theories predict that at $E_l = E_g$ the value of A will be 3 times higher than that in the gas one. In [91] is supposed that the increase of k for a bimolecular reaction proceeding in liquid phase will be about 5 times.

The same tendency is observed at comparing the experimental results from the ozonolysis of acyclic hydrocarbons in gas phase [2-10] and in an inert solvent - CCl_4 [47]. The ratio of the preexponential factors in them is in the range of 2-3.

The aforementioned considerations give us ground to introduce a coefficient for the ratio A_l/A_g equal to 2.8 (Table 1.5, column 11).

At comparing the experimental results with the theoretical data the experimental value of A is usually related to the number of the active centers. It is well known [6] that the hydrogen atoms in cycloparafins exist in various conformation in respect to the cycle planes they can be oriented in various positions from equatorial to axial and from endo- to exo-. In cyclohexane, for example, the three equatorial H atoms are 7 times more reactive that the three axial H atoms [74]. The H-atoms in the bigger cycles are even more nonequivalent since besides the axial and equatorial atoms in them there are also hydrogen atoms lying within the cycles for which a transanular interaction is characteristic. This can be clearly demonstrated by the models of Dreiding and the various stereochemical programs.The conformation analysis performed points out that the hydrogen atoms distribution in the cycles studied is as follows:

	C5	C6	C7	C8	C9	C10	C12
ax	10	6	8	8	10	10	16
eq	-	6	4	4	6	6	-
exo	-	-	2	4	2	4	8

In order to compare the experimental with the calculated values of A we relate the experimental values of A per the number of the equatorial H atoms (Table 1.6).

Table 1.6

The experimental and calculated values of A per one equatorial H-atom.

A, $M^{-1}.s^{-1}$	C5	C6	C7	C8	C9	C10	C12
$A_{exp}.10^{-7}$	0.32	3.34	0.36	0.29	2.48	22.8	12.9
$A_{cal}.10^{-7}$	1.65	6.40	2.06	0.86	0.77	1.34	9.90

The experimental values of A vary from $0.3.10^7$ to 23.10^7 $M^{-1}.s^{-1}$. It has been found that the values of A calculated for the activated complex with LC lie within this range $(0.8\text{-}10).10^7$, while those for the CC are in the range of 10^5-10^6 $M^{-1}.s^{-1}$. The lower values in the latter are related to the lack of free rotation (about 130 times) and to the more packed structure (2-4 times) of this complex. However in this case the eventual increase of A, on account of the occurrence of low frequency vibrations of the reaction center which according to us can bring about not more than 2-3 times increase, should be taken into account. Thus the calculated with CC values of A should be lower than the experimental ones by about 100 times. The coincidence of A calculated with LC with the experimental values of A becomes even better at their comparison for each cycle alone. In fact the differences in the values of A for all cycles do not exceed 3 which fact we will discuss further.

The results obtained show that the probability of the reaction to occurr via a cyclic transition state is negligible. Consequently, the probability of the ozone reaction with the C-H bonds to take place trough a mechanism of incorporation or formation of oxy-radicals in one step is also small. It follows from these two mechanisms that the reaction should take place minimum in two steps - the first is H-atom abstraction giving rise to ($R^\bullet$ + $^\bullet$OOOH) or to($R^\bullet$ + $HO^\bullet$ +O_2). Since the endothermity of the latter is lower by 16 kcal/mol it is thermodynamically more favorable.

In our opinion, one of the reasons for the existence of such a majority of schemes describing reaction RH + O_3 is due to the lack of entropy factors consideration. As a result of their taking into account it can be seen that the abstraction of H-atom is the limiting step in the ozonolisis of paraffins. By its own possibilities the ERM is method universal, because it can be applied to every typs of chemical reactions.

1.3. CYCLOHEXANE

The study of the relationship between the rate of ozone consumption and [ozone] and [cyclohexane] [52] in a wide range reveals a first order kinetics in regard to each reagent.

The typical kinetic curves of reaction products formation are given in Fig.1.6.

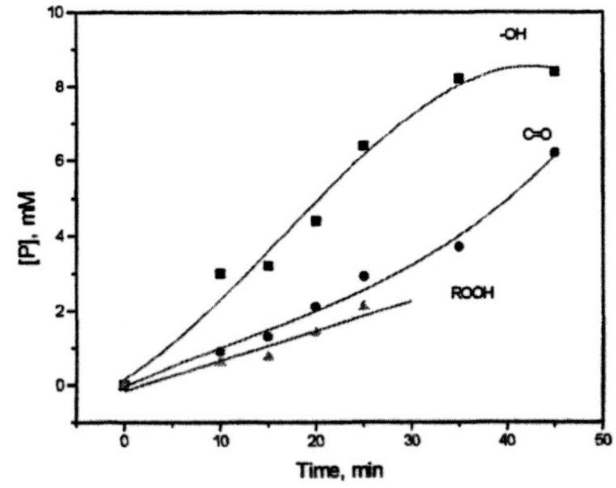

Fig.1.6. - Kinetics of basic products formation during cyclohexane ozonolysis at [RH] = 9.29 M, $[O_3]$ = 0.7 mM, 20^{0}C

The profile of the curves of products formation shows that cyclohexanone, peroxide and acids including adipic acid (the latter is not shown in the figure because of its extremely low concentration) are obtained in parallel reactions. The initial rate of alcohol formation is by about 3 times higher that that of the ketone. The same ratio for this reaction has been also reported by other authors [9, 67, 97].

The ratio between the rates of peroxides formation is W_{al}:W_{ket}:W_{ROOH} = 4:1:1. The nature of the peroxides formed in the course of the reaction has been clarified by studying the kinetics of iodine liberation during the oxidation reaction with HI solution in $CHCl_3$:CH_3COOH. The rate of iodine evolution was measured at λ= 470 nm. In order to ignore the formation of peroxide compounds arising form the further oxidation of the reaction products (alcohol and ketone), we have used reaction mixtures of very small conversion degrees, ca. 0.07 %. The results obtained (Fig.1.7.) justify the probable formation of H_2O_2, cyclohexylhydroperoxide (ROOH) and dicyclohexylperoxide (ROOR) in the reaction mixture, which react with different rates with HI (KI) [76].

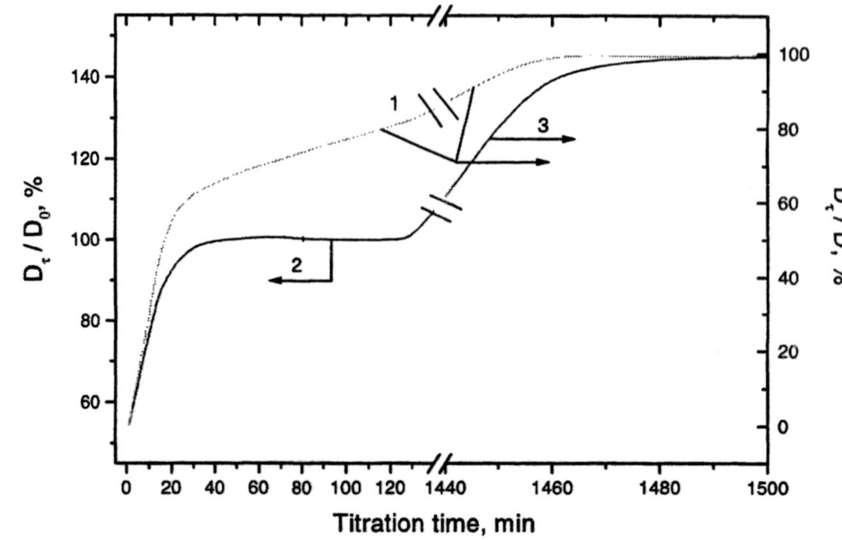

Fig.1.7. Dependence of the relative optical intensities D_τ/D (%) on the titration duration : D_τ, D - determined at time τ and after 24h at 470 nm; 1 - 0.07 % conversion; 2, 3 - reference mixture (solution) - *tert*-butyl hydroperoxide (2) /di-*tert*-butylperoxide (3) = 1:1

The formation of the peroxides abovementioned was confirmed by the iodometric procedure modified by us. 0.5 ml saturated water solution of KI and 0.5 ml CH_3COOH were added to 0.5 ml oxidate. The latter was kept at dark and room temperature for 2 h. The quantity of the iodine liberated after 30 min corresponds to H_2O_2 and ROOH, and this after 24 h to ROOR. Further, the separate determination of H_2O_2 and ROOH has been carried out by treatment of the oxidate sample with aqueous solution of catalasa (a specific enzyme decomposing H_2O_2) [98]. Thus the titration of the iodine liberated after 30 min gives the concentration of ROOH and by the difference between the two titration - the content of H_2O_2, respectively.

The duration of the quantitative iodine liberation was defined using a model system comprising H_2O_2, *tert*-butyl hydroperoxide and di-*tert* butylperoxide and it was found to be 30 min for H_2O_2 and *tert*-butyl-OOH and 24 h for di-*tert*-butylperoxide. These time durations at dark treatment correspond to well known literature data for their rates of interaction with HI (KI) [99-102]. The possibilities of the proposed procedure for the separate identification of H_2O_2, alkylhydroperoxides and dialkylperoxides are illustrated by the data in Table 1.7.

Table 1.7.

Concentrations of H_2O_2, t-BuOOH and t-BuOOBu-t x 10^2 M.

Compound	0 min	30 min	24 h
H_2O_2	1	1	-
H_2O_2 *	1	0	-
t.-BuOOH	1	1.13	-
t.-BuOOH*	1	1.2	-
t.-BuOOBu-t.	1	0.07	1.0
t.-BuROOBu-t.*	1	0.08	1.0
H_2O_2 : t.-BuOOH=1:1	1=0.5+0.5	0.49	1.08

The yield of peroxides was found to be 1-15 %. The authors of Ref. 78 reported a 45 % yield but they, in fact, do not discuss and have no explanation of the reasons for this high yield. In our opinion, most likely because of the higher experimental temperature the chain radical oxidation might generate higher quantities of peroxides.

The ratio of the initial rates of peroxides formation as shown by the kinetics of H_2O_2, ROOH and ROOR accumulation is 1:1:4. The relationships of the rates of products formation on the ozone content depicted in Fig.1.8 have a linear pattern

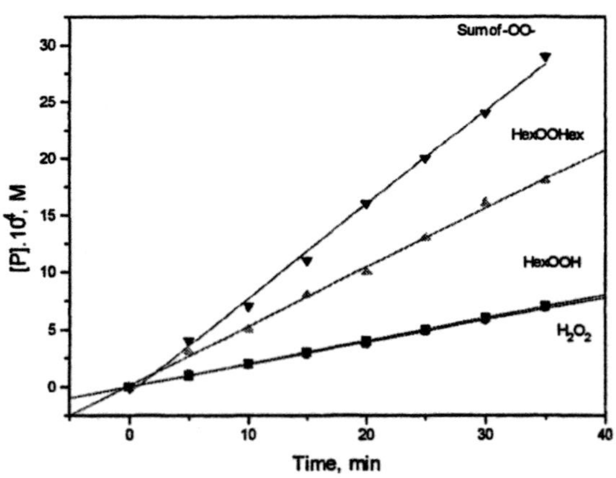

Fig.1.8. Kinetics of peroxides formation. [RH] = 9.26m, $W_{O3} = 1.10^{-5}$ $M.s^{-1}$, 20^0C

The acid numbers of the oxidates determined by titration with aqueous-alcohol solution of NaOH and indicator phenolphthalein is of an order higher than the content of adipic acid as determined by GC. This observation demonstrates the complex composition of the acids. In [97] is considered, however, that the acid number is completely defined by the presence of adipic acid. But this statement obviously contradicts the chromatographic data obtained by us. The latter shows that adipic acid comprises a very small percent of the total acid content.

In a number of experiments we have found out that the content of water is about 50 % of the total yield. The same water yield was obtained at tetradecane ozonolysis [1]. The most plausible mechanism of reaction products formation, in our opinion is consistent with the following Scheme:

Scheme 1.4.

$$RH + O_3 = [R^\bullet + O_2 + {}^\bullet OH]_{cage} \begin{cases} \rightarrow (\varphi)\ R^\bullet + O_2 + {}^\bullet OH \\ \rightarrow (1-\varphi)\ ROH + O_2 \end{cases} \quad (1)$$

$$RH + {}^\bullet OH = R^\bullet + HOH \quad (2)$$

$$R^\bullet + O_2 = RO_2^\bullet \quad (3)$$

$$RO_2^\bullet + RO_2^\bullet \begin{cases} \rightarrow (\alpha)\ ROH + R'C{=}O + O_2 \\ \rightarrow (1-\alpha)\ ROOR + O_2 \end{cases} \quad (6)$$

where: [] is the kinetic "cage"; φ - coefficient accounting to the number of radicals

in the bulk solution; (1-φ) - the part of radical, that recombine in the cage with alcohol formation

In the Scheme are not included the following reactions:

$$RO_2^\bullet + RH = ROOH + R^\bullet$$
$$ROOH = RO^\bullet + HO^\bullet,$$

as the concentration of the hydroperoxides in the reaction mixture is about only 2 % and the experiment temperature is below 40^0C and the chain radical reactions are of minor importance in the oxidation process.

The kinetic analysis of Scheme lead to the following expressions for the rates of formation of alcohol, ketone and ROOR:

$$W_{OH} = (1 - \varphi)\, W_{O3} + \alpha\, k_6\, [RO_2^{\bullet}]^2 = (1 - \varphi - 2\alpha\varphi) W_{O3} \quad \text{(Eq.1.35)}$$
$$W_{C=O} = \alpha\, k_6\, [RO_2^{\bullet}]^2 = 2\alpha\varphi\, W_{O3} \quad \text{(Eq.1.36)}$$
$$W_{ROOR} = (1 - \alpha)\, k_6\, [RO_2^{\bullet}]^2 = (1 - \alpha)\varphi W_{O3}, \quad \text{(Eq.1.37)}$$

where: W_{O3} - the rate of ozone absorption

The experimentally obtained dependencies of the accumulation rates of the alcohol, ketone and ROOR are in a good agreement with the equations derived from the Scheme (Fig. 1.9).

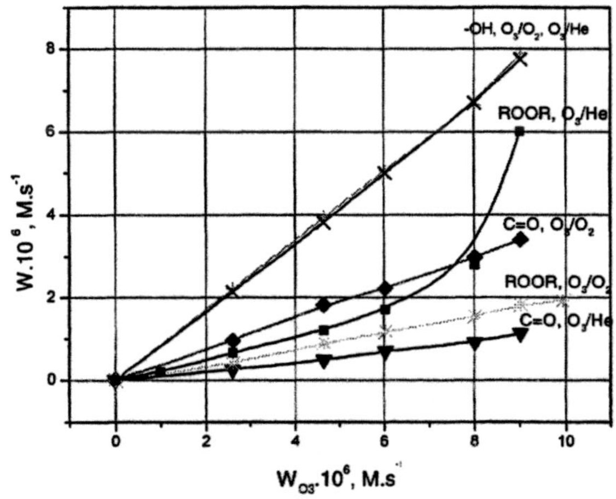

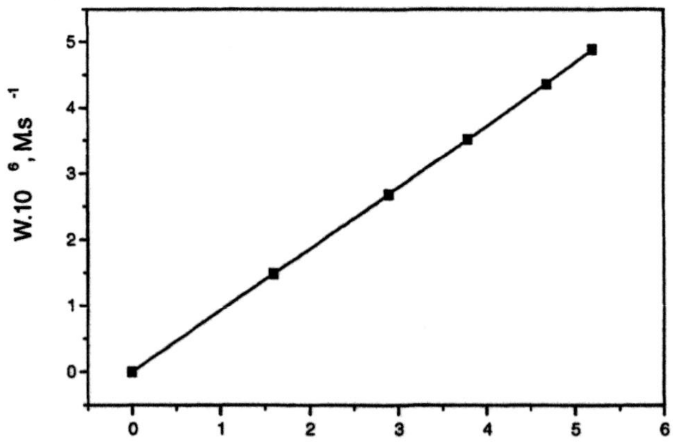

Fig.1.9. Dependence of the rates of products formation during cyclohexane ozonolysis on W_{O3} (20^0C).

On the basis of Eqns. 1.35-1.37 the value of α was determined as 0.7 (it stands for this part of alcohol and ketone which has been obtained via the recombination of $RO_2^{\bullet}$- radicals) and φ has been found to be 0.3 (the part of the radicals in the solvent bulk) and from the rates of reaction products accumulation was determined the stationary concentration of the peroxide radicals as $1.5.10^{-6}$ M.

It has been supposed in Refs. 11, 18 that ozone reacts with $RO_2^{\bullet}$ - radicals via reaction:

$$RO_2^{\bullet} + O_3 = RO^{\bullet} + 2O_2 \quad (4)$$

and then the alcohol will be obtained through reaction:

$$RO^{\bullet} + RH = ROH + R^{\bullet} \qquad (5)$$

However, in this case the dependence of W_{OH} on W_{O3} should be linear in coordinates $W_{OH}/(W_{O3})^{1/2}$, which is nor experimentally observed (Fig.1.9.). In addition W must be greater than the sum of the rates of all reaction products formation, fact which is also not confirmed be the experimental results. These two facts suggest that reactions (4) and (5) could be ignored.

If we accept that similarly to cumene oxidation [14] the square termination step generates RO-radicals:

$$RO_2^{\bullet} + RO_2^{\bullet} = 2RO^{\bullet} + O_2$$

during this reaction, then the additional quantity of alcohol will be provided by reaction (5). Thus it may be expected that the dilution of cyclohexane with an inert solvent should decrease the rate of reaction (5), consequently the ratio alcohol:ketone. But the experimental results show that the ratio is not changed at the cyclohexane dilution with tetrachloromethane thus rejecting the generation of oxy radicals from the peroxide radicals recombination.

The yield of H_2O_2 is 2-3 %. The formation of a such insignificant quantity of H_2O_2 could be explained by its yield via the following reactions [97]:

$$HO^{\bullet} + O_3 \rightarrow HO_2^{\bullet} + O_2$$
$$HO_2^{\bullet} + RH \rightarrow H_2O_2 + R^{\bullet}$$

Adipic acid is obtained at the monomolecular decomposition of $RO^{\bullet}$-radicals:

$$RO^{\bullet} \rightarrow {}^{\bullet}CH_2(CH_2)_4CHO \rightarrow HOOOC(CH_2)_4COOH$$

In addition one should also consider that in the total yield are also included, of course in tenths of percent, some products of the ozone reaction with cyclohexanone, cyclohexanol or cyclohexylperoxide. This shows that the reaction with $RO^{\bullet}$ - radicals participation has an insignificant contribution in the total yield.

We have also carried out cyclohexane ozonolysis at reduced concentrations of O_2 by its replacement with He. In this case, at equal initial ozone concentration its consumption rate increases two-fold which could be due to the proceeding of the following reaction:

$$R^{\bullet} + O_3 = RO^{\bullet} + O_2 \qquad (7)$$

and the increase of the contribution of reaction 5.

Considering reactions 7 and 5 W_{OH} and $W_{C=O}$ will depend on W_{O3} in the following manner:

$$W_{O3}^{obs} = W_{O3}^{RH} + W_{O3}^{R\bullet} \qquad \text{(Eq.1.38)}$$
$$W_{O3}^{R\bullet} = k_7.[R^{\bullet}][O_3] = 2\varphi\, k_7\,[O_3]\, W_{O3}^{RH}\,(k_2\,[O_2])^{-1} \qquad \text{(Eq.1.39)}$$
$$W_{OH} = \{1-\varphi+2\alpha\varphi+2\varphi k_7[O_3](k_2[O_2])^{-1}\}W_{O3}^{RH} \qquad \text{(Eq.1.40)}$$
$$W_{C=O} = 2\alpha\varphi W_{O3}^{RH} = 2\alpha\varphi k_2\,[O_2]\{k_2\,[O_2]+2\varphi k_7[O_3]\}^{-1}W_{O3}^{RH} \qquad \text{(Eq.1.41)}$$

The dependencies of the rates of product formation on W_{O3} using O_3/He and O_3/O_2 demonstrate that at equal rate of reaction (1) the rates of product formation vary with the replacement of O_2 by He (Fig.1.9.). This fact could be due to the occurrence of reaction (7) which apparently reduces the efficiency of ozone reaction. The comparison of the equations derived for the processes with and without reaction (7) provides a reasonable explanation for the relationships given in Table 1.8.

Table 1.8

The rates of cyclohexanol and cyclohexanone formation at various rates of ozone absorption ($W_{O3}.10^6$, $M^{-1}.s^{-1}$)

Rate	O_3/He, $W_{O3} = 7$	O_3/O_2, $W_{O3} = 3.5$	O_3/He, $W_{O3} = 5$	O_3/O_2, $W_{O3} = 2.5$
Cyclohexanol	6.0	3.0	4.8	2.2
Cyclohexanone	0.8	1.2	0.6	0.8
RH	3.6	3.5	2.4	2.5
$R^•$	3.4	-	2.5	-

At low O_2 concentration the rates of alcohol formation will be :

$$W_{OH}^{He} = (1 - \varphi) W_{O3}^{RH} + W_{C=O}^{He} + k_5[RO^•][RH] \qquad \text{(Eq.1.42)}$$

The rate of ozone uptake in its reaction with R -radicals can be estimated solving the following equation:

$$W_{O3}^{R•} = W_{OH}^{He} - W_{OH}^{O2} + W_{C=O}^{O2} - W_{C=O}^{He} \qquad \text{(Eq.1.43)}$$

The values obtained for W_{O3}^{RH} and $W_{O3}^{R•}$ are in a good agreement with the experimentally observed two-fold rise of W_{O3}.

Under ozonolysis by mixtures O_3/O_2 ROOR are obtained via reaction 6 (Fig.1.9), and with O3/He, because of the higher stationary concentration of $RO_2^•$ - radicals, the rate of the square termination of the alkoxy radicals increases and hence $W_{C=O}$ (Fig.1.9). In the chromatograms of the oxidates resulting from the ozonation of cyclohexane solutions in CCl_4 we have identified chlorocyclohexane. This fact confirms the generation of $R^•$ - radicals which can further abstract Cl - atom from the solvent molecule via reaction 8:

$$R^• + CCl_4 = RCl + CCl_3^• \qquad (8)$$

According to Ref.14 the rate constant of the reaction at 20°C amounts to 1.59 $M^{-1}.s^{-1}$.

We have estimated the stationary concentration of $R^•$ as 10^{-7} M from the dependence of the rate of cyclohexane formation on the CCl_4 concentration.

The mechanism of ozonolysis includes a stage of radical generation which has been also assumed by other authors [10]. ESR signals, however, have been registered only at ozonolysis of polyoleffins and cumene [50], which are characterized by relatively low values of k_6 and high $[RO_2^•]_{st}$.

We have confirmed the existence of radical steps in the paraffins ozonolysis by carrying the reaction directly in the ESR resonator. For this purpose we have constructed and worked out a special ESR cuvette (Fig.1.10)[50]:

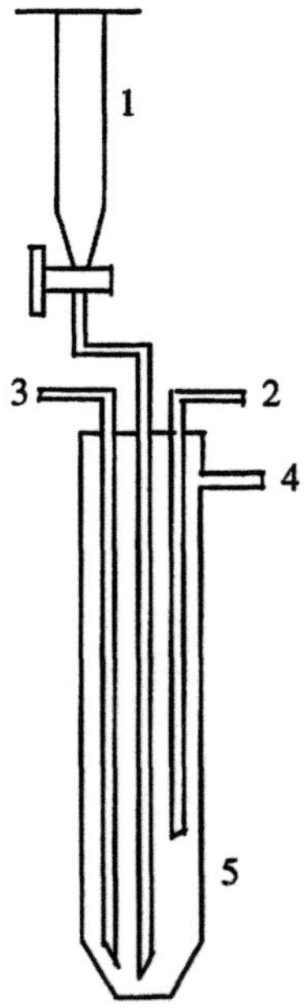

Fig.1.10. A device for carrying out chemical reactions in ESR-spectrometer
1 - separating funnel with capillary for liquid reagent feeding; 3 - capillary connected with the vacuum pump ensuring the necessary level of the liquid in the cuvette; 3 - capillary for the gas inlet; 4 - gas outlet; 5 - ESR-cuvette.

The ESR signals registered are shown in Fig. 1.11. They have been assigned to by comparison with those obtained in the initiated oxidation of n-decane, cumene and polyvinylcyclohexane (PVC) with oxygen. On the basis of the similarities in shape, g-factors and signals width we have assigned them to $RO_2^{\bullet}$-radicals.

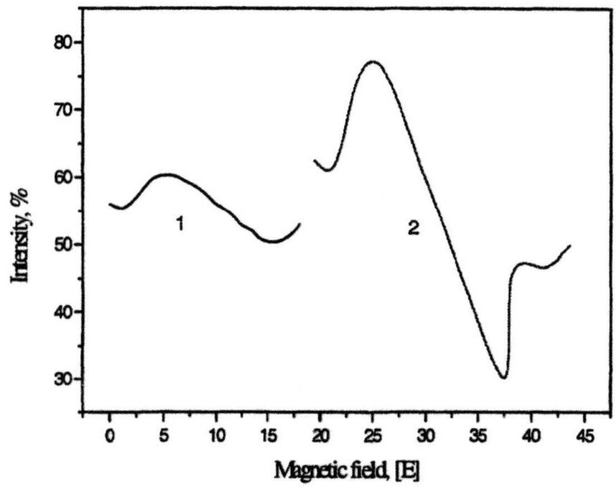

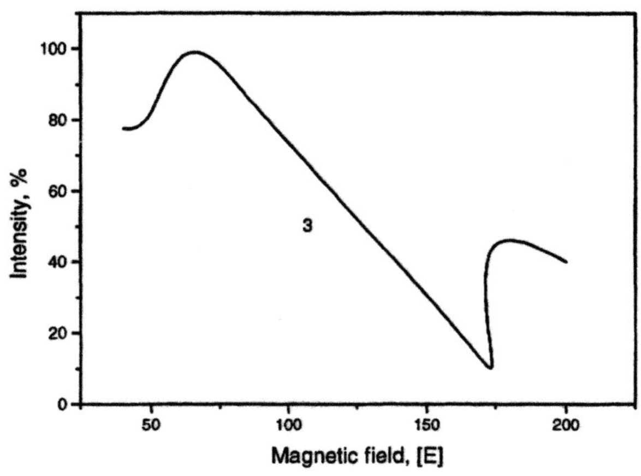

Fig.1.11. ESR -spectra; 1 - n-decane; 2 - cumene; 3 and 4 - PVC at ambient temperature.

The kinetics of ESR signals intensities during the PVCy ozonolysis is presented in Fig.1.12.

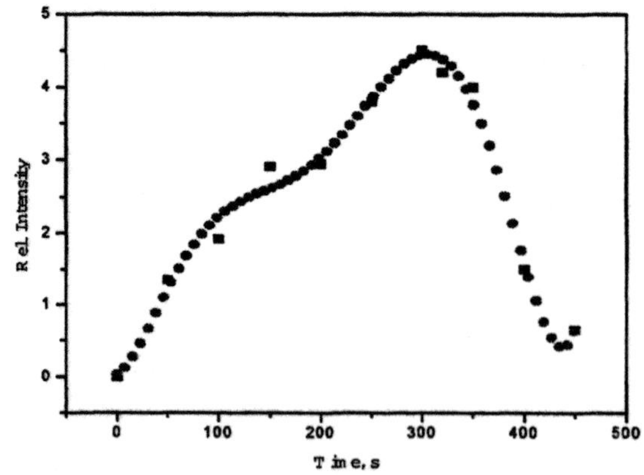

Fig.1.12. Kinetics of $RO_2^{\bullet}$ - radicals formation at the ozonolysis of powder PVC at 20^0C.

The rise of the intensity of $RO_2^•$ - singlet at the beginning has related to the disclosure of new surfaces and ozone interaction with the intermediate products of the reaction. The latter [50] has been considered as one of the reasons for the observed course of the CL - intensity in the ozonolysis of 2,7-dimethyloctane which has been found to be similar.

The drop off of the ESR intensity is due to the consumption of the initial compound (100 % conversion after 7 min). When O_3/He mixture was used, as it should be expected, the ESR signal intensity is of an order lower as a result of the active proceeding of reaction 7.

In Fig.1.13 is given the kinetics of $RO_2^•$-radicals formation during cumene ozonolysis. The absolute concentration of $RO_2^•$-radicals was determined using rubine as reference.

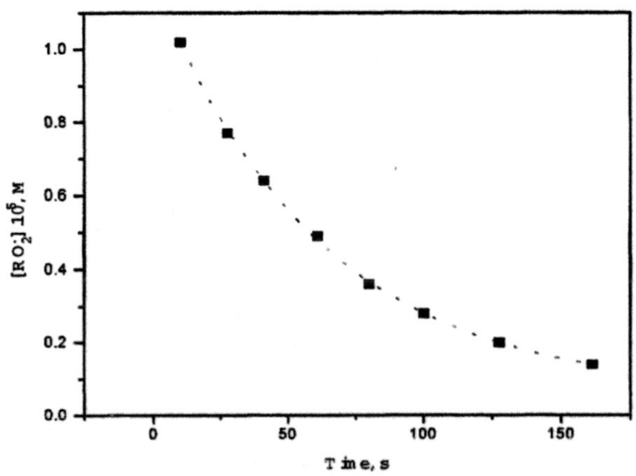

Fig.1.13. Kinetics of $RO_2^•$-radicals formation during cumene ozonolysis: W = $3.5.10^{-4}$ $M.s^{-1}$, 20^0C

The decrease in the intensity of the ESR signal with time is connected to the consumption of cumene because of the high rate of ozone absorption. As the first point in Fig 1.13 is registered at 0.6 % conversion , the stationary concentration of $RO_2^•$-radicals was determined by extrapolation to zero time and amounts to $[RO_2^•]_{st}$ = $1.4.10^{-5}$M. Under the experimental conditions applied the rate of the chain oxidation reaction is very low and thus $RO_2^•$-radicals would be consumed exclusively via square termination reaction. The rate of the later as estimated according to $[RO^•_2]_{st}$ and k_6 = $1.26.10^4$ $M^{-1}.s^{-1}$ [88] is $5.4.10^{-6}$ $M.s^{-1}$. This value is 64 times lower than that calculated at the assumption that W_{O3} = W_i (initiation rate). This fact confirms that ozone is apparently not an initiator in the studied reaction.

1.4. Cumene

It has been found that the O_2-oxidation of cumene takes place through a chain-radical route yielding cumylhydroperoxide (CHP), dicumylhydroperoxide (DCHP), acetophenone (AcP), dimethylphenylcarbinol (DMPC), α-methylstyrene, acetone, methanol, formic acid and benzoic acid [14] as reaction products.

The kinetic curves of products formation during cumene ozonolysis are shown in Fig.1.14.

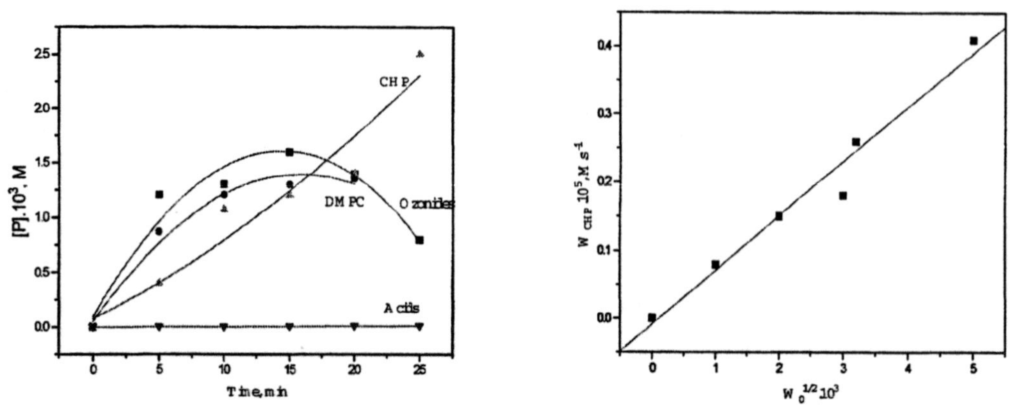

Fig.1.15. a. Dependence of the rates of DMPC and OZ formation on W_{O3}, 20°C.; b. Dependence of the rate of CHP formation on $(W_{O3})^{1/2}$, 20°C.

It is seen that while the rates of DMPC and OZ formation depend linearly on W_{O3}, the rate of CHP formation is proportional to $(W_{O3})^{1/2}$. These results suggest the occurrence of parallel steps of the formation of the reaction products as well as of radical stages in the reaction mechanism.

The chemiluminiscent kinetic curve shown in Fig.1.16, resulting from the $RO_2^{\bullet}$ - radicals recombination [55, 105] becomes a straight line in coordinates $\ln[(I_0^{1/2}+I^{1/2})/(I_0^{1/2}-I^{1/2})]/t.$, where I_o is the intensity of the chemiluminiscent signal, I - the current intensity. The tangent of the curves slope is equal to $0.5.(W_i.k_6)^{-1/2}$, where W_i is the rate of the initiation equal to αW_{O3} (α<1); k_6 - the rate constant of the square termination.

The results obtained could be described by the following mechanism:

Scheme 1.5

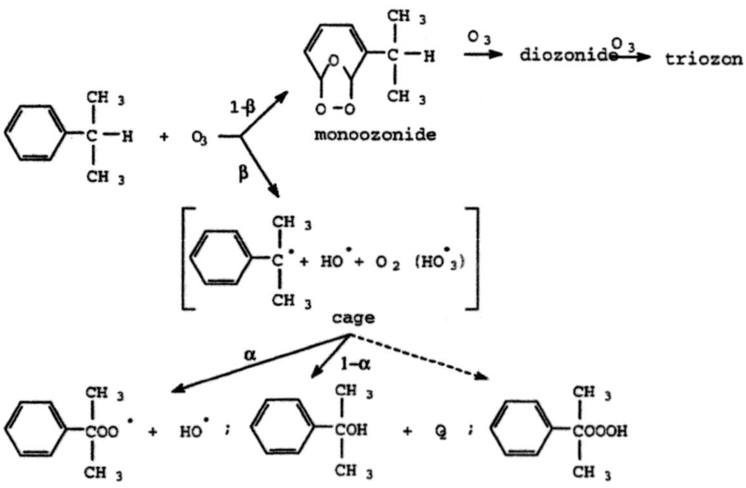

$$RH + HO^{\bullet} = R^{\bullet} + H_2O \ (1)$$
$$ROO^{\bullet} + RH = ROOH + R^{\bullet} \ (2)$$
$$2ROO^{\bullet} = 2RO^{\bullet} + O_2 \ (6)$$
$$RO^{\bullet} + RH = ROH + R^{\bullet} \ (5)$$

At quasi-stationary approximation [14, 28, 104] the rates of products formation will be presented as:

$$W_{DMPC} = (1 + 3\alpha).\beta.W_0,$$ (Eq.1.44)

$$W_{CHP} = k_2 (2\alpha\beta W_0/k_6)^{1/2} [RH]$$ (Eq.1.45

$$W_0 = W_{DMPC} + W^{I}_{DMPC} + W_{CHP} + W_{O3},$$ (Eq.1.46)

where: W'_{DMPC} is the rate of DMPC formation at square termination.

The values of α and β have been determined from the slopes of the kinetic curves in Fig.1.15a and b and on the basis of the literature values for the rate constants k_2 and k_6 [95] and they are 0.04 and 0.23, respectively. It follows form the tangent of the angle on the OZ curve (Fig.1.15a) that 75 % of the effective collisions between ozone and cumene are with the benzene ring and only 25 % are with the alkyl chain of cumene molecule, which fits very well the calculated value of β from Fig.1.16.

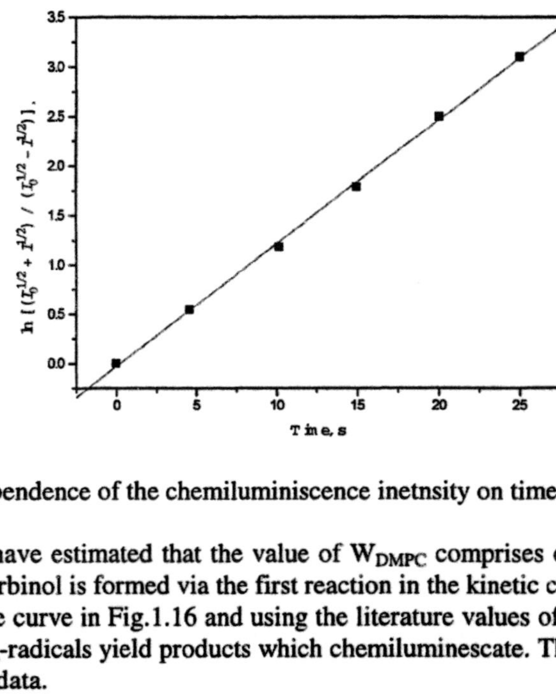

Fig.1.16. Dependence of the chemiluminiscence inetnsity on time at $W_o = 4.2.10^{-5}$ $M.s^{-1}$.

From Fig.1.15a we have estimated that the value of W_{DMPC} comprises only 7 % from that of W_{DMPC}, hence the basic part of carbinol is formed via the first reaction in the kinetic cage.

From the slope of the curve in Fig.1.16 and using the literature values of k_6 we have found out that 0.3 % of W_{O3} and 1/6 of RO_2-radicals yield products which chemiluminescate. These results correlate well with the aforementioned ESR data.

1.4.1.Ozonolysis in the Presence of Transition Metal Compounds

The data on the ozonolysis of cumene in the presence of transition metals are very scant [106, 107]. This reaction has been studied mainly in sulfuric acid medium and low cumene concentrations. Having this in mind we have carried out the ozonolysis reaction both in pure and 50 % solutions of acetic acid [57].

As it has been shown the main products of cumene ozonolysis are OZ, DMPC and CHP. In acetic acid, the product composition remains the same, but the amount of the products is decreased by about 1.6 times. The rates of CHP and OZ formation are lower by about 8.8 and 1.6 fold, respectively, and the rate of DMC accumulation grows 1.6 times. The ratio of the products is found to be 2.1:1.9:1 = OZ:DMPC:CHP. The rate constant (k) characterizing the ozone-acetic interaction at ambient temperature amounts to $1.5.10^{-5}$ $M.s^{-1}$ that is by about 3.5.10-4 smaller that that found for the ozone reaction with cumene ($k = 0.35$ $M^{-1}.s^{-1}$). It is known that AcOH deactivates ozone, inhibits $RO_2^{\bullet}$-radicals and protects the benzene ring [9-101]. The deactivation of ozone in AcOH is also confirmed by the change of the molar coefficient in the UV-spectrum as compared with that in CCl_4.

The catalytic properties of Co^{2+} in the hydrocarbon oxidation have been the subject of intensive investigations [107]. It has been established that during the cumene-AcOH ozonolysis in 1:1 (v:v) in the presence of $Co(AcO)_2$ the oxidation reaction is accelerated (Fig.1.17).In contrast to the noncatalysed process in the catalyzed by transition metal salts the ozonolysis is characterized by: 1) absence of ozonides formation that is indicative of the absence of ozone interaction with the phenyl ring and 2) the main product is DMPC, the accumulation rate of which proportional to the concentration of Co^{2+} after the 10 min. The initial rates of CHP formation do not vary with the changes in Co^{2+} but after the 15 min the rates increase with $[Co^{2+}]$. It can be seen from Table 1.9. that if we assume the ozonolysis of pure cumene as a reference then the addition of AcOH results in autoretardation of the oxidation rate and to reduction of the products yield. The ratio $[\Sigma P]/[O_3\}$ reaches value of 6.9.

Table 1.9

Kinetics of formation of the sum of products (ΣP) and their ratio per unit absorbed ozone $[\Sigma P]/[O_3]$ for the noncatalysed and catalyses reaction in the presence of AcOH (cumene:AcOH = 1:1) at ambient temperature

	Time, min	Cumene	Cumene+ AcOH	$[Co^{2+}]$= 1.10^3M	$[Co^{2+}]$= 2.10^3M	$[Co^{2+}]$= 3.10^3M	$[Co^{2+}]$= 4.10^3M	$[Co^{2+}]$= 5.10^3M
$[\Sigma P].10^3$ M	5	0.3	0.2	0.8	0.8	0.9	1.0	1.1
$[\Sigma P]/[O_3]$.		1	0.7	2.7	2.7	3.0	3.3	3.7
$[\Sigma P].10^3$ M	10	0.7	0.4	1.4	1.4	1.8	2.2	3.3
$[\Sigma P]/[O_3]$.		1	0.6	2.0	2.0	2.6	3.1	4.7
$[\Sigma P].10^3$ M	15	1.0	0.6	2.1	2.1	2.9	4.0	5.9
$[\Sigma P]/[O_3]$.		1	0.6	2.1	2.1	2.9	4.0	5.9
$[\Sigma P].10^3$ M	20	1.4	0.8	2.8	2.8	4.5	6.1	8.7
$[\Sigma P]/[O_3]$.		1	0.6	2.0	2.0	3.2	4.3	6.2
$[\Sigma P].10^3$ M	25	1.7	1.0	3.5	3.5	6.2	8.1	11.7
$[\Sigma P]/[O_3]$.		1	0.6	2.1	2.1	3.6	4.8	6.9
$[\Sigma P].10^3$ M	30	2.0	-	4.2	4.2	8.0	10.1	-
$[\Sigma P]/[O_3]$.		1	-	2.1	2.1	4.0	5.1	-

These data indicate the participation of oxygen accompanying ozone in the formation of the reaction products. The course of the kinetic curves of DMPC formation on the $Co(AcO)_2$ concentration is connected with its formation in reactions with participation of $Co(AcO)_2$ (Fig.1.17).

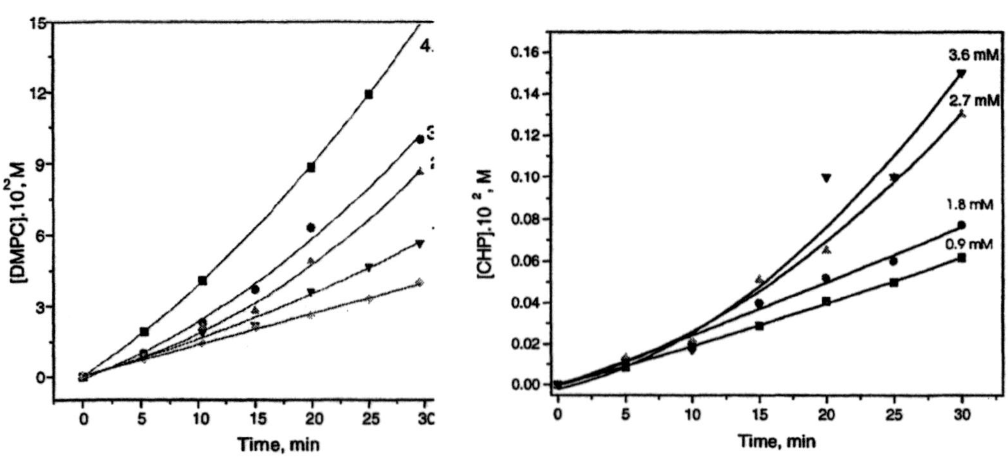

Fig.1.17. Kinetics of DMPC and CHP formation on $[Co(AcO)_2]$, 20°C.

The catalytic properties of the metal ions under other equal conditions depend on their rate constant with ozone [102]. For that reason we have determined the values of k in the presence of different transition metals in cumene-AcOH medium. In Table 1.10 are presented the rate constants of ozone reaction with some transition metals ions at 20°C in cumene: AcOH solution (1:1, v:v).

Table 1.10

Rate constants of the zone reaction with some transition metal ions

Reagent	$[Me^{+n}]$, mM	k, $M^{-1}.s^{-1}$
Co^{2+}	1	500
Mn^{2+}	1	1500
Cr^{3+}	0.5	64
Cumene	$6.52.10^3$	0.032

Upon ozonolysis of cume
ne in AcOH the value of W_{O3} is 2.10^{-6} $M.s^{-1}$ and at addition of $Co(AcO)_2$ in concentrations of (0.9, 1.8, 2.7 and 4.5).10^{-3} its value is 0.6, 1.1, 1.7 and 2.9 $M.s^{-1}$, respectively. This clearly shows that first of all ozone reacts with the Co^{2+} ions.

In conformity with the data obtained we have proposed the following scheme of ozonolysis;

Scheme 1.6

$$M^{n+}L + O_3 \rightarrow M^{n+1}L(O_3^-)$$
(0)
$$M^{n+1}L(O_3^-) + RH \rightarrow R^\bullet + M^{n+1}L\,(\hat{I}H) + O_2$$
(1)
$$M^{n+1}L(OH) + RH \rightarrow M^{n+}L + R^\bullet + H_2O$$
(1')
$$R^\bullet + O_2 \rightarrow RO_2^\bullet$$
(2')
$$RO_2^\bullet + RH \rightarrow ROOH + R^\bullet$$
(2)
$$R^\bullet + O_3 \rightarrow RO^\bullet + O_2$$
(3)
$$RO_2^\bullet + RO_2^\bullet \Big| \begin{array}{l} \rightarrow (\alpha)\ 2\ RO^\bullet + O_2 \\ \rightarrow (1-\alpha)\ ROOR + O_2 \end{array}$$
(6)
$$RO^\bullet + RH \rightarrow ROH + R^\bullet$$
(4)

Applying the quasi-state approximation approach in relation to WDMPC we will obtain:

$$W_{DMPC} = 2.[\alpha/(1-\alpha)].k_0\,.[Me^{n+}L].[O_3].[RH]$$
(Eq.1.47)

The last equation (1.47) has been derived without considering reactions 2 and 3 because of their insignificant contribution. It follows from eq.1.47 that W_{DMPC} is proportional to the concentration of the catalyst applied.

By applying literature values for k_2 and k_6 in cumene oxidation [14, 103-105] and $[RO_2]_{st}$ as determined from Fig.1.17 the value α is found to be 0.63 that is in a good agreement with that reported by other authors [14]. The analysis of Fig.1.18, Table 1.10 and considering the concentrations of ozone gives the same value of α. The good correlation of the kinetic results with the proposed Scheme 1.16 testifies its validity.

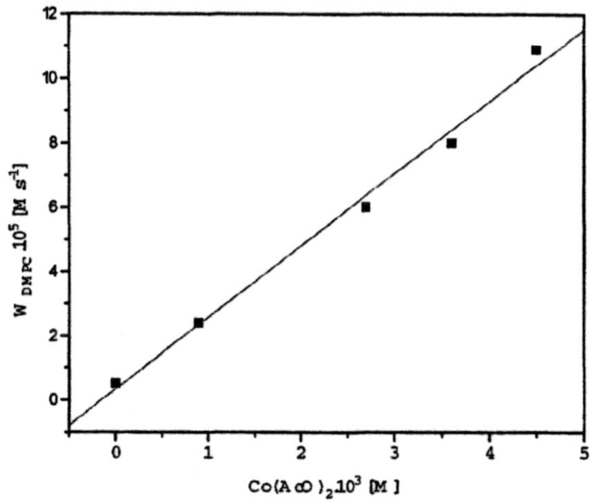

Fig.1.18. Dependence of W_{DMPC} on $[Co(AcO)_2]$

The addition of Co and Mn acetates to the reaction mixture changes the general features of the products formation kinetics (1.19). Thus W_{OZ} increases and acetophenone has been identified (AcP). The sum of the products exceeds by 7.3-fold that at cumene ozonolysis. OZ are obtained at the interaction of ozone with the benzene ring, and AcP is provided by the monomolecular decomposition reaction of $RO^{\bullet}$-radicals. Most likely the main role of Mn is in accelerating of these two reactions. The catalytic properties of the metal salts studied are confirmed by the ratio of the amount of the products formed in the catalyzed and noncatalysed processes per unit of absorbed ozone. At Co:Mn=5:1 this ratio becomes equal to 7.3 and is greater by about 6 % than that in the absence of Mn. The cumene conversion is increased but the selectivity of the process is reduced. The contribution of OZ and ApC to the total sum of the products formed is 27 %. The synergism of the simultaneous action of the both salts (Fig.1.19) can be associated with the occurrence of the following reaction:

$$Mn^{3+}(AcO)_2(O_3^-) + Co(AcO)_2 = Co^{3+}(AcO)_2(O_3^-) + Mn(AcO)_2$$

The authors of Ref. 108 stated that with increasing the Mn concentration the catalytic effect should decrease mainly on account of its noneffective interaction with the oxy radicals. The data of Fig.1.19, however, show that such a trend is not observed as the variation in the concentrations of DMPC, CHP and AcP are insignificant.

1.4.2. Ozonolysis in the Presence of NiO

The ozonolysis of cumene in the presence of heterogeneous additives which are effective in ozone decomposition [14] is of interest from two points of view: a) for elaboration of the mechanism of the ozone interaction with the heterogeneous surface and b) to increase the oxidation selectivity.

At bubbling of ozone (0.1 l/min) through a powder NiO (33 m^2/g) depending on the oxide concentration a partial or complete ozone decomposition is observed (Fig.1.20). The ozone concentration becomes zero at NiO concentrations . 140 mg.

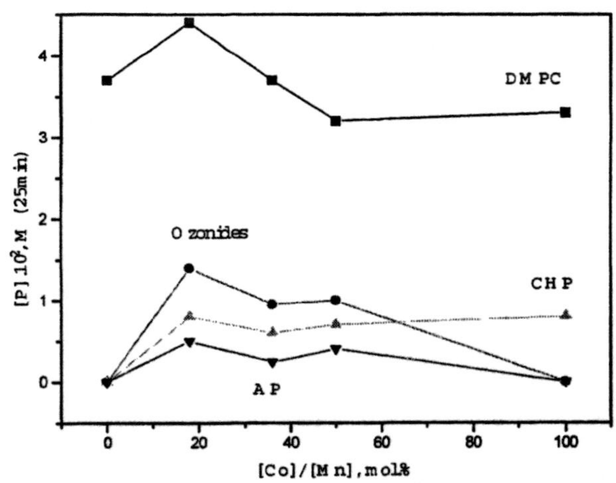

Fig.1.19. Reaction products formation depending on the ratio Co/Mn.

The profiles of the curves is explained by the oxy-reduction interactions of the catalysts and with the radical intermediates.

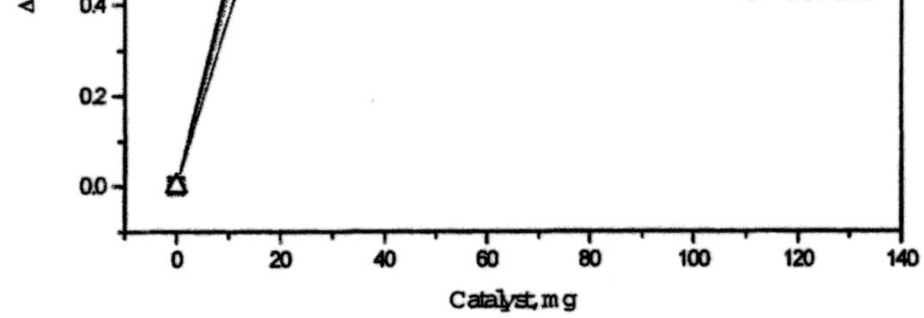

Fig.1.20. Dependence of the ozone decomposed quantity $\Delta D/D_{ot}$ on the NiO amount. $\Delta D=D_o-D_t$; D_o - ozone concentration at the reactor inlet at flow rate of 6 l/h

In the ESR spectrum a bad-resolved singlet is registered which, however, disappears in the course of the reaction. This signal was assigned to O_3^-anion radicals. The UV spectra have a maximum at 320 nm which is attributed to the electron transfers in Ni^2. The intensity of the latter goes down in the process of ozone decomposition (the gray color of the fresh NiO turns black during the reaction proceeding). We have estimated that the catalytic coefficient, i.e. the number of decomposed ozone molecules per one NiO molecule is about 20. By special experiments it has been established that on one Ni^{2+} of the gas phase consequently are absorbed about 133 ozone molecules, but the probability of Ni^{2+} to be transformed to Ni^{3+}

is 1/7, i.e. only one from seven ions is oxidized. These results indicate that the reaction of ozone with Ni^{2+} proceeds probably in two ways:

$$NiO + 2O_3 \rightarrow NiO + 3\ O_2 + Q$$
$$NiO + O_3 \rightarrow ONi^{3+}O_3^-$$

with the formation of surface compounds of $ONiO_3^+O_3^-$ type. They can further react with a new ozone molecule from a neighboring center, with NiO or with the oxides:

$$ONi^{3+}O_3^- + O_3 \rightarrow NiO_2 + 2.5O_2$$
$$ONi^{3+}O_3^- + ONi^{3+}O_3^- \rightarrow Ni_2O_3 + 2O_2$$
$$ONi^{3+}O_3^- + NiO \rightarrow Ni_2O_3 + O_2$$
$$NiO_2 + NiO \rightarrow Ni_2O_3$$

If a hydrocarbon is added to this system a proceeding of oxidation with participation of $ONiO_3^+O_3^-$ might be expected. For this purpose the cumene ozonolysis was carried out in the presence of NiO.

Fig.1.21 illustrates that OZ, DMPC and CHP are the main products of the cumene ozonolysis in the presence of NiO.

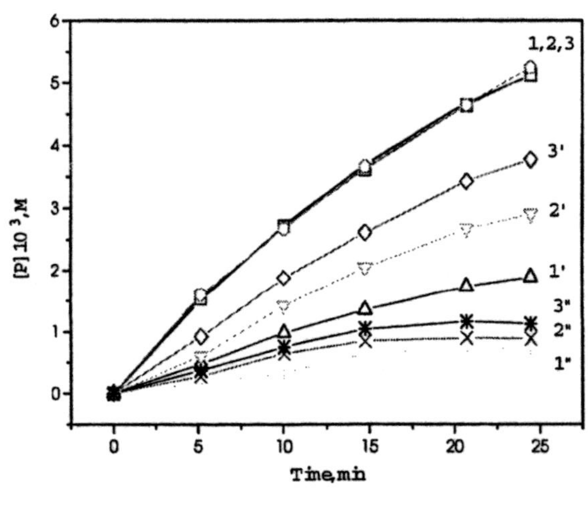

Fig.1.21. Kinetics of products formation of the ozone reaction with cumene in the presence of NiO at 20°C. 1, 2, 3-OZ; 1, 2, 3 - DMPC, 1, 2, 3 - CHP. 1-0.05, 2-0.325, 3-0.635 % wt NiO

It can be seen that the rate of OZ formation is the highest one and does not depend on [NiO] while those of DMPC and CHP are relatively lower and increase with [NiO]. As we have mentioned before the ozonides are obtained from the ozone interaction with C=C bonds in the benzene ring, and the ozonolysis of the hydrocarbon part takes place via a hydrogen atom abstraction mechanism. Our explanation for these experimental facts is related to the absence of ozonides formation on the surface and to the dominating homogeneous reactions as well as to heterogeneous reactions at ozonolysis of the alkyl part. The latter is related to the ozone activation on the NiO surface via the formation of the surface compound Ni3+O3- which according to the scheme:

$$Ni^{3+}O_3^- \rightarrow Ni^{3+} + O_3^-$$
$$O_3^- + RH \rightarrow R^{\bullet} + HO^- + O_2$$
$$Ni^{3+} + RH \rightarrow R^{\bullet} + Ni^{3+} + H^+$$

oxidizes the alkyl part of cumene.

The contribution of these reactions to the formation of the reaction products, is however small, as the total rate of products formation is only from 5 to 17 % higher as compared to the rate of ozone consumption.

1.4.3. Ozonolysis in the Presence of Mo and V Oxides

We have studied ozonolizis of cumene catalyzed by Mo-V catalysts. The reactions have been performed at ambient temperature in a glass reactor with a centrifugal pump with 33 ml cumene in the presence of 0.35 % wt mixtures of MoO_2 and V_2O_5 on SiO_2 (Fig.1.22) at various ozone concentrations.

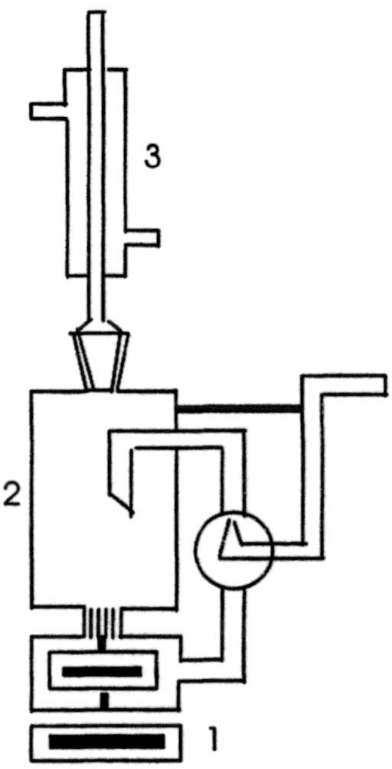

Fig.1.22. Reactor for kinetic studies; 1- magnetic stirrer; 2-reactor; 3-cooler; 4 - centrifugal pump with rotation of 100-1100 r/min

The catalysts presented in Table 1.11 were prepared by soaking of SiO_2 with water solutions of ammonium paramolibdate and ammonium vanadate with subsequent treatment at 400°C.

Table 1.11

Ccatalysts sample at total amount of oxides 10%.

No.	$MoO_2 : V_2O_5$	S, m^2/g
1.	0 : 1	13
2.	1 : 99	10
3.	1 : 9	11
4.	1 : 3	5
5.	1 : 1	4
6.	9 : 1	5
7.	1 : 0	10
8.	0 : 0	14

Fig.1.23 a-c illustrates the kinetic curves of major products formation - CHP, DMPC and OZ . The rate of CHP formation for sample 5 with Mo:V = 1:1 ratio amounts to $1.6.10^{-5}$ $M.s^{-1}$. The values of W_{CHP} for samples Nos 7 and 8 are similar to that value. In case of sample No1 the initial rate of CHP formation is to $0.4.10^{-5}$ $M.s^{-1}$ and is the same as that for the noncatalysed reaction (curve K). The untreated SiO_2 (sample 8) exerts also some catalytic properties. The integral areas below (under) the kinetic curves corresponding to the amount of the products formed vary in the presence and in the absence of the catalyst. For example, the area below the kinetic curve of CHP formation in the catalyzed reaction is greater as compared to that of the noncatalysed one. The stationary concentration of CHP depends on the type of the catalyst and in relation to their catalytic activity the sample shows the following order 5>8>7>1>K. Curve K is characterized by a S-shape form whereby three periods are distinguished: 1) the start of the product accumulation; 2) stationary regime in relation to the product; 3) autocatalytic process of product formation (Fig.1.23, A).

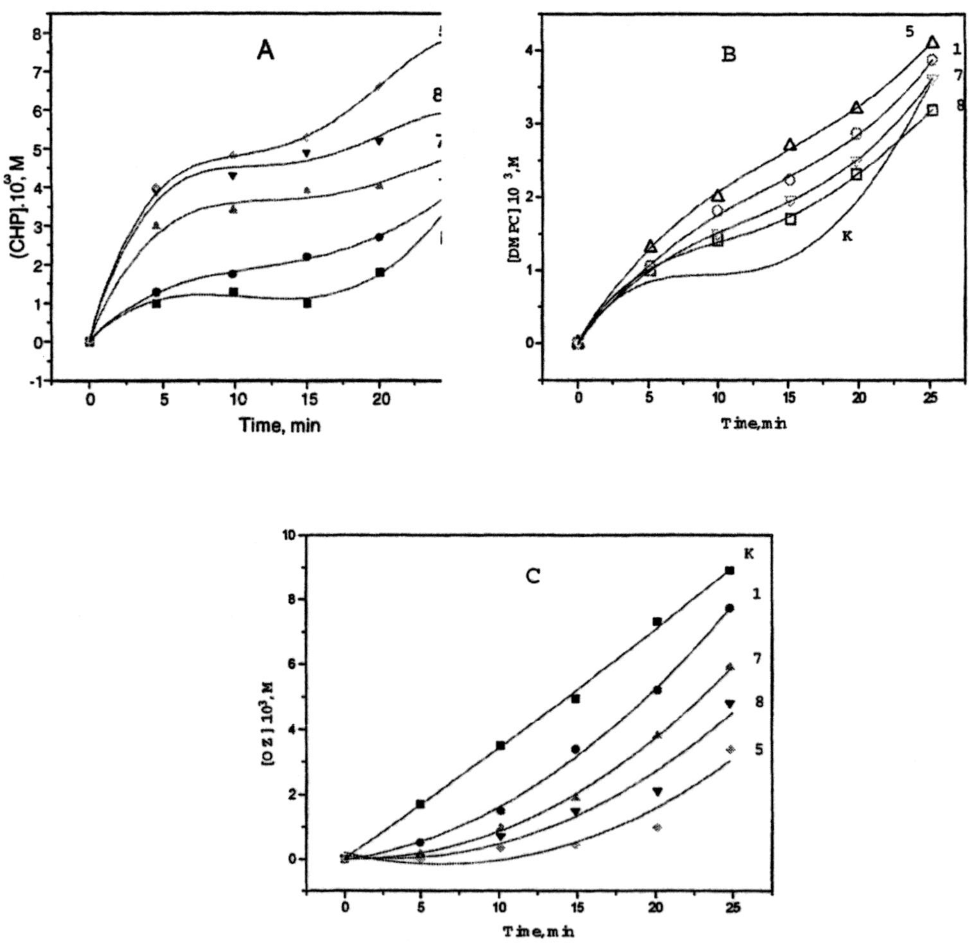

Fig.1.23.A-C. Kinetic curves of products formation during the catalyzed cumene ozonolysis. A-CHP, B-DMPC, C-OZ; 1, 5, 7 and 8 - the number of catalysts according to Table 1.11; K- in the absence of catalyst, $W_{O3} = 1.10^{-5}$ $M.s^{-1}$.

Upon DMPC formation (Fig.1.23, B) one can also notice some difference in the course of the kinetic curves in the presence (curves 1, 5, 7 and 8) and in the absence (curve K) of the catalyst. It has been found that the rate of DMPC formation is one and the same for all samples studied and it is $4.2.10^{-6}$ $M.s^{-1}$. After 10 min curve K becomes constant and after 15 min it again rises, i.e. its profile is typical for an autocatalytic process. The curves of DMPC formation in the presence of catalyst (1, 5, 7 and 8) are characterized by two sections - an initial one with a bigger slope and a second one with a smaller slope and the transition point is in the interval 5-10 min. Obviously, W_{DMPC} is slowed down in the advanced reaction because of the occurrence of secondary reactions associated to the DMPC depletion or are due to the catalyst deactivation. With respect to their activity the samples have the following sequence: 5>1>7>8>K.

The kinetics of OZ formation is entirely different from those discussed before (Fig.1.23, C). The initial section of the curves is characterized by the appearance of induction period up to 15 min. These results unambiguously point out that the catalyst direct selectively the process to the oxidation of the isopropyl substituent in the cumene molecule and only after the accumulation of a definite amount of products, which probably block the catalytic surface, the ozone attacs the benzene ring. In this case the order of activity of the samples is just opposite to that of CHP formation, i.e. 5<8<7<1<K.

The effect of the catalyst type on the amounts of products formed (20 min) is demonstrated in Fig. 1.24.

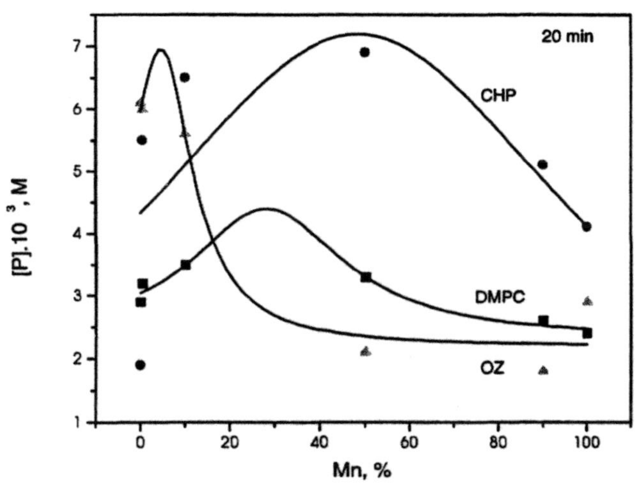

Fig.1.24. Kinetic curves of products formation depending on the Mo:V ratio

The increase of the Mo content in the catalyst leads to changes in the amounts of DMPC and CHP. The latter rise and reach maximum value at Mo content of 28.2 and 48.5 %, respectively. Simultaneously, the OZ yield is reduced with the rise of Mo content.

The catalytic effect of the samples studied is also demonstrated by the increase of the initial rate of oxidation that results in rise of CHP yield. This increase in time can be presented by the following equation:

$$[CHP] = a.t^2 + b.W_i^{1/2}.t \qquad \text{(Eq.1.48)}$$

For example, the CHP amount formed in the presence of catalyst is 2.4-fold higher as compared with that of the noncatalysed process. The selectivity of the catalysts studied consists in directing the process to oxidation of the isopropyl substituent in the cumene molecule.

1.4.4. Cumenehydroperoxide

In order to assess the effect of secondary reactions occurring during cumene ozonolysis we have investigated its reaction in the presence of CHP. The dependence of W_{O3} and $[O_3]$ is studied by means of stop-flow techniques at mixing time of 0.5 s.

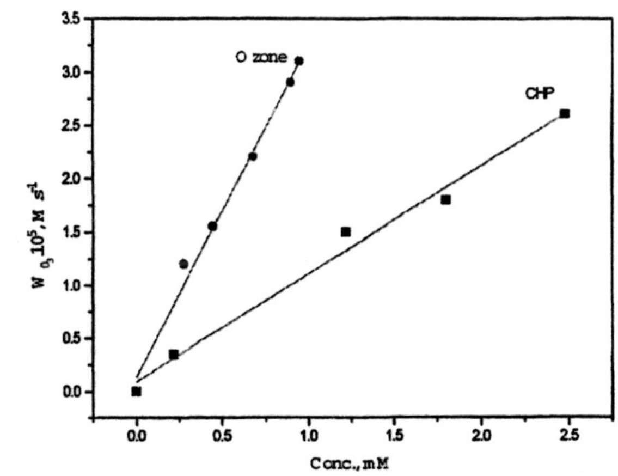

Fig.1.25. Dependence of the rate of ozone absorption on $[O_3]$ ([CHP] = 2.5 M) and [CHP] ($[O_3]$ = 0.68 mM), 20°C, CCl_4.

As seen the rate depends linearly on the concentration of the both reagents, thus, showing a first order kinetics with respect to each of them. The rate constant k was determined from the slope and it amounts to to 14.5±0.2 $M^{-1}.s^{-1}$. The main reaction products were found to be DMPC and OZ, while AcP was obtained in much smaller amounts (Fig.1.26).

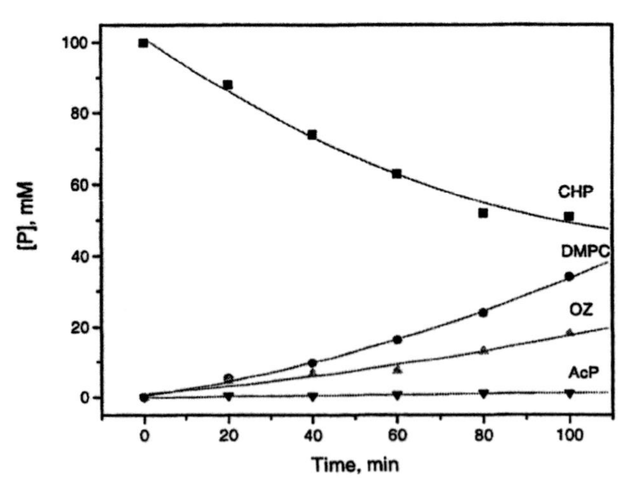

Fig.1.26 Kinetics of products formation during ozonolysis of CHP in CCl_4, $[O_3]$=0.36 mM.

Ozone reacts simultaneously with both the alkyl and phenyl moiety in the CHP molecule yielding DMPC, AcP and OZ. In the former case the attack on the OOH-group is much more kinetically beneficial resulting in the formation of cumeneperoxide and hydroxy peroxide radical. These radicals according to the classical concepts of oxidation, further undergo transformation giving rise to DMPC and AcP. Upon the attack on the phenyl ring, ozone is incorporated into the double bond forming mono-, di- and triozonides. The formation of monoozonide is the limiting step as that moment the aromatic character of the ring is destroyed while the ozone interaction with dienes and olefins occur with high rates and low activation energy [1, 25]. In view of this we have proposed the operation of the following mechanism for this reaction:

Scheme 1.7.

$$ROOH + O_3 \rightarrow \alpha\,(OZ) + \beta\,(RO_2^{\bullet} + HO^{\bullet} + O_2) + \gamma\,(RO^{\bullet} + HO_2^{\bullet}) \quad (1)$$
$$RO_2^{\bullet} + RO_2^{\bullet} \rightarrow 2\delta\,(RO^{\bullet}) + (1\text{-}\delta)\,(ROOR) \quad (6)$$
$$RO^{\bullet} + ROOH \rightarrow DMPC + RO_2^{\bullet} \quad (4)$$
$$RO^{\bullet} \rightarrow AcP + CH_3^{\bullet} \quad (5)$$

On the basis of the above pointed Scheme and the available value of k for reactions 4 and 5 the theoretical ratio W_{DMPC}: W_{AcP} is 25÷30:1. The experimentally observed value is 25:1 which is in a very good fit with the theoretical one. Another interesting ratio has been obtained for W_{DMPC}:W_{OZ} = 2.5:1 (Fig.1.16) and the values of α=0.4 and ($\beta+\gamma$)=0.6 have been found. The good correlation between the theoretical and experimental values of the W_{DMPC} :W_{AcP} ratio suggests that γ<<1, i.e. the free-radicals formation is insignificant. The rate constant of ozone reaction with DMPC at ambient temperature has been determined in a separate kinetic experiment and $k=3.4\pm0.1\ M^{-1}\ s^{-1}$. At this value of k the rate of end products formation becomes commensurable at [CHP]:[DMPC]=1:4.3. The linear profile of the kinetic curves of DMPC, OZ and AcP formation depending on time duration and the absence of induction periods can be regarded as a kinetic evidence for the independent formation of the products in parallel reactions (Fig.1.26). Upon ozonolysis of cumene I mol absorbed ozone yields 1 mol products, hence the reaction is not a chain one.

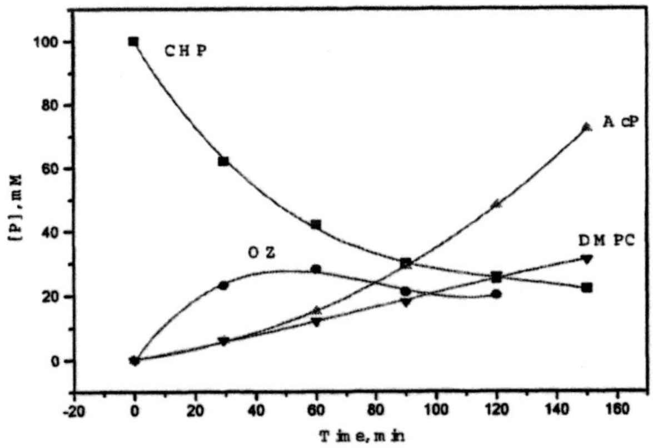

Fig.1.27 Kinetics of products formation during CHP ozonolysis in AcOH, $[O_3]$=0.8 mM

The main products in acetic acid medium (Fig.1.27) are OZ, DMPC and AcP. The curve of CHP consumption is found to decrease exponentially (it becomes a straight line in semilogarithmic coordinates), the curve of OZ formation passes through a maximum, DMPC is accumulated linearly with the time and that of AcP formation is strongly accelerated after 50 min. The ratio of the initial rates of DMPC:AcP and DMPC:OZ formation are 1:1 and 1:5, respectively. Phenol (PhOH)which have been identified in the oxidate could be obtained from the well known heterogeneous decomposition of CHP in acid medium giving rise to phenol and acetone.

Scheme 1.8

$$Ph–C(Me)_2–OOH + H^+ \Leftrightarrow Ph–C(Me)_2–O–^+OH_2$$
$$Ph–C(Me)_2–O–^+OH_2 \rightarrow Ph–C(Me)_2–O^+ + H_2O$$

$$Ph{-}C(Me)_2{-}O^+ \rightarrow Ph{-}^+O{=}C(Me)_2$$
$$Ph{-}^+O{=}C(Me)_2 + H_2O \rightarrow Ph{-}O{-}C(Me)_2{-}^+OH_2$$
$$Ph{-}O{-}C(Me)_2{-}^+OH_2 \Leftrightarrow Ph{-}O{-}C(Me)_2{-}OH + H^+$$
$$Ph{-}O{-}C(Me)_2{-}OH\ (H^+) \rightarrow PhOH + (Me)_2C{=}O$$

The composition of products has been found to be different in ozonolysis in $CHCl_3$ as compared with that in pure cumene and in AcOH solution (Fig.1.28).

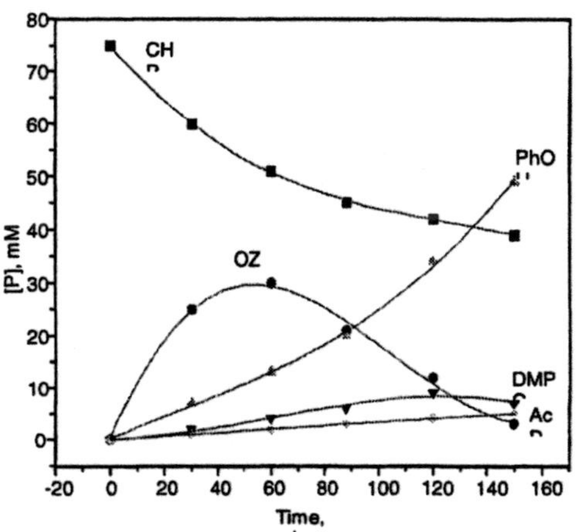

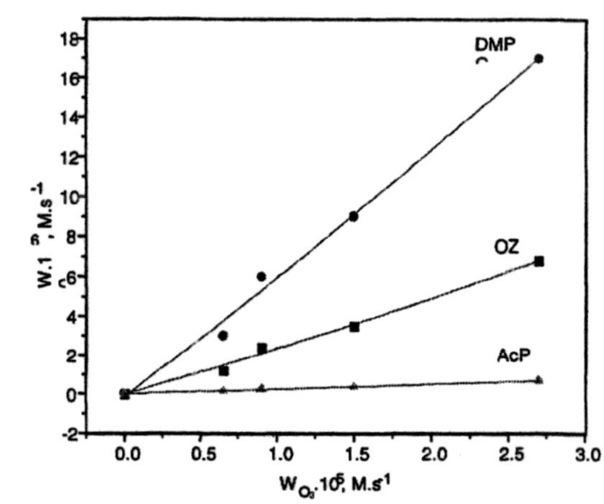

Fig.1.28. Kinetics of products formation at CHP ozonolysis in $CHCl_3$, $[O_3]$=0.8 mM
Fig.1.29. Dependence of the initial rates of products formation (W) on the rate of ozone absorption (W_{O3}) in CCl_4, 20°C.

It has been established that the main products are again OZ, DMPC and AcP but higher amounts of phenol are identified. In addition, the ozone reaction with $CHCl_3$ yields inorganic chlorocontaining acids and the pH of the medium approaches values of 1-2. These acids (probably HCl and HOCl) are efficient catalysts for the CHP decomposition to phenol. The ratios of the rates are W_{DMPC}:W_{AcP} = 2.5:1 and W_{DMPC}:W_{OZ}=1:14.

The kinetic curve of ozonides formation is indicative of its formation in a consecutive reactions.

The rates of AcP and PhOH formation rise just at the moment of the drop on the kinetic curve of OZ formation (Fig.1.27 and 1.28) thus suggesting the participation of OZ in their formation.

The dependence of the initial rate of DMPC, OZ and AcP on W_{O3} in the cumene ozonolysis in CCl_4 is presented in Fig.1.29. The values of α =0.4, β=0.6, γ<<1 and δ=0.04 calculated from this figure are in a good agreement with those obtained from the analysis of Scheme 1.7 and Fig.1.26.

1.5. Polyethylene and Polypropylene

We have extended our investigations on ozonation of the C-H bonds in polymer analogs of paraffins - polyethylene, polypropylene, their mixtures and polystyrene, with the aim of elaborating the effect of the polymer structure on this reaction.

The ozonolysis of polyolefins leads to change of the structural-physical and dynamic parameters of the polymer matrix such as: the degree of crystallinity (κ); density of the amorphous phase; orientation factor and segmental mobility of the macromolecules [109-110]. The variations of these parameters in polymer mixtures have been scarcely studied. This provokes us to study the effect of ozone reaction with oriented polyethylene high density (PE), oriented isotactic polypropylene (PP) and their mixtures on the melting point of crystallites (T_t), the relative melting heat, and thus on the degree of crystallinity and the zise distribution in the crystallites of the two components. The polymer blends were prepared from noninhibited PP samples with average weight mass (M_w) of $2.86.10^5$, average number mass (M_n) of $6.23.10^4$ and the ratio between them (M_w/M_n) of 4.6 and for PE samples with the corresponding values of $4.15.10^4$, $2.71.10^4$ and 1.53. The isotropic films were prepared by pressing of granules at 463 K and under pressure of 150 MPa on a cellophane support and tempered with water at 373 K. The oriented films of 12±2 μm thickness and stretching degree (λ=9) were obtained by local heating up to 373-383 K and 393-403 K. The melting point and the degree of crystallinity were evaluated by means of DSC method (Fig.1.30).

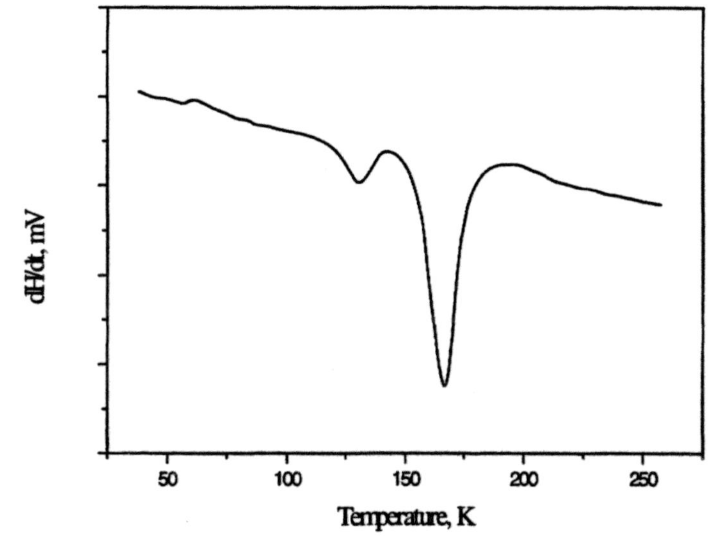

Fig.1.30. DSC curves of the PE:PP = 90:10 blend at λ=9

The values of (κ) for PP and PE were estimated using the following values of the relative heats of crystallites melting , i.e. 284.2 and 135.9 kJ/g, respectively. The polydispersity of the crystallites was characterized by the half width of its peak of melting Δh. The oriented samples were subjected to ozonolysis in a thermostated ozone chamber. The
reaction kinetics was followed by monitoring the variations of the IR spectral intensity at ν=1715 cm-1, characteristic of C=O. As an internal reference the frequency at 1455 cm^{-1} which remains constant during the course of the experiment, was used.

The dependencies of the oxidation level per unit thickness of the sample on the mixture composition at various oxidation times are presented in Fig.1.31.

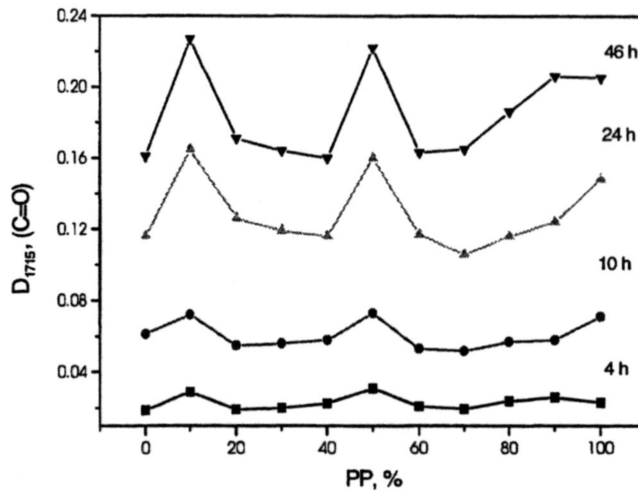

Fig.1.31. Kinetics of C=O accumulation in oriented films form PE/PP mixtures with λ=9, $[O_3]$ = 1 mM.

After 4 h in the IR spectra of the samples containing 10 and 50 % PP are registered two weak maxima. The appearance of the latter could be associated to the more porous structure leading to the better dissolving of ozone in them. This also suggests that the rate of oxidation should increase with increasing the time of ineteraction. We will discuss further the changes of T_T and κ of PE and PP in the ozonized and fresh samples in dependence on the mixture composition. The relationship between T_T of the crystallites in the initial and those ozonized for 45 h at ambient temperature are shown in Fig.1.32.

As seen T_T goes down for the PE and PP samples, irrespective of the composition and duration of treatment. For PE this decrease is 2-6°C depending on the composition of the mixture and in the case of PP it is considerably higher - 12°C for the initial PP and 24°C for the 1:1 mixture. As for the ozonized samples of PP content within 0-30%, PP crystallites are not observed and for those with PP wt 40 % their amount is within the limits of accuracy of the DSC method.

The lower melting point of the crystallites films points to a reduced ordering of the crystal structure of PP and PE or for increase of the relative free surface energy of the crystallites sides (σ_e) as compared with that of the nonozonized ones. The disturbance of the crystallites ordering might be ascribed to the penetration of ozone in them. At PE ozonation the ozone penetration in the crystallites depth is much weaker since T_T in the ozonized and nonozonized samples is similar (Fig.1.32) or it reacts only with the folded surface of the crystallites [115, 116]. The data in Ref.117 support the latter supposition for it has been found that the treatment by 100 % nitric acid of the oriented PE samples of stretching degree λ=10 leads to 5-8°C reduction of T_T of the crystallites. This decrease is related to the increase of σ_e of the crystallites walls under this treatment. A similar decrease of T_T (2-6°C) has been also established upon ozonation of oriented PE with λ=9 (Fig.1.32) which gives us ground to assume that similar to nitric acid, ozone reduces T_T also on account of σ_e rise.

The ozone ability to penetrate the PP crystallites is accelerated with the decrease of its content from 90 to 50 % in the mixtures (Fig.1.32).

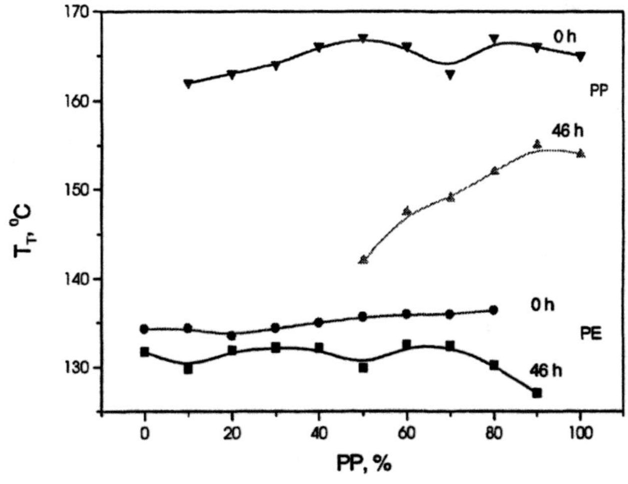

Fig.1.32. Variation of the melting points of PE and PP crystallites in mixtures

The marked decrease of m.p. (12-24°C) upon PP ozonolysis could not be explained by the increase of σ_e of the crystallites as they simply disappear. More probably the amorphization is connected to the destroy of crystallites as a result of the ozone action. Actually, in the DSC curves of the ozonized mixtures with 0-40 % PP content crystallites are not observed while in the IR spectra the amount of the CH_3 groups is not altered. Ozone reacts with the tertiary hydrogen bong in the amorphous and crystal part. The former leads to increase of the number of defects which facilitates the subsequent ozone access. Thus the crystallites are destroyed and further get amorphized.

The melting point appears to be a very sensitive parameter in investigating the behavior of the ozonized samples. In Figs 1.31 and 1.32 is observed an antibat dependence between the curves illustrating the dependence of the oxidation level on composition and T_T of the crystallites in blends with various PE content. The periods of acceleration, retardation and of maximum occurrence correspond to decrease, increase and minimum appearance depending on T_T of the PE crystallites in the mixture composition.

Fig.1.33 demonstrates the effect of mixtures composition on the degree of crystallinity(κ).

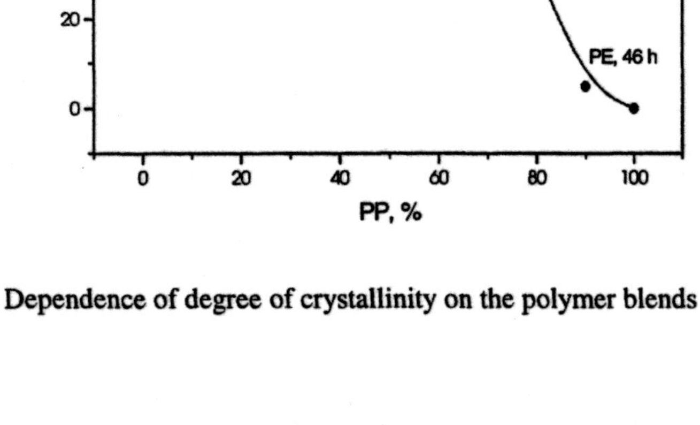

Fig.1.33. Dependence of degree of crystallinity on the polymer blends composition

First of all we have to note the decrease in the values of κ for PE and PP with the decrease of their content in the mixture both for the nonozonized and in the ozonized samples. However, in the case of the nonozonized samples the profile of κ is characterized by a range whereby it is constant.

The value of κ for PE in mixtures with PE content varying from 0 to 50 % amounts to 60 % and in those of PP content of 50-70 % is 64 %. The observation of the decrease of crystallinity degree of the components in the system PE/PP is also confirmed by other authors [117].

In the case of PE the values of κ for the ozonized samples are found to be higher than those of the nonozonized ones. An exception are the samples containing 10 and 20 % PE for which κ falls down abruptly. The increase, probably could be explained by the fact that ozone does not penetrate into the PE crystallites but rather reacts with PE in the transition zone on their surface. This reaction leads to break of the macromolecules in the transition zone and thus the crystallinity is enhanced on account of the section that provided by the cut-off molecules in the transition zone. The authors of Ref.118 have also observed a similar increase of κ for PE during its ozonation. These changes of κ are also demonstrated by the curves in Fig.1.33.

It has been found that the values of κ in the PP crystallites of ozonized samples are always higher as compared to that in the nonozonized ones (Fig.1.33).This observation could be related to the fact that ozone penetrates into the crystallites and reacts further with tert C-H bonds.Thus they are gradually crushed and transformed in an amorphous state. This phenomena is characterized by a decrease of T_T and κ in the ozonized samples whereby the initial content of PP was lower than 30 % and the crystal structure is completely destroyed.

It is well known that the polydispersity of crystallites can be characterized by the halfwidth of the melting peaks provided by the DSC method. The changes of this parameter for the two components are illustrated in Fig.1.34.

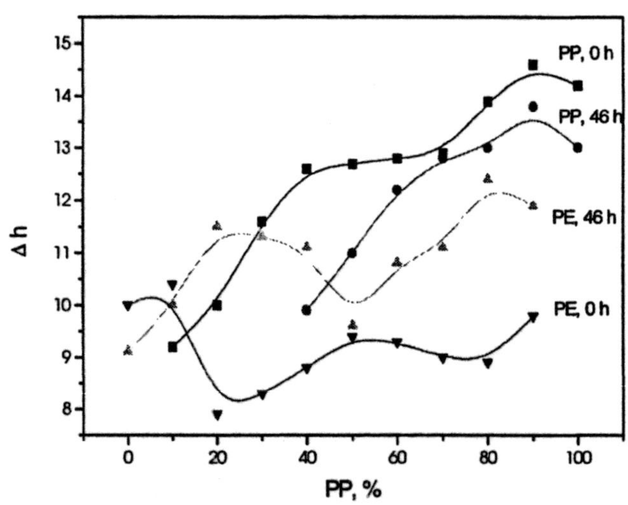

Fig.1.34. Dependence of halfwidth of melting peaks on mixture composition.

For PE, Δh in the ozonized samples is higher than that of the nonozonized ones in the whole range of the mixtures composition with an exception of the pure PE and the mixture with 1:9 content. The comparison of the relationship between Δh and κ allow us to assume that the increase in polydispersity of PE crystallites can be explained by physical reasons stipulated by the mixture composition. Thus, in mixtures with high PE content ranging from 30 to 100 % is observed a simultaneous increase of Δh and κ. The polydispersity of the crystallites is resulting from the rebuilding of the destructed parts of the macromolecules from the transition zones. In mixtures with lower PE content, below 30 %, the rise of polydispersity is accompanied by a drop of the crystallinity degree.

It has been found that the polydispersity of PP crystallites monotonously decrease with the decrease of PP content in the mixtures (Fig.1.34). In each case the polydispersity and κ in the ozonized samples are always lower than those in the nonozonized ones in the whole range of mixtures composition. Apparently, the decrease in the PP polydispersity is associated with its initial reaction with the smallest crystallites.

Upon more prolonged ozonation of PE and PP their amorphous phases get considerably altered as a result of the accumulation of oxygencontaining compounds of various molecular mass which can also destroy them.

1.6. POLYSTYRENE

When bubbling ozone through a polystyrene (PS) solution in CCl_4 it is absorbed. [58].

The rate constant at ambient temperature k has been determined according to eq.1.14. and it is 0.37 $M^{-1}.s^{-1}$. The rate of ozone absorption (W_{O3}) remains constant up to the uptake of 3.2 mol ozone per a monomer unit. Further, at consumption of 4.5 mol ozone it has been reduced 10-times. W_{O3} does not depend on the ozone flow rate in it but increases linearly with the increase of ozone and PS concentration (Table 1.12).

Table 1.12

Dependence of the rate constant on ozone and PS concentration and on the ozone flow rate (v) at 25°C

$v.10^3$, l/s	0.6	0.8	0.8	1.0	1.0	1.0	1.0	1.0	1.0	1.4	2.1	2.8
[PS], M	0.01	0.027	0.054	0.027	0.027	0.027	0.027	0.027	0.027	0.068	0.12	0.072
$[O_3].10^4$, M	0.75	0.75	0.81	0.09	0.23	0.75	1.7	2.9	7.1	0.72	0.78	0.71
$W.10^7$, $M.s^{-1}$	4.2	8.4	34	1.2	3.9	11.6	39	35.2	131	35	80	30.4
k, $M^{-1}.s^{-1}$	0.31	0.23	0.45	0.27	0.35	0.32	0.47	0.25	0.38	0.40	0.48	0.33

The analysis of the IR spectra of the ozonized samples (Table 1.13) shows that the content of t-CH_2-groups practically does not change during ozonolysis as the intensity of the bands at 700 and 2930 cm^{-1} do not change essentially.

Table 1.13

The intensity of the IR-absorption bands of the products of PS ozonolysis: [PS]=27.4 mM; $[O_3]$ = 0.88 mM; ambient temperature

Time, min	700 cm^{-1}, %	1500 cm^{-1}, %	2930 cm^{-1}, %	$-lg(I_t/I_0)$ at 1500 cm^{-1}
0	36.1	23.0	36.0	0
10	35.4	21.2	35.8	0.035
15	36.0	20.1	35.9	0.059
25	34.5	19.3	35.1	0.076
30	34.9	18.0	35.8	0.106

From the other hand the intensity at 1500 cm^{-1} which is assigned to the aromatic ring vibrations decreases in the course of the reaction. This fact confirms the ozone interaction with the benzene ring giving rise to ozonides. The latter have been identified by the occurrence of C-O vibrations in the range of 1100-1200 cm^{-1}, as well as by the iodometric titration of the oxidates. It is known that the stoichiometry of the ozone reaction with benzene is 1:3. The kinetic curves presented in Fig.1.35 demonstrate a linear kinetics of ozonides formation.

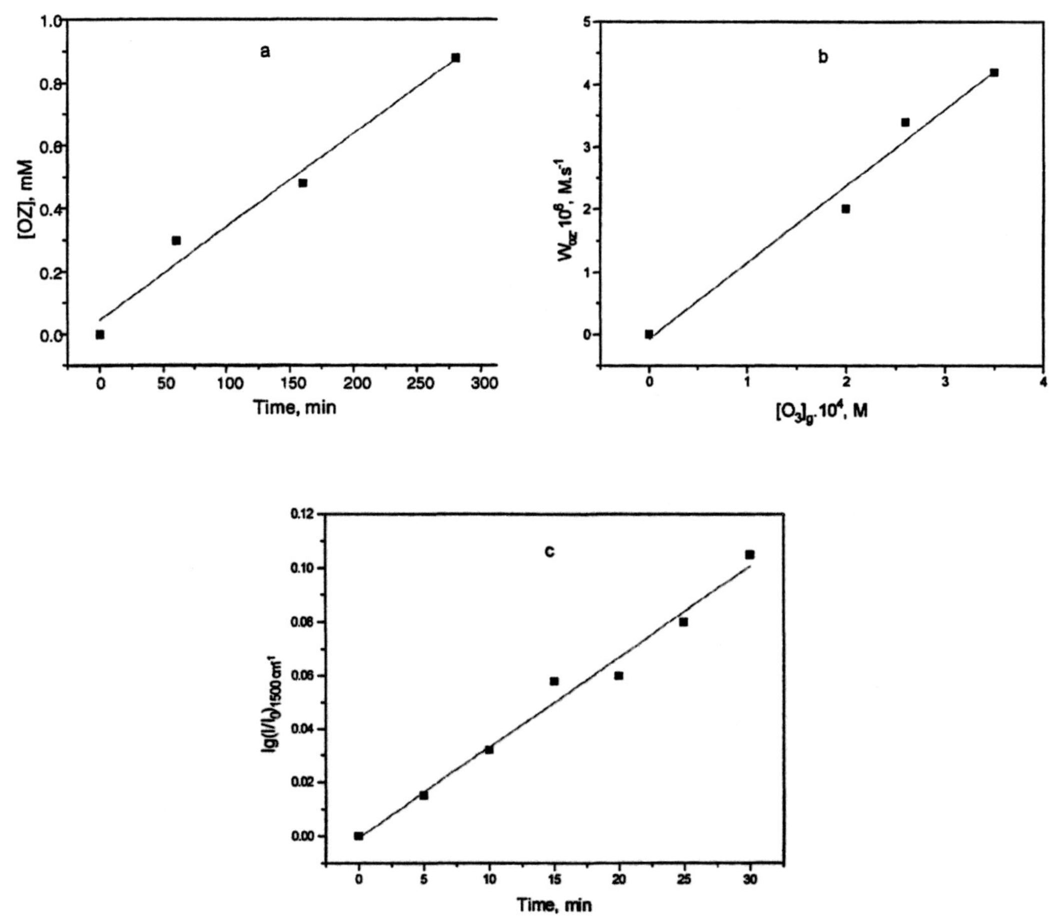

Fig.1.35. Kinetics of ozonides formation (a)-25°C, [PS] = 24.7 mM; $[O_3]_g$=1.2.10^{-5} M; (b) - dependence of the rate of ozonides formation on ozone concentration at the reactor outlet, 25°C; [PS] = 27.4 mM; (c) variation of the band intensity at 1500 cm^{-1}, 25°C; [PS] = 27.4 mM; $[O_3]_g$ =8.8.10^{-4} M.

The rate of the ozonides formation ate 25oC has a magnitude of 0.29±0.01 M^{-1}.s^{-1}. This value correlates well with the rate constant of benzene ring consumption for cumene which is 0.28±0.02 M^{-1}.s^{-1}.

The kinetics of hydroperoxides accumulation is illustrated by the curves in Fig.1.36. Its profile suggests the intermediate character of the hydroperoxides formed in this reaction.

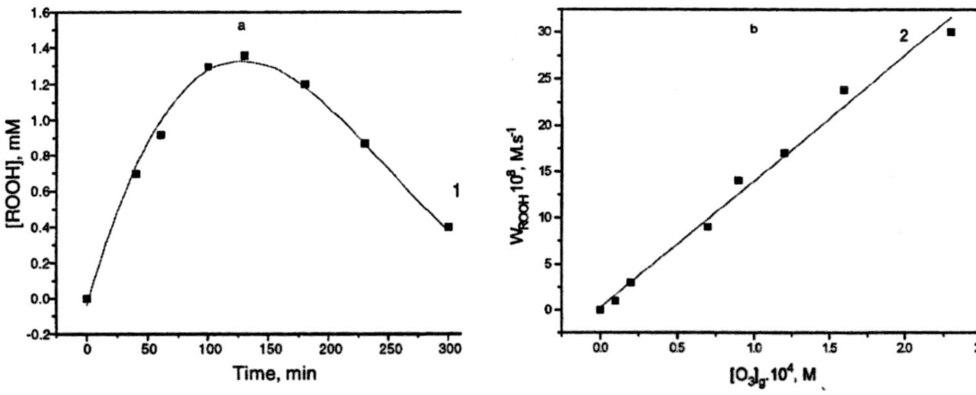

Fig.1.36. (a) Kinetics of ROOH formation (1), 25°C; $[PS]_0$=60 mM; $[O_3]_g$=1.4.10^{-4} M and (b) Dependence of the initial rate of ROOH formation on the ozone concentration at reactor outlet (2), 25°C; $[PS]_o$=27.4 mM.

It is seen that at the initial stage of the reaction [ROOH] increases and W_{ROOH} depends linearly on $[O_3]$ (Fig.1.36, curve 2). The ROOH content at 40 % conversion is about 6 % irrespective of [PS] which is in the range 0.1-1.2 %.

The kinetic curves of the variation of PS molecular weight at [PS]=1.5 % passes through a maximum (Fig.1.37, 1).

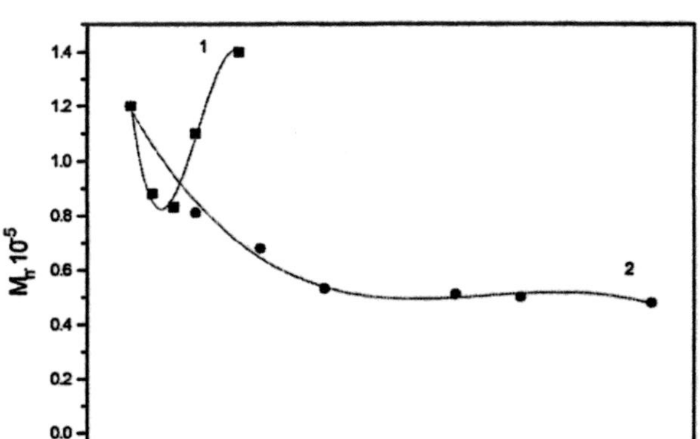

Fig.1.37. Kinetics of PS molecular weight variation with average number molecular mass 1.2.10^5 upon ozonolysis (1) [PS]=3.5 %; (2)-]PS] =0.3 %, W_{O3} = 2.10^{-4} mol/(kg.s); 25°C

It has been established that carbonyloxides take part in cross-linking processes [1].

For this reason we have studied the PS degradation in solutions with concentrations below 1.5 %. Upon ozonolysis of dilutes PS solutions with concentrations below 0.5 % we have observed a decrease of the average molecular weight (Fig.1.37, 2). The rate of degradation is constant up to a conversion degree of 30 %.

$$W_d = (1/M - 1/M_0).t^{-1}.[PS]_0^{-1} \qquad \text{(Eq.1.49)}$$

It is linearly dependent on the ozone uptake rate and temperature (Fig.1.38).

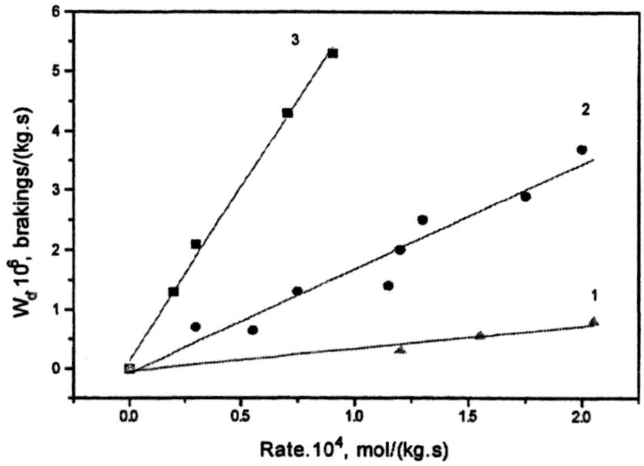

Fig.1.38. Dependence of the rate of PS degradation on the rate of ozone consumption. 1-0 °C; 2-25°C; 3-60°C; [PS]<1 %.

The number of chain breaks in diluted solutions related to 1 mol absorbed ozone $n=W_d/W_i$ does not depend on the polymer concentration but rises with temperature (Table 1.14).

Table 1.14

Dependence of the rate of PS degradation on the solution composition and temperature

t, ^{0}C	0	0	25	25	25	25	25	25	25	60	60
[PS], %	1.5	0.7	0.19	0.30	0.42	0.71	1.5	6.5	10.4	1.5	0.7
$W_d.10^6$	0.88	1.6	3.4	3.8	5.0	1.4	5.8	1.4	0.4	6.0	7.7
$W.10^4$	2.2	3.1	1.2	2.0	2.2	1.2	3.4	1.7	2.1	1.0	1.2
$n. 10^2$	0.4	0.54	2.8	1.9	2.3	1.2	1.7	0.8	0.2	6.0	6.4
$n_{av}.10^2$	0.47	±0.07		1.8	±0.6			0.9±0.8		6.2	±0.2

Under these conditions the reactions leading to increase of the molecular mass do not exhibit any essential influence. This is also confirmed by the profiles of the molecular mass distribution (MMD) of low and high molecular components (Fig.1.39).

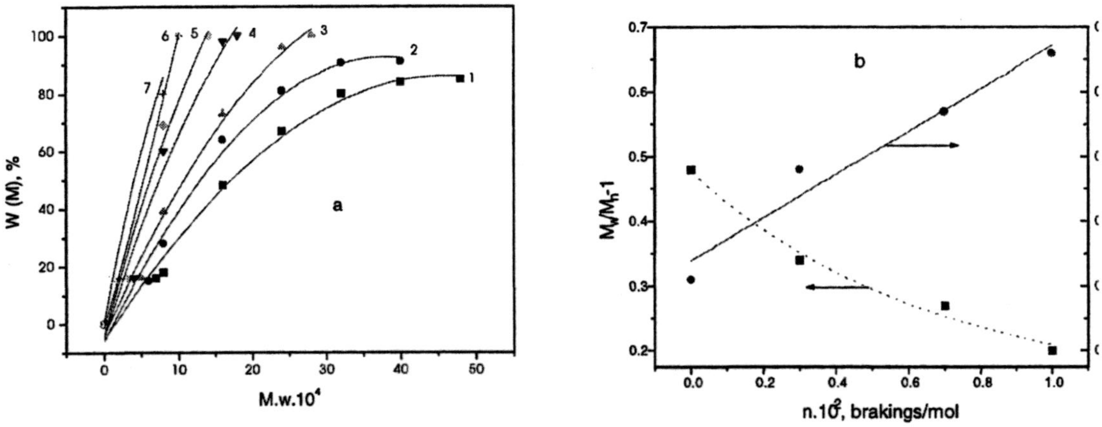

Fig.1.39 (a) Integral curves of MMD at 25°C, W_{O3}=6.6.10^{-5} mol/(kg.s), [PS] = 1% and ozonation time of:1-0 min, 2 - 15 min, 3 - 30 min, 4 - 45 min, 5- 60 min, 6 - 90 min and 7 - 120 min; (b) MMD in the course of ozonation.

The decrease of the molecular mass and the rise of the number of chain breaking (n) is accompanied by narrowing of the MMD (Fig.1.39 a) and the ratio M_W/M_n approaches a value equal to 1 (Fig.1.39b).

By means of ESR it has been established that the ozonolysis of PS in solid phase generates peroxide radicals [126] which arise not by the direct ozone reaction with PS but are also secondary products. The kinetic scheme of the reaction, proposed by us, is based on the experimentally obtained data and can be presented as follows:

Scheme 1.9

$(\text{-}CH_2\text{-}(Ph)CH\text{-})_m + O_3 \rightarrow \alpha P_1 + (1\text{-}\alpha)(P_2 + P_3 + \ldots)$ (1)

$P_1 + O_3 \rightarrow RO_2^{\bullet}$ (2)

$(\text{-}CH_2\text{-}(Ph)CH\text{-})_m + RO_2^{\bullet} + O_2 \rightarrow ROOH + RO_2^{\bullet}$ (3)

$RO_2^{\bullet} + RO_2^{\bullet} \rightarrow \beta(2RO^{\bullet})$ (6')$+ (1\text{-}\beta)\ ROOH + O_2$, (6")

where; P_1, P_2 and P_3 are intermediate products

The reactions describing the ozonides formation and the ozone interaction with ROOH are included in the scheme pointed above:

$(\text{-}R\text{-})_mPh + O_3 \rightarrow$ monoozonide (slow)

monoozonide + $O_3 \rightarrow$ diozonide

diozonide + $O_3 \rightarrow$ triozonide (4)

$ROOH + O_3 \rightarrow \gamma ROH + 2O_2$ (5')

$ROOH + O_3 \rightarrow (1\text{-}\gamma)\ (RO_2^{\bullet} + HO^{\bullet} + O_2^{\bullet})$ (5")

The linear dependence of the rate of ozone absorption on its concentration as well as the stoichiometric coefficient of the reaction in the range of 4.2-4.5 can be regarded as evidences for the absence of ozone absorption in a chain reaction. According to Scheme 1.9 the reaction of ozone consumption will be presented as:

$$W = (k_1 + 3k_4)[PS][O_3] + k_2[P_1][O_3] \quad \text{(Eq.1.50)}$$

and

$$[P_1] = \alpha.[PS]_0.k_1.(k_2 - k_1)^{-1}.\{\exp[-k_1[O_3].t - \exp[-k_2[O_3].t\} \quad \text{(Eq.1.51)}$$

It is apparent that reaction 2 has an insignificant effect in the total balance of ozone consumption since only 6 % of ROOH are obtained at the absorption of 1 mol ozone. The value of the observed rate constant k_{obs} is equal to $k_1 + k_4$ and then $k_1 = 0.13\ M^{-1}.s^{-1}$.

The insignificant amounts of ROOH testify that actually the oxidation of tert.-H bonds does not take place via a chain reaction. At the initial stage of ozonation $[P_1] = \alpha.k_1.[PS]_o[O_3].t$, since $[RO_2] \sim O_3$ and $W_{ROOH} = k_3[RO_2].[PS] \sim O_3$.This is also observed experimentally (Fig.1.36, 2). The results suggest that the RO_2 radicals formation in PS is not dependent on the phase state of the reaction occurrence. The recombination of RO_2-radicals according to reaction 6" gives rise to RO-radicals. The macromolecule destruction takes place via a monomolecular decomposition of RO-radicals:

$$RO^{\bullet} \rightarrow r^{\bullet} + (R')_2C{=}O \text{ (degradation)} \qquad (7)$$

In case the degradation process takes place via reaction 7, then $W_d \sim W_{O3}$. In order to elaborate the mechanism of degradation process we have carried out an initiated oxidation of PS {[AIBN]=2%, 60°C, W_i=1.10-^{5} mol/(kg.s)}, where AIBN - azoisobutyronitrile. For the inhibited oxidation W_d has been found as 9.10^{-7} breaks/(kg.s) and upon ozonation a similar value for W_d has been obtained at $W_{O3}=1.5.10^{-5}$ mol/(kg.s) (Fig.1.38, 3). If we assume that the initiated oxidation and the ozonolysis differ only in the pathway of RO_2-radicals generation, then the mechanism of destruction and the kinetic relationships should be similar. The experimentally found values of the ratio W_i/W_{O3} accounting for the occurrence of radical steps in the ozonolysis are: 0.05 at 0°C, 0.2-25°C and 0.69 - 60°C (Fig.1.38, 1-3). The activation energy of degradation E_a calculated on the basis of Fig.1.38 amounts to 7.6 kcal/mol and correlates well with the activation energy of RO-radical izomerisation E_a [120]. The ratio W_i/W could be estimated using the values of k_1 and k_4 and the quasistationary conditions in relation to P_1:

$$W_i/W = 0.43\alpha\beta \qquad \text{(Eq.1.52)}$$

From the experimentally measured ratio W_i/W= 0.2 at 25°C it follows that $\alpha\beta$=0.47. Because of lack of any data for the magnitude of β for PS we have used the known value of β for cumene oxidation (low molecular analog of PS) that is 0.86 and then we have obtained 0.55 for α. The latter demonstrates that the major part of RO-radicals have been generated not via reaction 6" but as a result of the direct ozone attack on the PS macromolecule.

Thus it can be reasonably expected that the degradation will occur via C=C bonds, for which is known [1, 25] the rate of their reaction with ozone , i.e. $10^5 M^{-1}.s^{-1}$ and similar to polydienes they would be responsible for the PS degradation. However the amount of C=C bonds in PS is of order of $(3\pm0.5).10^{-2}$ mol/kg and at average number mass of PS - 10-4 mol/(kg.s) which means that one macromolecule comprises 1-2 double bonds. Upon statistical distribution of C=C bonds along the macromolecule length, degradation at $W_{O3} \sim 10^{-4}$ mol/(kg.s) should finish in 5-10 min. However we failed to observe a marked decrease of the molecular weight within this time. Further, W_d should begin to go down. Actually the rate of degradation remains constant even after 150 min when the conversion is already 30 %. Apparently, the C=C bonds in PS are not effective in this, mainly because they are terminal double bonds. This supposition has been confirmed by measuring the concentration of C=C bonds which during ozonation decreases with the simultaneous rise of PS molecular mass.

If we take into account that the degradation proceeds via the random law then the equilibrium width of MMD should approach value of 2 only after 2-3 breaks. However this is in contrast to the experimentally obtained data illustrated in Fig.1.39 b. It is seen that MMD depends on the conversion degree and approaches to 1. This effect could be explainded by the variation of the ratio of the values of the elementary constants which characterize the breaking of C-C bonds in the macromolecule. A decrease of W_d has been registered at the transition to lower molecular polymers as is the case with fractionated PS and with polymers with broad MMD. This might be the explanation of the decrease of the effective W_d upon ozonolysis and the narrowing of the MMD.

The mechanism of alkane ozonolysis will be further discussed again from quantum chemical point of view in Chapter 5.

REFERENCES

1. Razumovskii, S.D., S.K. Rakovski, D.M. Shopov, G.E. Zaikov. *Ozone and Its Reactions with Organic Compounds (in Russian)*. Publ. House of Bulgarian Academy of Sciences, Sofia, 1983.
2. Klejmanov, NA, AB Nalbandjan. *Dokl. Akad. Nauk SSSR*, 122, 103 (1958).
3. Schubert, CC, RN Pease. The Oxidation of -Lower Paraffin Hydrocarbons. I. Room Temperature Reaction of Methane, Propane, n-Butane and Isobutane with Ozonized Ohygen. *J. Am. Chem. Soc.* , 78. 2044-2048 (1956).
4. Schubert, CC, RN Pease. Reaction of Paraffin Hydrocarbons with Ozonized Oxygen: Possible Role of Ozone in Normal Combastion. *J. Chem. Phys.*, 24, 919-920 (1956).
5. Dillemuth, FJ, DR Skidmore, CC Schubert. The Reaction of Ozone with Methane. *J. Phys. Chem.*, 64, 1496-1499 (1960).
6. Morrissey, RJ, CC Schubert. The reaction of Ozone with Propane and Ethane. *Comb. Flame*, 1, 263268 (1963).
7. Dardin, VJ, LN Albright.Partial Oxidation of Propane Initiated by Ozone. *Ind. Eng. Chem. Proc. Des. Dev.*, 4, 61 (1965).
8. Grewer, TH.Chapter IV.4.3. *Oxidation of Hydrocarbons by Ozone* in: Low Temperature Oxidation (Ed. W. Jost), 1, 186-188 (1964).
9. Hamilton, GA, BS Ribner, TM Hellman. On The Mechanism of Alkane Oxidation by Ozone. *Oxidation of Organic Compounds. Adv. Chem. Ser.*, 77, 15 (1969).
10. Shljapintoh VYa, AA Kefeli, VI Gol'denberg, SD Razumovskii. *Dokl. Akad. Nauk SSSR*, 168, 1132 (1969).
11. Razumovskii SD, AA Kefeli, GE Zaikov. *Zurnal Org. Khim.*, 7, 2045 (1971).
12. Durland, J. R., H. Adkins. *J. Am. Chem. Soc.*, 61, 429 (1939).
13. Geiseler, F. Asinger, H. Wien. *Chem. Ber.*, 92, 958 (1959).
14. Emanuel', NM, ET Denisov, ZK Majzus. *Hudrocarbon Chain Oxidation Reactions in Liquid Phase*. Nauka, 1965, p. 202.
15. Whiting, MC, AJN Bolt, JH Parish.The Reaction between Ozone and Saturated Compounds. *Int.Symp.Oxidation, Preprints*, 2, 267-287 (1967), San Francisco.
16. Tjutjunnikov, N., AI Ljashenko, AM Shiman, IS Suhoterin, GI Moskwina, AS Drozdov, GF Sanchenko. *Khim. Promish.*: 653 (1971).
17. Williamson, DG, RJ Cvetanovic. Rates of Ozone Reactions in Carbon Tetrachloride Solution. *J. Am. Chem. Soc.*, 92, 2949 (1970).
18. Razumovskii, SD, AA Kefeli, GR Trubnikov, GE Zaikov. *Dokl. Akad. Nauk SSSR*, 192, 1313 (1970).
19. ET Kang, KG Neoh, X Zhang, KL Tan, DJ Liaw. Surface modification of electroactive polymer films by ozone treatment. *Surface and Interface Analysis*, 24: 1, 51-58 (1996).
20. VA Basiuk. Organic reactions on silica surface: Synthetic applications. *Uspekhi Khimii*, 64: 11, 1073-1090 (1995).
21. Horwath, M., L. Vitzky, G. Huttner. *Ozone*. Budapest Academy Kiadoi, 1985.
22. Razumovskii, SD, GE Zaikov. *Izv. Akad. NAuk SSSR ser. khim.*: 2244 (1973).
23. Tedder M, *Quart. Revs.*, 14, 336 (1960).
24. Herron, T, R. E. Huie. *J. Phys. Chem* , 73, 3327 (1969).
25. Bailey, P.S. "*Ozonattion in Organic Chemistry*", Academic Press, New York, v.I, 1978, v.II, 1982, *Adv. Chem. Ser. - A*, (Ed. HH Wasserman), 39 -I, 39 - II, 1978, 1982.
26. Denisov, ET, TG Denisova. Reaction Ability of Ozone in the Reactions with C-H Bonds in Hydrocarbons, Alcohols, and Ketones: Analyst in the Frame of Parabolic Model. *Kinet. Katal.*, 37, 51 (1996).
27. Nangia, PS, SW Benson. Thermochemistry and Kinetics of Ozonation Reactions. *J. Am. Chem. Soc.*, 102, 3105-3115 (1980).
28. Benson, SW. *The Foundations of Chemical Kinetics*. Moscow, Mir, 1964.
29. Atkinson, R. Gas-phase tropospheric chemistry of volatile organic compounds. 1. Alkanes and alkenes. *J Phys Chem Ref Data*, 26, 2, 215-290 (1997).

30. Reiser, O. Oxidation of Weakly Activated C-H Bonds. *Angewandte Chemie - International Edition in English,* 33: 1, 69-72 (1994).
31. Razumovskii, SD. *Zurnal Priklod. Khim.*, 42, 2118 (1969).
32. Razumovskii, SD. *Khim. Promishl.*: 491 (1967).
33. Hatakeyama, S, H Lai, S Gao, K Murano. Production of Hydrogen Peroxide and Organic Hydroperoxides in the Reaction of Ozone with Natural Hydrocarbons in Air. *Chemistry Letters* : 8, 1287-1290, (1993).
34. A Hudson. Applications of Electron Spin Resonance to Environmental Problems. *Research on Chemical Intermediates,* 19: 7, 635-641 (1993).
35. Kulik, EA, MI Ivanchenko, K Kato, S Sano, Y Ikada. Peroxide generation and decomposition on polymer surface. *Journal of Polymer Science, Part A - Polymer Chemistry,* 33: 2, 323-330 (1995).
36. Jacob, P., Wehling, B., Hill, W., Klockow, D. Feasibility study of Raman spectroscopy as a tool to investigate the liquid-phase chemistry of aliphatic organic peroxides. *Appl Spectrosc,* 51, 1, 74-80 (1997).
37. Gross, Z., Simkhovich, L. Ozone as primary oxidant in iron(III). porphyrin catalyzed hydroxylation of hydrocarbons. *J Mol Catal A-Chem,* 117, 1-3, 243-248 (1997).
38. Atkinson, R. Gas-phase tropospheric chemistry of volatile organic compounds. 1. Alkanes and alkenes. *J Phys Chem Ref Data,* 26, 2, 215-290 (1997).
39. J Cerkovnik, B Plesnicar. Characterization and Reactivity of Hydrogen Trioxide, (HOOOH). - A Reactive Intermediate Formed in the Low- Temperature Ozonation of 2-Ethylanthrahydroquinone. *J.Am.Chem.Soc.*, 115: 25, 12169-12170 (1993).
40. Marshall, JA, AW Garofalo Oxidative Cleavage of Monosubstituted, Disubstituted, and Trisubstituted Olefins to Methyl Esters Through Ozonolysis in Methanolic NaOH. *J.Org.Chem,* 58: 14, 3675-3680 (1993).
41. Dunkin, IR, A Mccluskey Tetrachlorocyclopentadienone O-Oxide, a Facile Oxygen Atom Transfer Reagent - The Oxidation of Cyclohexane. *Journal of Photochemistry and Photobiology A - Chemistry,* 74: 2-3, 159-164 (1993).
42. Dhandapani, B., Oyama, S.T. Gas phase ozone decomposition catalysts. *Appl Catal B-Environ,* 11, 2, 129-166 (1997).
43. Encinar, JM, FJ Beltran, JM Frades. Liquid Phase Ozonation of Cyclohexanol Using Acetic Acid as Solvent. *Anales de Quimica,* 89: 7-8, 682-690 (1993).
44. Galstyan, GA, TM Galstyan, LI Mikulenko. The Kinetics and Mechanism of the Catalytic Reaction of Ozone with Methylbenzenes in Acetic Acid. *Kinetics and Catalysis,* 35: 2, 231-236 (1994).
45. Rakovsky, SK, DR Cherneva, M Deneva. Ozone Reactions with Aliphatic Ethers in CCl4 - Kinetics and Mechanism. *International Journal of Chemical Kinetics,* 27: 2, 153-165 (1995).
46. Deninno, MP. "Anomalous" ozonolysis of cyclic allylic alcohols: Mechanism and synthetic utility. *Journal of the American Chemical Society,* 117: 39, 9927-9928 (1995).
47. Razumovskii, SD, SK Rakovski, GE Zaikov. Influence of the Paraffine Hydrocarbon Structure on its Reaction Rate with Ozone. *Izv. Akad. Nauk, ser. khim.*: 9, 1963-1967 (1975).
48. Rakovski, SK, AA Popov, SD Razumovskii, DM Shopov, GE Zaikov. On the Mechanism of the reaction of Cycloparaffins with Ozone. *Comm. Dep. Chem - Bulgarian Academy of Sciences,* 8 (3) 544-557 (1975).
49. Popov, AA, SK Rakovski, SD Razumovskii, DM Shopov, GE Zaikov. Steric Strain Effects on the Kinetic of Chemical Reactions of Cycloparaffins with Ozone. *Comm. Dep. Chem - Bulgarian Academy of Sciences,* 8 (3) 571-577 (1975).
50. Rakovsky SK, SD Razumovskii, GE Zaikov. Investigation of the Ozone Paraffine Reactions by ESR. *Izv. Akad. Nauk, ser. khim.*: 3, 701- 403 (1976).
51. Popov, AA, SK Rakovsky, DM Shopov, LV Ruban. Mechanism of the Ozone Reactions with Saturated Hydrocarbons. *Izv. Akad. Nauk, ser. Khim.*, 40, (5), 982-990 (1976)
52. Ruban, LV, SK Rakovski, AA Popov. Kinetics and Mechanism of Cyclohexane Oxidation by Ozone. *Izv. Akad. Nauk, ser. Khim.*:5,1950-1956 (1976).
53. Ruban, LV, SK Rakovski, SD Razumovskii, GE Zaikov. On the Composition of Peroxide Products during Cyclohexane Ozonation. *Izv. Akad. Nauk, ser. Khim.*:5,2104-2106 (1976).

54. Rakovski SK, DR Cherneva, DM Shopov, SD Razumovskiy. Application of the Barbotage Method in Investigating the Kinetics of Ozone Reactions with Organic Compounds. *Comm. Dep. Chem - Bulgarian Academy of Sciences*, 9 (4) 711-717 (1976).
55. Rakovski SK, DR Cherneva, DM Shopov, VM Parfenov. Kinetics and Mechanism of Cumene Ozonation. *Comm. Dep. Chem - Bulgarian Academy of Sciences*, 11 (1) 153-159 (1978).
56. Rakovski SK, LK Nenchev, DR Cherneva, DM Shopov. Mechanism of the Reaction of Ozone with Cumene in the Presence of Silica-Supported MoO3 and V2O5 Catalysts. *Comm. Dep. Chem - Bulgarian Academy of Sciences*, 12 (2) 254-1257 (1979).
57. Rakovsky SK, DR Cherneva, DM Shopov. Cumene Oxidation by Ozone-Oxygen Mixture in the Presence of Transition Metal Salts. *Izv. Akad. Nauk, ser. Khim.*:9,1991-1995 (1979).
58. Kefely A.A., S.K. Rakovski, D.M. Shopov, S.D. Razumovskii, K.S. Rakovski, G.E. Zaikov. Kinetic Relationships and Mechanism of Ozone Reaction with Polystyrene in CCl_4 Solution. *J. Polymer Science: Polymer Chem. Edition*, 19, (6), 2175 - 2184 (1981).
59. Razumovskii S.D., S.K. Rakovski, D.M. Shopov, G.E. Zaikov. Ozone and Its Reactions with Organic Compounds (monograph in Russian). *Publishing House of the Bulgarian Academy of Sciences*, 1 - 278, 1983.
60. Rakovski S.K., D.R. Cherneva, D.M. Shopov. Ozonolysis of Cumene Hydroperoxide in Different Solvents. *Oxid. Commun.*, 7, (1), 181 - 190 (1984).
61. Blinov, NN, AA Popov, SK Rakovski, AK Stoyanov, DM Shopov, GE Zaikiv. Change of the Melting Point, Polymolecularity and Polycristalinityof Components of High Density Polyethtylene-Polypropylene Blend under Deep Ozone Oxidaton. *Visoko Molekul. Soedin. A*, 31, 2212-2218 (1989).
62. Rakovski, SK, DM Shopov, DR Cherneva. The Investigation of Ozone Reaction with Cumene Hydroperoxide as a Model of Polyolefinic Degradation. 25-th Microsymposium "Processing and Lohg-Term Stabilities of Hydrocarbon Polymers", Prague, July 18-21, 1983, P33 p.1-4.
63. Rakovsky, SK, LK Nenchev, DR Cherneva, DM Shopov. Catalytic Degradation of Ozone over NiO., *7-th Intern. Symp. Heter. Catal.*, Varna, 1979, v.2, p.231-236.
64. Rakovsky, SK, DR Cherneva, AA Spozhakina, K Novakov, Catalytic Decomposition of Ozone over Suported Metal Oxide Catalysts. 8-Th Intern. Symp. Heter. Catal., Varana, 5-9 October 1996, p.407-413.
65. Martinez, RI. The Mechanism of O_3-Aldehyde Reactions. *Int. J. Chem.Kinet.*, 14, 237-249 (1982).
66. Komissarovva, IN, VD Komissarov, ET Denisov. *Izv. Akad. Nauk ser. khim.*: 9, 1991 (1978).
67. Hellman, TM, GA Hamilton. *J. Am. Chem. Soc.*, 96, 1530 (1974).
68. Stary, FE, Emge, DE, RW Murray. Ozonation of Organic Substrates. Hydrotrioxide Formatin and Decompozition to Give Singlet Oxygen. J. Am. Chem Soc., 98, 1880-1884 (1976).
69. Vedeneev, I., LV Gurvich, VN Kondrat'ev, VA Medvedev, EL Frankevich. *Energy Breaking of the Chemical Bonds , Ionization Potencials, and Electron Affinity*. Publ. House Akad. Nauk SSSR, Moscow, 1962, p. 90.
70. Voevodskii, VV. *Physics and Chemistry of Elementary Chemical Processes*. Nauka, Moscow, 1969.
71. Frank-Kamenetzkii, DA. *Diffusion and Heat Transfer in Chemical Kinetics*. Nauka, Moscow, 1987.
72. Sherwood, TK, RL Pigford, CR Wilke. *Mass Transfer*. McGraw Hill, New York, 1975.
73. Danckwerst, PV. *Gas-Liquid Reactions*. McGraw Hill Book Comp., New York, 1970.
74. Astarita, G. *Mass Transfer with Chemical Reaction*. Elsevier, Amsterdam, 1967.
75. Hammond, GS. *J. Am. Chem. Soc.*, 77, 334 (1955).
76. Hisatsune, IC, LH Kolopajlo, J. Heicklen. *J. Am. Chem. Soc.*, 99, 3704 (1977).
77. Alcock, WG, B Mile. *J. Chem. Soc. Chem. Comm.*:5 (1976).
78. Bukert, U, NL Allinger. Molecular Mechanics. ASC Monograph 177, Washington, D.C., 1982.
79. KS Chen, NL Allinger A Molecular Mechanics Study of Alkyl Peroxides. *Journal of Computational Chemistry*, 14: 7, 755-768 (1993).
80. Eliel, EL, NL Allingerĺ, SJ Angyal, GA Morrison. *Conformational Analysis*. Moscow, Mir, 1969.
81. Eyring, H, SH Lin, SM Lin. *Basic Chemical Kinetics*. J. Wiley, New York, 1980
82. Levine, IN. *Physical Chemistry*.McGraw Hill, New York, 1988.
83. Benson, SW. *The Foundations of Chemical Kinetics*. Moscow, Mir, 1964.

84. Tyutyulkov, N. *Quantum Chemistry*. Nauka i Izkustvo, Sofia 1978.
85. Klark T. *A Handbook of Coputational Chemistry*, J. Wiley, New York, 1985.
86. Dobbs, KD , DA Dixon, A Komornicki. Ab initio Prediction of the Barrier Height for Abstraction of H from CH_4 by OH. *J.Chem.Phys.*, 98: 11, 8852-8858, (1993).
87. Dewar, MJS, JC Hwang, DR Kuhn. *An AM1 Study of the reactions with Ozone with Ethylene and 2-Butene. J. Am. Chem. Soc.*, 113, 3, 735-741 (1991).
88. Meredith, C, GE Quelch, HF Schaefeer, III. Open-Chain and Cyclyc Protonated Ozone: The Ground-State Potencial-Energy Hypersurface. *J. Am. Chem. Soc.*, 113, 1186-1193 (1991).
89. Karadakov, P, SK Rakovsky. *Computer Program. - Calculation of Pre-exponents According to the Activated Complex Theory and Collision Theory*. Inst. Org. Chem., Inst Catlal., 1988.
90. Herzberg, G Molecular Spectra and Molecular Structure (two volumes), New York, 1954/1957.
91. Benson, S. *Thermochemical Kinetics*. Moscow, Mir, 1971.
92. Pitzer, K. S. *J. Chem. Phys.*, 14, 239 (1946).
93. Bazilevskii, MV. Molecular *Orbitals Method and Reaction Ability of Organic Molecules*. Moscow, Khimia, 1969.
94. Bazilevskii, MV, ZT Trosman. *Uspehi Khimii*, 41, 3 (1960).
95. Radinovitch, E. *Trans. Farady Soc.*, 33, 1225 (1937).
96. OEvans, M. G., M. Polany. *Trans. Farady Soc.*, 32, 1333 (1936).
97. Galimova, LG, VD Komissarov, ET Denisov. *Izv. Akad. Nauk SSSR, ser. khim.*: 307 (1973).
98. Dioxin, M, E Ebb. *Ferments*. Mir, Moscow, 1966, p. 277.
99. Karnogitzkii, V. *Organic Peroxides*. IL, Moscow, 1961, p. 115.
100. Kritchfild, F. *Analysis of Basic Functional Groups in Organic Compounds*. Mir, 1965, p. 188.
101. Hawkins, EJE. *Organic Peroxides*. Khimia, Moscow, 1965.
102. Johnson, RM, JW Siddiqi. *The Determination of Organic Peroxides*. Pergamon Press, New York, 1970, p 15.
103. Denisov ET. *Rate Constants of the Homolytical Liquid-Phase Reactions*. Nauka, Moscow, 1971.
104. Denisov ET. *Kinetic of Homogeneous Chemical Reactions*. Visshaya Shkola, 1988.
105. Shlyapintoh, VYa, ON Karpuhin, LM Postnikov, IV Zaharov, AA Vichutinskii, VF Tzepalov. *Chemiluminescent Methods of Investigation of the Slow Chemical Proceses*. Nauka, Moscow, 1966.
106. Yakubi, VA, BA Ponomarev, NF Tupalo, GA Galstyan, LP Petrenko, SM Tyutyunnik, PP Karpuhin. *Theory and Practis of Liquid Phase Oxidation*. Ed. NM Emanuel', Nauka, Moscow, 1974, p 303.
107. Galstyan GA, *Pross. II-nd Ozone Conference*, MGU, Moscow, 1977, p.40.
108. Nenchev, LK, DM Shopov. *Neftekhimia*, 18, 2, 256 (1978).
109. Karpova, SG, AA Popov, SN Chvalun, YuA Zubov, GE Zaikov. *Visoko Molekul. Soed.- B*, 27, 9, 686 (1985).
110. Karpova, SG, AA Popov, SN Chvalun, YuA Zubov, GE Zaikov. *Visoko Molekul. Soed.- A*, 28, 7, 1404 (1986).
111. Popov, AA, AV Rusak, OA Ledneva, GE Zaikov. *Visoko Molekul. Soed.- A*, 28, 9, 1836 (1986).
112. Blinov, NN, AA Popov, ES Popova, AV Butusov, AN Neverov, GE Zaikov, G Irinyi. *Polym. Deg. Stab.*, 22, 7 (1988).
113. Cipitoin, A, A Constantinesku, V. Dobresku. IUPAC Macro'83, Budapest, 1983, Sec.4, P.727.
114. Patel, KHH. *J. Polym. Sci. Polym. Phys.*, 13, 2, 303 (1975).
115. Popov, AA, NN Blinov, GE Zaikov. *Dokl.Akad. Nauk SSSR*, 286, 4, 922 (1986).
116. Selihova, VI, LA Ozerina, LN Ozerin, NF Bakeev. *Visoko Molekul. Soed.- A*, 28, 2, 342 (1986).
117. Priest, DJ. *J. Polym. Sci.A-2*, 9, 10, 1777 (1971).
118. Karpova, SG. *PhD Theses*. Inst. Chem. Phys., Moscow, 1985.
119. Gaponova, IS, VM Gol'dberg, GE Zaikov, AA Kefely, GB Pariiskii, SD Razumovskii, VYa Toptigin. *Visoko Molekul. Soed.- A, 20*, 2037 (1978).
120. Howard, JA, KV Ingold. Can. J. Chem., 47, 3797 (1969).

Chapter 2

OZONOLYSIS OF OXYGEN CONTAINING ORGANIC COMPOUNDS

The ozonolysis of oxygen containing compounds is a promising reaction that takes place under mild conditions and yields compounds of higher oxidation level that the starting ones. It may find various applications in chemical and pharmaceuticals industries, fine organic synthesis, etc. [1-2] Moreover, this reaction is extremely important from ecological point of view for utilization and purification of waste industrial waters from hydroxybenzene production, through their partial or complete oxidation [3-19]. The importance of this reaction for the theory and practice gave us establishment to make a system investigations [20-28].

2.1. ALCOHOLS

Current concepts on the ozonolysis of alcohols are discussed in several monographs [1, 2, 20, 29]. During the past years a number of authors tried to elucidate the kinetics and mechanism of this reaction and its application for the selective preparation of ketones and aldehydes in high yields under mild conditions [29-34]. It has been established that the rate of ethanol ozonolysis does not change when the hydroxyl group is deuterated [33] and its value is 4.17 times higher than when the methylene group is deuterated. On this basis, together with the data from the analysis of the products composition and the kinetics of their formation the authors, have suggested that the H-atom abstraction from ozone is the rate determining step of the reaction. This conclusion is confirmed by the data in Table 2.1 as reported by various authors [31-33].

Table 2.1.

Kinetic parameters of the ozone reaction with aliphatic alcohols at 25^0C

Alcohols	k, $M^{-1}.s^{-1}$	lg A	E, kcal/mol	Ref.
t.-Butanol	0.01	4.6	9.0	32
t.-Butanol	0.05	-	-	31
Ethanol	0.35	6.7	9.8	32
Ethanol	0.25 (22^0C)	6.7	9.8	33
n-Butanol	0.54	7.3	10.3	32
n-Butanol	0.39			31
iso-Propanol	0.89	7.3	10.0	32
Cyclopentanol	1.35			31

The ratio of the relative reactivities of tertiary:secondary:primary alcohols is 1:12:241 according to Ref. 31, or 1:156:817 according to Ref.[6]. On the basis of the studied on the reaction of ozone with methyl, ethyl and 2-propyl alcohol, the following reaction mechanism is proposed [30]:

Scheme 2.1

$$CH_3OH + O_3 \Leftrightarrow [HOC^{\bullet}H_2...HO_3^{\bullet}]^{\#} \rightarrow HOCH_2OOOH,$$

and in the presence of a base:

$$CH_3OH + O_3 + (B) \rightarrow HOCHO + {}^{-}O_2H + BH^{+}$$
$$CH_3CH_2OH + O_3 \Leftrightarrow [CH_3HC^{\bullet}(OH)...HO_3^{\bullet}]^{\#} \rightarrow CH_3HC(OOOH)OH \rightarrow$$
$$CH_3OOH + H_2O_2 \text{ and/or } CH_3CHO + O_2 + H_2O$$
$$(CH_3)_2C(OH)H + O_3 \Leftrightarrow [(CH_3)_2C^{\bullet}(OH)...HO_3^{\bullet}]^{\#} \rightarrow (CH_3)_2C(OH)OOOH$$
$$\downarrow -O_2$$
$$CH_3C(OH){=}CH_2 + H_2O + O_2\ (CH_3)_2CO + H_2O$$
$$\downarrow +O_3$$
$$CH_3COOH + HCHO$$

It is supposed that an intermediate ion or radical pair is formed whose recombination in the kinetic cage yields α-hydroxy-hydrotrioxide. The latter leaves the cage and passes into the solution volume. The authors of Ref. 30 have found that the product composition of 2-propanol ozonolysis depends on the way of propanol purification. It has been established that the ozonolysis of 21 alcohols leads always to the formation of the corresponding aldehydes and ketones in 53-83 % yields [34].

Thought the interesting results given above and the assumptions for the reaction pathway, in our opinion, the mechanism is still unclear and new data should be provided for its elaboration.

In this connection we have carried out intensive experimental and theoretical studied of this reaction, the results of which are summarized in Fig.2.1-2.3 and Tables 2.2-2.4.

Fig.2.1 demonstrates the results from ozone consumption in solutions of methanol and iso-propanol by means of stopped flow technique with time of mixing below 0.2 s. It is seen that irrespective of the type of alcohol being ozonized, the kinetic curves coincide and a first order kinetics is observed under the experimental conditions applied.

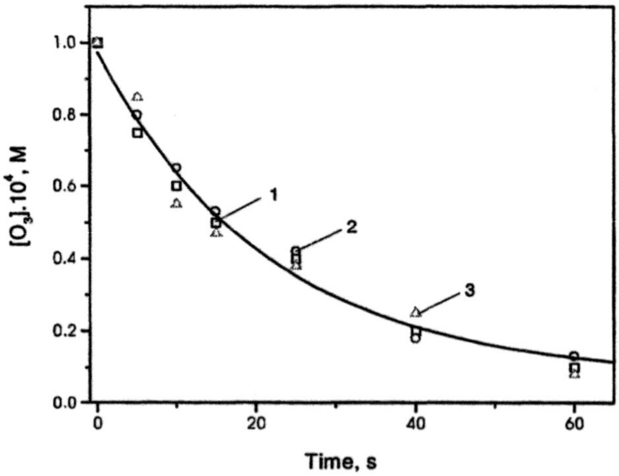

Fig.2.1. Kinetics of ozone consumption in reaction of ozone with 1-methanol, 22°C, 0.74 M; 2-ethanol, 3°C, 0.24 M and 3 - *iso*-propanol, 3.5°C, 0.037 M.

Calculations of the kinetic curves of ozone reaction with MeOH, EtOH and i-PrOH gave the following values of the rate constants: 0.057, 0.17 and 1.13 $M^{-1}.s^{-1}$, respectively.

The same values were also obtained at carrying out the reaction in a bubble reactor. The values of k calculated according to Eq.1.4 are given in Table 2.2 and their graphic presentation in Figs.2.2 and 2.3.

Table 2.2.

Kinetic parameters of ozone reaction with MeOH in CCl_4 and pure MeOH solutions: 22°C, ω=0.167 s^{-1}; v = $1.67.10^{-3}$ l/s, maximum rate of ozone inlet - $1.67.10^{5}$ $M.s^{-1}$

[MeOH], M	$[O_3]_0.10^5$, M	$[O_3]_g.10^5$, M	$\Delta[O_3].10^5$, M	$W.10^6$, $M.s^{-1}$	k, $M^{-1}.s^{-1}$
0	10	0	10	0	-
0.247	10	8.32	1.68	2.80	0.057
0.439	10	7.35	2.65	4.43	0.056
0.618	10	6.64	3.36	5.61	0.058
0.740	10	6.22	3.78	6.31	0.055
0.987	10	5.53	4.47	7.46	0.057
1.0	10	5.49	4.51	7.53	0.056
2.0	10	3.79	6.21	10.37	0.055
3.0	10	2.89	7.11	11.87	0.057
4.0	10	2.34	7.66	12.79	0.058
5.0	10	1.96	8.04	13.42	0.054
10.0	10	1.09	8.91	14.88	0.056
24.7	10	0.47	9.53	15.91	0.058
0.740	8	4.98	3.02	5.04	0.057
0.740	6	3.73	2.27	3.79	0.055
0.740	4	2.49	1.51	2.52	0.056
0.740	2	1.24	0.76	1.27	0.058

Processing the data of Table 2.2, columns 5, 2 and 3, according to Eq.1.3 yields the following linear dependencies (Fig.2. a and b):

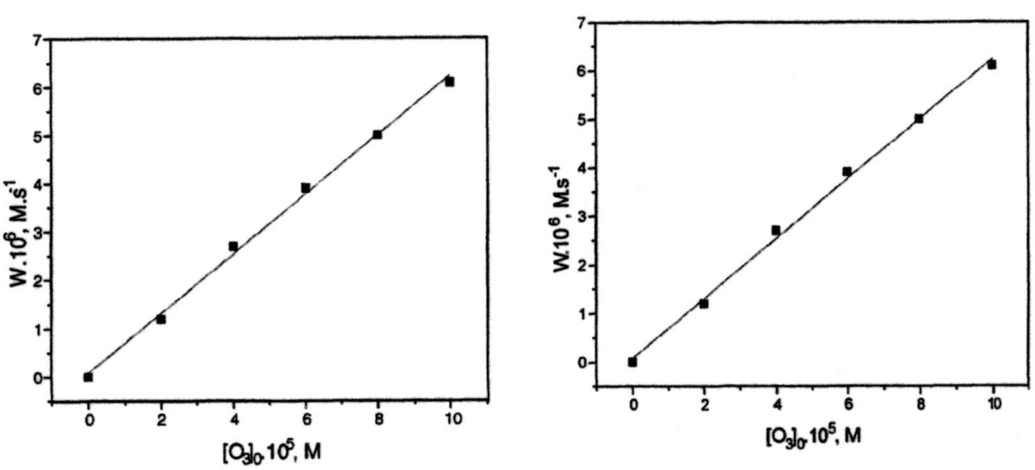

Fig.2.2. Dependence of the rate of ozone consumption on the inlet (a) and outlet (b) ozone concentration

The rate of W on ΔO_3 has been found to be linear too.

The dependence of W on [ROH] (column 1) is a curve which approached a limit at W=$1.67.10^{-5}$ $M.s^{-1}$. The nonlinear character of this relationship is in accordance with the complexity of Eq.1.3 (Fig.2.3) since when [ROH] $\rightarrow\infty$, $[O_3] \rightarrow 0$ and $\Delta O_3 \rightarrow [O_3]_o$.

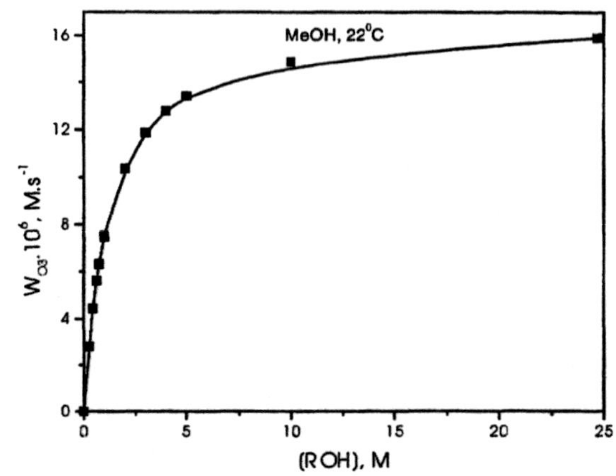

Fig.2.3. Dependence of the rate of ozone consumption on methanol concentration according to Eq.1.4.

The values of k calculated by Eq.1.4 for each point of the curve are equal to 0.057. Hence, at using the bubbling method when the rate is of first order in relation to ozone the dependencies of W on $[O_3]_o$, $[O_3]_g$ and $\Delta[O_3]$ should be linear and that on [ROH] should have the profile of the curve in Fig.2.3.

From the analysis of the data obtained (Table 2.3 and 2.4) it follows that the rate constant and the activation energies are strongly dependent on the alcohol structure.

Table 2.3.

Dependence of k on temperature in ^{0}C for ozone reaction with three types of alcohols

k, $M^{-1}.s^{-1}$	0^0C	10^0C	20^0C	25^0C	30^0C
MeOH	0.008	0.021	0.049	0.072	0.108
t.-BuOH	0.005	0.013	0.029	0.045	0.064
EtOH	0.14	0.28	0.54	0.74	1.10
n-PrOH	0.19	0.36	0.67	0.89	1.18
n-BuOH	0.15	0.30	0.56	0.76	1.10
i-PrOH	0.93	1.61	2.71	3.46	4.39
s.-BuOH	0.88	1.54	2.58	3.29	4.18
c-HexOH	0.92	1.59	2.65	3.37	4.27

The interaction of ozone with MeOH possessing primary-α-H-atoms and tert-BuOH with only primary-C-H bonds has been found to be the slowest and the value of k at 20°C per one α-H-atom in the first case it is $1.62.10^{-2}$ $M^{-1}.s^{-1}$, and in the second case related to one primary-H-atom is $3.22.10^{-3}$ $M^{-1}.s^{-1}$. The difference in the values of these constants is due to the fact that while the -OH group in MeOH affects directly the α-H-atom, for the tert.BuOH that does not possess any α-H-atoms, the effect of the OH- groups is carried through one σ-bond and it is considerably weaker. As a result of this the reactivity of tert.BuOH becomes similar to that of methane and neopentane. The interaction of ozone with EtOH, n-PrOH, n-BuOH takes place at higher rate and the values of k per one α-H-atom amount to: 0.27, 0.34 and 0.28, respectively. The higher rate of ozone reaction with these alcohols is associated with the presence of secondary -α-H-atoms in their molecule, which have lower bond energy that the primary ones. The enhanced reactivity of n-PrOH as compared to that of EtOH could be ascribed to the donor effect of the second -CH_3-group while the donor effect of the C_2H_5-group in n-BuOH is weaker than that of the methyl group and thus k is lower. It has been found that the reaction of ozone with i-PrOH, s-BuOH and s-HexOH alcohols possessing tert.-H-atoms with lowest energy bond is the fastest with k equal to 2.71, 2.58 and 2.65 , respectively, i.e. in fact they are the same. The ratio of the reduced values of k at 25°C for methanol:ethanol;sec.butanol are 1:15:137. Simultaneously the values of E_a decrease with the decrease of the α-H-atoms bond energy (D) and their ratio is 1:0.78:0.61 [11, 14]. This fact can be regarded as an important evidence for the mechanism of -α-C-H-atoms abstraction by ozone.

Table 2.4.

Kinetic parameters of ozone reaction with some alcohols at 25°C

	MeOH	t.-BuOH	EtOH	n-PrOH	n-BuOH	i-PrOH	s.-BuOH	c-HexOH
$k.10^2$, $M^{-1}.s^{-1}$	7.2	4.5	74	89	76	346	329	337
n, α-C-H	3	9 (β-C-H)	2	2	2	1	1	1
$k.10^2/n$, $M^{-1}.s^{-1}$	2.4	0.5	37	44	38	346	329	337
E_a, kcal/mol	13.9	13.7	10.9	10.1	10.4	8.5	8.5	8.4
$A.10^{-7}$, $M^{-1}.s^{-1}$	41	5.8	3.9	2.4	1.7	0.62	0.59	0.51

2.1.1. Application of ERM

We have tried to elaborate the mechanism of alcohols ozonolysis by means of ORM described in Chapter I. The calculations of A were performed considering the two possible structures of the activated complex: linear (LC) and cyclic(CC) (Fig.2.4).

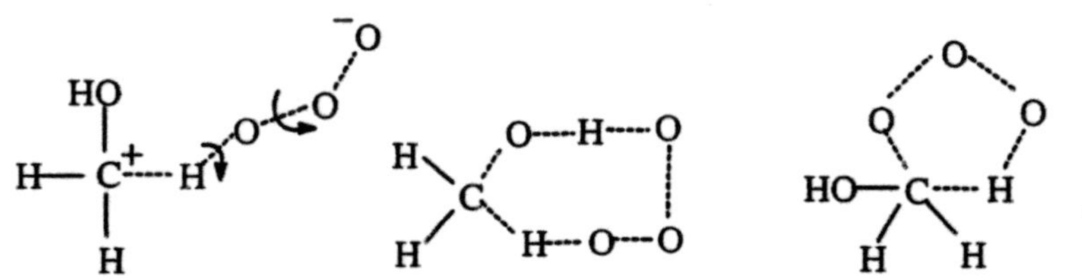

Fig.2.4. Structure of the probable activated complexes in the reaction of ozone with aliphatic alcohols: LC-linear with free fragment rotation and CC' and CC-cyclic complexes without free rotation

The good agreement between the experimental and theoretical values for A allows to determine the AC structure and on this basis to select the most possible mechanism.

Simultaneously the value of the preexponential was calculated by the collision method and at comparing its value with that calculated by the AC theory the steric factor -p was determined [36-40]:

$$k = p.Z_0.\exp(-E_a/RT), \qquad \text{(Eq.2.1)}$$

where: $Z_0 = \pi.(r_A+r_B)^2.(kT/\pi m^*)^{1/2}$; Z_o is the collision factor; r_A and r_B are the van der Waals radii of the reagents; k - Boltzmann constant; T-absolute temperature ; m* is the reduced mass.

The preexponentials for the reaction of ozone with alcohols, calculated according to the activated complex method (ACT) and collision theory (CT) are presented in Table 2.5.

Table 2.5.

The values of A calculated by ACT with LC and CC and by CT, the sums (R_{fr}) and energy (E_{fr}) of inner rotation, steric factors (p) VDW-radii of the molecules (r) and the ratio between the calculated and experimental values of preexponentials $A^{LC}{}_{cal}/A_{obs}$.

	MeOH	EtOH	n-PrOH	i-PrOH	n-BuOH	s.-BuOH	t.-BuOH	c-HexOH
$A^{CC}.10^{-4}$, $M^{-1}.s^{-1}$	52	8.3	3.7	3.5	2.4	2.2	6.3	1.6
$A^{LC}.10^{-4}$, $M^{-1}.s^{-1}$	56	15	6.7	6.3	4.4	4.0	4.9	2.9
$F_{fr}.10^{-2}$	7.1	9.8	12	11	13.8	12.6	12.9	12.6
$A^{LC}{}_{cal}.10^{-7}$, $M^{-1}.s^{-1}$	40	15	8.1	6.8	6.1	5.0	6.3	3.6
r, Å	2.24	2.57	2.91	2.88	3.13	3.23	3.13	3.23
$A^{CT}.10^{-11}$, $M^{-1}.s^{-1}$	2.6	2.6	2.7	2.7	2.8	2.8	2.8	2.7
$p.10^{5}$	22	5.8	2.5	2.3	1.6	1.4	1.8	4.1
$A^{LC}{}_{cal}/A_{obs}$	1.0	3.8	3.4	11	3.1	8.3	1.1	7.2
A_{fr}, cal/mol	0	730	670	1368	701	1172	52	1081

Note: ALC is the preexponential calculated with LC without free fragment rotation (it is absent at $E_{fr} \geq 4$ kcal); A_{CC} -calculated with CC; ALC $_{calc}$ = at free fragment rotation, i.e. when exp ($-E_{fr}/RT$)=1 or E_{fr} =0 kcal; A_{ct} - calculated according to the collision theory at 300K and r - calculated by the Alinger method (pcmod4 program) at radius of the ozone molecule of 2A; E_{fr} - the calculated values of the rotation energy at ratio of $A^{LC}{}_{cal}/A_{obs}$=1.

An interesting fact is the good agreement between the values of A^{LC} and A^{CC}. The values of A^{CC} with five and seven member cyclic form of AC are actually similar as they differ not more than by 1%. However the free rotation produces an increase in $A^{LC}{}_{cal}$ by 3 orders of magnitude as compared to those of A^{CC}. The steric factor has a value which is in agreement with the liquid phase reaction. The ratio between the calculated and experimental values of A^{LC} which increases in the sequence primary:secondary:tertiary alcohols is 1:3.4:8.8, i.e. the coincidence in this sequence decreases. This is connected with the fact that E_{fr} is not zero and they increase with the changes in the alcohol structure in the same sequence. It has been found that the values of $A^{LC}{}_{calc}$. and A_{obs}. coincide when the values of rotation energy have the values in line 10. We have obtained the same values by means of *mopac6* program and therefore the values in line 10 can be also considered as calculated ones.

The good agreement between $A^{LC}{}_{calc}$ and A_{obs}. suggests that the mechanism of the rate determining step of the ozone reaction with alcohols is associated with the formation of linear AC and abstraction of α-H-atom from the alcohol.

The experimental and theoretical results conform well with the following scheme:

Scheme 2.2.

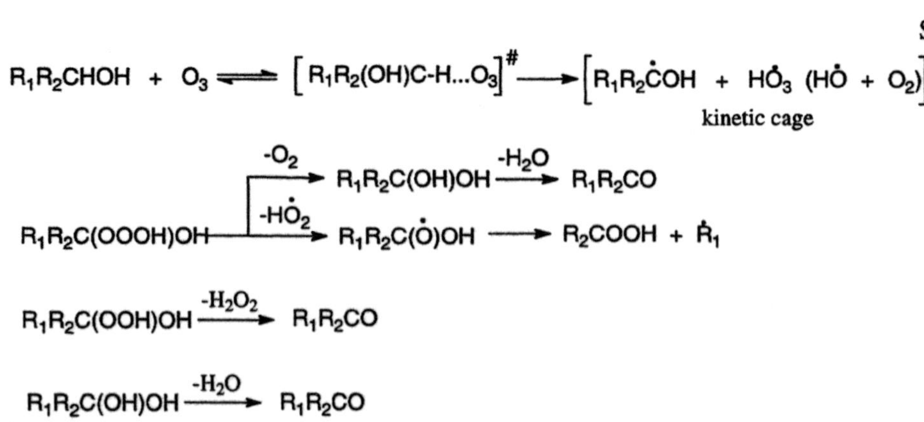

Ozone attacks the α-H-atom forming a LC which further undergoes decomposition to radical (or ion) pair in one kinetic cage. The α-hydroxyalcohol, α-hydroxyperoxyalcohol and α-hydroxytrioxyalcohol being unstable leave the cage and decompose rapidly to the corresponding aldehyde or ketone liberating water, hydroperoxide and oxygen or lead to the formation of hydroperoxy and alkoxy radicals. The latter can further undergo monomolecular decomposition.

2.2. KETONES

The investigations of the reaction of ozone with ketones are of interest in relation to the theory of reactivity, the ozone chemistry, preparation of oxygen-containing compounds, degradation and stabilization of organic materials. In particular, the ozonation of methylethylketone [41-43] yielded acetic acid, diacetyl and hydroperoxide as the main reaction products in CCl_4. It has been assumed that the process proceeds as a non-chain radical reaction (RO_2). Similar products are also formed when the reaction takes place in water. However, in this case ozone interacts with the enol form of MEK [41]. The addition of nitric acid accelerates enolization that is evidenced by the increase of the oxidation rate [43]. The decrease of the oxygen content in the ozone-oxygen mixture results in an increase of the oxidation rate that is due to the change of the radical leading the oxidation, from $RO_2^{\bullet}$ to $RO^{\bullet}$. The ozonation of isomeric decanones indicates that the reactivity of ketones depends on the length and structure of the alkyl chain. In this case the main products are α-ketohydroperoxides, monocarboxylic acids, keto alcohols and alkylbutyric lactones. However, it should be *Note*d that the conclusions in these papers were based on products analysis at very high degrees of conversion whereby the secondary reactions play an important role in the products formation. This gives us ground to doubt whether the proposed mechanism describes adequately the initial steps of the reaction. The rate constant and the activation energy of interaction between cyclohexanone and ozone are $1.6.10^{-2}$ $M^{-1}.s^{-1}$ and 11.2 kcal/mol, respectively. The reactivity of ozonized methyl derivatives of cyclohexanone is considerably higher than that of the H-form. An increase in rate of interaction between the above derivatives and ozone is also observed depending on the site of substitution and it follows the sequence: 3-Me< 4-M2< 2-Me [44-46].

It is evident from the foregoing that so far no systematic investigations on the ozonolysis of ketones have been carried out.

In our studied on the ozonolysis of ketones we have combined the kinetic approach and products analysis with theoretical methods. The results from the ozonolysis of some aliphatic ketones are shown in Table 2.6

Table 2.6.

Kinetic parameters of the reaction of ozone with some aliphatic ketones at 21°C

Ketone	$k_0.10^3$, $M^{-1}.s^{-1}$	$k_{st}10^3$, $M^{-1}.s^{-1}$	k_{rel}	lg A	E_a, kcal/mol
CH_3COCH_3	3.5	7.7	0.2	10.823	18
$CH_3COC_2H_5$	12	24	0.7	10.63	17
$CH_3COC_5H_{11}$	23	67	1	10.54	16.5
$CH_3COC_6H_{13}$	45	58	2.4	10.46	16
$CH_3COC_7H_{15}$	448	54	2.3	10.31	15.8
$CH_3COC_{10}H_{21}$	42	113	2.2	10.58	16.2
$C_2H_5COC_5H_{11}$	20	74	1	11.48	15.5
$CH_3COCH_2COCH_3$	29000	-	-		-
$c\text{-}C_7H_{12}O$	87	80	4.1	9.00	17
CH_3CO-2-Naph	22000	-	-	5.77	6
4-CH_3-c-C_6H_9O	117	101	6.2	8.97	12
4-t.-Bu-c-C_6H_9O	166	123	8.1	7.93	11.8

Note: k_o is calculated according to Eq.1.4; k_{rel} - relative rate constant determined by GC relative to ethylpentylketone.

The following six types of ketones have been the subject of our investigations: 1-acetone containing only α-C-H bonds; 2 - ketones with general formula $CH_3CO(CH_2)_nCH_3$, where n=1, 4, 5, 6 and 9, which contain secondary α, β, γ C-H bonds; 3 -ketones with general formula formula $CH_3(CH_2)_m CO(CH_2)_nCH_3$, where m=1, n=4, which contain α, α'-secondary C-H bonds and cyclic - cycloheptanone which contains α, β and γ secondary C-H bonds; 4-4-methyl or 4-tert-butyl cyclohexanone with γ-tert. C-H bond; 5 -2-naphthylmethylketone with various substituents and 6 - acetylacetone containing α-C-H bond activated by two keto groups. The rate constant of keto-enol tautomerism for the latter is considerably higher than that of the other ketones studied.

$$CH_3\text{-}CO\text{-}CH_2\text{-}CO\text{-}CH_3 \Leftrightarrow CH_3\text{-}CO\text{-}CH{=}C(OH)\text{-}CH_3$$

The values of k depending on the ketone structure, particularly on the content of primary, secondary or tertiary α-C-H bonds, vary in a wide range. The ratio of their reactivity is 1:3.4÷24.8:33.4÷47.4. In contrast to paraffins these values are more similar which is associated to the activation of α-C-H-bonds by the keto group. Ozone is sensitive not only to the activation to the α-C-H-bonds but also to the keto-enol equilibrium. The rate of its reaction with acetylacetone is by three orders of magnitude higher than that with monoketones. Ketones with different substituents, for example, 2-naphthylmethylketones show also higher reactivity than the aliphatic ketones. It is concluded that the reaction center in this case is the naphthyl ring and not the α-C-H-bond in the methyl group.

Another interesting observation are the higher values of E_a as compared to those of paraffins (by 2-3 kcal/mol). Since ozone is an electrophilic agent and the activation of the α-C-H-bonds from the keto group favors the nucleophilic attack, the increase of E_a should be reasonably expected.

In Fig.2.5 is shown a typical kinetic curve illustrating the change of ozone concentration at the reactor outlet during ozonolysis of ketones in solution.

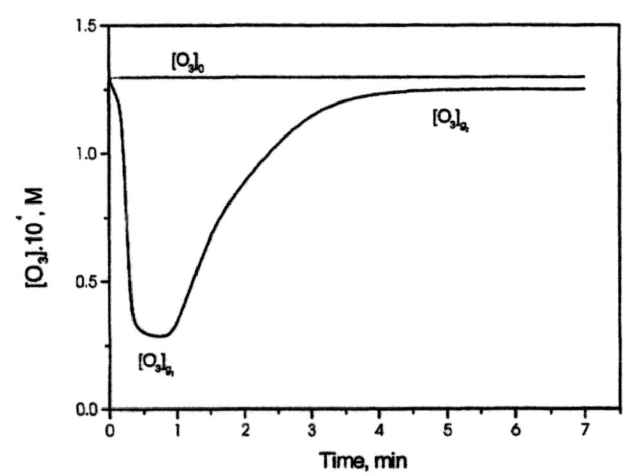

Fig.2.5. Ozonolysis of 0.712 mmol cycloheptylketone in 10 ml CCl_4; 21°C; v-0.1 l/min.

The horizontal line describes the inlet ozone concentration and the curve corresponds to the change of its concentration at the reactor outlet. It is seen that the addition of ketone results in abrupt fall of the ozone concentration followed by a steady-state part ($[O_3]_{g1}$). After some time the ozone concentration rises again and a second steady-state region is attained ($[O_3]_{g2}$).

This profile of the kinetic curve ozone-gas suggests ozone interaction with more than one compound [1, 21, 22]. The accuracy of ozone titration is determined by the difference in the values of the rate constants and when it is about an order then in the curves appear well defined separate steps. The areas over the steps correspond to the amount of absorbed ozone and at known stoichiometry, to the quantity of the reacting substance, respectively.

This type of the kinetic curve for ketones is associated with the keto-enolic equilibrium and to the ability of ozone to react with the C=C bonds 106 times faster than with C-H bonds with nearly zero activation energy [1].

The processes corresponding to the curve ozone-gas (Fig.1.5) are the following: 1) ozone starts its reaction with ketone (probably with the enol form) and an abrupt fall of its concentration is registered; 2) when the rate of ozone supply becomes equal to the rate of the chemical reaction, the first step is formed; 3) the end of the step is connected with the consumption of the enol form and the ozone concentration begins to rise; 4) at this moment ozone begins to react with the keto form and the second step is formed whereby the rate of ozone supply has become equal to the rate of the second reaction. The sharp transition between the two steady-state region testifies that the rate of restoration of the keto-enol equilibrium is considerably lower than the rate of ozone interaction with the keto form.

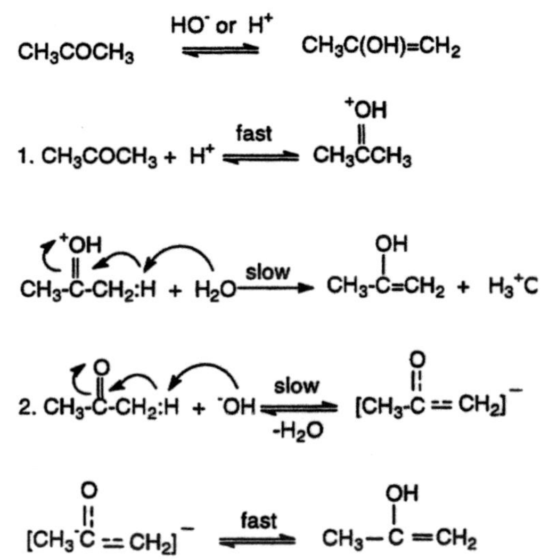

The concentration of the enolic form can be found from the area below the kinetic curve in Fig.2.5 and then the equilibrium constant of enolization could be measured. It is known that the stoichiometry of the reaction of ozone with C=C is 1:1 and thus the amount of the enolic form can be judged by the amount of absorbed ozone, i.e. in this particular case the concentration of the enolic form is 0.134 mM. Since the initial concentration of ketone is known, the equilibrium one will be the difference between the initial and that of the enolic form. Then the equilibrium constant can be easily calculated using the formula:

$$K_e=[Enol]_e/[Keton]_e,$$

and it amounts to 1.9.10-2.

From the curve in Fig.2.5 one can also determine the rate constants of ozone interaction with the two tautomeric forms. On the basis of the parameters of the first steady-state part and by using Eq.1.4, the value of the rate constant with the enol form is found to be k_E= 2.9.10^2 $M^{-1}.s^{-1}$ and from that second -the rate constant with the keto form k_K= 0.048 $M^{-1}.s^{-1}$.

This demonstrates the importance of the method of ozone titration that makes possible the measurement of the equilibrium constant and the rate constants of ozone interaction with the two tautomeric forms in one kinetic experiment.

It is evident that the rate constants of ozone interaction with ketone as measured by the steady-state method (Fig.2.6) through mixing of ozone and ketone solutions at $[K]_o>>[O_3]_o$ are higher than those found by the bubbling method (Table 2.6).This is not difficult to explain because in the former case the values obtained represent the total effective constant of interaction of the enol and the keto form.

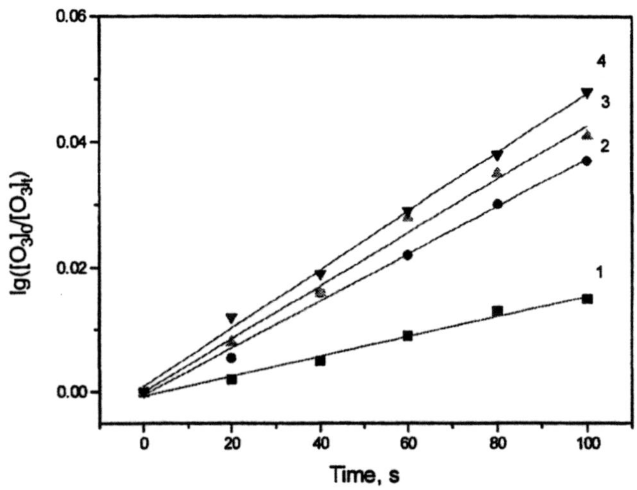

Fig.2.6. Semilogarithmic anamorphosis of the kinetic curve of ozonolysis of 1- methylethylketone; 2-methylpentylketone; 3-ethylpentylketone and 4-cycloheptylketone. 21°C, ketone concentration - 28 mM.

The results from the calculation of the equilibrium constants of keto-enol tautomerism for some aliphatic ketones in CCl_4 are given in Table 2.7. In the Table are not presented the data for acetylacetone and 2-naphtylmethylketone because in the former case the rate of reaching the equilibrium is commensurable with the rate of ozone interaction and in the latter case the ozone reacts with the double bonds in the naphthyl ring. The equilibrium constants do not differ from those found within the temperature range of 21°C to 3°C and agree with data from literature [47-50].

Table 2.7.

Equilibrium constants of keto-enol tautomerism of some ketones in CCl_4 solution at 21°C determined by titration with ozone.

Keton	CH_3COCH_3	$CH_3COC_2H_5$	$CH_3COC_5H_{11}$	$CH_3COC_6H_{13}$	$CH_3COC_7H_{15}$	$CH_3COC_{10}H_{21}$	c-$C_7H_{12}O$
$K_e.10^4$,	1.2	1.2	46	29	96	130	190

The kinetics of ozone reactions with ketones are also determined gas-chromatographically. The relative rate constants shown in Table 2.6, column 4 demonstrate that only acetone and methylketone possess lower reactivity that the standard. It is seen that the rate constants calculated from the relative ones and the value of the standard constant correspond to those found by the bubbling method.

The main products of ozone interaction with methylethylketone are 2-hydroxymethylethylketone, diacetyl, peroxides -alkyl and hydro, acetaldehyde and acetic acid.

On the basis of the kinetic results obtained and the product analysis we suggest the following scheme for the proceeding of ketone ozonolysis:

Scheme 2.3

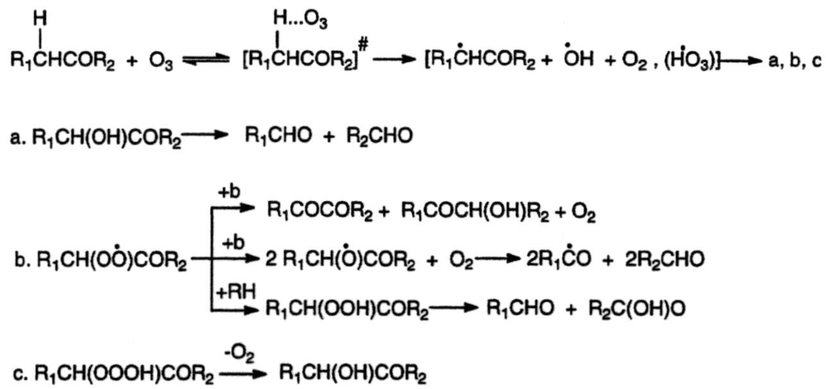

Ozone attacks the α-H-atom forming a LC which further undergoes decomposition to radical (or ion) pair in one kinetic cage. or lead to the formation of hydroperoxy and alkoxy radical. The latter can further undergo monomolecular decomposition.

The intermediate formation of LC is assumed in the first stage, followed by breaking of α-C-H leading to the formation of radical (or ion) pair. Then α-hydroxyketone, α-peroxyketone radical and α-hydroxytrioxyketone through C-C bond breaking can decompose to two aldehydes. The α-peroxy ketone radical reacting with the initial ketone can be transformed into α-hydroxyperoxyketone which is decomposed through breaking of C-C bond to aldehyde and acid or two radicals recombinate giving rise to diketone and α-hydroxyketone or two α-oxy ketone radicals. The latter reacting with the initial ketone are transformed into α-hydroxyketone or through a monomolecular decomposition and breaking of C-C bond to alkoxy radical and aldehyde. α-Hydrotrioxyketone is decomposed to α-hydroxyketone and it in its turn to two aldehydes.

2.3. Hydrotrioxides

Hydrotrioxides (HTO) are defined as compounds containing HOOO-group. Hydrotrioxides of oxygen-containing compounds have been studied and described in the literature in detail. They include those of ethers, acetals, alcohols and aldehydes [51-63]. They are obtained by ozonation of the corresponding organic compounds, for example: benzaldehyde, 2-methyltetrahydrofuran, methylisopropyl ether, di-iso-propyl ether, etc., at low temperatures (< -60°C - -80°C). HTO are unstable compounds and at heating they decompose and liberate oxygen The NMR method has been used to confirm their structure including the unstable HOOO-group.[53].

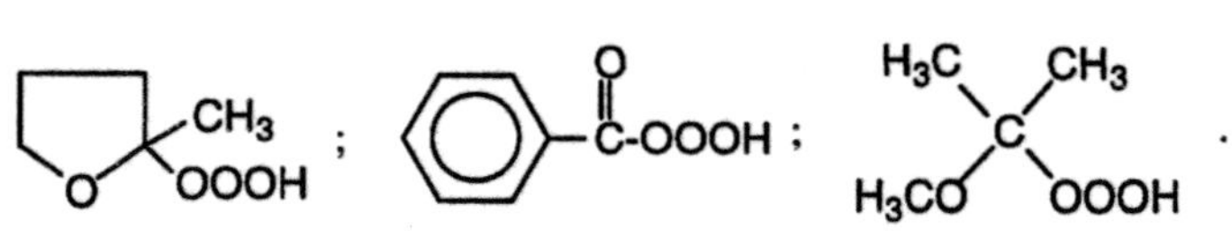

The NMR signal of the proton in HOOO-group appears at δ=13.1 ppm, while that in HOO-group at δ=9 ppm [53, 57]. Benzoic acid and singlet oxygen were the main products of decomposition of benzaldehyde HTO. This reaction due to the high yield of $^1O_2$ appears to be a suitable source for its preparation [55].

The proton signals of the HOOO-group in the pointed below acetals:

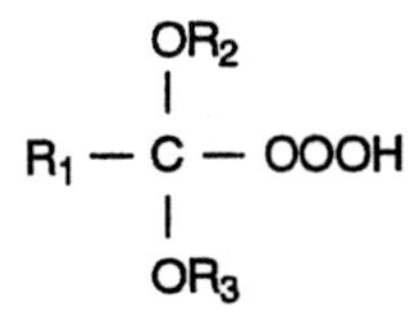

$R_1=R_2=R_3=CH_3$; $R_1=CH_3$, $R_2=R_3=C_2H_5$; $R_1=Ph$, $R_2=R_3=CH_3$

appear in the vicinity of δ=12-14 ppm [30, 31] and the proton signal of HTO of ethanol at 13.7 ppm is a doublet [62].

The kinetic parameters of decomposition of HTO of benzaldehyde, 2-methyltetrahydrofuran and methyl-iso-propylether are found to be dependent on whether the decomposition is performed in a solvent or in a pure substrate [53]. Thus in pure HTO a slight downfield shift (by <0.15) ppm of the peak of HOOO is observed [64-66]. The doublet at <-50°C is transformed into a singlet [56, 57, 62] with the rise of temperature to ambient. This is explained by the participation of proton in exchange reaction with hydrogen bonds formation. The lower values of A and E_a found for the decomposition of pure HTO are resulting from the hydrogen bonds formation and HTO stabilization in TS:

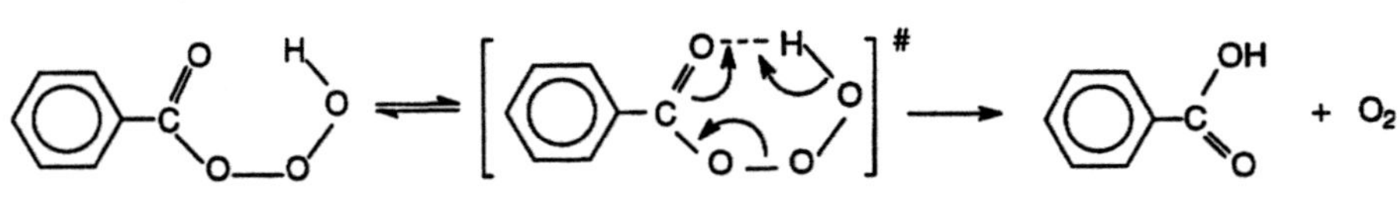

The kinetic parameters of the decomposition of 23 HTO obtained by different authors are given in Table 2.8 [53-71].

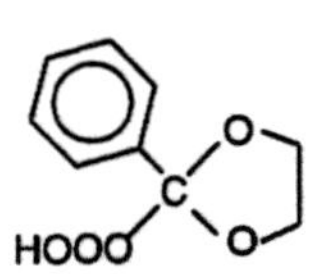
1

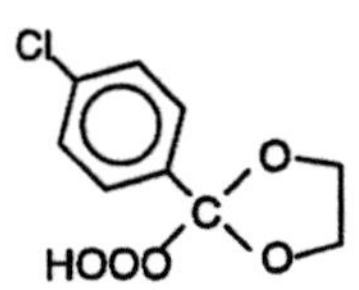
2

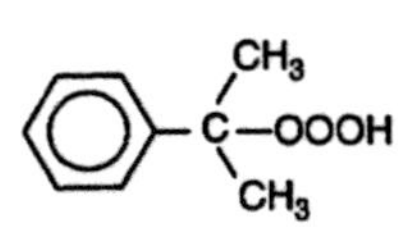
3

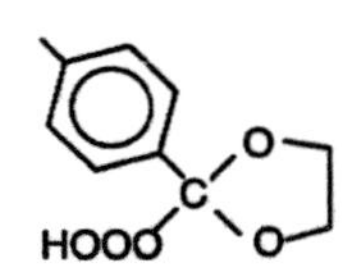
4

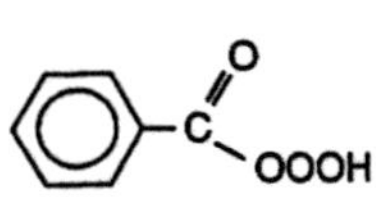
5

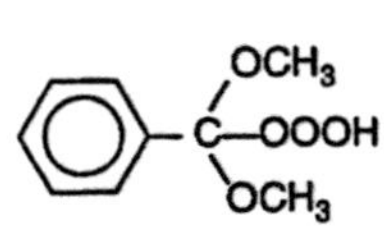
6

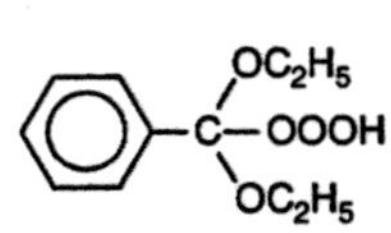
7

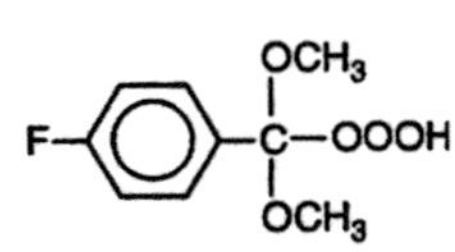
8

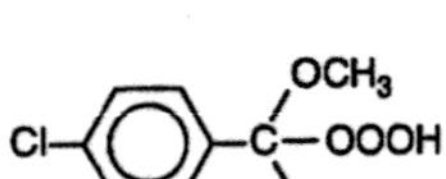
9

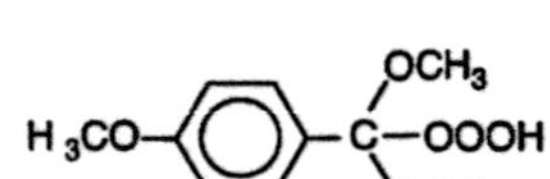
10

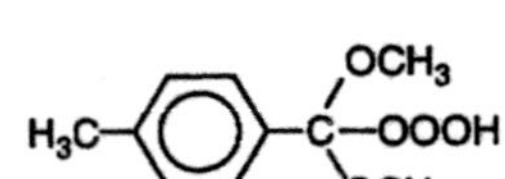
11

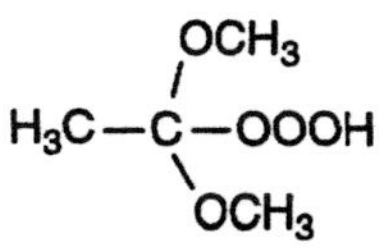
12

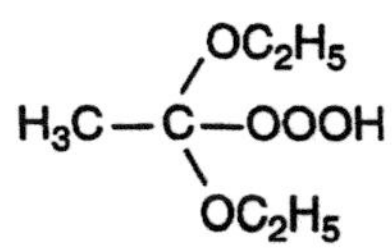
13

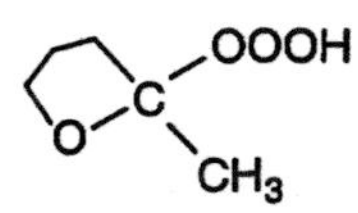
14

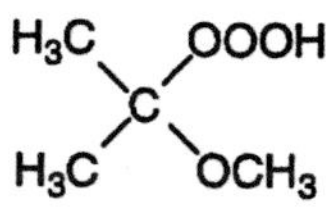
15

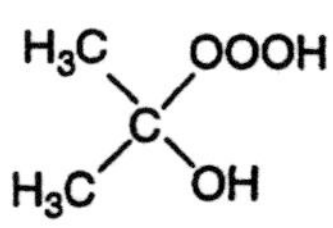
16

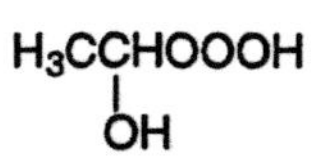
17

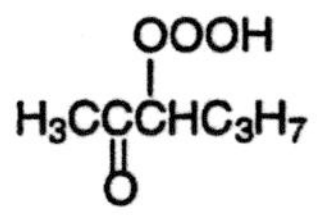
18

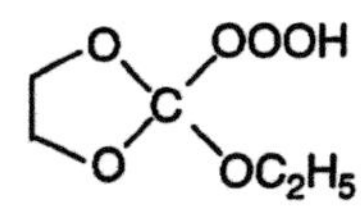
19

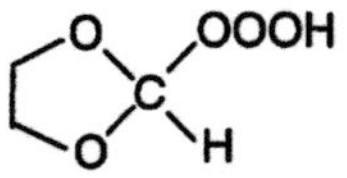
20

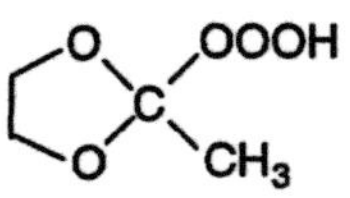
21

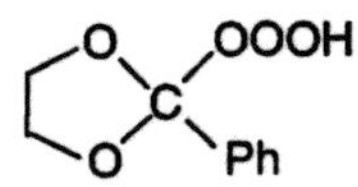
22

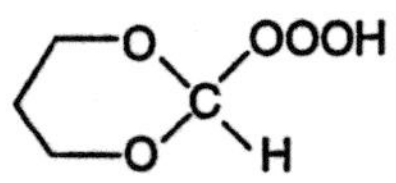
23

Table 2.8

Kinetic and activation parameters of the decomposition reaction of HTO in pure substrate (-), ether and dichloromethane (DCM)

HTO	Solvent	$\tau_{1/2}$, min	t, ^{0}C	k.10^4, s^{-1}	lgA	E_a, kcal/mol
1	Ether	43	-40	160	14.8	12.2
			-30	660		
	DChM	58	-20	640	20.6	16.6
			-30	120		
2	Ether	-	-40	270	-	-
3	d_6-Acetone	162	-35	0.7	16	10.4
		50	-23	2.3		
		37	-20	3.1		
		21	-17	5.4		
		13	-12	10		
4	DChM	85	-20	380	18.6	14.7
			-30	80		
	Ether	63	-40	110		
5	-	14	0	8.6	10.8	5.4
	-	33	-9	3.5		
	-	82	-2-	1.7		
	-	180	-34	4.6		
	Ether	41	-27	2.8	16.5	11.1
		82	-33	1.4		
		216	-39	5.3		
6	Ether	58	-30	660	18.9	15.9
			-35	290		
			-40	120		
7	Ether	53	-30	760	19.8	16.7
			-35	280		
			-40	130		
8	Ether		-30	630	19.6	16.4
			-35	310		
		63	-40	110		
9	Ether		-35	320	18.4	15.4
		49	-40	140		

10	Ether		-25	440	17.2	13.7
			-30	200		
		69	-35	100		
11	Ether		-20	2100	21.5	17.8
			-25	810		
			-30	310		
		49	-35	140		
12	-		-20	270	20.8	15.6
	-		-10	560		
	DChM		-10	400		
	Ether		-10	290		
		84	-20	90		
13	-		-10	1100	14.6	11.1
	-		-20	370		
	Ether		-10	760	13.1	9.8
			-20	280		
	-		-50	0.1	22.7	18.2
	-		-40	2		
	-		-29	120		
	-		-20	370		
	-		-10	1000		
14	-	3	-6	40	7.9	4.1
	-	17	-30	6.9		
	-	58	-50	2		
	-	330	-68	0.35		
	Ether	8	-12	10	17.4	11.7
		18	-19	6.5		
		60	-27	1.9		
		138	-33	0.83		
15	-	2.6	-10	50	16.5	11.5
	-	13	-22	8.8		
	-	16	-26	7.3		
	-	52	-33	2.2		
	-	114	-38	1		
16	-	150	--30	0.76	11.7	6.4
	-	18	-10	6.2		
	-	7	0	20		
17	-	19	-30		6.9	3.8
	-	29	-40	4		
	-	46	-45	2.5		
	-	120	-60	0.9		
	-	288	-70	0.4		
18	-	2	0	60	11.5	6.9
	-	3	-10	40		
	-	10	-20	10		
	-	23	-31	5		
19*	-	3	-10	40	12.4	8.0
	-	10	-20	10		
	-	23	-30	5		
	-	38	-40	3		
20*	-	168	-30	0.7	-	-
21*	-	84	-30	1.4	-	-
22*	DChM	12	-30	10	-	-
23*	-	38	-30	3	-	-

Note: *- our data

In studying the decomposition of α-methylbenzylmethylether HTO Pleshnichar has established that the main products are acetophenone, methanol, methylbenzoate and traces of water and hydroperoxide [53, 54]. He has explained the formation of these products by the following scheme

Scheme 2.4.

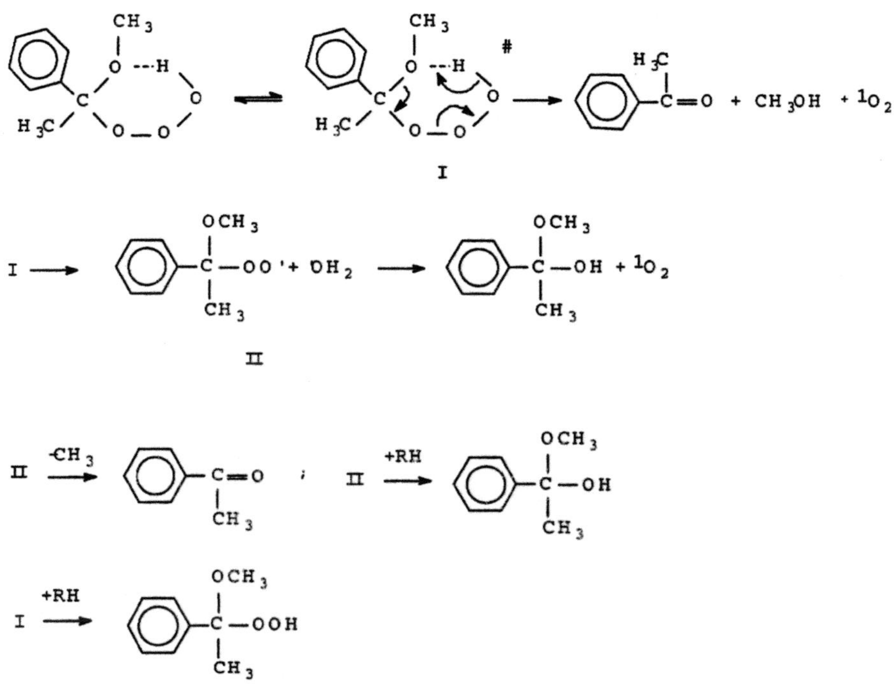

We have established that the decomposition reaction of HTO in the temperature range of (-70°C +)°C obeys the following equation:

$$-d\,[HTO]/dt = k_d \cdot [HTO], \qquad \text{(Eq.2.3)}$$

where: k_p is the decomposition rate constant

The values of the rate constants of the HTO decomposition indicated that electron-accepting substituents to the benzene ring affect the process in a rather complex way (Table 2.9).

Table 2.9.

Activation parameters of the decomposition of hydrotrioxides of some α-methylbenzylalkylethers

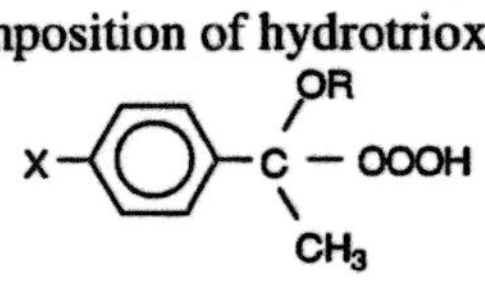

HTO	Solvent	$k_d.10^2$ (-30°C), s^{-1}	E_a, kcal/mol	lg A
X=H; R=CH_3	EtAc	0.5	15.2	11.4
	Ether	8.9	17.7	14.9
Cl; CH_3	Ether	12	16.7	14.1
Br; CH_3	Ether	7.6	16.8	14.1
CH_3;; CH_3	Ether	49	18.1	16
Br; C_2H_5	Ether	6.9	17.6	14.7

The decomposition reaction of the synthesized HTO of 1,1-diethoxyethane yields ethylacetate (0.75-080 mol per 1 mol of absorbed ozone), ethanol (0.80-0.85mol), water (0.06-0.12 mol), acetaldehyde, acetic acid ethylformiate, diethylcarbonate, dissolved gases, probably methane and ethane, and traces of 2-hydroxydiethylether and α,α-diethoxydiethylperoxide and hydroperoxide. It was supposed that ethylacetate, ethanol, and singlet oxygen are formed as follows:

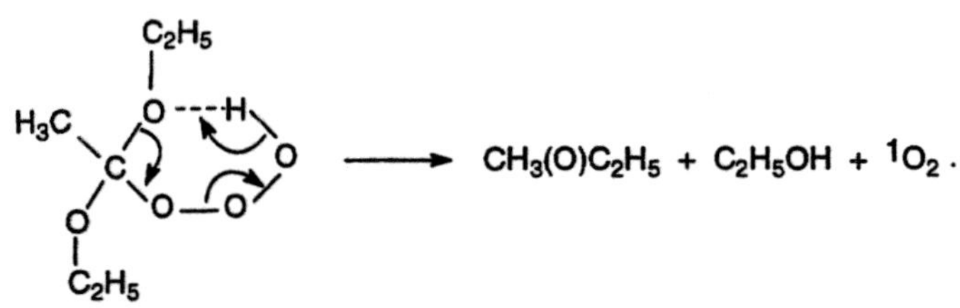

The rest of the decomposition products were supposed to be formed by the interaction of the generated in the decomposition reaction radicals with the initial and intermediate reaction products.

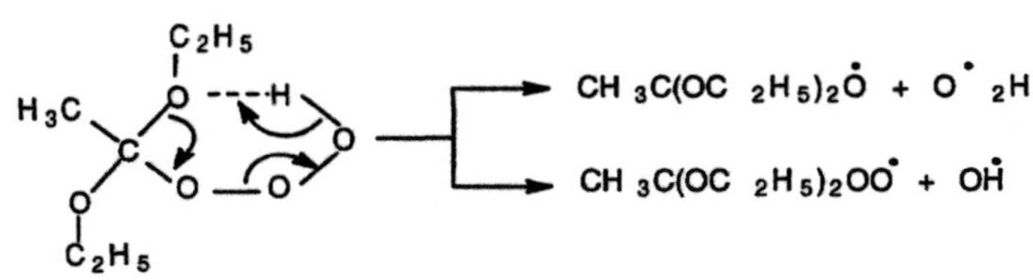

$$CH_3C(OC_2H_5)_2OO^{\bullet} + RH \longrightarrow CH_3C(OC_2H_5)_2OOH + R^{\bullet}$$

$$2CH_3C(OC_2H_5)_2OO^{\bullet} \longrightarrow [CH_3C(OC_2H_5)_2O]_2 + O_2$$

$$CH_3C(OC_2H_5)_2O^{\bullet} + RH \longrightarrow CH_3C(OC_2H_5)_2OH + R^{\bullet}$$

$$CH_3C(OC_2H_5)_2OH \longrightarrow CH_3C(O)OC_2H_5 + C_2H_5OH$$

The comparatively low values of the activation energy of the thermal decomposition of some HTO (Table 2.9) can be explained by the fact that the reaction takes place simultaneously in several directions. The reaction step of radical decomposition was suppressed and became negligible with the decrease in temperature. Ozonolysis of 1,1-diethoxyethane in the -20°C-+60°C range lead mainly to the formation of ethylacetate, ethanol, the corresponding HTO and water. The radical steps of the process yielded ROOH and water and hence, they were responsible for the formation of the latter about 40 % at 50°C whereas practically it was below 1 % at 5°C.

Table 2.10.

`Effect of temperature on the formation of water and HTO in the ozonolysis of 1,1-diethoxyethane (1) and 2-methyl-1,3-dioxolane at 12 mM absorbed ozone and initial concentration of acetal 0.1M

Acetal	t, ^{0}C	HTO, mM	H_2O, mM	HTO/O_3	H_2O/O_3
1	60	5.7	6.1	0.48	0.51
	50	4.5	5.0	0.38	0.42
	40	3.6	4.2	0.3	0.35
	20	2.3	2.2	0.19	0.19
	5	0.8	0 7	0.07	0.06
	-20	0	0	0	0
2	50	4	4.3	0.33	0.36
	40	3	3.3	0.25	0.27
	20	2.1	2.2	0.18	0.19
	5	0	0	0	0
	-20	0	0	0	0

These results were also confirmed by the data in Ref.34, 37. According to these authors the share of the free radical route is 1.2 % at -60°C for the HTO of 1,1-diethoxyethane while it varies in the range of 0.3-3.7 between -30°C-0°C for the decomposition of HTO of 2-propanol.

Formation of nitroxyl radicals has been observed upon decomposition of HTO of ethanol, isopropanol, 1,1-diethoxyethane and benzaldehyde in the presence of diphenylamine and 2,2,6,6-tetramethylpiperydine [66].

According to Pryor and co. [70] the values of the activation parameters of cumene HTO decomposition (E_a and lgA) are 16 kcal/mol and 10, respectively, whereas other authors reported 24 kcal/mol and 16, respectively [67]. These parameters are very close to those of benzaldehyde HTO [53] and acetals [56, 57]. The main products of the cumene HTO decomposition are dimethylphenylcarbinol (DMPC), acetophenone (AcP), singlet oxygen and hydroperoxide (ROOH). The assumption that the decomposition process takes place via chain-radical route was confirmed with the help of 2,5-di-tert.butyl-4-methylphenol (Ionol). In this case the activation energy is strongly reduced. The products formation could be explained by the following scheme [70].

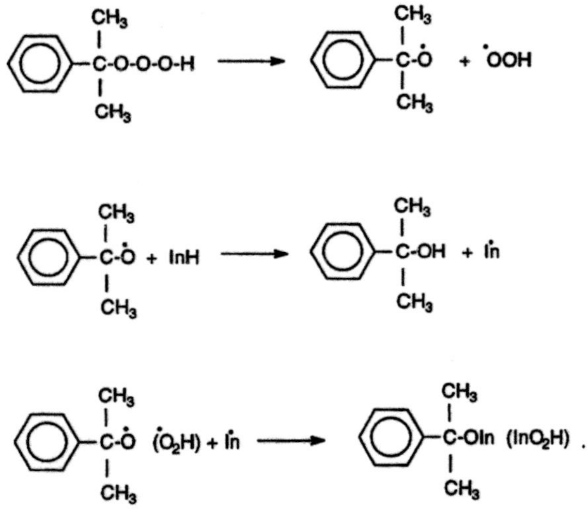

Then the process occurs without an inhibitor:

$$RO^{\bullet} + ROOOH \rightarrow ROH + ROOO^{\bullet}$$
$$ROOOH + {}^{\bullet}O_2H \rightarrow ROOO^{\bullet} + H_2O_2$$

$$ROOO^{\bullet} \rightarrow RO^{\bullet} + {}^{1}O_2$$
$$RO^{\bullet} \rightarrow AcP + {}^{\bullet}CH_3$$

We have found that the thermal decomposition of HTO is accompanied by an intensive chemiluminescence in the visible and infrared regions [51, 60, 65], which fact has been also observed by other authors. Carbonyl compounds and singlet oxygen are supposed to be chemiluminescent emitters. The rate constant of the thermal decomposition determined by chemiluminescence (k_{CL}) turned out to be very close to that measured by the NMR (k_{NMR}) method through monitoring the disappearance of the signal at δ=13 ppm (Table 2.11).

Table 2.11.

Rate constants of decomposition determined by chemiluminescence and by NMR for some HTO

HTO	t, ^{0}C	$k_{CL}.10^3$, s^{-1}	$k_{NMR}.10^3$, s^{-1}
$(CH_3)_2C(OOOH)OH$	0	2	2
$(CH_3)_2C(OOOH)OCH(CH_3)_2$	-26	12	18
$CH_3C(OOOH)(OC_2H_5)_2$	0	1.3	0.9

It should be *Noted*, however, that the decomposition rate determined by chemiluminescence in the infrared range of the spectrum is lower than that in the visible region [60]. The reason for this difference is not yet clear.

Our attempts and those of other investigators to use HTO as oxidizing agents have given satisfactory results in a number of cases [1, 2, 73-78].

It is known that the oxidation of sulfides to respective sulfoxides using HTO of ethanol, 2-propanol, 1,1-diethoxyethane and benzaldehyde takes place at a high rate and selectivity.. Among them the latter two compounds were the most active [68]

$$R_2S \xrightarrow{+R'OOOH} R_2S{=}O$$

$$R = CH_3, C_4H_9, C_7H_{15}$$

It has been found that dimethylsulfide (DMS) was readily oxidized by the HTO of 1,3-dioxolane to dimethyl sulfoxide (DMSO) (Table 2.12).

Table 2.12.

Oxidation of dimethylsulphide by HTO of 1,3-dioxolane (1) and 2-methyl-1,3-dioxolane (2), at -30°C and mol ratio [ROOH]/[DMC] = 1:4

HTO	$[ROOOH]_0$, mM	[DMSO], mM	Yeld, %
1	0.7	0.6	86
	0.6	0.6	100
2	0.5	0.3	60
	0.6	0.4	67

Initially the hydrotrioxide forms quantitatively a monoether. It was also demonstrated in the case of dimethyl- and dibutyl sulphoxides that HTO can oxidize sulphoxides to the respective sulphones:

$$R_2S{=}O + R'OOOH \rightarrow R_2SO + R'OOH$$

The HTO of benzaldehyde and 2-propanol were the most effective. The yield of sulphone per 1 mol of converted HTO was increased with the increase of the thermal stability of HTO (Table 2.13).

Table 2.13.

Oxidation of dialkylsulphoxides by HTO

HTO	$[HTO]_0 \cdot 10^2$, M	R_2SO	$[R_2SO]_0 \cdot 10^2$, M	$[HTO]_0/[R_2SO]_0$	$[R_2SO_2] \cdot 10^2$, M	$[R_2SO_2]/[R_2SO]_0$	$[R_2SO_2]/[HTO]_0$
$CH_3CH(OH)(OOOH)$	13	$(C_4H_9)_2SO$	13.8	0.9	1.6	0.12	0.12
	20		6.9	2.9	1.0	0.15	0.05
$(CH_3)_2C(OH)(OOOH)$	92	$(CH_3)_2SO$	100	1.1	19	0.18	0.21
	63		19	3.3	16.4	0.86	0.26
	25	$(C_4H_9)_2SO$	21	1.2	6.5	0.31	0.16
	30		5.3	5.7	5.4	1	0.18
$(CH_3)_2CHOC(CH_3)_2(OOOH)$	50	$(C_4H_9)_2SO$	44	1.1	7.6	0.17	0.15
	32		11	1.2	1.2	0.11	0.04
$CH_3CH(OC_2H_5)(OOOH)$	26	$(C_4H_9)_2SO$	27	1	2.5	0.09	0.1
	46		9.2	5	0.8	0.087	0.02
C6H5C(O)(OOOH)	29	$(C_4H_9)_2SO$	30.6	0.9	21.6	0.67	0.74
	30		9	3.3	8.2	1	0.27

Another direction in HTO application is the oxidation reaction of phosphines. According to Ref. 76 HTO of alcohols, ethers and acetals convert readily, selectively and quantitatively triaryl- and trialkyl-phosphines to respective phosphites:

$$(RO)_3P + R'OOOH \rightarrow (RO)_3P{=}O + R'OOH$$

The data from the oxidation process of triphenylphosphine by HTO of ethanol and propanol , obtained by the authors of [50] are listed in Table 2.14.

Table 2.14.

Stoichiometric ratios at the oxidation of triphenylphosphine by HTO of 2-propanol and ethanol in ether solution at -50°C [76].

HTO	$[HTO]_0$, M	$[HTO]_f$, M	$[Ph_3P]_0$, M	$[Ph_3P]_f$, M	$\Delta[Ph_3P]/\Delta[HTO]$
2-Propanol	0.52	0.37	0.14	0	0.9
	0.19	0.08	0.14	0	1.3
	0.09	0	0.018	0.01	1.9
	0.1	0	0.33	0.1	2.0
Ethanol	0.10	0	0.23	0.04	1.9
	0.08	0	0.29	0.13	2.0

Note: in 2-propanol R in HTO is $(CH_3)_2C(OH)$ and in ethanol R - $CH_3CH(OH)$; index f - final concentration

It is evident that the reaction stoichiometry depends on the initial concentration of the reagents and it attains its ultimate value of 2 at low HTO concentration and excess of phosphine. In this case HTO is converted to hydroperoxide which in its turn yields alcohol.

It has been found that the titration with phosphines can be used as an analytical method for HTO determination. Its reliability is supported by the close values of the rate constant of α-hydroxyhydrotrioxide decomposition determined by the above method (-30°C, k_d=6.10^{-4} s^{-1}) and by the NMR method (-30°C, k_d=4.10^{-4} s^{-1}).

It has been shown that pyridine is oxidized by HTO at -60^{0}C to N-oxide (60-80 % yield). The oxidation of other organic compounds to corresponding products can also be accomplished by HTO. In this case, however, catalysts and higher temperatures should be applied.

Upon ozonation of 2-methyl-1,3-dioxolane in the temperature range of -60°C - +50 °C it has been found that one mol ozone gives 2 moles of glycol monoester. It was presumed that the intermediate HTO can also oxidize the substrate (Table 2.15).

Table 2.15.

Synthesis (-60°C- +50°C) and decomposition (-30°C) of HTO of 2-methyl-1,3-dioxolanes in ethylacetate

Acetal	$[Ac]_0$, mM	$[Ac]_f$, mM	Time, min	Δ[Ac], mM	$[O_3]$, mM	[Monoester], mM
2-Methyl-	22.4	18.0	45	4.4	1.8	4.2
1,3-dioxolan	11.2	9.4	10	1.8	0.7	1.2
	11.2	9.4	15	1.8	0.8	1.6
	11.2	9.7	10	1.5	0.6	1.5
	5.6[a]	3.8	15	1.8	0.9	1.5
	3.7[a]	1.9	20	1.8	1.0	1.8

The decomposition reaction of the HTO pointed above proceeds via the following way:

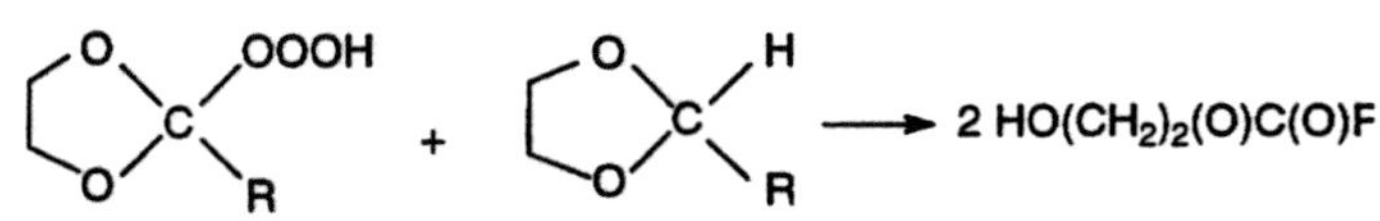

R = H, CH_3

This suggestion was confirmed in the oxidation of 2-methyl-1,3-dioxolane by the HTO of 1,3-dioxolane. Besides ethyleneglycol-monoformiate, ethyleneglycol-monoacetate was found in the mixture of reaction products.

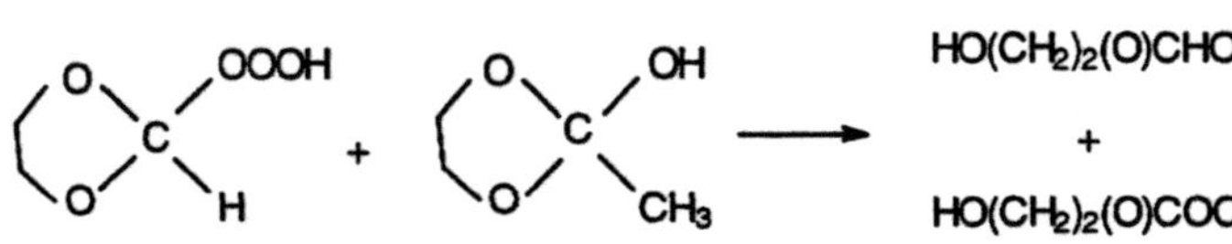

Ring opening upon the decomposition of HTO formed by the ozonolysis of nonsymmetric dioxanes (2-methyl, 2,4-dimethyl and 4,4-dimethyl-1,3-dioxane) may take place via two routes resulting in the formation of two isomeric esters:

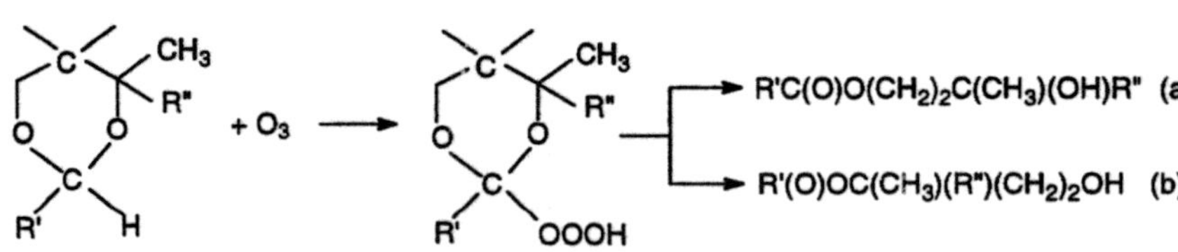

R' = R" = H; R' = CH_3, R" = H; R' = H, R" = CH_3

The ratio of the products of type (a) and (b) upon ozonolysis of 1,3-dioxycyclanes at various temperatures are shown in Table 2.16.

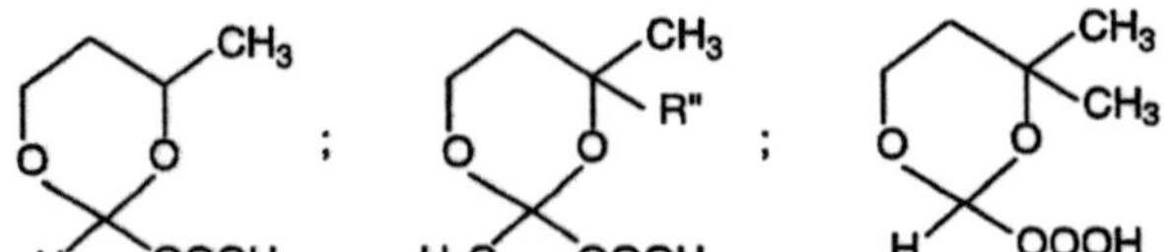

1 ; 2 ; 3

Table 2.16.
Products of ozonation of 4-methyl-1,3-dioxane (1), 2,4-dimethyl-1,3-dioxane (2) and 4,4-dimethyl-1,3-dioxane (3) at various temperatures

HTO	a/b, -50^{0}C	a/b, +15^{0}C	a/b, +50^{0}C
1	2:1	2:1	2:1
2	2.5:1	2:1	2:1
3	7:1	3:1	2:1

Note: (a) -monoesters with acylated primary groups

The main products of the reaction were monoesters with acylated primary groups. The more selective decomposition of the HTO of 4,4-dimethyl-1,3-dioxane (Table 2.16) is due to the steric repulsion between the OOOH-group and the axial CH_3 group at the 4C-atom of the ring in the intermediate HTO:

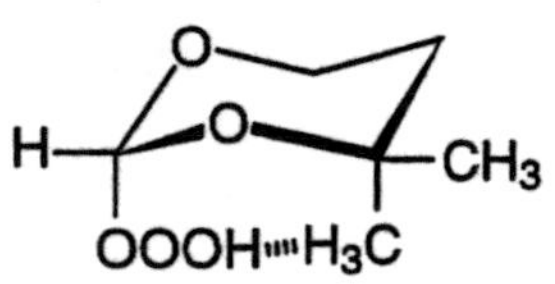

The above discussion shows that the chemistry of organic trioxides is a new rapid-developing branch of the organic chemistry and any progress in this field is essential for the development and improvement of the theory or reactivity and methods of organic synthesis [79-96].

1,3-Dioxolanes, in particular those substituted in the 4th position, have found wide application as solvents, plasitifiers and biologically active substances. In this connection we have tried to develop a new method for the preparation of the above compounds in satisfactory yields and to obtain a number of new derivatives. The two phase catalytic: "liquid-liquid" and "liquid-solid" systems were applied.

2.4. Synthesis of Oxolanes

2.4.1. Derivatives of 4-Hydroxymethyl-1,3-Dioxolane

A four-necked flask, supplied with a reflux condenser, a magnetic stirrer, thermometer and a funnel was charged with 10 ml aqueous NaOH and 5 mM Katamine AB and under continuos stirring 0.05 M 4-hydroxymethyl-1,3-dioxolane was added dropwise [24]. The mixture was heated up to 15°C and the corresponding chloride (0.05 M) was added dropwise for 30 min. Then the mixture was diluted with water and extracted with ether. The ethereal solution was dried over $MgSO_4$, the diethyl ether removed and the mixture subjected to vacuum distillation. Thus the following oxolanes were obtained:

1. 4-Propyloxymethyl-1,3-dioxolane ; NMR spectrum (CCl_4, δ ppm of HMDS): 0.7÷1.0 (t) CH_3; 1.3÷1.7 (qt) CH_3-CH_2; 3.2÷3.5 (m) CH_2-O-CH_2; 3.6-4.2 (m) 5H, -OCH_2-CH_2-CH_2O; 4.7 (s) -O-CH_2-O-ring.

2. 4-Benzyloxymethyl-1,3-dioxolane; (Ccl4, δ ppm of HMDS): 3.6÷4.2 (m) 5H, -OCH_2-CH_2-CH_2O ; 4.35 (s) OCH_2-Ph; 4.7÷4.8 (s) O-CH_2-O- ring; 7.0÷7.3 (m) C_6H_5.

3. Phenylsulphone of 4-hydroxymethyl-1,3-dioxolane: (Ccl4, δ ppm of HMDS): 3.6÷4.4 (m) 5H -OCH_2-CH_2-CH_2O ; 4.7÷4.9 (s) 2H, O-CH_2-O-ring; 7.5-8.0 (m) C_6H_5.

2.4.2. Phenyl Ethers of 4-Hydroxymethyl-1,3-Dioxolane

A four-necked flask, provided with a reflux condenser, a magnetic stirrer, thermometer and a funnel was charged with the respective phenol, 0.1 M of solid KOH, a catalytic quantity of 18-crown-6 and acetonitrile or DMSO as solvent and under continuos stirring 0.05 M 4-hydroxymethyl-1,3-dioxolane was added dropwise. The mixture was stirred for 3 h at 70°C followed by ether extraction and the extract was dried over $MgSO_4$. The diethyl ether was distilled off and the residual mixture subjected to vacuum distillation. Thus 4-(2,5-dichlorophenoxymethyl)-1,3-dioxolane was obtained with the following NMR characteristics: (CCl_4, δ ppm of HMDS):3.2÷4 (m) 5H -CH_2-CH_2-CH_2O; 4.8÷4.9 (s) 2H, -CH_2-O-ring; 6.6÷7.2 (m) C_6H_5.

2.4.3. Alkyl Ethers of 4-Hydroxymethyl-1,3-Dioxolane

The synthesis of alkyl ethers of 4-hydroxymethyl-1,3-dioxolane (B) was carried out by the interaction of 4-hydroxymethyl-1,3-dioxolane (A) with alkylhalides (RX): propylchloride, butylchloride, octylchloride, butylbromide and hexylbromide. In the presence of an aqueous alkaline solution with the application of interphase catalyst -Katamin AB the following products were obtained:

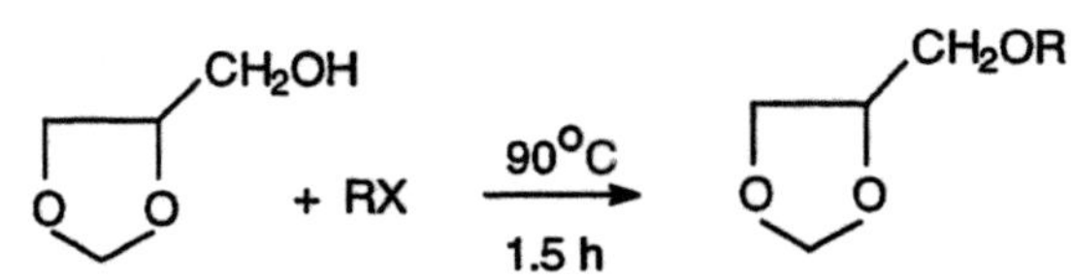

where: R=C_3H_7 (I); C_4H_9 (II); C_6H_{13} (III); C_8H_{17} (IV) with the following physical constants: boiling point t in °C at mm Hg., n_{D20} and yield in %:42(5), 1.4313, 55; 58(4), 1.4320, 57; 80(6), 1.4359, 50 and 132(4), 1.5213, 44, respectively.

In the series of normal alkychlorides the yields of monoethers decrease with the increase of the molecular mass of the substituents. The substitution of chlorides by the corresponding bromides leads also to increase of the yields up to 65-70 %.

The arylmethylethers of 4-hydroxy-1,3-dioxolane were obtained undre similar conditions. Benzyl and p-chlorobenzylchloride react with 4-hydroxy-1,3-dioxolane to give the corresponding ethers:

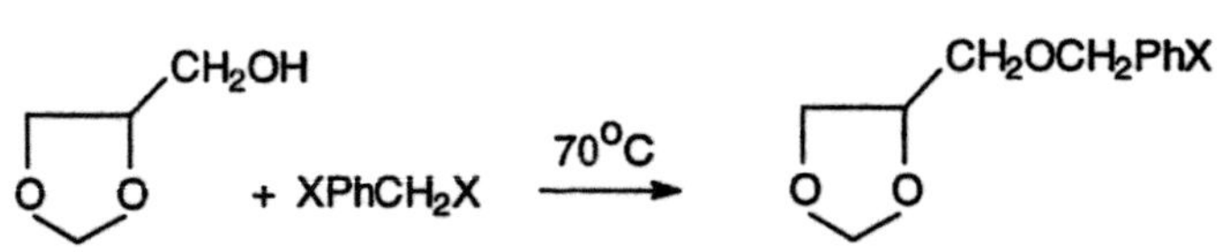

where: X=H(V) or Cl(VI) in yields of 84 and 70 %, respectively.

The physical constants for V are 132(4), 1.5231, ρ=1.1832 g/cm^3, and for VI 160(7), 1.5242, 1.0119, respectively.

In some case it seems more preferable to use epychlorohydrine derivatives. In interphase catalysis (solid KOH, catalyst - tetrabutylammonium bromide and 90°C) it became possible to obtain benzyl ether of 4-hydroxymethyl-1,3-dioxolane:

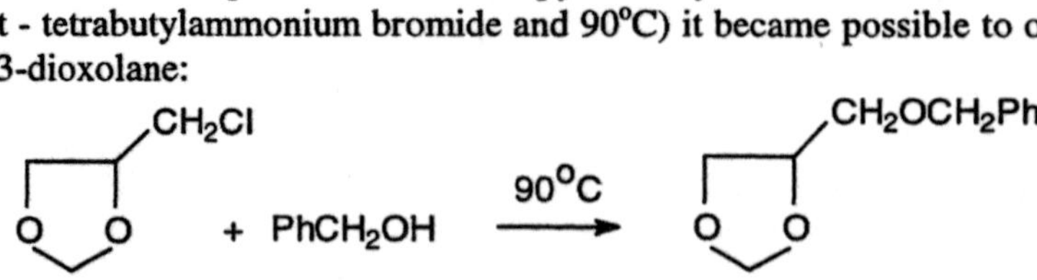

The yield, however, does not exceed 25 %.

A series of aryl ethers were obtained by the reaction of 4-chloromethyl-1,3-dioxolane with substituted phenols in the presence of various solvents:

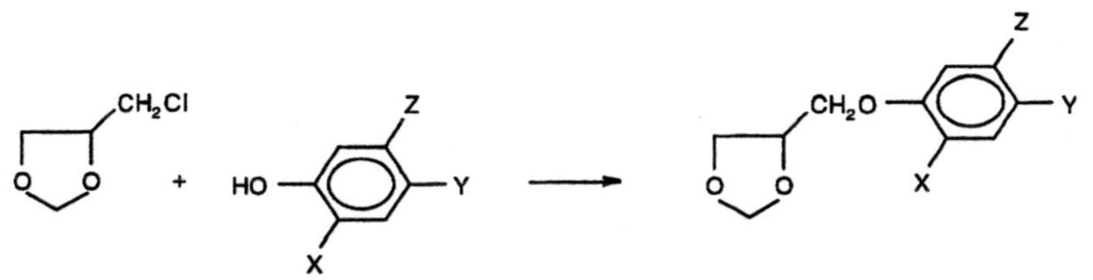

where: X, Y and Z are: CH3, Cl, H (VII); H, H, H (VIII); Cl, H, Cl(X); Cl, H, H (XI); H, CHI, H (XII); and H, Cl, H (XIII).

Depending on the conditions and the type of the substituent the yield of monoethers varies from 10 to 80 % (Table 2.17).

Table 2.17.

Synthesis of aryl derivatives

Substrate	Solvent	Catalyst	b.p., ^{0}C (mm Hg)	ρ, g/cm^3	n_D^{20}	Yield, %
VII	NaOH/H_2O	Katamin AB	154(5)	1.2381	1.5330	10
VIII	NaOH/H_2O	Katamin AB	112(7)	1.0548	1.5150	42
IX	CH_3CN	18-crown-6	165(4)	-	1.5445	15
X	CH_3CN	18-crown-6	150(5)	-	1.5341	10
XI	DMSO	18-crown-6	118(5)	-	1.5194	50
XII	DMSO	18-crown-6	75-77 ðîï.		-	20
XIII	DMSO	18-crown-6	113(3)	1.1832	1.5231	80

The C_4-C_8 aliphatic alcohols do not react with 4-chloromethyl-1,3-dioxolane under interphase catalysis conditions. The results obtained show that it is better to conduct the synthesis of 4-alkoxymethyl-1,3-dioxolanes on the basis of 4-hydroxymethyl-1,3-dioxolane and the corresponding alkyl halides. Derivatives of 4-hydroxymethyl-1,3-dioxolane-arylsulphonates were obtained by the interaction of 4-hydroxymethyl-1,3-dioxolane with arylsulphochrorides (aqueous-alkaline solution, Katamin AB):

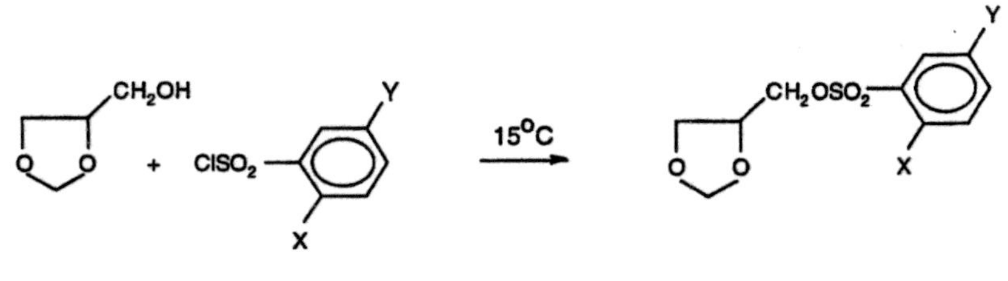

where: X and Y are: H, H (XIV); CH_3, H (XV); CH_3, CH_3 (XVI); Cl, H (XVII). The yields of the compounds in (%) and their melting points in (^{0}C) are as follows: 88, 55-57; 80, 75; 50, 90-92 and 30, 37.

Compounds XIV-XVI are new. The data reveals that 4-hydroxymethyl and 4-chloromethyl-1,3-dioxolanes can be successfully applied in the synthesis of cyclic acetals, containing alkyl and aromatic substituents.

2.4.4. -5-NONYLENE-1,2,4-TRIOXOLANE

1,2,4 - Trioxolanes are the main products at the ozonolysis of olefins [1, 2]. In contrast to peroxides [100], and 1,2,3,4, - tetraoxolanes their thermal properties are not investigated enough.

The synthesis of 5-nonylene, 1,3,4-trioxalane - 1-deceneozonide was accomplished by ozonation of 10 ml solution of 0.5M 1-decene at temperature -70°C [25]. The yields of the ozonide was 60 %. Its structure is:

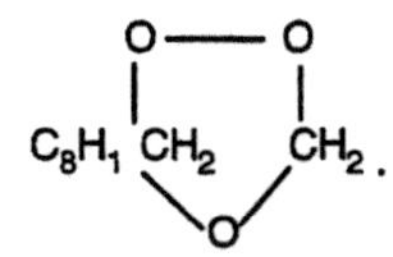

Its thermal properties were investigated by means of differential scanning calorimetry (DSC). A highly intensive exthermal peak with maximum (TM) in the range of 384.4-403 K was observed on the thermogram of 1-decene ozonide. It has been found that this peak depends on the heating rate. The analysis of the thermograms data are presented in Table 2.18.

Table 2.18.

DSC analysis of the thermal decomposition of 1-decene ozonide

Heating, ^{0}C/min	T_m, K	ΔH, kJ/mol	E_a, kJ/mol	n
3	384.4	341	129	1.02
5	390.0	349	127	0.98
10	396.0	353	131	1.01
15	401.0	345	129	0.99
0	403.0	347	128	0.98
Average value	-	349±9	129±4	1±0.06

Note: ΔH, E_a and n are the enthalpy and the activation energy of decomposition and the order of the reaction, respectively.

It is seen that ΔH for the ozonide is 349 kJ/mol which is about 62 % higher that the respective value for dicumene peroxide- 215 kJ/mol [110] (Table 2.19).

Table 2.19.

Activation parameters of the reaction of thermal decomposition of organic peroxides

Substrate	E_a, kJ/mol	lgA	Ref.
RO-OR	155±5	15.3±0.5	115-118
Cumene-DPO	140±8	-	101
Acetone-DPO	153±5	13.8±1.1	101
Benzoil-DPO	128	-	101
Pynacoline-DPO	128	12.6±1.3	101
Ethylene-ozonide	115.1	13.6	105
Hexene-1-ozonide	99.2	10.8	105
Heptene-1-ozonide	95.4	10.2	105

Note: R= Me, Et, n-Pr, i-Pr, sec.-Bu, t-Bu radicals; PO - peroxide; DPO- diperoxide.

A modified version of the method of Ellerstein [111] Is used to determine the values of the activation energy and the kinetic order of the reaction. Calculations were made according to the equation proposed by Crane et al. [112].

$$T^2(S/h) = (E_a/R) - nT^2(h/r), \qquad \text{(Eq.2.3)}$$

where: $S = dH^2/dT^2$, $h = dH/\varphi dt = dH/dT$, $r = \Delta H - H_p(T)$, ΔH - is the total enthalpy of the reaction, $H_p(T)$ - the partial enthalpy evolved up to T, t- reaction time, φ - heating rate.

From the linear plot in coordinates of Eq.2.4 (correlation coefficient of 96 %) the values of E_a and n were determined form its intercept and slope, respectively.

The Kissinger relationship between the peak maximum temperature (T_m) and E_a for data obtained at different heating rates was used as an alternative method for E_a determination 112, 113]:

$$d(\ln\varphi)/d(1/T_m) = -E_a/R - 2T_m \qquad \text{(Eq.2.4)}$$

In this case the plot in coordinates $\varphi/(1/T_m)$ gives a straight line with a slope E_a/R when $E_a/R >> 2T_m$. The value of E_a was determined to 132 kJ/mol with a correlation coefficient of 0.999.

Taking into account the first-order kinetics of the thermal decomposition of 1-decene, the value of A in the Arrhenius equation was calculated from the following expression [114]:

$$A = E_a.\varphi.n.(RT_m{}^2)^{-1}.\exp(E_a/RT_m), \qquad \text{(Eq.2.5)}$$

according to which $A = 1.7.10^{15}\ s^{-1}$. Thus the Arrhenius equation for the thermal decomposition reaction of 1-decene can be written as:

$$k = 1.7.10^{15}.\exp(-129000/RT),\ s^{-1} \qquad \text{(Eq.2.6)}$$

The thermal data allows to analyse the scheme for the decomposition of the ozonides given below:

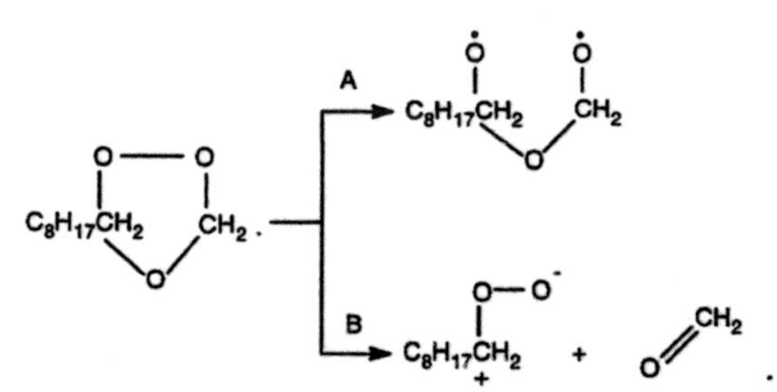

Kinetic data on the thermolysis of three different classes of peroxide compounds are summarized in Table 2.19. It is known that the homolytic scission of the peroxide bond [59, 64, 65] is the rate-limiting step in the thermal decomposition of dialkylperoxide and ketodiperoxides. In the case of 1,2,4-trioxalanes, it has been reported that they can decompose through a synchronous step along the ether and peroxide C-O bonds yielding carbonyloxide and carbonyl compounds [66, 67]. One of the arguments in favor of this suggestion (route B) are the lower values of E_a and A for ozonides than those for peroxides. However, the proximity of the E_a and A values of 1-decene ozonide with those of alkyl peroxides corroborate with the hypothesis of homolytic decomposition (route A). Thus it can be concluded that depending on the conditions and ozonide type either of the two possible routes for the decomposition reaction can dominate, but always being in a dynamic equilibrium.

2.5. Ethers

Investigations of ozone reactions with ethers have started as early as the last century [119-123]. Results of these studied are summarized and reported by Bailley in his review [74]. The main products formed during ozonolysis of aliphatic ethers are alcohols, aldehydes, esters, acids, hydrotrioxides, hydrogen and organic peroxides, singlet oxygen and water. Price et al. [125] propose the so called "insertion" mechanism, according to which ozone is inserted into the α-C-H bonds at the first step through a 1,3-dipolar addition, thus forming unstable hydrotrioxide. This mechanism was also supported by the data of Erickson and Bailey [126, 2], Murray et al. [127] who measured the activation parameters of the decomposition of a series of hydrotrioxides obtained by the ozonolysis of some ethers. Giamalva et al. [128] summarized the known so far possible mechanisms, i.e. interaction with the ether oxygen atom, 1,3-dipolar insertion of ozone into the α-C-H-bonds, homolytic abstraction of α-H-atom and heterolytic abstraction with carboanion and carbocation formation. These authors clearly indicate the predominance of the one-step mechanisms with transfer of hydrogen atom, hydride anion or cation.

As it has been mentioned above the low temperature ozonolysis of ethers yields hydrotrioxides which however are stable at very low temperatures, At normal temperatures the composition of the products is found to be different from that at the decomposition of hydrotrioxides. These two reasons provoke us to carry out extended studied on the ozone reaction with ethers in order to establish the real mechanism of this reactions at normal temperatures.

It was found from Fig.2.7 that the kinetic constant of diethyl ether (DEE) ozonation at -5^{0}C is $M^{-1}.s^{-1}$ and the stoichiometric coefficient is 1.

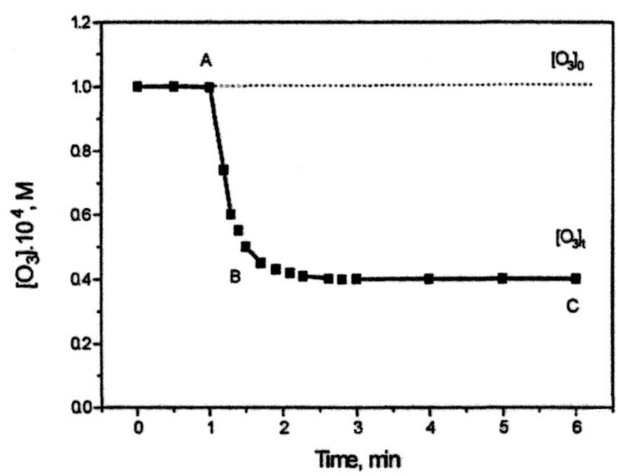

Fig.2.7. Kinetic curve of ozone concentration at the reactor outlet 10 ml, 92 mM DEE at -5°C.

The calculated stoichiometric coefficient of the ozonation reaction with other ethers was unity, too.

The kinetics of ozonolysis was studied on the example of n-dibutylether (DBE). The kinetic curves of the reaction products formation and the initial ether consumption are shown in Fig.2.8

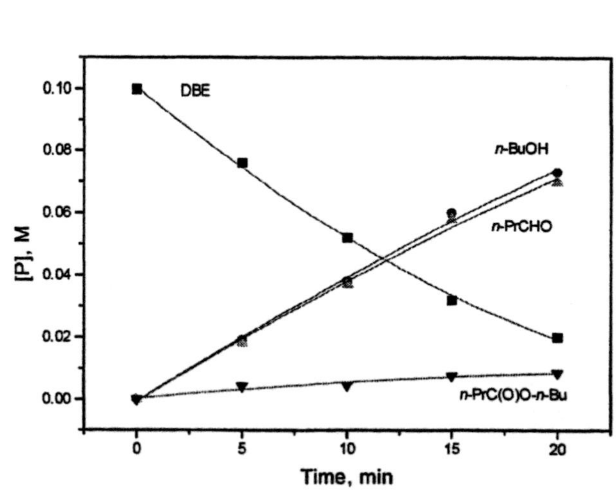

Fig.2.8. Kinetics of products formation and consumption of DBE during ozonolysis at ambient temperature. $[O_3]_o = 2.34.10^{-4}$ M

The following products were identified by GC-MS in the reaction mixture after 10 min:

130 * 57 41 87 56 55 101 39 130 43 45 * *n*-Dibutylether;
74 * 56 41 43 42 55 39 57 45 40 41 * *n*-Butanol;
72 * 44 43 41 72 57 42 38 37 40 71 * Butanal;
88 * 60 73 41 42 43 45 39 55 61 88 * Butiryc acid;
144 * 71 89 56 43 41 57 60 73 55 42 * Butyl butyrate and
164 *& 247 * - Chloro containing compounds like:

$CH_3(CH_2)_2CH(Cl)O(CH_2)_3CH_3$ (M.w.164.4) and
$CH_3(CH_2)_2CH(CCl_3)O(CH_2)_3CH_3$ (M.w. 247)

The rate of DBE decomposition ($7.4.10^{-5}$ $M.s^{-1}$ determined from Fig.2.8) is almost equal to the sum of butanal ($6.5.10^{-5}$ $M.s^{-1}$), butanol -$6.71.10^{-5}$ $M.s^{-1}$ and butylbutyrate ($1.0.10^{-5}$ $M.s^{-1}$) formation rates. The kinetic curves of butanol and butanal formation have almost the same slope and they start without any induction period. This could mean that they are formed in parallel reactions from a common precursor. The rate of butylbutyrate accumulation, as evidenced by its kinetic curve, is ca. 7 times lower, thus indicating its formation in parallel reaction to butanol (butanal) formation, most probably from the same precursor . The ratio product amount:initial DBE after 10 min is found to be 1:1. This means that the share of the autooxidation process is negligible though we have identified some amount of butyric acid by its IR spectrum (1765 cm^{-1} (monomer) and ε=1470 $M^{-1}.cm^{-1}$) resulting from the butanal oxidation.

The possible existence of common precursor raises the question about its nature. Such a precursor could be either α-hydroxyether (EOH) or α-hydrotrioxyether (EOOOH) which through intermolecular disproportionation can further produce simultaneously aldehyde and alcohol:

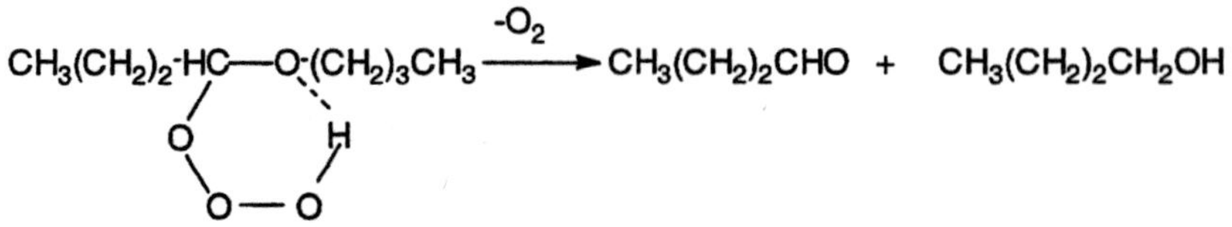
(2)

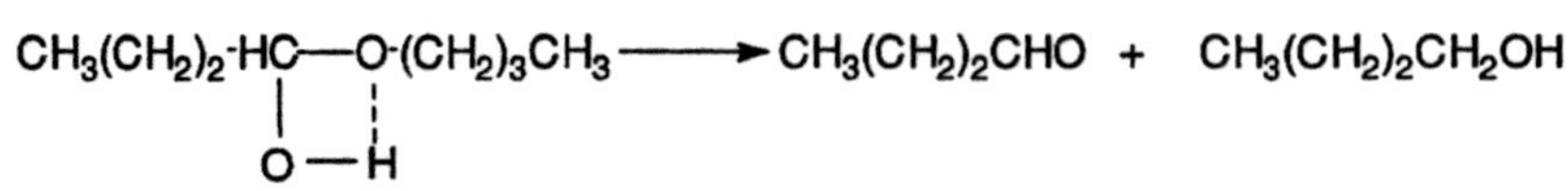
(2')

Probably, the intermolecular reaction of EOOOH disproportionation is preferable, because of the six-member ring transition state, while EOH disproportionation occurs via a strained four-member-ring transition state, which is energetically unfavourable. If EOH is supposed to be the precursor then butylbutyrate should be the major reaction product as a result of the rapid oxidation of EOH by ozone;

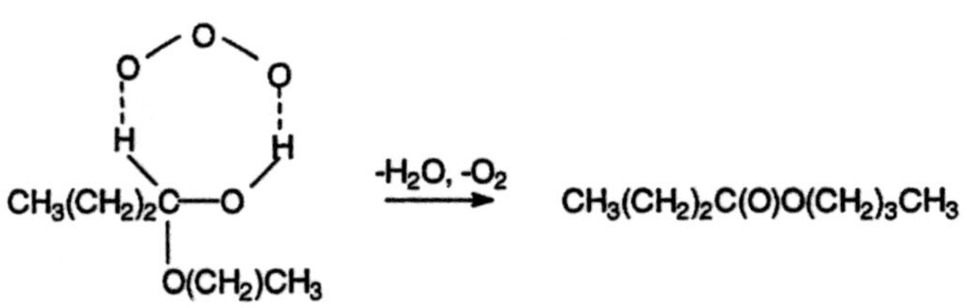
(3')

which is not experimentally observed.

When EOH is supposed to be a precursor, it is difficult to imagine its one-step transformation into an ester without any additional assumptions. The solvent - CCl_4 could mot also affect this transformation having in mind its weak oxidizing properties and nonspecifity.

The formation of the ester from EOOOH in one step can be easily presented through a four-member ring transition state:

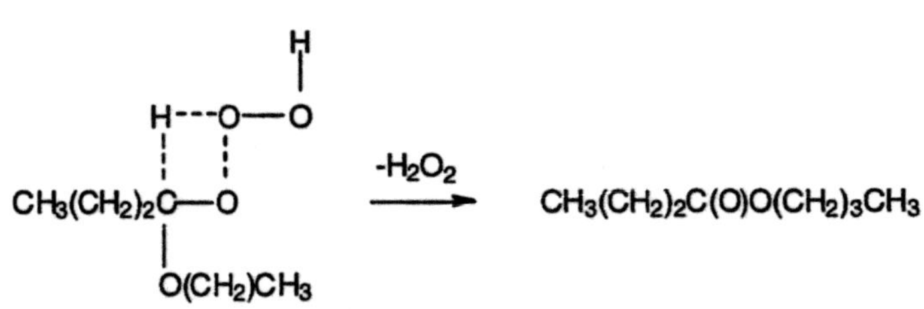
(3)

The formation of four-member transition state will be more unfavorable than the 6-member transition state. This conclusion is in agreement with the kinetic data, namely, the rate of ester formation is ca. 7 times lower than that of alcohol and aldehyde formation. Direct evidence for EOOOH formation was found only after prolonged ozonation for 24h at -78°C [129]. The NMR spectrum of the oxidate has a signal at δ= 13.52 ppm , which is attributed to OOOH.

The order of the reaction in [DBE] and $[O_3]$ is unity (Fig.2.9) and the rate law can be written as : $W = k.[DBE][O_3]$.

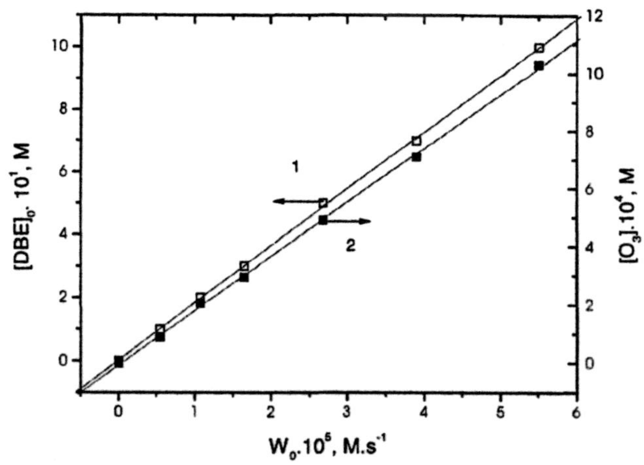

Fig.2.9. Dependence of the reaction rate of DBE ozonation (21°C) on the concentration of ozone(1) at [DBE]=10 mM and DBE (2) at $[O_3]$ =0.1 mM

Based on the results discussed above, a scheme of the ozone reaction with aliphatic ethers is proposed:

Scheme 2.5.

Ether + $O_3 \rightarrow$ EOOOH (k_1)
EOOOH $\rightarrow$ Alcohol + Aldehyde + O_2 (k_2)
EOOOH $\rightarrow$ Ester + H_2O_2 (k_3)

The formation of chrlorine-containing compounds (1-2%) can be explained by the presence of radical intermediates in the reaction mixture. This means that EOOOH will also decompose via radical route:

EOOOH $\rightarrow EO^\bullet + {}^\bullet O_2H$ (k_4)
$EO^\bullet + EH \rightarrow EOH + E^\bullet$ (k_5)
$E^\bullet + O_2 \rightarrow EO_2^\bullet$ $(k_{p'})$
$E^\bullet + CCl_4 \rightarrow ECl + CCl_3^\bullet$ $(k_{p'})$
$EO_2^\bullet + EH \rightarrow EOOH + E^\bullet$ (k_p)
$E^\bullet + CCl_3^\bullet \rightarrow ECCl_3$ $(k_{t'})$
$2EO_2^\bullet \rightarrow$ nonradical products (k_t)

The validity of the mechanism pointed above was confirmed by the good agreement between the experimental points on the DBE decomposition curve, and products accumulation and the theoretical curves calculated according to the scheme and given in Fig.2.8. For example, the curve describing the DBE consumption is obtained from equation:

$$[DBE]_t = [DBE]_0.\exp(-k't) \quad \text{(Eq.2.7)}$$

where: k' - constant measured by stop-flow method - $1.26.10^{-3}$ s^{-1}, the corresponding bimolecular constant is 6.3 M^{-1}.s^{-1}.

2.5.1. Application of the ERM

In order to elucidate the intimate mechanism of ozone interaction with ethers we have applied the developed by us experimentally-theoretical approach which has been described in Chapter I. It is based on comparison of the experimental and calculated values of A assuming linear (LC) and cyclic (CC) form of the activated complex (AC).

The kinetic constants of ozone reaction with diethylether (EtE), dichlorodiethylether (DClEtE), di-iso-propylether (i-PrE), di-n-butylether (n-ButE), di-iso-amylether (i-AmE) and di-n-amylether (n-AmE) in CCl_4 are given in Table 2.20.

Table 2.20.

Rate constants of ozone reaction with the ethers studied at various temperatures (°C)

			k ±	5%	$M^{-1}.s^{-1}$				
Ether	-15^0	-6^0	-5^0	3.5^0	4^0	10^0	13^0	21^0	24^0
EtE	0.5	0.9	1.0	1.2	1.6	2.2	2.4	3.1/3.0	3.6
DClEtE	-	-	0.009	0.016	-	-	0.028	0.035/0.04	-
i-PrE	-	-	1.6	1.9	-	-	3.4	5.4/5.1	-
n-ButE	-	-	1.9	2.1	-	-	4.3	6.3/6.3	-
i-AmE	-	-	1.9	2.5	-	-	4.1	7.8/7.0	-
n-AmE	-	-	2.2	2.8	-	-	4.2	7.8/7.5	-

Note: in column 9, after the slash are given the values of the constants determined by stop-flow method

The rate constants for EtE, n-ButE, i-AmE and n-AmE have similar values. All these ethers have close α-C-H bond energies and similar electronic environment. The value of the rate constant for i-PrE ozonation is also close to those mentioned above, while the constant for DClEtE is of 100 times lower value. In DIPE the presence of tert.C-H bonds, according to our discussion in Chapter I, should contribute to their higher reactivity because of the lower energy of these bonds. The presence of oxygen, however and the more difficult stabilization of the transition state due to steric factors, makes this interaction slower. The very low rate constant of DClEtE ozonation can be attributed to the strong electron-accepting properties of the Cl-atom.

Arrhenius parameters calculated on the basis of the data in Table 2.20 are summarized in Table 2.21.

Table 2.21.

Arrhenius parameters of ozone reaction with some aliphatic ethers

Ether	k, $M^{-1}.s^{-1}$	lg A	E_a, kcal/mol
EtE	3.1	6.4 (7.3)	7.9 (9.9)
DClEtE	0.035	4.9 (5.5)	8.5 (10.1)
i-PrE	5.4	6.3 (7.9)	7.6 (10.0)
n-ButE	6.3 (6.0)	6.6 (7.4)	7.8 (9.7)
i-AmE	6.8	6.6	7.8
n-AmE	7.8	6.5	7.6
THydP	(1.3)	(6.8)	(9.7)
THydF	(12.3)	(6.6)	(8.2)

Note: the values in brackets are taken from Ref.79; THydP and THydF -are tetrahydropyrane and tetrahydrofurane, respectively.

Comparison of the data obtained with reference data shows a good coincidence in relation to the values of the rate constants at 21°C. Only a slight difference in the activation energies was observed. Perhaps, it is the heat of ozone dissolution in CCl_4 that has to be taken into account as the reason for these differences.

All literature reaction schemes describing the ozonation of ethers proceed by two geometric forms of the activated complex- LC and CC:

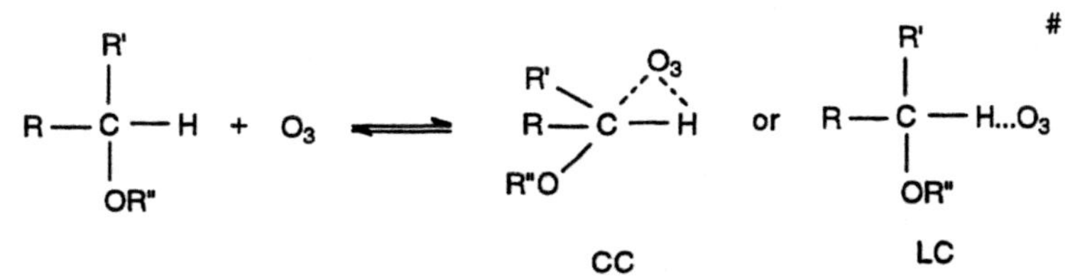

The CC has a more compact structure, without any possibility for free rotation. It can be a transition state for the following reaction:

$$EH + O_3 \rightarrow EOOOH \rightarrow \text{products} \qquad (a)$$

where EOOOH if formed in one step via 1,3-dipolar insertion.

The LC structure is an open one, allowing free rotation around H-O and O-O bonds. LC can be a transition state for the reactions:

$$EH + O_3 \rightarrow E^{\bullet} + HO_3^{\bullet} \rightarrow EOOOH \rightarrow \text{products} \qquad (b)$$
$$EH + O_3 \rightarrow E^{+} + HO_3^{-} \rightarrow EOOOH \rightarrow \text{products} \qquad (c),$$

in which an H-atom or hydride anion abstraction occurs and EOOOH is formed during the second step.

The necessary parameters for estimating the A values are presented in Table 2.22.

Table 2.22.

Heats of formation (ΔH) in kcal of initial, intermediate and final products, van der Waals radii of the ethers and free rotation energy (E_{fr}) around H-O bond

Ether	r, [Å]	ΔH, EH	ΔH, EOOOH	ΔH, $E^{\bullet}$	ΔH, E^{+}	E_a, kcal/mol
EtE	3.18	-68	-66	-32	135	1.3
dClEtE	3.51	-74	-73	-41	132	2.2
i-PrE	3.48	-76	-74	-42	125	1.4
n-ButE	4.03	-81	-79	-47	120	1.3
i-AmE	4.31	-87	-86	-53	115	1.3
n-AmE	4.45	-86	-85	-51	116	1.3

The calculated values of A are shown in Table 2.23

Table 2.23.

The calculated preexponents (A) and steric factors (p)

Ether	lg A, CT	lg A, CC	lg A, LC	p.10^6, CC	p.106, LC	$A_{èç÷}^{LC}/À_{àêñï}$
EtE	11.66	4.02	6.40	0.02	5.5	1
dClEtE	11.66	3.62	4.99	0.009	0.2	1.23
i-PrE	11.67	4.28	6.35	0.04	4.8	0.91
n-ButE	11.72	3.99	6.09	0.02	2.3	0.31
i-AmE	11.75	3.90	6.00	0.01	1.8	0.25
n-AmE	11.74	3.77	5.87	0.01	1.3	0.23

Note: the values of A in columns 3 and 4 are per one equivalent α-C-H atom

The good agreement between the calculated and experimental values for A confirms the conclusion that the geometry of the transition state is a linear (LC) and abstraction mechanisms a and c seem more probable.

In order to evaluate which mechanism occurs we have used the thermodynamic parameters of Table 2.22. The calculated heats of formation of the individual compounds fit well to the experimentally measured values [36, 37, 51, 66]. Using different expressions for the Hamiltonians, we have obtained the following values for ΔH: O_3 - 33.3 kcal against the literature data of (34.1); HOOOH -17.8 (-17.1); HOOO.9.2 (17.8); HOOO- -20.3(-24.9); HO- -5.2(-25.8) [66].

$$\Delta H = -\Delta H\,(EH) - \Delta H\,(O_3) + \Delta H\,(EOOOH) \quad (a)$$
$$\Delta H = -\Delta H\,(EH) - \Delta H\,(O_3) + \Delta H\,(E^{\bullet}) + \Delta H\,(HO_3^{\bullet}) \quad (b)$$
$$\Delta H = -\Delta H\,(EH) - \Delta H\,(O_3) + \Delta H\,(Å^{+}) + \Delta H\,(HO_3^{-}) - e^2/r_{ip} - E_s \quad (c)$$

where: e^2/r_{ip} = 5.4 eV and Ea= 17 kcal for tetrachloromethane were taken from [66]

The calculated values for ΔH ΔH are depicted in Table 2.24.

Table 2.24.

Calculated heats of investigated reactions according to mechanisms a, b and c

Ether	a, kcal/mol	b, kcal/mol	c, kcal/mol
EtE	-32	-16	0
dClEtE	-33	-19	6
i-PrE	-32	-18	1
n-ButE	-32	-18	1
i-AmE	-32	-18	2
n-AmE	-33	-17	2

From thermodynamic point of view , (a) is the preferable mechanism and further between (b) and (c), (b) is more possible.

2.6. HYDROXYBENZENES

The reaction of ozone with mono- and dihydroxybenzenes has provoked a particular interest [1, 2, 129-132], namely , because of its great importance for environment protection, chemical stabilization and theory of reactivity. The ozonation of phenol, pyrocatechol, resorcinol and hydroquinone has been studied in different solvents - aqueous and organic, aiming at elaboration of the kinetic parameters and products composition [133-139]. The rate constants at room temperature of phenol and resorcinol ozonation in water are $1.3.10^3$ $M^{-1}.s^{-1}$ and >3.10^5 $M^{-1}.s^{-1}$, respectively, whereas the rate constants of benzene, toluene and anisole ozonation in organic media are 2, 14 and $2.9.10^2$ $M^{-1}.s^{-1}$ [129]. Gurol et al. [130] have found that the relative rate of pyrocatechol/phenol and resorcinol/phenol ozonation in water medium are 220 and 70, respectively. Provided that the rate constant of phenol ozonation is known [129], the calculated values of the rate constants of pyrocatechol and resorcinol ozonation are $2.86.10^5$ $M^{-1}.s^{-1}$ and $9.1.10^4$ $M^{-1}.s^{-1}$, respectively. However, if the reaction is carried out in organic solvents the values are quite different. For example, in CCl_4 and at room temperature the. Following values have been obtained for: benzene - 0.06; ethylbenzene - 0.2; anisole - 1.1; phenol - 230 and pyrocatechol - $3.2.10^3$ $M^{-1}.s^{-1}$ [136-139]. One of the possible explanations for the different values of the rate constants of hydroxyphenols ozonation obtained by the various researches could be the great influence of the water on this reaction, for example for phenol they vary in the range 100-180 $M^{-1}.s^{-1}$, 500 $M^{-1}.s^{-1}$ for pyrocatechol and 300 $M^{-1}.s^{-1}$ for 3,6-di-tert-butylpyrocatechol. Pryor et al.[127] have reported that the rate constants of ozone reaction with α-tocopherol in CCl_4 and water are $5.5.10^3$ $M^{-1}.s^{-1}$ and 1.10^6 $M^{-1}.s^{-1}$, with α-tocopherol acetate - $1.45.10^2$ $M^{-1}.s^{-1}$ and that with α-tocopherolquinone it is $1.15.10^4$ $M^{-1}.s^{-1}$.

Product analysis of pyrocatechol ozonation in aqueous medium shows that three moles of ozone are readily absorbed to give CO_2 (24.8 %), CO (6 %), formic acid (32.5 %) and glyoxal (4.2 %). Since CO_2 and formic acid are the main reaction products, it seems very likely that most of the pyrocatechol undergoes anomalous ozonolysis [135]. Radical formation was observed during 2,6-di-tert-butylphenol ozonolysis. When the ratio of absorbed ozone to phenol = 0.8, 50 % viscous yellow oil and 3 a5 2,6-di-tert-butyl-o-quinone were identified in the products [138]. In 3,6-di-tert-butyl-pyrocatechol ozonation the corresponding quinone was found to be the main product after complete consumption of the initial substrate [139]. Side products like 2,5-di-tert-butyl-, 3-hydroxy-p-quinone and 3,6-di-tert-butyl-1,2-phenylacetal-5-hydroxy-3,6-di-tert-butyl-p-benzophenone were also found. The rate constant of this reaction amounts to 3.10^2 $M^{-1}.s^{-1}$. The same constant with the corresponding pyrocatechol with 0-hydroxy-acetylated groups is estimated at a two orders of magnitude lower value. A mechanism has been suggested involving the formation of either an 1,3-cyclic activated complex between two hydrogen atoms from the OH-groups and one ozone molecule or π- or σ-complexes with the benzene ring.

The different values of the literature constants and the various mechanisms proposed for this reaction testify the necessity of further research of its kinetics and reaction pathway. In this connection we have studied the ozonolisis of the following hydroxybenzenes (Scheme 2.6).

Scheme 2.6

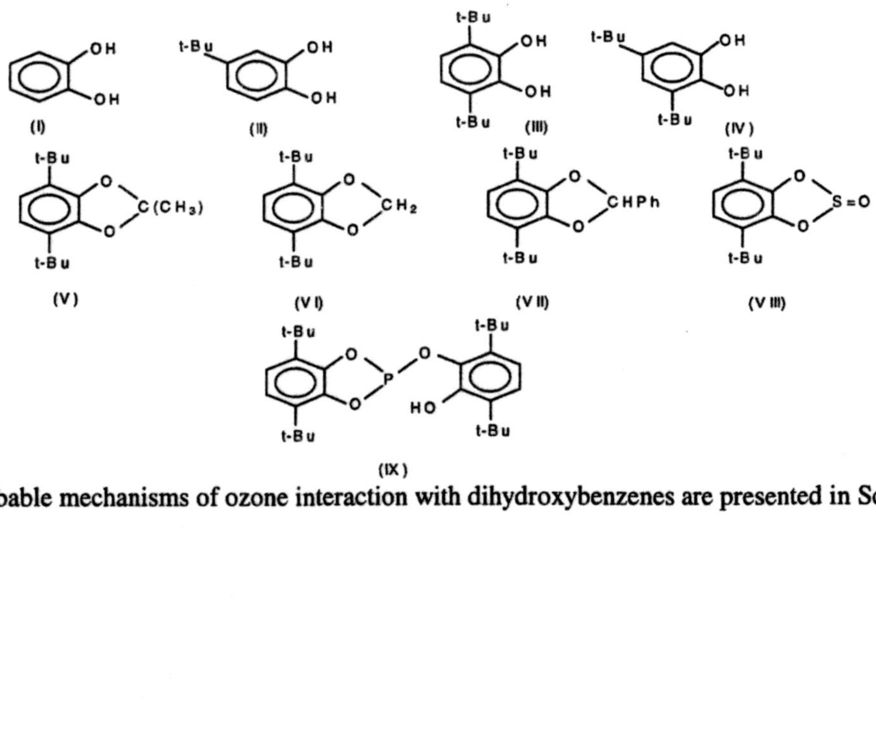

The probable mechanisms of ozone interaction with dihydroxybenzenes are presented in Scheme 2.7.

Scheme 2.7.

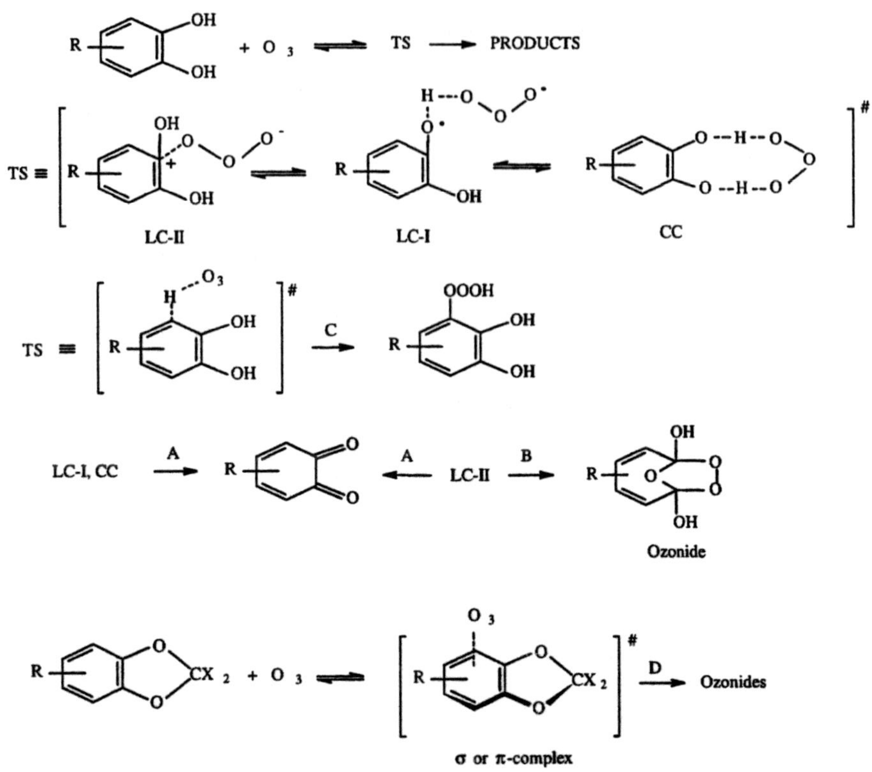

Mechanism A with a cyclic complex (CC) formation in the transition state was suggested by Razumovskii et al. [139]. Mechanism B with linear complex LC-II in the transition state was suggested and discussed by Biley [2]. The interaction of ozone with C-H bonds with the formation of trioxide [127] as a possible parallel reaction is indicated in mechanism C. The acetylated forms of dihydroxybenzenes can react only via attack on the benzene ring according to mechanism D or C. Formally, mechanism D can be regarded as an extended version of mechanism B, involving the formation of TS similar to π- or σ-complexes. We suggest a new mechanism, an extended version of the Razumovskii mechanism whereby the transition state is linear with LC-I structure.

Upon ozonation of any of the investigated catechols directly in the ESR cell or after freezing the reaction products at liquid nitrogen no signals were detected.

The kinetic curve of the change of ozone concentration at the bubbling reactor outlet (Fig.2.10, 1) is characterized by three different regions: AB - fast ozone consumption after the addition of pyrocatechol; BC - steady-state part when the rate of the chemical reaction becomes equal to the rate of ozone supply and CD - ozone concentration begins to rise due to the pyrocatechol consumption. BC part of the curve allows to calculate the rate constant, and from the area below curve ABCD - the stoichiometry of the reaction. The straight line designated {O3}0 is the ozone concentration at the reactor inlet. Curve 2 presents the o-quinone formation with the reaction time. Its profile suggests the intermediate formation of o-quinone.

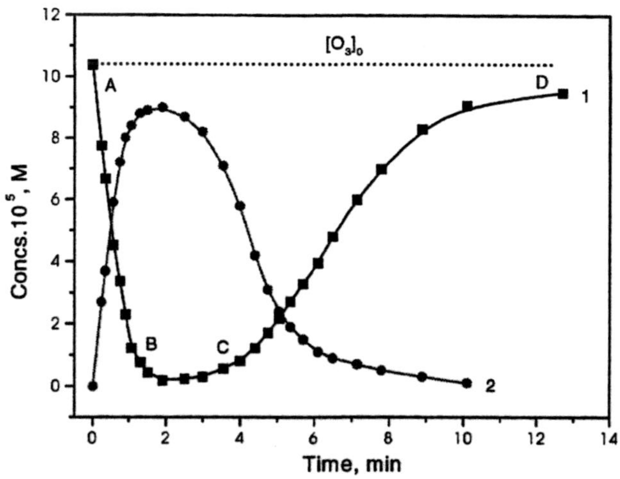

Fig.2.10. Kinetic curves of ozone absorption at reactor outlet (1) and *o*-quinone accumulation (2), at ambient temperature, [PC]= 0.227 mM in CCl_4 (10ml) durring bubbling of ozone with 0.1 l/min flow rate.

The kinetic curves of the products formation in pyrocatechol ozonolysis, its consumption and *o*-quinone consumption during its ozonation in a separate experiment are given in Fig.2.11.

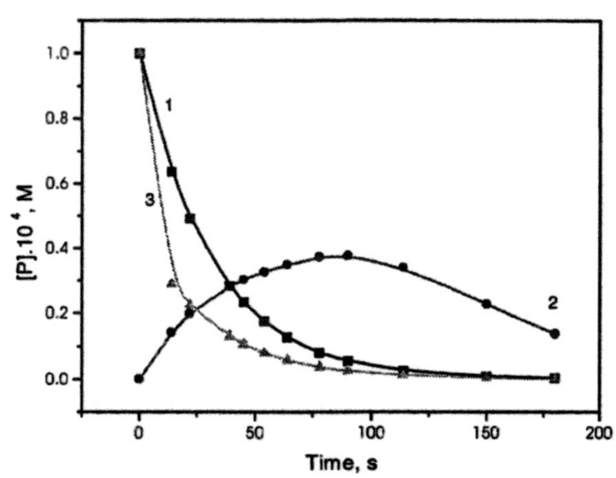

Fig.2.11. Kinetics of pyrocatechol consumption (1), o-quinone accumulation (2) and o-quinine consumption (3), ambient temperature, $[O_3]=1.10^{-5}$ M.

Calculated from the kinetic curves in Fig.2.11 rate constants of pyrocatechol and *o*-quinone consumption were $3.2.10^4$ $M^{-1}.s^{-1}$ and $7.1.10^4$ $M^{-1}.s^{-1}$, respectively. The initial rate rate of o-quinone formation had almost the same value as that of pyrocatechol consumption. The small variation of the constants is due to the participation of pyrocatechol in parallel reactions. Actually, during the reaction small amounts of open chain products were identified.

Pyrocatechol ozonation at 15 % conversion gave the following yields: o-quinone - 85 %, pyrogallol - 3%, ozonide - 10%, muconic acid -2%, maleic acid and fumaric acids and polymer products - 1%. The ratio between the amount of absorbed ozone to the consumed pyrocatechol was calculated as 6. Similar ratio values were obtained for the other hydroxybenzenes with free hydroxy groups.

It was found that with the increase of the conversion from 0 to 100% the amount of open chain products continuously increases while that of the rest products passes through a maximum. The individual compounds identified by ^{13}C-NMR are: muconic semialdehyde and acid - 20%, maleic and fumaric semialdehyde and acid - 40 %, glyoxal, formic acid, oxalic acid, CO2 and polymer products - 40 %. The reaction rate of dihydroxybenzenes ozonation follows first order kinetics in relation to each reagent (fig.2.12, A and B).

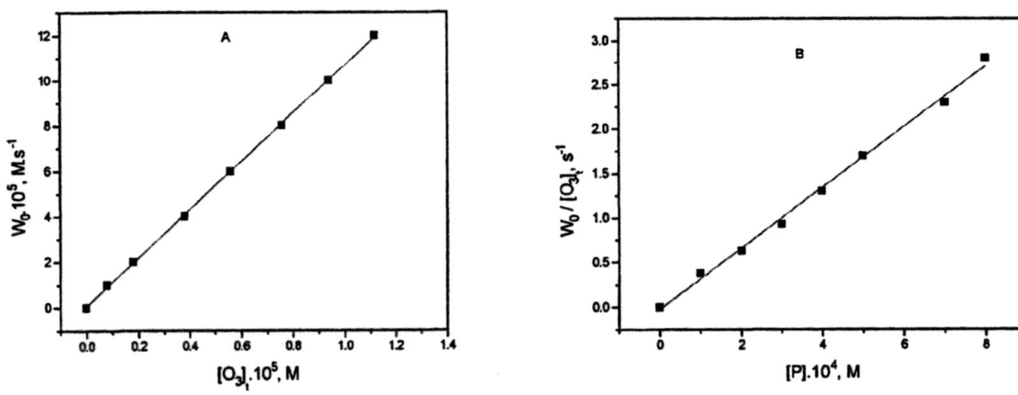

Fig.2.12 Dependence of the rate of pyrocatechol ozonation on the concentration of ozone (A) and pyrocatechol (B)

Table 2.25 presents experimental kinetic data on the ozonation of the investigated dihydroxybenzenes at various temperatures.

Table 2.25.

Kinetic data of dixydroxybenzenes ozonolysis in CCl_4, 20°C.

Substrate	k, $M^{-1}.s^{-1}$	lg A	E_a,, kcal/mol	Substrate	k, $M^{-1}.s^{-1}$	lg A	E_a, kcal/mol
I	3230	5.35	2.5	VIII	598	6.84	5.5
II	3100	4.91	1.9	IX	111	6.81	6.4
III	3100	5.07	2.1	Benzene	0.06	8.32	12.8
IV	3200	4.80	1.7	Toluene	0.40	7.05	10.3
V	500	7.14	6.0	Anisole	10	-	-
VI	749	7.17	5.8	Pheole	160	7.31	7.0
VII	828	7.06	5.6				

A drastic difference between the values of the kinetic parameters of catechols (I-IV) and (V-IX) in which the OH groups are acetalated is observed. The rate constants of catechols I-IV manifested 4-28 times higher values than those of catechols of V-IX. On the other hand the preexponentials demonstrated by about 2 orders of magnitude lower values with the former compounds. The activation energies are 2.05±0.5 and 5.9±0.5 kcal/mol, respectively or the acetalation of the HO-groups leads to increase in EA by about 4 kcal. Probably, ozone reacts predominantly with the hydrogen atoms of the HO-groups and, to a very small extent, with the benzene ring. The ratio of these ozone interactions varies from 94:4 to 80:20 depending on the hydroxybenzenes nature. The lower activation energies of catechols I-IV with respect to those of V-IX are more consistent with the formation of an activated complex of contact type in the transition state, i.e. with structure LC-I (Scheme 2.7). This assumption could also be confirmed by analysis of the kinetic parameters of the ozonation of benzenes, toluene, anizol and phenol. The reaction of ozone with benzene proceeds with a low specific rate of 0.06 $M^{-1}.s^{-1}$ and relatively high activation energy equal to 12.8 kcal/mol. The methyl derivatives of benzene, i.e. toluene, reacts 6.7 times faster and the E_a is decreased by 2.5 kcal/mol, and the rate of anizol ozonation is even more higher - 10 $M^{-1}.s^{-1}$. The exchange of kethyl group by a hydroxyl one leads to significant acceleration of the ozonation rate and approaches values of 160 $M^{-1}.s^{-1}$, and the activation energy is reduced by 5.8 kcal/mol. Probably, in this case the mechanism of the reaction is changed and ozone interacts predominantly with the hydrogen atoms of the OH groups. The decrease in the reactivity of anizol as compared to that of phenol (~16 times) supports this assumption.

Analysis of the kinetic parameters of catechols in acetalated form and those of benzene, toluene, anizol and phenol shows that the rate constants increase and the activation energies go down with increasing the number of electron donating substituents (Table 2.25). The high values of the preexponential factors could be associated with the high values of the energy of free rotation in the activated complex as a result of the steric hindrances caused by the presence of the bulky tert.butyl groups. This means that when the hydroxyl groups are acetalated, the activated complex should have a LC-II like structure and the ozonation would lead only to the formation of ozonides and open ring products.

The kinetic parameters of some dihydroxybenzenes were determined in ozonation in aqueous medium (for those which are water soluble - I-IV, VIII and IX). However, we found that because of diffusion limitations in the bubbling reactor we were not able to measure constants higher than 1.10^4 $M^{-1}.s^{-1}$. In fact, the literature values for some of these constants are of this order, but they have been measured by a modified stop-flow technique [27].

The higher reactivity of dihydroxybenzenes in aqueous medium is connected with the change of the ozonation mechanism from hidden radical to ion one. However, this question requires special study and discussion.

The analyses made so far show that the reaction path of ozone reaction with dihydroxybenezenes depends strongly on their nature and it proceeds via a transition state with activated complex - LC-I or LC-II

2.6.1. Application of ERM

For detailed analysis of the reaction mechanism of ozone reaction with catechols we have used the ERM method. The necessary parameters for calculations are presented in Table 2.26.

Table 2.26.

Symmetry numbers (σ), VdW radii (r), heats of formation (ΔH) of dihydroxybenzenes and the corresponding quinones and ozonides

Substrate	σ	r, Å	ΔH, kcal/mol	$\Delta H^{quinone}$, kcal/mol	$\Delta H^{ozonide}$, kcal/mol
I	2	3.02	-62(-85)	-39(-43)	-73
II	1	3.82	-88	-66	-99
III	2	4.46	-109	-86	-125
IV	1	4.39	-110	-90	-123
V	2	5.07	-116		-125
VI	2	4.56	-90		-103
VII	1	5.37	-63		-86
VIII	1	4.77	-73		-77
IX	1	5.30	-135		-140
Benzene	6	2.86	19(13)		21
Toluene	1	3.37	12(3)		11
Anisole	1	3.45	-16(-26)		-20
Phenol	1	2.96	-21(-38)		-24

Note: The experimental values are put in parenthesis [140].

In addition, we have calculated the heats of formation of pyrogallol - -117.6 kcal/mol (-129 kcal/mol [140]), 3-trihydroxy pyrocatechol - -59 kcal/mol, H_2O_3 - -21.2 kcal/mol (-17.7 kcal/mol [140]) and ozone - 34.1 kcal/mol.

Thus, the heats of the ozonation reaction according to the different mechanisms were calculated and amount as follows: A -33.1; B and D - 46.1 and C -32.1 kcal/mol. All the mechanisms are exothermic and, therefore thermodynamically favorable. In this case only the magnitude of the activation energy and the entropy benefits will determine the reactivity of these compounds and the reaction pathway.

Table 2.27.

Calculated preexponential values for ozone reaction with dihydroxybenzenes

Substrate	lg A TC	lg A, CC	lg A, LC	lg A, calc.	A_{calc}/A_{obs}	lg p
1	2	3	4	5	6	7
I	11.318	4.346	7.69	5.39	1.10	5.928
II	11.422	3.905	7.25	4.95	1.10	6.472
III	11.500	4.070	7.42	5.12	1.12	6.380
IV	11.490	3.780	7.12	4.82	1.05	6.670
V	11.572	4.017	7.36	7.36	1.66	4.212
VI	11.511	4.056	7.40	7.40	1.70	4.102
VII	11.801	3.934	7.28	7.28	1.66	4.521
VIII	11.533	3.717	7.06	7.06	1.66	4.473
IX	11.575	3.689	6.88	6.88	1.10	4.695
Benzene	11.319	5.124	8.47	8.47-	1.41	2.849
Phenol	11.322	4.196	7.54	7.54	1.70	3.782

Note: p - collision factor

Calculated preexponentials were compared with those obtained experimentally (Table 2.25) and it is seen that A have lower values (~10 times) for the cyclic form of the activated complex if compared with the experiment. At the same time the values of A for the linear complex are about 200 times higher. The values of A for CC are the highest as predicted by the theory and they are lower than the experimental ones. This suggests that the reaction takes place via a cyclic complex. The values of A calculated for LC in the Table were obtained without taking account of the energy of free rotation. The latter was calculated as a sum of the rotation around H-O and O-O axes (by *mopac6*) and amounts to 3.1 kcal. This means that the real values in column 4 are about 200 times smaller. The comparison between the corrected in this way A values and the experimental data for the compounds I-IV reveals a full coincidence.

The reaction pathway of the ozonation reaction of compounds V-IX is quite different since the transition state includes the formation of a σ or π-complex. For these compounds the formation of AC in the transition state is impeded due to the breaking of their aromatic character but simultaneously the free rotation is facilitated, which may be even zero, due to action of the principle of the lowest energy. If the energy of free rotation is assumed to be very low then the values of A calculated for LC-II would coincide with the experimentally found ones. Such a coincidence is observed for compound V in column 5 (Table 2.27).

The basic conclusion for the analysis of the results obtained for this reaction is that the kinetics and mechanism of the ozonation reaction of dihydroxybenzenes depend strongly on their structure and the type of the reaction medium.

2.7. Carbohydrates and Model Compounds

The selective oxidation of carbohydrates has received increasing attention in recent years [141-151]. This stems from interest in the transformation of carbohydrates to new valuable and ecologically friendly oxygen-containing compounds from available starting materials. Hence, a detailed clarification of the processes related to oxidation, the right choice of the catalysts and the definition of optimum conditions are of great importance [143, 145, 146, 148]. These reactions are invaluable for structure determination and transformation of polyhydrates and other natural products.

The formation of two carboxylic groups alters significantly the carbohydrate (polymer) properties making it water soluble and imparting it ion acceptor properties. This polymer can be used for preparing new classes of detergents and ion capture agents. The most intersting reaction is oxidative cleavage of C_2-C_3 - bonds [147]:

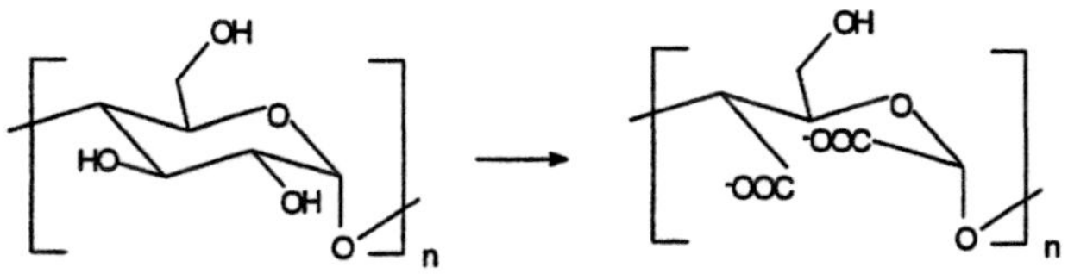

Formation of two carboxylic groups dramatically changed the properties of carbohydrates as it makes them water soluble and they become ion captures.

If combined with hydroxylation of the double bond the oxidation can be used as an alternative method for direct cleavage of double bonds by ozone [1, 2]. The oxidation of C_2-C_3 - bond is accomplished namely by applying stoichiometric oxidants like $Pb(AcO)_4$ in organic solvents and HIO_4 in aqueous media. Occasionally, other oxidizing agents have been used such as ceric ammonium nitrate, sodium bismuthate, chromium trioxide, and trivalent or pentavalent [152-161] iodine compounds. The mechanism of the C-C bond cleavage of vicinal diols with periodine acid can be represented as follows:

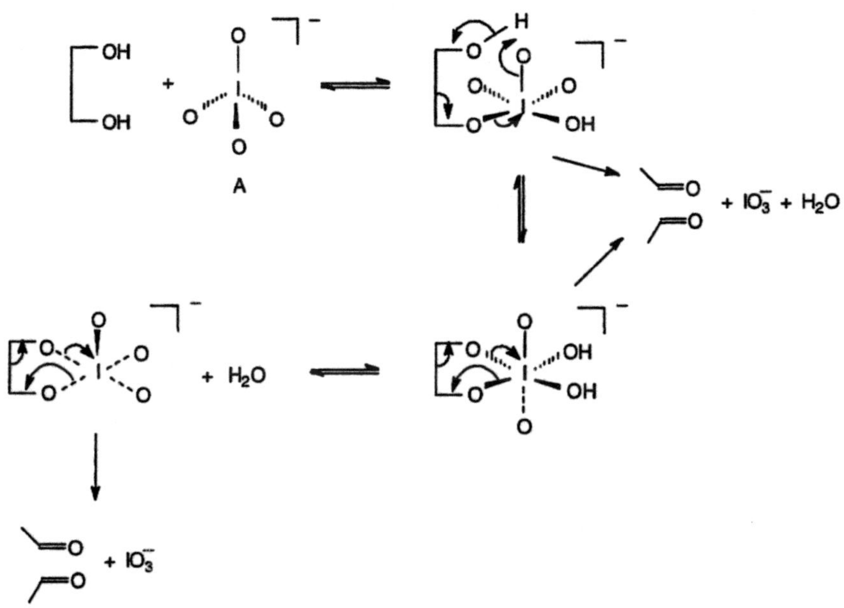

The cleavage of C-C bonds between two adjacent OH groups in the linear diols is a good source for aldehydes [162-164].

The efficiency and selectivity of the ozone reaction with C=C bonds in organic compounds is well known [1, 2]. Rate constants are of the order of 14 - 10^6 $M^{-1}.s^{-1}$, in contrast to 10^{-2} - 10 $M^{-1}.s^{-1}$ for the ozone reaction with C-H - bonds. The rate-determining step of the alcohols ozonolysis is the hydrogen atom abstraction in α position in relation to the OH group. The relative reactivity of the primary:secondary:tertiary alcohols with respect to ozone lies in the following sequence: 1:15:138 and corresponds to the C-H bonds energies. The main products of alcohol ozonation are the corresponding aldehydes or ketones. The carbonyl compounds are in equilibrium with their enol forms which brings about their further decomposition via ozone reaction with the C=C bond of the enol form according to Scheme 2.9

Scheme 2.9.

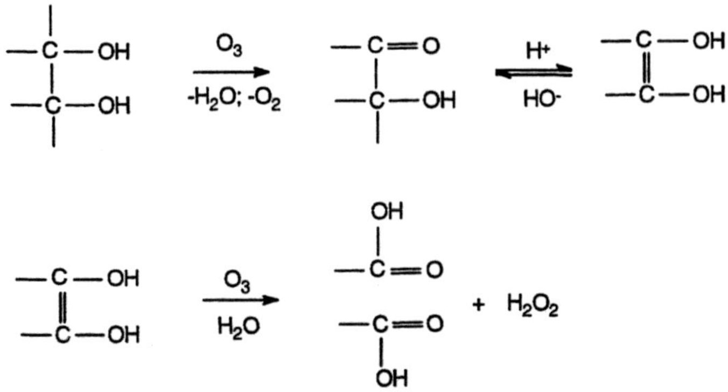

The monomer glucose unit contains the following α-hydrogen atoms: tert. C-H bond at C_2 and C_3 adjacent to OH-group, at C_4- adjacent to acetal group, at C_1- aldehyde hydrogen atom, at C_5-with alkylsubstituents and sec.-C-H bond at C_6. The energy of the aldehyde hydrogen atom is slightly lower than that of the tert-C-H bonds (difficult to make the difference) and considerably lower than that of the sec.C-H bond. Obviously, the hydrogen atom in the aldehyde group will be the most reactive. Among the tert.C-H bonds those with higher electron density will be more reactive. They will be presented in the following order of decreasing reactivity: $C_5>C_4>C_3=C_2$. The sec.C-H bond at C_6 appears to be the most inert in ozonolysis reaction.

The preliminary analysis reveals that ozone will attack predominantly the H-atom at C_1-atom in starch with the formation of carboxylic groups and opening of the glucoside ring, and to a lesser extent, give rise to carbonyl compounds at C_5, C_4, C_3 and C_2-atoms which can be easily converted to acids. However a selective cleavage of C_2-C_3 bond could not be expected to occur.

Conventionally, the oxidation reactions of carbohydrates are carried out in aqueous medium. It is known that the solubility of ozone in water is low and require very good stirring. Moreover, at pH above 7 ozone begins to react rapidly with the hydroxyl anions in the solution yielding HO-radicals [129, 165-167]. The latter reduces the selectivity of the oxidation process (Scheme 2.10):

Scheme 2.10.

$$HO_2^- \Leftrightarrow O_2^- + H^+ \quad (1)$$

Initiation: $$O_3 + HO^- \rightarrow HO_2^\bullet\,(O_2^-) + O_2^- \quad (2)$$

Propagation $$O_3 + (H^+)O_2^- \rightarrow (H^+)O_3^- + O_2 \quad (3)$$

$$O_3 + HO^\bullet \rightarrow HO_2^\bullet\,(O_2^-) + O_2 \quad (4)$$

Termination: Any recombination between $HO^\bullet$, $HO_2^\bullet$, O_2 (t)

where: (1) pK_a= 4.8; (2) k_2 = 48; 70 $M^{-1}.s^{-1}$.

$$O_3 + O_2^- \rightarrow O_3^- + O_2 \quad (5)$$

$$O_3^- + H^+ \Leftrightarrow HO_3^\bullet \quad (6)$$

$$HO_3^\bullet \rightarrow HO^\bullet + O_2, \quad (7)$$

where: $k_5 = 1.6.10^9\ M^{-1}.s^{-1}$; $k_6 = 5.2.10^{10}\ M^{-1}.s^{-1}$, $k_{-6} = 3.7.10^4\ s^{-1}$; $k_7 = 1.1.10^5\ s^{-1}$

The study of this reaction is impeded by the low solubility of polymeric carbohydrates (starch) in water and in this connection we have used model compounds for elucidation of the reaction mechanism.

The ozonolysis of methanol, ethanol, isopropanol and other alcohols proceeds via α-H-atom abstraction by ozone leading to the formation of carbonyl group which is subsequently oxidized to carboxylic one. It has been found that the ethanol ozonolysis is promoted by adding small amounts of hydroperoxide [168] because the steady concentration of the HO-radicals is increased.

Upon ozonolysis of iso-propanol in aqueous medium at pH=8, controlled by titration with NaOH, ambient temperature and 36 % conversion the following products in the HPLC chromatogram were identified: acetone - 25 %, acetic acid - 6 % and formic acid - 5 % (Fig.2.13).

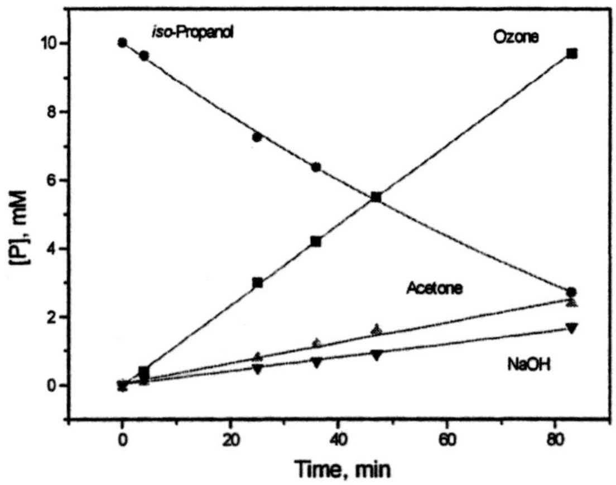

Fig.2.13. Kinetics of products formation during the ozone reaction with iso-propanol at pH=8. The products were identified by means of HPLC and ^{13}C-NMR.

The mechanism of the reaction could be expressed in the following way:

Scheme 2.11.

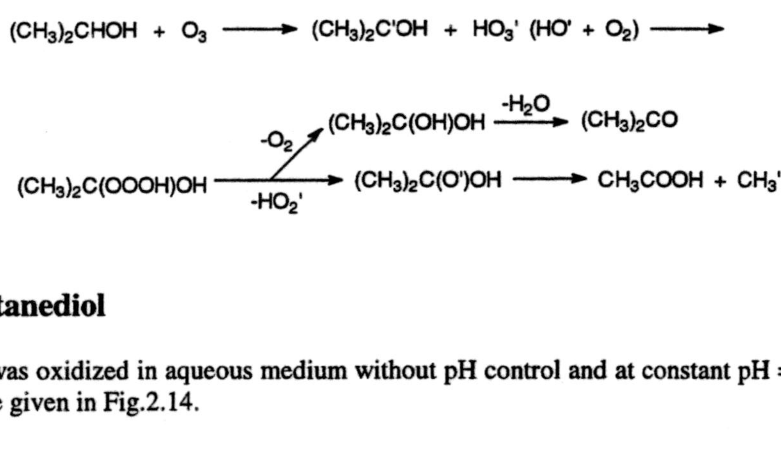

2.7.1. 2, 3-Butanediol

Butanediol was oxidized in aqueous medium without pH control and at constant pH = 10 and the main kinetic results are given in Fig.2.14.

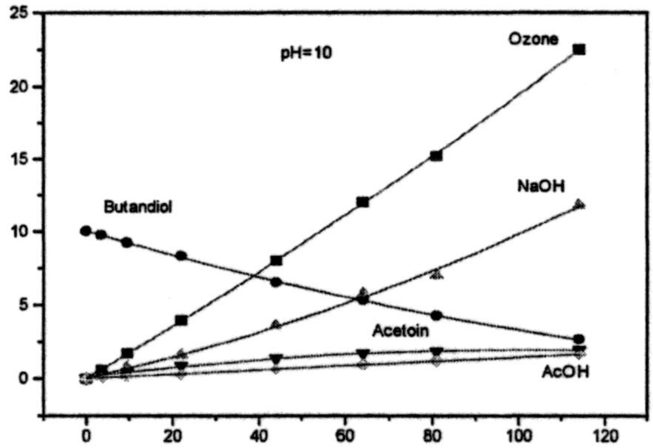

Fig.2.14. Kinetics of products formation during ozonolysis of 2,3-butanediol at pH=10. The products were identified by means of HPLC.

In the HPLC chromatogram of the starting material two peaks were recorded from the two stereo isomers. New peaks were found after ozonation: keto alcohol, acetoin, acetic acid, acetaldehyde, formaldehyde, and formic acid. At higher conversions the peaks of the keto alcohol and acetoin disappeared whereas the peak from acetic acid increased its intensity. The observed rate of BD consumption is higher that the sum of rates of products formation determined by HPLC. Probably at higher conversions ozone the HO radicals oxidized the reaction products to CO_2 and water. In this connection $CaCO_3$ was found in the trap. The stoichiometry of the ozonation reaction with respect to ozone is unity and the rate constant of the reaction is 0.2 $M^{-1}.s^{-1}$. At low conversions up to 10-120 % the main products of the reaction are acetoin and acetic acid. It has been found that at pH=8 the rate of acetoin formation is lower than that without pH control. The scheme proposed for the ozonation reaction is:

Scheme 2.12.

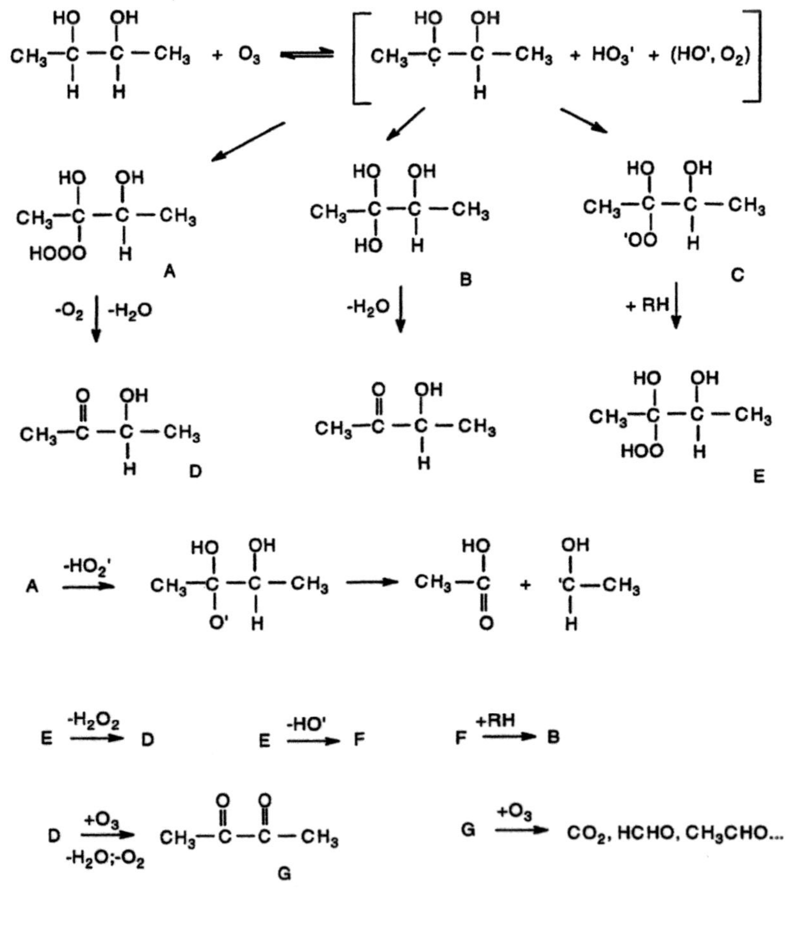

2.7.2. 1,2-Cyclohexanediol

100 ml 10 mM 1,2-cyclohexanediol were ozonized under stirring at ambient temperature. pH of the medium was kept constant at 7, 8, 9 and 11. In the experiments without pH control its change was measured during the reaction duration. In all cases formation of keto alcohol, diketone and open chain products was observed (Fig.2.15).

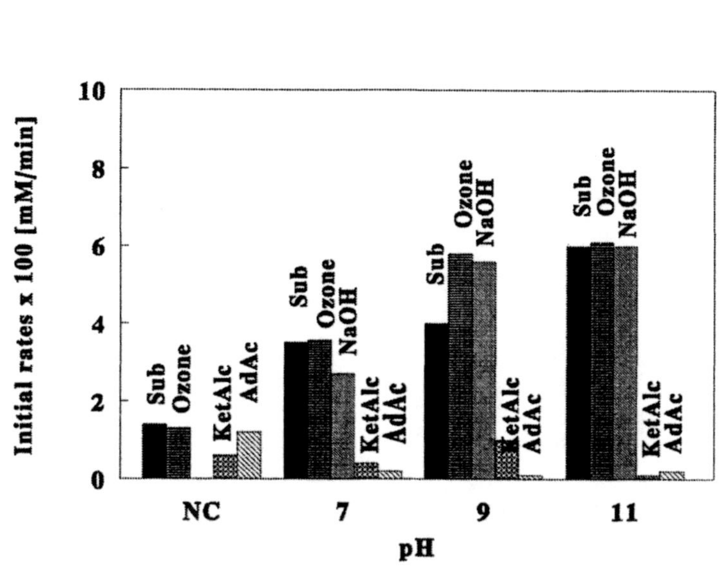

Fig.2.15. Initial rates of product consumption and formation upon 1,2-cyclohexanediol ozonation (Sub) at various pH and without pH control (NC). The products cyclohexanol-2-on (KetAlc) and adipic acid (AdAc) were identified by HPLC and ^{13}C-NMR, and the total acid number by NaOH.

Upon ozonolysis without pH control the main products were: keto alcohol and adipic acid. The value of pH decreases rapidly at the beginning from 7 to 2.2. In the ^{13}C-NMR spectra of the reaction mixture were recorded peaks at 179.5, 180.1 and 179.4 ppm which could be ascribed to adipic acid semialdehyde (179.5 and 179.4 ppm) and adipic acid (180.1 ppm). Moreover, at conversions up to 20 % other products with open chains like glutaric, succinic, malonic, acetic and formic acids were identified, too. The rate of cyclohexanediol is found to be 2-3 times higher that the rate of ozone uptake, the latter corresponding to the sum of the rates of keto alcohol and adipic acid formation. This fact suggests that its decomposition occurs via another route. The ratio between the rate of formation of cyclic and open chain products decreased from 2 to 0.5 with rising of pH from 4 to 11. Simultaneously the rate of alcohol consumption increases 4 times. When the pH was controlled from 7 to 11, the rate of consumption of NaOH and ozone corresponds to the rate of cyclohexanediol decomposition. In the NMR spectra the diversity of peaks corresponding to carboxylic carbons reaches its maximum at pH=11 and peaks at 167.8, 174.7, 184, 178.3 and 184.8 ppm were recorded. All above testifies that the mechanism of cyclohexanediol ozonolysis in aqua media is changed from stoichiometric α-C-H atom abstraction in acid-neutral media to a chain hydroxy-radical oxidation in alkaline media.

2.7.3. Mannitol and its Derivatives

The kinetics of products formation during the ozonolysis of mannitol and 1,2,5,6-di-O-iso-propyl-mannitol:

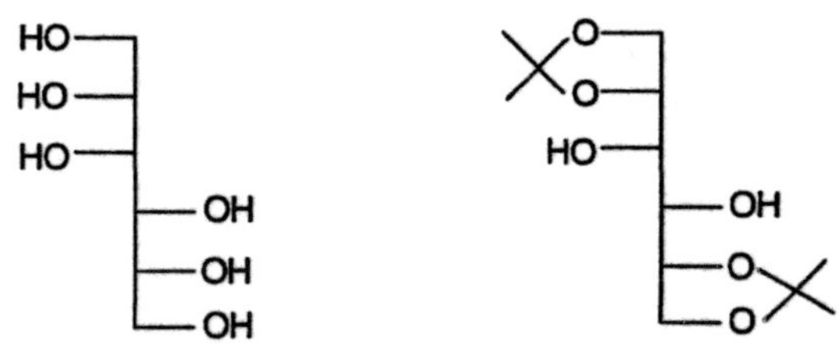

is shown in Figs 2.16 and 2.17. In the ozonation of mannitol at pH=7 the inner OH groups were converted to carbonyl ones or the C-C bonds breaks leading to the formation of products with reduced number of carbon atoms as compared with that of the starting compound. For example, carbonyl and carboxylic compounds, CO_2, H_2O were obtained. The rate of ozone absorption and that of mannitol consumption coincide, the sum of the rate of acids formation and keto-compounds is also close to these values.

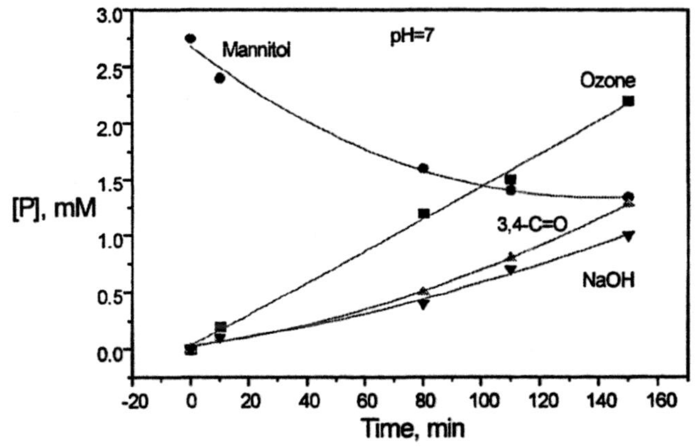

Fig.2.16. Kinetics of products formation during the ozone reaction with mannitol at pH=7. Products were identified by HPLC and ^{13}C-NMR.

The acetalation of the 1,2,5,6-OH groups at pH up to 7 causes hydrolyzation of the compound thus obtained to mannitol and acetone (Fig.2.17). At pH above these values the selectivity of ozone interaction with the third and fourth-α-H-atom is increased.

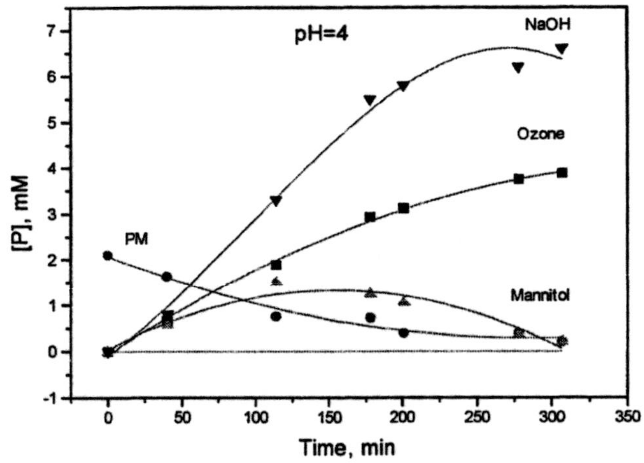

Fig.2.17. Kinetics of products accumulation during 1,2,5,6-O-di-iso-propylmannitol (PM) ozonolysis. The products were identified by HPLC and ^{13}C-NMR.

If the ozonation was carried out without pH control the pH changes from 7 at the beginning to 2.8 at 20 % conversion. A good correlation has been found between the rate of PM decomposition, ozone consumption and the rates of products (acetone, ketones and acids) formation. Ozonation at constant pH=4, kept constant through titration with 0.1N NaOH, shows similar kinetic characteristics, i.e. monoketone at 3 and 4th position is formed which is further converted into 3-4-diketone. It has been established that the rate of the reaction was accelerated by 4-5 times in alkaline medium.

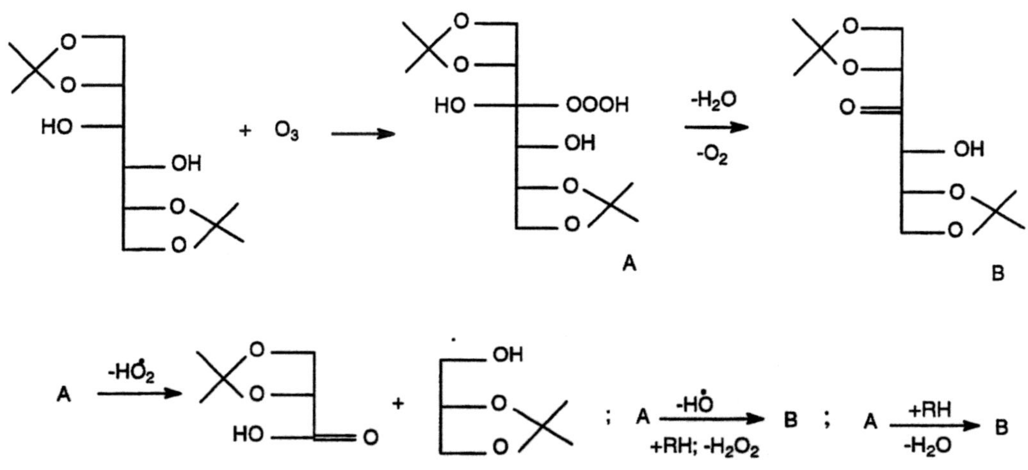

2.7.4. α-D-Glucose

It the ozonation of α-D-glucose it was found that firstly ozone attacks the H-atom of the aldehyde group at the first C-atom, and after that the H-atoms of the tert-C-H-bonds at C_2, C_3, C_4 and C_5. Upon ozonation without pH control, the values of pH changes from 7 to 2 in the course of the reaction (Fig.2.18). The yield of gluconic acid was 60 % at 5 % conversion (C_1 signal at 177.2 ppm). The rest of products at this moment was about 40 %: D-arabino-hexulose (C_2 - 178.4 ppm), D-ribo-hexos-3-ulose (C_3 - 177.8 ppm), D-xylo-hexos-4-ulose (C_4 - 176.9 ppm), D-xylo-hexos-5-ulose (C_5 - 176 ppm), D-gluco-hexodialdose (C_6 - 175.9 ppm), oxalic (167.2 ppm) and formic acids.

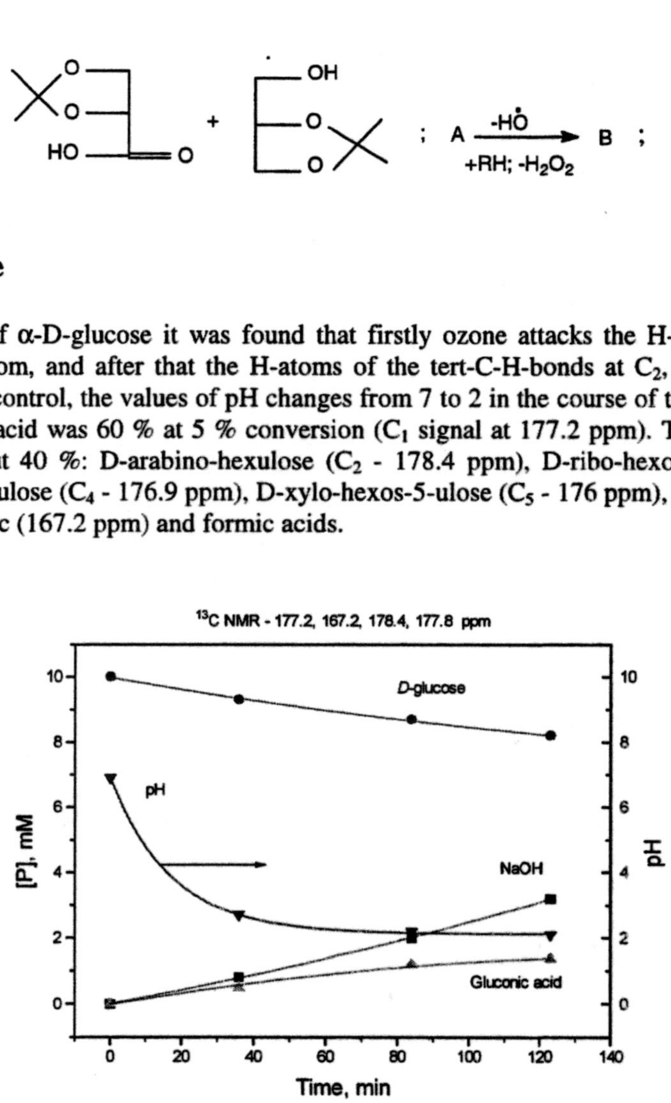

Fig.2.18 Kinetics of products formation during ozonolysis of D-glucose without pH control. Products were identified using HPLC and ^{13}C-NMR.

The pattern of products formation during glucose ozonolysis at pH from 7 to 9 was similar to that in acid medium, but with different ratio of products distribution. In that case the main product was gluconic acid reaching values up to 90-95 % (Fig.2.19).

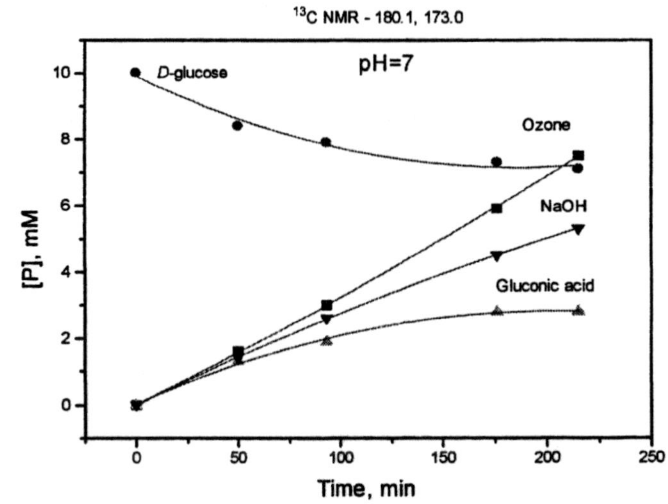

Fig.2.19 Kinetics of products formation during the reaction of ozone with D-glucose at pH=7. Products were identified by means of HPLC and ^{13}C-NMR.

In alkali medium sodium salts were formed and the peaks of carboxylic carbon atoms were slightly shifted. The rate of the reaction rises 1.5-2 times as compared with that in acid medium. Among the rest products were identified products from the ozone attack on: C_2, C_3, C_4 and C_5 - 177.4, 176.2, 174.7 and 172.6 ppm. Thus analyzing the results it can be assumed that ozone attacks all tert.C-H-bonds in the glucose molecule with almost equal probability but the rate of cleavage depends on the energy bond and the electron density. The variations in pH change the rate of ozone interaction with the hydroxyl anions and thus changes the oxidation carrier species from ozone to hydroxyl radical.

2.7.5. α-D-Methyl-Glucose

The change of H in the hydroxyl group at C1 for methyl group is used for protection of this group and for modeling of a part of the polymeric carbohydrate chain. From this point of view α-D-methyl_D-glucoside (MG) can be regarded as a suitable low molecular model for elaborating the reactivity and mechanism of ozonolysis. Fig. 2.20 demonstrates a pattern of the ozone reaction with MG at pH=8. The reaction was carried out in alkaline medium to prevent the hydrolyzation of MG in acid medium and the transformation of the methoxy group in a hydroxyl one.

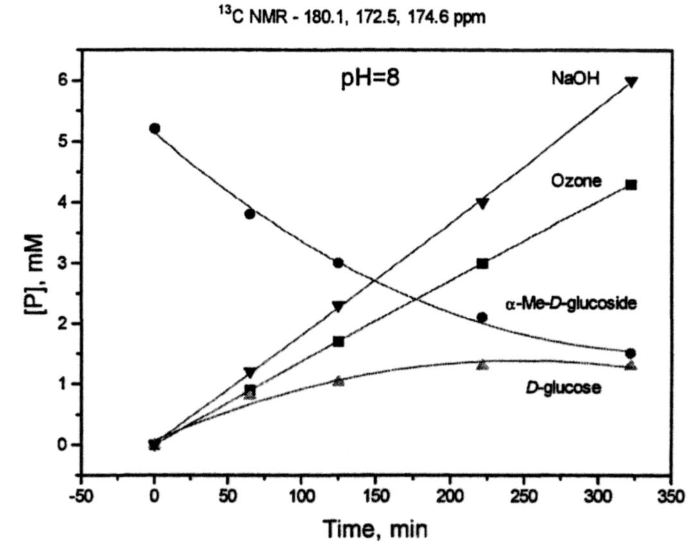

Fig.2.20. Kinetics of products accumulation during α-D-methyl-glucoside at pH=8. The products were identified by HPLC and ^{13}C-NMR.

It is seen that MG is almost quantitatively converted into glucose though the ozonation was carried out in alkaline medium. This could be attributed to the occurrence of oxidative hydrolyzation under the action of ozone. After then ozone an HO-radicals react with it yielding acids, identified by the ^{13}C-NMR signals (Fig.2.20.). We have determined that the rate of MG decomposition is $3.4.10^{-7}$ $M.s^{-1}$, the rate of acid groups accumulation as $3.12.10^{-7}$ $M.s^{-1}$, the rate of ozone absorption $-2.43.10^{-7}$ M.s-1 and the initial rate of glucose formation as $1.9.10^{7}$ $M.s^{-1}$.

2.7.6. β-Cyclodextrine and Starch

The ozonolysis of β-cyclodextrine and starch was carried out in solution and suspension and the kinetic results are presented in Figs 2.21 and 2.22.

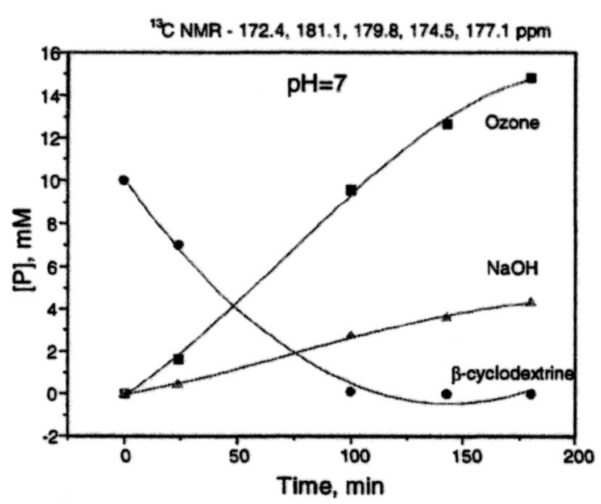

Fig.2.21. Kinetics of products formation during ozone reaction with β-cyclodextrine at pH=7. Products were identified by HPLC and ^{13}C-NMR.

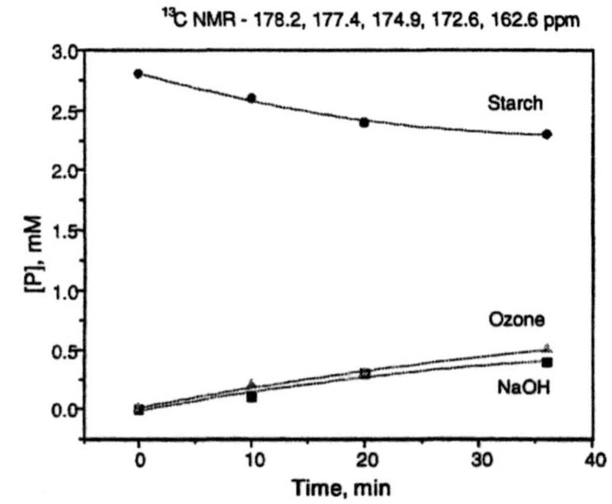

Fig.2.22. Kinetics of products accumulation during starch ozonolysis at pH=7. Products identification was carried out using HPLC and ^{13}C-NMR.

β-Cyclodextrine is an excellent model of polysaccharides that contains ten glucoside units arranged in one cycle, in which the OH-groups in 1 and 4 position are blocked (acetalated). The OH-groups at C_2, C_3 and C_6 remain free. Upon ozonolysis of aqueous β-cyclodextrine solution at pH=7 its consumption rate amounts to $1.67.10^{-7}$ $M.s^{-1}$. This value is almost equal to the rate of ozone consumption while the rate of titration by NaOH is found to be by about 25 % of these rates. Ozone converts the β-cyclodextrine molecule into a linear one with formation of carboxylic groups. The latter were identified by the appearance of chemical shifts at 172.4, 181.1, 179.8, 174.5 and 177.1 ppm in their ^{13}C-NMR spectra. The underlined values can be assigned to the carboxylic atoms at C_2-C_3 atoms.

At ozonation of starch suspension (Fig.2.22) firstly it is oxidized and only after that is dissolved. A series of peaks which could be ascribed to the carboxylic acid formation has been observed in the NMR spectra. A very good agreement was obtained between the rate of starch consumption, ozone uptake and rate of titration by NaOH.

The observation that the suspension is dissolved and the coincidence between the amount of absorbed ozone and the acid number of the oxidate evidence that the ozonation method can be used for transformation of natural polyhydrates to new valuable products.

The mechanism of the reaction of ozone with model compounds (vicinal alcohols), as shown above, is based on the interaction between ozone and α-H atom in the alcohol molecule, the cleavage of C-C bonds being a very rare and random process. The latter was observed during the monomolecular decomposition of oxy-radicals generated in the course of some of the conversions steps of intermediate products. The possibility of improving the selectivity of the C-C bond cleavage could be based on oxidation of carbohydrates using catalytic systems and specific reagents.

2.8. Catalytic Ozonolysis and Oxidation

The selectivity of the C-C bond cleavage at vicinal OH-groups is improved by applying specific reagents like RuO_4 [29]. Usually Ru(VIII) is obtained in cito via Ru(III) oxidation by suitable oxidants, for example NaOCl in alkaline medium. After that RuO4 interacts with the alcohol forming a Ru-diol complex according to the following scheme (Scheme 2.13):

Scheme 2.13.

$$RuCl_3 + 3NaOH \longrightarrow Ru^{+3} + 3OH^- + 3Na^+ + 3Cl^-$$

$$Ru^{+3} \xrightarrow[-O_3^-]{+O_3} Ru^{+4} \xrightarrow[-O_2]{+O_3} Ru^{+6} \xrightarrow[-O_2]{+O_3} Ru^{+8}$$

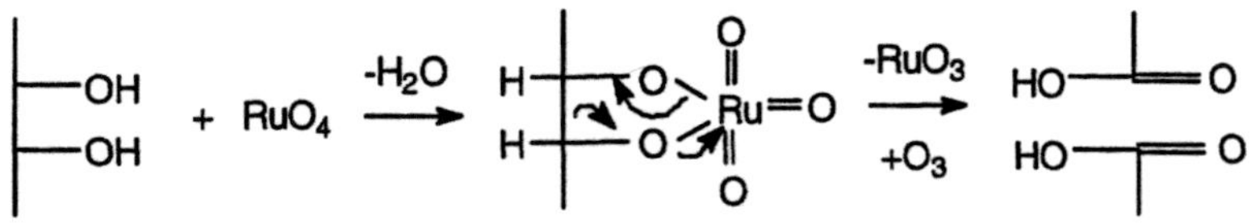

The hypochoride oxidizes the reduced by the alcohol or carbohydrates Ru(VIII) to Ru(VI), and further to Ru(VIII) and simultaneously the aldehyde to acid.

Our idea was to use ozone instead of hypochloride. In this connection we have measured the rate constant of Ru(III) oxidation by ozone which at ambient temperature at pH=7 has the value of 15 $M^{-1}.s^{-1}$. Unfortunately, from a kinetic point of view this value is extremely low to compensate the rate of RuO4 interaction with the alcohol or carbohydrate and to ensure high Ru(VIII) concentration. Because of this, the ozonolysis was carried out in the following manner: the black solution of 1 % wt. $RuCl_3.H_2O$ in 50 ml water was ozonized at pH=7 until it turned yellow ($RuCl_4$). After that the substrate solution was added dropwise. In this way all alcohols and carbohydrates pointed above, were ozonized.

Acetic acid was the main product of 2,3-butanediol ozonolysis (Fig.2.23)

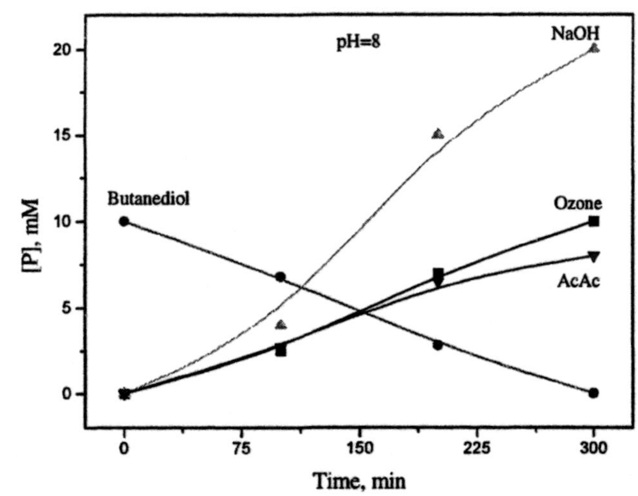

Fig.2.23. Kinetics of 2,3-butanediol ozonolysis (50 ml water), in the presence of 0.1 g $RuCl_3$.H2O, ozone flow rate - 0.1l/min and pH=8. Products identification was performed by HPLC and ^{13}C-NMR.

It is seen that (Fig.2.23) that the initial rates (up to 90 min) of butanediol consumption, ozone absorption, acetic acid formation (AcAc) and titration by NaOH are almost the same as evidenced by the equal slopes of the curves. This unambiguously support the proceeding of the reaction via the mechanism proposed in Scheme 2.8. However, after 90 min other products were also formed and the acidity of the reaction mixture begins to rise as evidenced by the sharp increase in the titration curve (by NaOH).

Upon ozonolysis of 1,2-cyclohexanediol, mannitol and its derivatives vie the method mentioned above, a selective C-C bonds cleavage between the vicinal HO-groups with carboxylic groups formation was observed only to 20-30 % conversions. At higher conversions the number of products is increased, namely because of the occurrence of secondary processes. In the cases with straight-chain alcohols two monoacids are formed from one alcohol molecule and with cyclic alcohols - one dicarboxylic acid, for example adipic acid at 1,2-cyclohexanediol ozonolysis or 2,3-dicarboxylic acid at glucose (its derivatives) ozonolysis. When the oxidation was carried out at lower pH up to 6.5, a hydrolysis of the protected HO-groups takes place. Such an example is the ozonolysis of 1,2,5,6-di-O-mannitol at pH=6.5 (Fig.2.24).

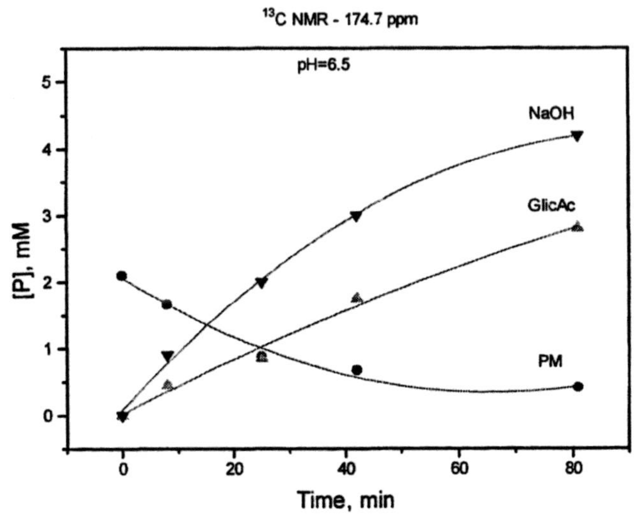

Fig.2.24 Kinetics of products formation during ozonolysis of 1,2,5,6-di-O-mannitol (PM) in the presence of 1 % wt. $RuCl_3.H_2O$.

The rate of glyceric acid formation (GlicAc - 174.7 ppm) is almost the same as that of PM consumption but is found to be lower that the rate of NaOH titration. This could be associated only with the partial hydrolyzation of PM and the formation of low molecular acids like formic and acetic ones.

However, it should be noted that the carbohydrate oxidation by ozone and Ru has some shortcomings, such as the removal of ruthenium from the reaction mixture, its volatility, high toxicity and price.

It is known that $NaIO_4$, similar to Ru(VIII), can cause the selective C-C-bonds cleavage at the vicinal hydroxy groups with two carbonyl compounds fromation. Simultanously it is reduced to $NaIO_3$. The idea was ozone to be used for reoxidation of iodate to periodate. Unfortunately, the rate of this process compared with that of reduction

$$IO_4^- + \text{Substrate} = IO_3^- + \text{Products (very fast)}$$
$$IO_3^- + O_3 = IO_4^- + O_2 \text{ (slow)}$$
$$\text{Products} + O_3 = \text{Acids} + O_2 \text{ (fast)}$$

is not sufficiently high. This requires special conditions for performing ozonolysis: maximum iodate concentration and the absence of organic products in the ozonolysis medium. Such a situation could be achieved if the products of $NaIO_4$ interaction with the substrate are oxidized and removed, and the ozonolysis proceeds further for the $NaIO_3$ oxidation to $NaIO_4$.

The oxidation of 1,2-cyclohexanediol (10 ml) by an equimolar amount of $NaIO_4$ under continuous stirring leads to the quantitative formation of adipic aldehyde which is further oxidized by ozone to semialdehyde and adipic acid. No oxidation of iodate to periodate was observed.

The application of a number of heterogeneous catalysts Ag_2O, $Co(AcO)_2$, $FeCl_3$, Pt/C, Pd/C, Adams catalyst, zeolite catalysts TS-1, ETAS and catalysts synthesized by us on the basis of ion exchange: Ce with silicalite - Ce-Silic, Ce and Ru with alumophosphates - Ce-APO and Ru-APO and phthalocyanine complexes: $PtAu_6Pc$, $PdAu_6Pc$ and MnPc for the selective ozonation of alcohols , model compounds of sacharides and natural carbohydrates, in all cases accelerate the process but did not make the process selective with respect to C-C bonds cleavage.

REFERENCES

1. Razumovskii SD, SK Rakovski, DM Shopov, GE Zaikov. Ozone and Its Reactions with Organic Compounds (monograph in Russian). *Publishing House of the Bulgarian Academy of Sciences*, 1 - 278, 1983.
2. Bailey, PS. *Ozonization in Organic Chemistry, v.II-Nonolefinic Compounds. Organic Chemistry* (A Series of Monographs), v.39-11, Acad. Press, 1982, p.255-328.
3. Otal, E., Mantzavinos, D., Delgado, M.V., Hellenbrand, R., Lebrato, J., Metcalfe, I.S., Livingston, A.G. Integrated wet air oxidation and biological treatment of polyethylene glycol-containing wastewaters. *J Chem Technol Biotechnol*, 70, 2, 147-156 (1997).
4. Benitez, F.J., Beltranheredia, J., Acero, J.L., Pinilla, M.L. Simultaneous photodegradation and ozonation plus UV radiation of phenolic acids - Major pollutants in agro-industrial wastewaters, *J Chem Technol Biotechnol*, 70, 3, 253-260 (1997).
5. Amalric, L, C Guillard, P Pichat. The GC-MS identification of some aliphatic intermediates from the TIO2 photocatalytic degradation of dimethoxybenzenes in water. *Research on Chemical Intermediates*, 21: 1, 33-46 (1995).
6. Heinzle, E, H Stockinger, M Stern, M Fahmy, OM Kut. Combined biological chemical, (ozone). treatment of wastewaters containing chloroguaiacols. *Journal of Chemical Technology and Biotechnology*, 62: 3, 241-252 (1995).
7. Beltran, FJ, PM Alvarez, B Legube, H Allemane. Chemical degradation of aldicarb in water using ozone. *Journal of Chemical Technology and Biotechnology*, 62: 3, 272-278 (1995).
8. Lee, MJ, H Arai, T Miyata. Decomposition of t-butanol in water by gamma- irradiation combined with ozone, Cupric ion effect. *Chemistry Letters* : 6, 1069-1070 (1994).
9. Benko, EM, VV Kovaleva, NA Mamleeva, AN Mitrofanova, VA Voblikova, AN Pryakhin, VV Lunin Ozone Oxidation of Lignins. *Zhurnal Fizicheskoi Khimii*, 68: 11, 1964-1967 (1994).
10. Artemev, YM, MA Artemeva, MG Vinogradov, TI Ilika Photocatalytic decomposition of methylene blue and 2, 4,5-trichlorophenol in ozonated aqueous TIO_2 and Nb_2O_5 suspensions. *Russian Journal of Applied Chemistry*, 67: 9 Part 2, 1354-1357 (1994).
11. Scheuer, C, B Wimmer, H Bischof, L Nguyen, J Maguhn, P Spitzauer, A Kettrup, D Wabner Oxidative decomposition of organic water pollutants with UV-activated hydrogen peroxide - Determination of anionic products by ion chromatography. *J.Chrom. A* 706: 1-2, 253-258 (1995).
12. Hoger, B. E Gilbert, SH Eberle. Influence of Chloride Ions in Water Treatment with Ozone Hydrogen Peroxide. *Vom Wasser*, 80, 187-191 (1993).
13. Ruck W.Ozonization of Biologically Purified Sewage Water. *Vom Wasser*, 80, 253-272 (1993).
14. Gab, S, WV Turner, S Wolff, KH Becker, L Ruppert, KJ Brockmann. Formation of alkyl and hydroxyalkyl hydroperoxides on ozonolysis in water and in air. *Atmospheric Environment*, 29: 18, 2401-2407 (1995).
15. Takeuchi, K, C Fushimi, K Nakamura, T Ibusuki. Determination of Ammonia Ammonium Ion in Water by Continuous-Flow Chemiluminescence After Derivatization to an Isoindole. *Bunseki Kagaku*, 42: 10, 625-629 (1993).
16. Pan, GY, CL Chen, JS Gratzl, HM Chang. Model compound studies on the cleavage of glycosidic bonds by ozone in aqueous solution. *Research on Chemical Intermediates*, 21: 3-5, 205-222 (1995).
17. Wada, H, T Naoi, Y Kuroda. Recycling of water from chromium(VI). waste water by ozone oxidation combined with UV radiation and ion exchange. *Nippon Kagaku Kaishi*, : 4, 306-313 (1995).
18. Marco, A, E Chamarro, S Esplugas. Formic Acid Degradation by O3/UV. *Afinidad*, 51: 452, 265-271 (1994).
19. Starek, J, A Zukal, J Rathousky. Comparison of the Adsorption of Humic Acids from Aqueous Solutions on Active Carbon and Activated Charcoal Cloths. *Carbon*, 32: 2, 207-211 (1994).
20. Rakovski, S.K., D.R. Cherneva. Kinetics and Mechanism of the Ozone Reaction with Aliphatic Alcohols. *Int.J.Chem.Kinet.*, **22**, (4), 321 - 329 (1990).

21. Rakovski SK, DR Cherneva, DM Shopov, SD Razumovskiy. Application of the Barbotage Method in Investigating the Kinetics of Ozone Reactions with Organic Compounds. *Comm. Dep. Chem - Bulgarian Academy of Sciences*, 9 (4) 711-717 (1976).
22. Rakovski, S.K., D.R. Cherneva. Kinetics and Mecanism of the Reaction of Ozone with Organic Compounds. I. Interaction of Aliphatic Ketones with Ozone. *Oxidation Communications,* **12,** (3), 108 - 115 (1989).
23. Rakovski SK, LG Kulak, EM Kuramshin, SS Zlotsky, DL Rakhmankulov. A Review of the Reaction of Ozone with Saturated Hydrocarbons. Formation, Properties and Reactions of Organic Hydrotrioxides. *Comm. Depart. Chem. (Bulgaria)*, **22,** (3), 722 - 736 (1989).
24. Rakhmankulov D.L., S.S. Zlotsky, L.Z. Rudnik, G.T. Teregulova, S.K. Rakovski. Synthesis of 4-Hydroximethyl-1,3-dioxolane ether and its Derivatives *Comm. Depart. Chem. (Bulgaria)*, **22,** (3), 652 - 655 (1989).
25. Anachkov M.P., S.K. Rakovsky, A.K. Stoyanov, R.K. Fotty. DSC Study of the Thermal Decomposition of 1-Decene Ozonide. *Thermochemical Acta*, **237**, 213 - 217 (1994). KW: ozonide, ozone.
26. Rakovsky S.K., D.R. Cherneva, M. Deneva. Ozone Reactions with Aliphatic Ethers in CCl_4. Kinetics and Mechanism. *Int. J. Chem. Kinet.* **27,** (2), 153 - 165 (1995). KW: ozone, ethers.
27. Rakovsky S.K., D.R. Cherneva, M. Deneva, V.V. Ershov. Ozonation of Catechols. Kinetic and Mechanism. *Oxid. Commun.*, **20,** (2), 169-178 (1997). KW: ozone, catechols.
28. Rakovsky S.K., R.A. Sheldon, F.v. Rantwijk. Carbohydrate Ozonation. Selective Cleavage of the 2,3- Bond in Carbohydrates. *Oxid. Commun.*, **19,** (4), 482 - 498 (1996). KW: ozonation, oxidation, alcohols, carbohydrates.
29. Denisov, ET, NI Mickeevich, VE Agabekov. *Mechanism of Liquid-Phase Oxidation of Oxygen Containing Compounds*. Publ. House "Science and Technique ", Minsk. 1975.
30. Whiting, M, A Bolt, J Parish. *Adv.Chem.Ser.*, 77, 4 (1968).
31. Williamson, D, RJ Cvetanovich. *J.Am.Chem.Soc.*, 92, 2949 (1970).
32. Gerchikov, AYa, ER. Kuznetzov, ET Denissov. *Kin. Kat.*, 15, 109 (1970).
33. Shereshovetz, VV, N. Shafikov, VD. Komissarov. *Kin. Kat.*, 21, 1596 (1980).
34. Waters, M, A. Rollin. *J.Org.Chem.*, 42, 889 (1976).
35. Rakovski, SK, PhD Theses. *Moscow, Institute of Chemical Physics, Akad. Sci. USSR*, 1975.
36. Kondratiev, VN (Ch. Ed.). *Bond Energy, Ionization Potential and Electron Affinity*, Nauka, Moscow, 83, 1974.
37. Gordon, AJ, RA Ford. *The Chemists Companion*, Publ. House Mir, Moscow, 1976 (Russian).
38. Eiring,G, SG Lin, SM Lin. *The Principles of Chemical Kinetics*, Publ. House Mir, Moscow, 1983 (Russian).
39. Karadakov K, SK Rakovsky. Software: Calculation of Pre-exponents ver.1.0. Copyright, 1987, based on the Howbert, J. Software: The Molecular Animator, vers. 1-3, Copyright, 1983.
40. Thiel W. Software, IBM, Program MNDOC, QCPE ver., Fachbereich Physikalische Chemie der Philipps-Universitat, auf den Zahbergen, D-3550, Marburg, Germany, based on Dewar M.Y.S, W. Thiel, *J.Amer.Chem.Soc.*, 99, 4899 (1977).
41. Gerchikov, AYa, VD Komissarov, ET Denisov, GB Kochemassov. *Kinet. Katal.*, 13, 1126 (1972); 15, 509 (1974).
42. Gerchikov, AYa, VD Komissarov, ET Denisov. *Kinet. Katal.*, 15, 230 (1974).
43. Gerchikov, AYa, VD Komissarov, LG Galimova, ET Denisov. *Dokl. Akad. Nauk SSSR*, 213, 881 (1973).
44. Korotkova, NP, AM Syroezhko, VA Proskuryakov. *Zh.Prikl.Khim.*, 54, 885 (1981).
45. Komissarov, LG, Galimova, ET Denisov. *Kinet. Katal.*, 15, 1063 (1974).
46. Syroezhko, AM, NP Korotkova, AA Vihorev, VA Proskuryakov. *Zh.Prikl.Khim.*, 15, 2562, (1978).
47. Barton, D, VD Li (Eds). *General Organic Chemistry*, v.2, Khim., Moscow, 577-80 (1982).
48. House HO. *"Modern Synthetic Reactions"*, Benjamin, Menlo Park, 2-nd ed. (1972).
49. Zabicky J (Ed.). *"The Chemistry of Carbonyl Groups"*, Interscience, London, v.2 (1970).
50. Alcais, P, R. Brouillard. *J.Chem.Soc. Perkin II*, 1214 (1972).

51. Houben-Weyl. *Methoden der organischen Chemie, Organische Peroxo-Verbindungen.* H. Kropf, G (Ed.), Thieme Verlag, Stuttgart, N. York, 1988, Band E 13/Teil 1 und 2.
52. Batterbee, JE, PS Bailey, *J. Org. Chem.*, 32, 3899 (1967).
53. Kovac, F, B Pleshnichar. *J. Chem. Soc., Chem. Commun.*, 122 (1978).
54. Kovac, F, B Pleshnichar. *J. Amer. Chem. Soc.*, 101, 2677 (1979).
55. Pryor, WA, N Ohto, DF Church. *J. Amer. Chem. Soc.*, 104, 5813 (1982).
56. Brudnikar BM, EM Kuramshin, Pross. Vsesouznaya Konferentzia po khim. tekhn. acetal. heteroanal. Pross,, Ufa, 1981, 112.
57. Shereshovetz VV, VD Komissarov, SI Maslennikov, NN Kabelanov. *Izv. AN SSSR ser. chim.*, 2631(1982).
58. Kuramshin, EM, LG Kulak, SS Zlotskii, DL Rahmankulov. *Izv. Chim., Bulg. Akad. Sci.*, 11, 2631 (1982).
59. Pleshnichar, B. *Organic peroxides* (The chemistry of functional groups, Peroxides, Ed. S. Patai), J. Wiley, New York, 1983.
60. Taillefer, RJ, SE Thomas, Y Nadeau, H Bierbeck. *Can. J. Chem.*, 57, 3041 (1979).
61. Taillefer, RJ, SE Thomas, Y Nadeau, S Fliscar, H Henzy, *Can. J. Chem.*, 58, 1138 (1980).
62. Rakovski, SK, DR Cherneva, DM Shopov, *Int. Microsymp. Oxidation of Organic Compounds*, Tallin, 30, (1987).
63. Rakovski, SK. *Catalytic Oxidation of Alcohols.* Plenary Lecture, High All-Military School "V. Levski" - V. Tarnowo, Bulgaria, 1997. Nangia, PS, SW Benson. *J. Amer. Chem. Soc.*, 102, 3105 (1980).
64. Denisov, ET. *Kinet. Katal.*, 9, 95 (1981).
65. Rahmankulov, DL, Kuramshin, EM, SS Zlotskii, *Usp. chim.*, 54, 923 (1985).
66. Kuramshin, EM, UB Ivashov, SS Zlotskii, DL Rahmankulov. Izv. VUZ, khim. khim. tekhn., 27, 13 (1984).
67. Rahmankulov, DL, RA Karahanov, SS Zlotzkii, EA Kantor, UV Imashev, AM Sirkin. Chemistry and Technology of 1,3 - Dioxolanes. *Itogi Nauki i Tekhniki*, Moscow, 5, 202 (1979).
68. Brudnik, BM, EM Kuramshin, UB Ivashov, SS Zlotskii, DL Rahmankulov. *Zurn. Org. Khim.*, 17, 700 (1981).
69. Brudnik, B.M., V.V. Shereshovetz, E.M.Kuramshin, U.B. Imashev, S.S. Zlotsky, D.L Rakhmankulov, *React. Kinet. Catal. Lett.*, **13**, 97 (1980).
70. Kuramshin, EM, RH Sadeeva, BK Gumerova, SS Zlotskii, DL Rahmankulov. *Zurn. Org. Khim.*, 19, 149 (1983).
71. Kuramshin, EM, BK Gumerova, LG Kulak, SS Zlotskii, DL Rahmankulov. *Zurn. Obsh. Khim.*, 55, 1611 (1985).
72. Bailey, P.S., D.A. Lerdal, *J. Amer. Chem. Soc.*, **100,** 5820 (1978).
73. Erickson, R.E., R.T. Hansen, J. Harkins, *J. Amer. Chem. Soc.*, **90,** 6777 (1968).
74. Stary, F.E., D.E. Emge, R.W. Murray, *J. Amer. Chem. Soc.*, **98,** 1880 (1976).
75. Price, C.C., A.L. Tumolo, *J. Amer. Chem. Soc.*, **86,** 4691 (1964).
76. Murray, R.W.,W.S. Lumma, J.W. Lin, *J. Amer. Chem. Soc.*, **92,** 3205 (1970).
77. Shereshovetz, VV, FA Galieva, NN Kabanov, NA Shishlov, RA Sadikov, VD Komissarov, GA Tolstikov. *Izv. Akad. Nauk ser. khim.*: 317 (1986).
78. Deslongshamps, P., *Tetrahedron*, **31**, 2463 (1975).
79. Shafikov, NYa, RA Sadikov, VV Shereshovetz, AA Panasenko, VD Komissarov., *Izv. Akad. Nauk ser. khim.*: 1923 (1981).
80. Shereshovetz, VV, FA Galieva, VD Komissarov. *Izv. Akad. Nauk ser. khim.*: 1668 (1984).
81. Rahimov, AI. Chemistry and Technology of the Organic Peroxides. Moscow, 1979. p. 392.
82. Shereshovetz, VV, NN Kabanov, VD Komissarov, GA Tolstikov. *Izv. Akad. Nauk ser. khim.*: 1660 (1985).
83. Shereshovetz, VV, RA Sadikov, LA Bahanov, AA Panasenco, VD Komissarov, GA Tolstikov. *Izv. Akad. Nauk ser. khim.*: 1675 (1982).
84. Shereshovetz, VV, VD Komissarov, NYa Shafikov, LA Bahanova, VS Kolonicin, YuE Nikitin, GA Tolstikov. *Zurn. Org. Khim.*, 19, 225 (1983).
85. Kuramshin, EM, LG Kulak, SS Zlotskii, DL Rahmankulov. *Zurn. Org. Khim.*, 22, 1986 (1983).

86. Shereshovetz, VV, LA Bakhanova, VD Komissarov, GA Tolstikov. *Zurn. Org. Khim.*, 21, 482(1985).
87. Shereshovetz, VV, VD Komissarov, LA Bakhanova, RSh, Burkhanova, GA Tolstikov. *Zurn. Org. Khim.*, 19, 892(1983).
88. Shereshovetz, VV, FA Galieva, NYa Shafikov, RA Sadikov, AA Panasenko, VD Komissarov, GA Tolstikov. *Izv. Akad. Nauk ser. khim.*: 1177 (1982).
89. Shereshovetz, VV, LA Bakhanova, VD Komissarov, NS Vostrikov, GA Tolstikov. *Izv. Akad. Nauk ser. khim.*: 1922 (1982).
90. Lorenz, O. *Anal. Chem.*, 37, 101 (1965).
91. Stein, RA, U Slawson. *Anal. Chem.*, 35, 1008 (1963).
92. Tolstikov, GA. Reactions of Hydroproxide Oxidation, Nauka, Moscow, 1976, p.243.
93. Fizer, L, M Fizer. Reagents for the Organic Syntheses. Mir, Moscow, 3, 66 (1970).
94. Karakulova, EN. Chemistry of Oil Sulfides, Nauka, Moscow, 1970, 202.
95. Brudnik, BM, LV Spirikhina, EM Kuramshin, UB Imasheva, SS Zlotskii, DL Rahmankulov. *Zurn. Org. Khim.*, 16, 1281 (1980).
96. Clerici, A, O Porta. *Synthesis Commun.*,18, 2281 (1988).
97. Weber, V, G Hokel. *Interphase Catalysis in Organic Syntheses*. Mir, Moscow, 1980, p. 327.
98. Rakhmankulov, DL, RA Karakhanov, SS Zlotzkii, EA Kantor, UB Imashev. Chemistry and Technology of 1,3 - Dioxocyclanes. *Itogi Nauki - Khim. Tekhm.*, 5, 248 (1979).
99. Baldin, A.C. Thermochemistry of Peroxides. in S.Patai (Ed.), *The Chemistry of Functional Groups, Peroxides*, J Wiley & Sons Ltd., 1983, p. 97.
100. Cafferata, LFR, GN Eyler, EL Svartman, AI Canizo, EJ Borkowski. *J. Org. Chem.* 55,1058 (1990).
101. Jefford, CW, A Jaber, J Boukouvalas, P Tissot. *Thermochim. Acta*, **188**, 337 (1991).
102. Bernatek, E, M Hvatum. *Acta Chem. Scand.* 14, 836 (1960).
103. Privett, OS, EC Nickel. *J. Am. Oil. Chem. Soc.* 43, 393 (1966).
104. Boldenkov, NI, SD Razumovskii. *Khim. Promst.*:2, 100 (1977).
105. Criegee, R, A Kerckow, H Zinke. *Chem. Ber.*, 88,1878 (1955).
106. Razumovskii, SD, YuN Yur'ev. *Zhurn. Org. Khim.*, 3, 251 (1967); *J. Org. Chem. USSR* (Engl. Transl.), 3, 238 (1967).
107. Story, PR, TK Hall, Morrison lll, JC Farine. *Tetrahedron Lett.*: 5397 (1968).
108. Hull, LA, IC Hisatsune, J Heicklen. *J. Phys. Chem.* 76, 2659 (1972).
109. Brazier, DW, NV Schwartz. *Thermochim. Acta*, 39, 7 (1980).
110. Ellerstein, SM in *Analytical Calorimetry* - RS Porter and JO Johnson (eds.), Plenum Press, N.Y., 1968, p.279.
111. Crane, LW, PJ Dynes, DH Kaelble. *J. Polym. Sci., Polym. Letters*, 11, 533 (1973).
112. Kissinger, H. *Anal. Chem.*, 21, 1702 (1957).
113. Rogers, RN, LC Smith. *Thermochim. Acta*, 1, 1 (1970).
114. Batt, L, HO McCullough. *Int. J. Chem. Kinet.*, 8, 911 (1976).
115. Batt, L, K Christie, RT Milne, AJ Summers. *Int. J. Chem. Kinet.*, 6, 877 (1974).
116. Walker RF, L Philips. *J. Chem. Soc. (A)* : 2103 (1968).
117. Lewis, DK. *Can. J. Chem.*, 54, 581 (1976).
118. *Pract. Chem.*, 66, 273, (1855).
119. L von Balo. *Ann.*, 140, 348 (1866).
120. Berthlot, M. *Compl. Rend.*, 92, 895 (1881).
121. Harries, V. *Ann.*, 343, 311 (1905).
122. Fisher, FG. *Ann. Chem.*, 476, 233 (1929).
123. Bailey, PS. *Chem. Rev.*, 58, 925 (1958).
124. Price, CC, AL Tumolo. *J. Am. Chem. Soc.*, 86, 4691 (1964).
125. Ericson, RE, D Bakalik, C Richards, M Seanlon. *J. Org. Chem.*, 31, 461 (1966).
126. Stary, FE, DE Emge, RW Murray. *J. Am. Chem. Soc.*, 98, 1880 (1976).
127. Giamalva, DH, DF Church, WA Pryor. *J. Am. Chem. Soc.*, 108, 7678 (1986).
128. Pryor, WA, N Ohto, DF Church, *J. Am. Chem. Soc.*, 105, 3614 (1983).
129. Hoigne, J et al. *Water Res.*, 17, 173, 185 (1983); 19, 993 (1985).
130. Gurol, MD, S Nekoulnalnl. *Ind. Eng. Chem. Fundam.*, 23, 54 (1984).

131. Bernatek, E, C Frangen, *Acta Chem. Scand.*, 15, 471 (1961).
132. Bernatek, E, J Moskeland, K Valen. *Acta Chem. Scand.*, 15, 1454 (1961).
133. Bernatek, E, C Frangen. *Acta Chem. Scand.*, 16, 2421 (1962).
134. Bernatek, E, A Vincze. *Acta Chem. Scand.*, 19, 2007 (1965).
135. Jarret, M, A Bermond, C Ducauze. *Analysis*, 14, 185 (1983).
136. Razumovskii, SD, GA Nikiforov, GM Globenko, AA Kefely, YaA Gurvich, NA Karelin, GE Zaikov. *Neftekhimia*, 12, 376 (1972).
137. Konstantinova, ML, SD Razumovskii, GE Zaikov. *Izv. Akad. Nauk SSSR, ser. khim.*: 324 (1991).
138. Razumovskii, SD, ML Konstantinova, GE Zaikov, *Izv. Akad. Nauk SSSR, ser. khim.*: 1203 (1992).
139. Konstantinova, ML, VB Vol'eva, SD Razumovskii. *Izv. Akad. Nauk SSSR, ser. khim.*: 1443 (1992).
140. *Handbook of Chemistry and Physics*, 66th Edition, 1985-86, CPC Press, p. D278.
141. Sheldon, R.A. *Heterogeneous catalytic oxidation and fine chemicals. Heterogeneous Catalysis and Fine Chemicals II*, 1991, Elsevier Sci.Publ., B.V., Amsterdam.
142. Sheldon, R.A.; Kochi, J.K. *Methal-Catalyzed Oxidation of Organic Compounds*. Acad.Press, N.Y., 1981.
143. Emons, C. *Carbohydrate-Based Synthesis of C -Chirons. PhD theses*, January 31, 1992, Eindhoven, The Netherlands.
144. Floor, M.; Schenke, K.M.; Kieboom, A.P.G.; Bekkum, H.v. *4th Europian Carbohydrate Symposium*, July 12-17,1987, Darmstadt, FRG, D-27.
145. Wit, D.de. *Oxidation of Sucrose and Glucose of Chiral Building locks. PhD theses*, 1992, Delft, The Netherlands.
146. Floor, M. *Glycol-cleavage Oxidation of Polysaccharides and Model Compounds. PhD theses*, 1989, Delft, The Netherlands.
147. Davidson, E.A. *Carbohydrate Chemistry*. Holt, Rinchart and Winston, Inc., N.Y., 1967.
148. *Abstracts of XV-th Int.Carbohydrate Symp.*, July 5-10, 1992, Paris, France.
149. Bochkov, A.F.; Zaikov, G.E. *Chemistry of the O-Glicosidic Bond: Formation and Cleavage*. Perg.Press, Oxford, 1979.
150. Andreozzi, R.; Inspla, A.; Caprio, V.; D'Amore, M. *Water Res.*, 26, 917 (1992).
151. Kiviniemi, A.-M.; Virtanen, P. *J.Carbohydrate Chem.*, 11, 195 (1992).
152. Hudlicky, M. *Oxidation in Organic Chemistry*. ACS Monograph 186, Washington DC, 1990.
153. Wiesner, K.;Chan, K.K.; Demerson, C. *Tetrahedron Lett.* : 2893 (1965).
154. Trahanovsky, W.S.;Young, L.H.;Bierman, M.H. *J.Org.Chem.*, 34, 869 (1969).
155. Uskokovic, M.; Gut, M.; Trachtenberg, E.N.;Klyne, W.; Dorfman, R.I. *J.Am.Chem.Soc.*, 82, 4965 (1960).
156. Rigby, W. m.*Chem.Soc.* : 1907 (1950).
157. Tavares, D.F.;Borger, J.P. *Can.J.Chem.*, 44, 1323 (1966).
158. Rocek, J.; Westheimer, F.H. *J.Am.Chem.Soc.*, 84, 2241 (1962).
159. Ohloff, G.; Giersch, W. *Angew.Chem.* 85, 401 (1973).
160. Cambie, R.C.; Chambres, D.; Rutledge, P.S.; Woodgate, P.D. *J. Chem. Soc. Perkin Trans. I* : 1483 (1978).
161. Criegee, R.; Beucker, H. *Justus Liebigs Ann.Chem.* : 541, 218 (1939).
162. Barton, D, et al. *Tetrahedron Lett.*, 23, 957 (1982).
163. Rakovski, S.K.; Cherneva, D.R.; Minchev, L. *VII-th National Conference on Chromatography, B-35*, Blagoevgrad, November 8-12, 1990.
164. Plesnicar, B, Kovac F, Schara M. *J. Am. Chem. Soc.*, 110, 2114 (1988).
165. Augustin, RL (Ed.) *Oxidation (in two volums). Techniques and Applications in Organic Synthesis*. Marcel Dekker, Inc., N.Y., 1969.
166. Schuchmann, MN, Cv Sontag. *JWSRT-Aqua*, 38, 311 (1989).
167. Lamberti et al. *US Patent 3873614/March 25*, 1975.
168. Namba, K, S Nakayama. *Bull. Chem. Soc. Jap.*, 55, 3339 (1982).

Chapter 3

OZONOLYSIS OF ALKENES IN LIQUID PHASE

3.1. OLEFINS

The reactions of ozone with alkenes are the best-known of all reactions of ozone. These reactions take place with high rates, low activation energy and are a basic source for ozonides preparation, i.e. cyclic 1,2,4-trioxalanes [1-7]. The interaction of ozone with the C=C bonds is a powerful method for structural studies of polydienes, functionalizing of complex organic compounds at low temperatures, and synthesis of valuable oxygencontaining compounds [8-20]. This reaction plays an important role in atmosphere [23-30] during degradation of rubber materials [31, 32]. All this makes the study of the theoretical and practical aspects of these reactions attractive and relevant for many researchers [1-13]. The great interest stems both for elucidating the mechanism and reactivity of alkenes and for ozone application in organic synthesis, rubber products protection against ozone degradation and creation of novel efficient and economical technologies [1, 2, 8, 13, 15, 25-31]. We have carried out detailed and systematic studies on the ozonolysis of olefins and their polymeric analogues [35-50].

3.1.1 Mechanisms

The reactions of ozone with alkenes are a subject of interest and study since ozone discovery in 1840 [51-78]. The first experimental research and conclusions concerning these reactions dated from the beginning of this century [51-53]. Harries observed that certain olefinic compounds reacted with ozone to give peroxide oils to which he firstly assigned structure (1) and later he changed the structural assignment to structure (2) as shown in Scheme 3.1 [54-58]:

Scheme 3.1

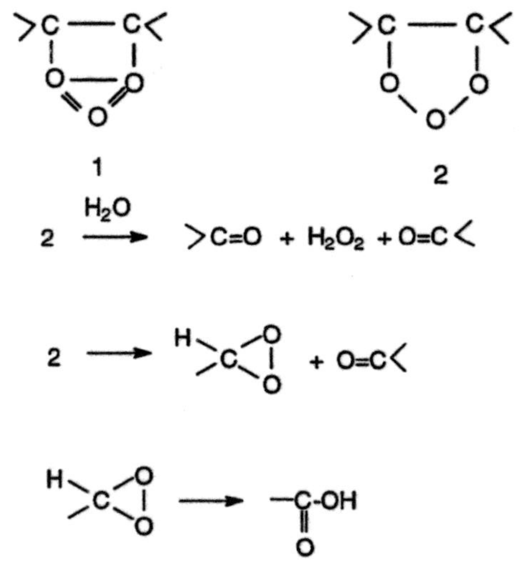

Harries proposed that the compound with structure (2) decomposed to hydrogen peroxide and carbonyl compounds. However, the possibility for self decomposition of (2) to 1,2-dioxalane rearranging further to a carboxylic acid and a carbonyl compound could not be excluded.

Staudinger [59] suggested a more complicated mechanism of olefin ozonolysis (Scheme 3.2) than Harries:

Scheme 3.2

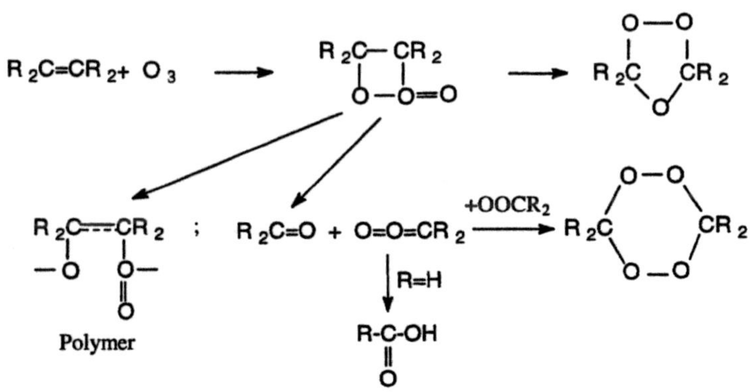

He assumed that ozone reacts with olefin producing an intermediate adduct with a four-member ring, which undergoes rearrangement to 1,2,3-trioxalane, diperoxide, carbonyl compound or an acid or polymerizes. Later it has been established that this scheme failed to explain a number of experimental observations [61].

In 1949-51 Criegee proposed another mechanism of ozonolysis which can be considered as classical [61-66]. This mechanism is outlined in Scheme 3.3

Scheme 3.3

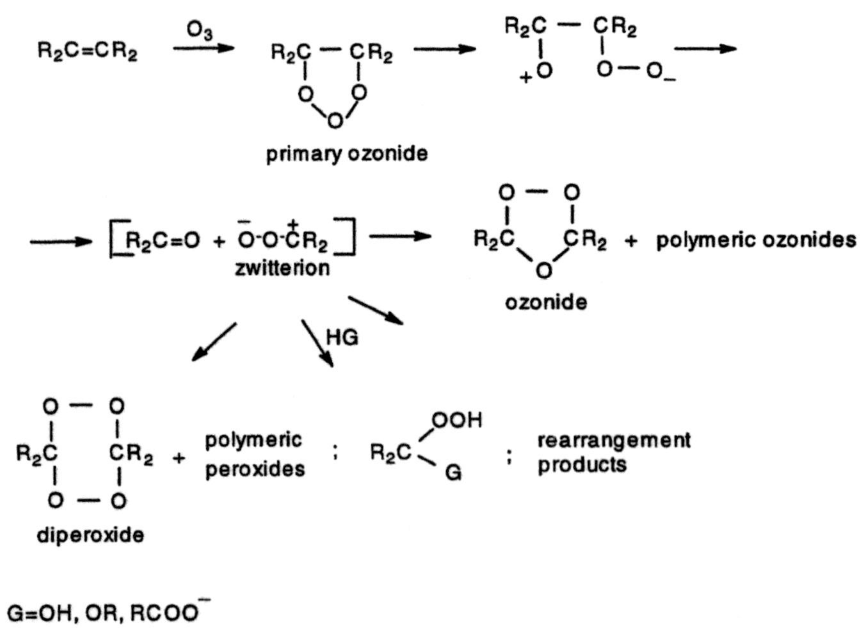

According to Scheme 3.3 ozone reacts with C=C bonds forming a primary ozonide (PO) - 1,2,3-trioxalane. The reaction is exothermic with a heat effect more than 50 kcal/mol. PO is very unstable and rapidly decomposes to zwitterion and carbonyl compound. There are suggested at least four ways through which the zwitterion can stabilize itself: 1) reaction with its own ketone or aldehyde to give ozonide (OZ) (1,2,4-trioxalane); 2) reaction with the wall of solvent cage, composed mainly of solvents molecules thus producing hydroperides; 3) reaction with other zwitterions outside the cage yielding diperoxide - 1,2,4,5-tetraoxalane and polymeric peroxides - -(-O-O-C-O-O-)$_n$-, reactions with carbonyl compounds resulting from the decomposition of another PO and formation of "cross" ozonides, reactions with preliminary added aldehydes, ketones, alcohols, water, etc. leading to the formation of "foreign" ozonides or peroxide compounds and 4) undergoes monomolecular rearrangement giving 1,2-dioxalane which isomerizes further to acid.

The proposed mechanism has been confirmed by numerous experimental data carried out by various researchers [1, 2]. The low temperature reduction of olefins ozonolysis products, obtained at low temperatures, gives 1,2 -alcohol with 1,2,3-trioxalane as a precursor [81-83]. The formation of zwitterion is evidenced by the formation of: 1) " foreign" ozonide [84]; 2) methoxy hydroperoxides in methanol solutions [85-88]; 3) "cross" ozonides [89-91]; 4) ozonide during the photochemical decomposition of diazocompounds solutions in the presence of oxygen and an aldehyde [92]:

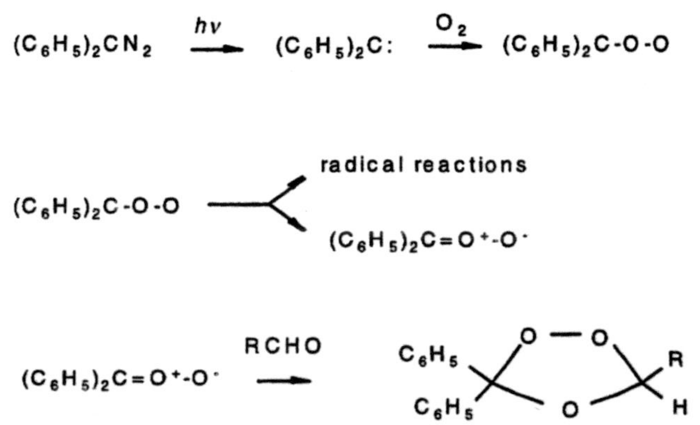

Rieche [68, 69] has confirmed the ozonide structure through a counter synthesis, and later it was established by different spectral and theoretical methods [69-77].

The Criegee mechanism is very close to the contemporary concepts for the mechanism of olefins ozonolysis. The additional studies carried by Criegee accounting for the effect of the number, size, spatial arrangement, electronic properties of the substituents, conformational state and electronic structure of the zwitterion [78-80] on the alkene ozonolysis allow to throw light on the mechanism of the "anomalous" ozonolysis of some olefins with more particular structure and substituents [81-93].

The ozonide stereochemistry can be predicted if the configuration of the zwitterion - *anti* or *syn* is known, which is strongly dependent on the nature and size of the substituents and the experimental conditions [93, 94]:

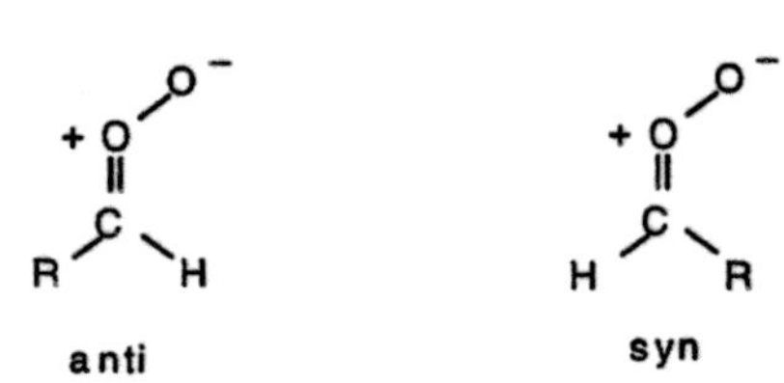

The studies on olefin ozonolysis up to 1958 were summarized by Bailey [3]. Many other authors have carried out research on separate problems regarding of these reactions [4-16]. These works stimulate a number of new studies on the mechanism and stereochemistry of the reaction, on the ozonolysis of new types of olefins, the preparation of new products and the application of novel methods of analysis [52-55, 66-70].

Concurrently with the accumulation of the convincing evidences favoring the Criegee mechanism, other evidences began to appear which could not be explained by the classical Criegee mechanism. For example, upon ozonolysis of *cis*- and *trans*-1-arylpropenes the ratios of ozonide to polymer peroxides and/or free aldehydes are quite different for the two isomers [71-73]. It was also reported that the *cis:trans* ratios of cross ozonides [74-82] obtained from *cis* and *trans* unsymmetrical olefins, often differ whereas according to the simple Criegee mechanism *cis*-olefin should yield *cis*-ozonide and *trans*-olefin *trans*-ozonide, respectively.

For this reason, Story et al. [93] proposed a scheme which explains the formation of *trans*- and *cis*-ozonide from *cis*- and *trans*-olefin, respectively (Scheme 3.4):

Scheme 3.4.

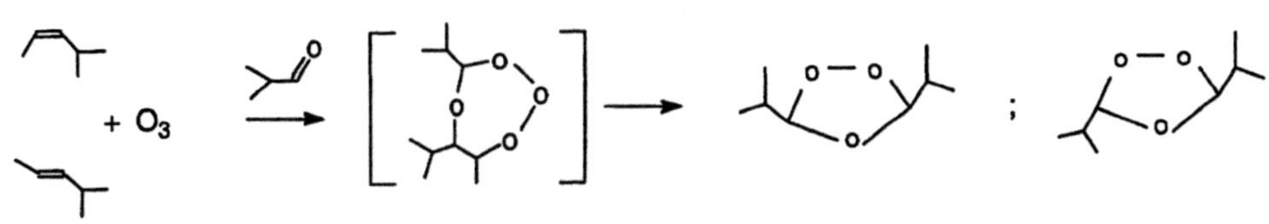

The mechanism pointed above includes the formation of seven-member ring intermediate with conformation responsible for the final ozonide conformation. Upon ozonolysis of *trans*-olefins the *cis:trans*-ozonides ratio should be close to 1, and higher than 1 upon *cis*-olefins ozonolysis. In case of hindered olefins, the formation of σ-complex rearranging further to ozonide, is considered as most probable.

This mechanism lost its importance very quickly as many new experimental results emerged which could be hardly explained by it. Thus the ozonolysis of *trans*-di-tert-butylethylene yields *trans*-ozonide [94] while according to the Story mechanism it should give *cis*-ozonides. Moreover, the Story mechanism could not explain the epoxides formation which are the major products in ozonolysis of highly hindered olefins. It was found that the foreign aldehyde with deuterated ^{18}O-atom is incorporated into the zwitterion in preference to the epoxy-orientation of the ozonide cycle [95-104]. This observation is in consonance with the classical concept of the Criegee mechanism whereas the Story mechansim supposes the preference of peroxy bridge orientation [93]

Bauld, Bailey et al. [94] suggested refinements for the Criegee mechanism in order to account for the strong and varied experimental facts on the ozone reaction with olefins. For this purpose, they accepted that the PO exists in C-C half-chair conformation; the zwitterion (carbonyloxide -CO) exists in *syn* and *anti* configurations depending on the geometry of the reagent and the experimental conditions; the OZ preferred conformation is C-O half-chair; the addition of ozone to C=C bond is 1,3-*cis*-stereoselective yielding five-member ring - 1,2,3-trioxolane; the concerted cleavage of PO gives zwitterion and carbonyl compounds; which via 1,3-bipolar cycloaddition results in ozonide formation in the cage, and in the volume, the products suggested above by Criegee; the reaction of the zwitterion and the carbonyl compound occurs stereoselectively depending on the zwitterion configuration and the preferred conformation of the end ozonide. Thus, Bauld and Bailey formulated the following stereochemical rules: 1) equatorial (*trans*) substituents in PO are preferentially converted into *anti*, and axial (*cis*) substituents into *syn* carbonyl oxide; 2) the equatorial substituent in the PO is incorporated into a carbonyl oxide zwitterion in preference to an axial substituent; 3) aldehydes interact preferentially with anti carbonyl oxides so as to orient bulky substituents into a,e-conformation, i.e. *cis* ozonide formation, and with *syn*-carbonyl oxides so as to orient bulky substituent into a,a or e,e-conformation (*trans*). Thus Bailey succeeded in explaining: the cis:trans ratios of normal and cross ozonides, particularly those obtained from olefins with bulky substituents; trans-olefins yield trans-ozonides and cis olefins - cis ones, predominantly.

These suggestions, however, are true only for olefins with bulky substituents such as, *iso*-C_3H_7 or *tert*-C_4H_9, but not for olefins with smaller substituents, i.e. CH_3, C_2H_5 [105-108]. On the basis of microwave spectra Kuczkowski et al., [104] have shown that ethylene, propylene and *trans*-2-buteneozonide have an oxygen-oxygen half-chair rather than carbon-oxygen chair conformation as suggested by the Bailey mechanism. This means that the e,e- and a,s-substituents are located in *trans*-position and a,e-substituents in *cis*-position [109-115]. The studies on the conformation of PO, OZ and zwitterions reveal that Bailey' s rule 2 is not always valid and should be refined. It was shown that during the decomposition of *cis*-2-butene PO the anti carbonyl oxide yields *cis*-ozonide and the *syn*-carbonyl oxide *trans* ozonides, regardless of the OZ conformation. The small substituents - CH_3, C_2H_5 and C_3H_7 in the PO of 1,2-disubstituted ethylenes are oriented in e,e-position and the bulky ones - *iso*- C_3H_7 and *tert*-C_4H_9 in a,a-position [116-121].

According to Ref. 104 the ozonolysis of olefins takes place as a *supra-supra*-cycloaddition of $4n$-$2n$-systems via a five-member ring formation. The cycloreversible decomposition of PO, depending on the configuration, results in: 1) *syn*-carbonyl oxide with a-orientation of the R substituent; 2) *anti*-carbonyl oxide in the case of e-orientation of the substituent. The stereochemistry of the olefins ozonolysis is shown in Table 3.1.

Table 3.1.

Stereochemical course of OZ formation decomposition of PO after the mechanism of Bauld-Bailey (before the slash) and Kuczkowski (after the slash)

Olefin	PO, C-C half chair/ O-envelope	COX	OZ, O-O-half chair/-	OZ
1-Alkene	*e/e*	*anti/anti*	*e,a/-*	*cis/cis*[3]
Trans[1]	*a, a/e*	*syn/syn*	*a, a/-*	*trans/trans*
Cis[2]	*e, a/e*	*anti/anti*	*e, a/-*	*cis/cis*
Cis[3]	*-/a*	*-/syn*	-	*-/trans*

Note: 1 - bulky substituents; 2- bulky substituents; 3 - small substituents.

Later, Criegee summarizes the available data and modified the mechanism of olefin ozonolysis taking account of the different conformations of the carbonyl oxide [117-119].

Murray, Hagen and Bailey have established that the *cis : trans*-ozonides ratios vary at: 1) addition of various complex-forming reagents like toluene, o-xylene, isodurene, 1-mesithyl-1-phenylmethane, hexamethylbenzene; 2) change of the heating rate during PO decomposition [120-122]:

Scheme 3.5

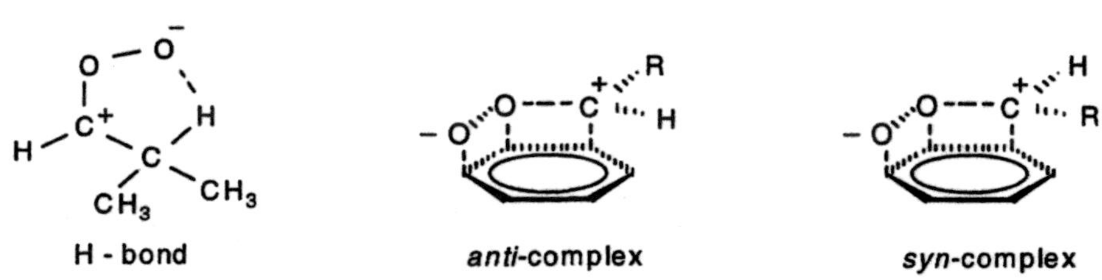

The more stable transition state of PO would give *anti*-carbonyl oxide predominantly, being dependent on the size of the substituents. Thus, *trans*-1,2-di-*iso*-propylethylene in its reaction with ozone in the presence of complexing agents, gives *cis* : *trans*-ozonides ratio >1. It has been found that this ratio remains the same in the absence of complexation agent and at slow heating from (-155°C) which, however, is against the stereochemical predictions. This fact could be explained by the occurrence of an equilibrium between the *syn*- and *anti*-isomers of CO, which takes place at high rate even at -155°C passing through a carbonyl oxide formation.

Scheme 3.6

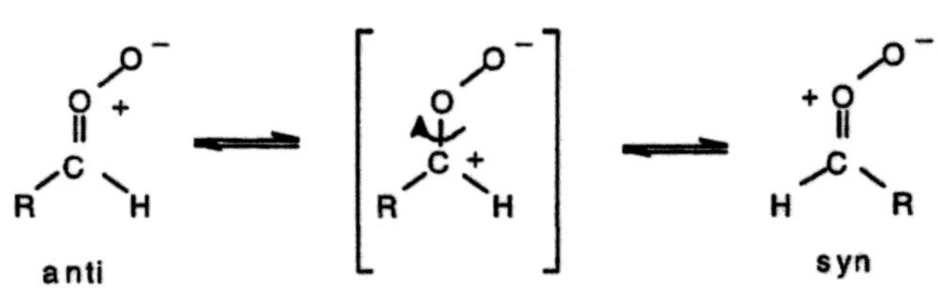

This hinders the complexation between the CO and the complexing agent. These results have been observed during the study on the low temperature ozonolysis of *cis*-olefins.

On the basis of conformational considerations, the *anti*-isomer of CO appears to be more stable than the *syn*-one although in some cases the latter form can be stabilized on account of H-bonds formation (Scheme 3.5). From the other hand, it can be assumed that CO preferentially leads to *anti*-CO-complex formation. Its stability will increase with increase of the substituents size and then the equilibrium *syn-anti* will be shifted to the latter. This will result in rise of the *cis*-ozonide yield as predicted by the third rule of Bailey [121].

On the basis of experimental and theoretical data [122-125], and applying the rule of least motion whereby the PO decomposition occurs via a minimum change of atoms coordinates, Bailey [122] has demonstrated that the CO conformation depends rather on the rate of PO decomposition than on the thermodynamic stability of the stereoisomers. On this ground he has proposed some additional refinements to the mechanism of olefins ozonolysis. The *cis-trans* ozonides ratio depends on: 1) the decomposition of the thermodynamically preferable PO conformer; 2) the decomposition of the kinetically favorable PO conformer; 3) the competition in regard to the second and probably the third rule of Bailey for olefins with small substituents and 4) the equilibrium of *syn-/anti-* CO conformers prior their recombination with the carbonyl moiety. The slow heating of *e,e*-conformer, which most likely has a lower activation energy of decomposition than the *a,a*-conformer, would produce preferentially *anti*-CO and the *cis*-ozonide in a smaller amount. In the case of rapid heating the decomposition would occur so quickly that it would not affect the equilibrium of the two conformers as it is shown below:

$$anti\text{-COX} \leftarrow e,e\text{-PO} \leftrightarrow a,a\text{-PO} \rightarrow syn\text{-COX}$$

In Ref. 121, 122 it is reported that rule 3 should be altered so as to conform to the O-O half-chair conformation of the end ozonide [103, 104, 125]:

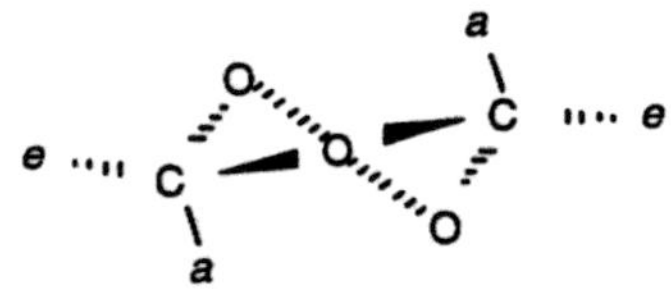

trans - a,a or e,e
cis - a,e

Thus the revised rule 3 should be: upon the interaction of the aldehydes with *anti*-CO the bulky substituents are oriented equatorially -axially (e,a) producing *cis*-ozonide, while aldehydes react with *syn*-CO to orient them diaxially (a,a) producing *trans*-ozonide. Bulky saubstituents in PO are oriented diaxially (a,a) resulting in *trans*-ozonide formation.

The mechanism of ozone reaction with olefins has been studied by means of thermochemical and quantum chemical methods in a series of contributions [126-131]. Harding and Goddard III [128] have found that the exothermity of the reaction between ozone and ethylene amounts to 92-96 kcal/mol while it is 101.7 kcal/mol according to Ref. 11. Goddard III states that the ozonolysis of *cis-trans* olefins takes place through a biradical pathway (Scheme 3.7):

Scheme 3.7

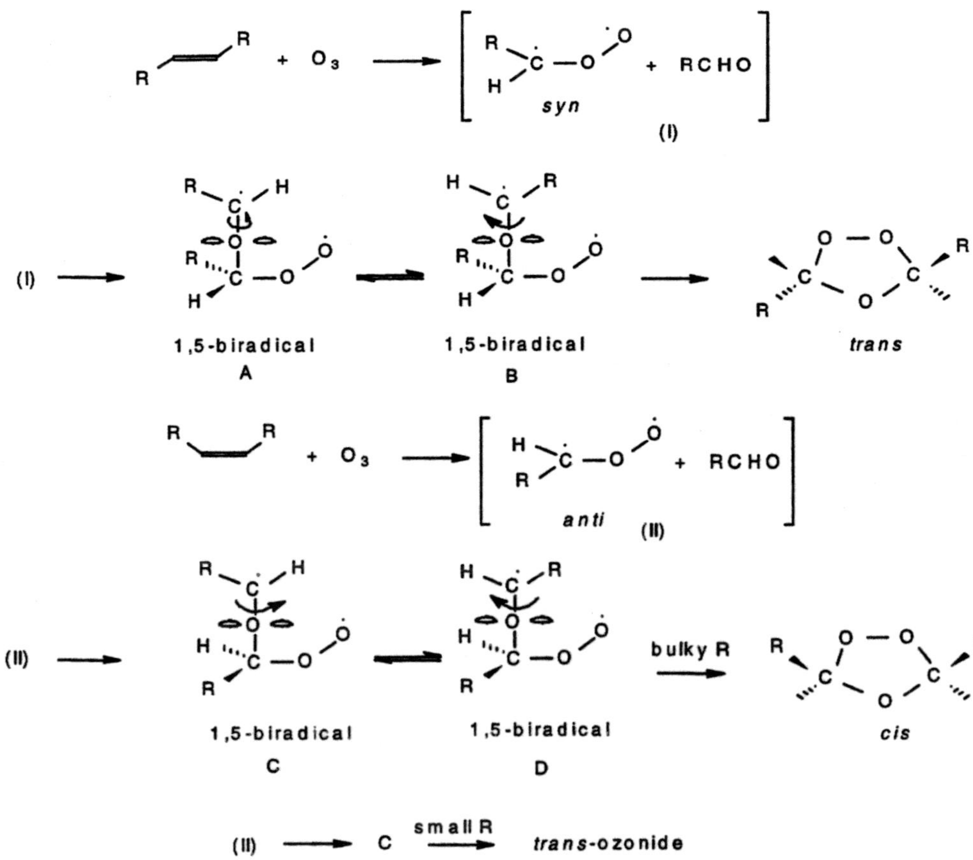

In the transition states A, B, C and D, the orientation of the carbonyl oxygen adjacent to the carbon atom of the peroxymethylene allows the stabilization of the lone electron pair on the oxygen *p*-orbital through exchange with the C-O bond. The arrows around the biradical C-O bond denote the most favorable orientations for rotation leading to cyclization (the bulky substituents should be kept away from their neighbours in their movement) . The orbital phase consideration predicts *supra*-surface addition whereas the rotation around the biradical bond should be clockwise in the transition states B and D, and counter-clockwise in A and C. The pathway of ozonides formation from olefins with bulky substituents *(tert*-butyl) is denoted by arrows. In case of olefins with small substituents, i.e. CH_3, the transition state C might be assumed to follow the orbital permissible suprasurface addition even when the other transition states (A, B and D) follow the steric directions. In the light of the biradical mechanism the authors interpret the Baileys rules in the following manner: 1) an equilibrium between the *syn*- and *anti*-peroxymethylenes (not carbonyloxides) in the absence of complexing agent, does not occur or take place to a very small extent, because of the high conversion barrier of 29 kcal/mol; 2) the complexing agents accelerate the rate of erquilibration which leads to increase of the *anti*-peroxymethylene concentration, respectively of the *cis*-ozonide, which is in contradiction with the data of Ref. 131; 3) at higher temperature the cyclization reaction is less stereoselective which would favor the formation of the more stable *trans*-product, fact which is experimentally confirmed; 4) the increase of the solvents polarity increases the lifetime of the 1,5-biradical species and thus reduces the stereoselectivity of the reaction.

On the basis of quantum chemical calculations Cremer]118-121] has proposed a revised rules for determination of the stereoselective pathways of olefins ozonolysis (Table 3.2).

Table 3.2.

Stereochemical courses of ozonides formation in dependence on the olefin conformation at ozone 1,3-cycloaddition and 1,3-cycloconversion of PO to CO at early (x<0.5) and later (x>0.5) transition state

Olefin	Size of R_1, & R_2	COX x<0.5	COX. x>0.5	OZ x<0.5	OZ x>0.5
cis-	small	*anti-*	*syn-*	*cis-*	*trans-*
trans	small	*anti-*	*syn-*	*cis-*	*trans-*
cis	bulk	*anti-*	*ànti-*	*cis-*	*cis-*
cis	bulk	*syn-*	*syn-*	*trans-*	*trans-*

It should be noted that the works discussed above and the conclusions drawn in them are based mainly on the analysis of the products obtained and on the activation energies calculations of the intermediate and stable products. However the kinetic methods could also provide essential evidence for the elucidation of the mechanism of this reaction.

3.1.2. Kinetics

The reaction of ozone with olefins usually proceeds with high rate and low activation energy at temperatures ranging from -100°C to 300°C [89, 134, 137].

3.1.2.1. Gas Phase

Cadle and Schadt were the first who have determined the values of *k* for the ozonolysis of ethylene and hexene-2 in gas phase [138]. Later the ozonolysis kinetics of various olefins has been studied by many investigators [137-148]. Most of them reported second-order kinetics of the reaction with *k* values (20°C) in the range of $0.8\text{-}450.10^3\ M^{-1}.s^{-1}$. (Table 3.3).

Table 3.3

Arrhenius parameters for the ozone reaction with olefins in gas phase

No.	Olefin	$k.10^{-3}$ (20°C), $M^{-1}.s^{-1}$	lg A	E_a, kcal/mol
1.	Ethene	1.6±0.2 1.6; 1.8; 0.8; 1	2.2; 6.2; 6.73; 5.5	4.2±0.4; 4.2; 2.6; 4.9
2.	Propene	7.1	5.46	3.9
3.	Butene-1	6.2; 3.9	6.2	1.7
4.	*trans*-Butene-2	29; 260; 13	5.55; 5.41; 6.93	0.2±0.3; 2.3; 1.1
5.	*cis*-Butene-2	200; 17	6.28	0.96
6.	*iso*-Butene	14±2	6.15; 6.28	2.8±0.4; 1.7
7.	2,3-Dimethylbutene	-	6.23	0.82
8.	Pentene-1	3.9; 3.2	-	-
9.	Hexene-1	4.5	-	-
10.	2-Methylbutene	450	-	-
11	Trimethylethene	12	-	-
12.	Cyclopentene	6.1; 4.6	-	-
13.	1,1-Dichloroethene	2.2	-	-
14.	1,2-Difluoroethene	0.24	-	-
15.	Tetrafluoroethene	81	-	-
16.	Hexafluoropropene	13	-	-
17.	Octafluorobutene	1.1	-	-

The various investigators have reported close values for k for the reaction of ozone with ethene, but the values of A differ more than four fold [138, 139].

There appears that the values of A should be similar if one takes into account the structure of *trans*-butene-2 and *iso*-butene.with lower E_a for *iso*-butene. In fact, the values of A and E_a differ essentially and the magnitude of E_a for *iso*-butene is higher (Table 3.3).

It has been found that the ozonides yield, the major products in liquid phase, is very low. The reaction mixture contains both lower and higher molecular compounds than the initial olefin, such as hydrocarbons, acids and aldehydes, etc. These products are obtained from the consequent reactions of the C=C bonds cleavage products [144, 145]. Compounds with mass numbers up to 5 times of order higher than the initial compounds have been identified by means of mass-spectrometry (for butenes with molecular mass greater than 200) [145]. It has been reported that epoxides were the major products of halogenated olefins ozonolysis [146].

The ozonolysis in gas phase is accompanied by chemiluminescence, fact, which has found practical application for the production of analytical equipment for ozone analysis in the atmosphere. The light emission is due to the electronic transfer of various excited species such as: H_2COX ($^1A''$) and OH ($X^2\pi_i$; $v' \leq 9$; $A^2\Sigma^+$).

In the reaction schemes outlined below the various products composition and the formation of the intermediate excited species are associated with the occurrence of radical steps. Due to the breakdown of the O-O bond of the primary ozone [144] (Scheme 3.8)

Scheme 3.8

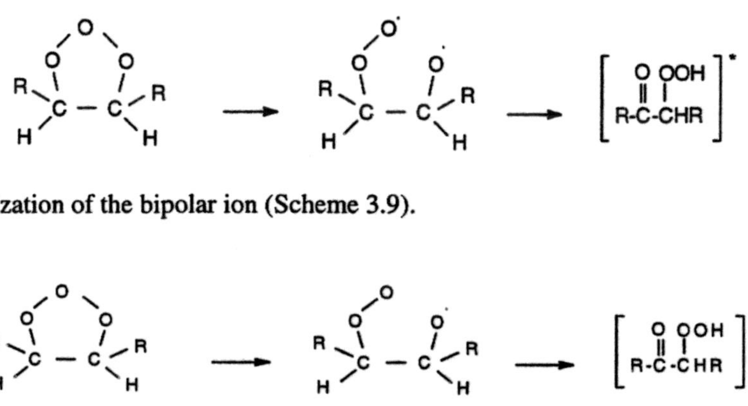

and further izomerization of the bipolar ion (Scheme 3.9).

Scheme 3.9

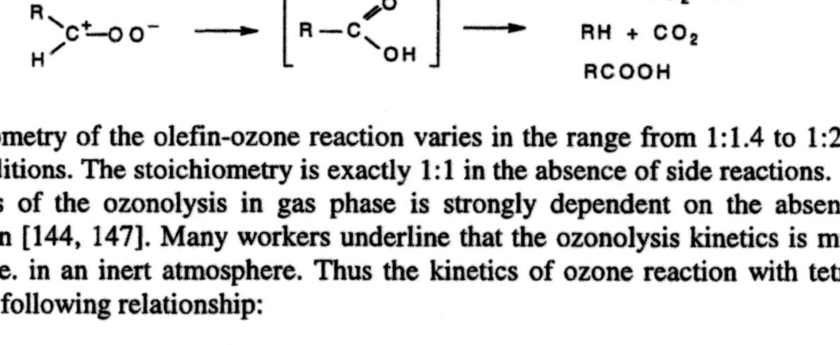

The stoichiometry of the olefin-ozone reaction varies in the range from 1:1.4 to 1:2 in dependence on the reaction conditions. The stoichiometry is exactly 1:1 in the absence of side reactions.

The kinetics of the ozonolysis in gas phase is strongly dependent on the absence or presence of molecular oxygen [144, 147]. Many workers underline that the ozonolysis kinetics is much more complex in its absence, i.e. in an inert atmosphere. Thus the kinetics of ozone reaction with tetrafluoroethylene is described by the following relationship:

$$d[O_3]/dt = k_1[O_3][C_2F_4] + k_2 [O_3]^2.[C_2F_4], \quad \text{(Eq.3.1)}$$

$$\lg k_1 = (8.2\ (0.5) - [(9.5\pm0.7)/2.3.RT] \quad \text{(Eq.3.2)}$$

$$\lg k_2 = (14.6\ (0.5) - [(10.1\pm0.6)/2.3.RT \quad \text{(Eq.3.3)}$$

and the reaction of ozone with allene [150]:

$$d[O_3]/dt = k_1[O_3][C_3H_4] + k_2 [C_3H_4].([O_3]^2/[O_3]_0) \quad \text{(Eq.3.4)}$$
$$\lg k_1 = (6.0 (0.7) - [(5.5\pm1)/2.3.RT] \quad \text{(Eq.3.5)}$$
$$\lg k_2 = (6.9(0.7) - [(6.2\pm0.8)/2.3.RT] \quad \text{(Eq.3.6)}$$

The foregoing discussion shows that the analysis of the kinetic data in gas phase requires particular attention. Thus one should bear in mind that probably some of the reported constants for gas phase ozonolysis are actually products, summations or partial quotients of parallel reactions constants.

3.1.2.2. Liquid Phase

The study of the kinetics of olefins ozonolysis in liquid phase appears to be a more complicated task than in gas phase because of the high reaction rate and more complex analysing equipment [136, 137]. The first studies in this respect were devoted to determination only of the relative rates in regard to a standard olefin-cyclohexene [153]. The first values of k were reported in 1969-1970. It has been found that olefins ozonolysis in liquid phase follows also second order kinetics [154, 155].

We have established that the outlet ozone concentration is proportional to the change of the olefin concentration during acrylic acid ozonolysis (Fig.3.1):

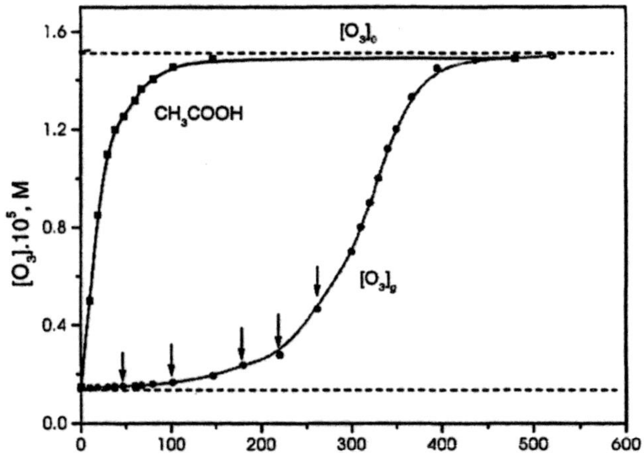

Fig.3.1. Kinetics of ozone concentration at reactor outlet upon ozonolysis of 0.02M acrylic acid in acetic acid medium at 200°C. The arrows denote the points of stoichiometry and rate constant estimation.

The outlet ozone concentration falls down abruptly after the start of the reaction and it remains low, close to zero, almost up to the complete olefin depletion (up to 4 min). Then it begins to rise rapidly and approaches the ozone concentration at the reactor inlet. The pattern of the kinetic curve depends on k it becomes rectangular at high values of k and at low values of k the ozone concentration in the stationary part rises, the time necessary for reaching the inlet concentration becomes longer and the curve form is changed dramatically. A 1:1 stoichiometry has been found. After the addition of 1 mol ozone to 1 mol olefin the reaction rate is reduced by a factor of about 6-8 . This fact suggests that the bipolar ion and the carbonyl compounds practically do not interact with ozone. The major part of the products (more than 95 %) are generated via one single reaction pathway at low and moderate temperatures. The values of A are found to amount to 10^6 $M^{-1}.s^{-1}$ which are typical for simple liquid phase reactions. The methods for kinetic studies are based on determination of the consumption rate of one of the reagent. In the absence of equilibrium intermediate steps, the rate of the first reaction step can be directly found as follows:

are based on determination of the consumption rate of one of the reagent. In the absence of equilibrium intermediate steps, the rate of the first reaction step can be directly found as follows:

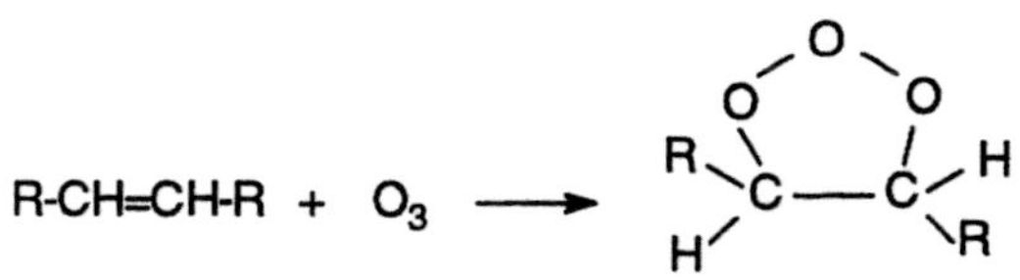

The absence of a reverse reaction is testified by the following two facts: 1) the primary ozonide formation is an exothermic process - 38 kcal/mol [134]; 2) the accumulation of significant amounts of PO at temperatures - 110-120°C. If then an inert gas is blown through the reaction mixture is liberated only the dissolved ozone. If a reverse reaction occurs:

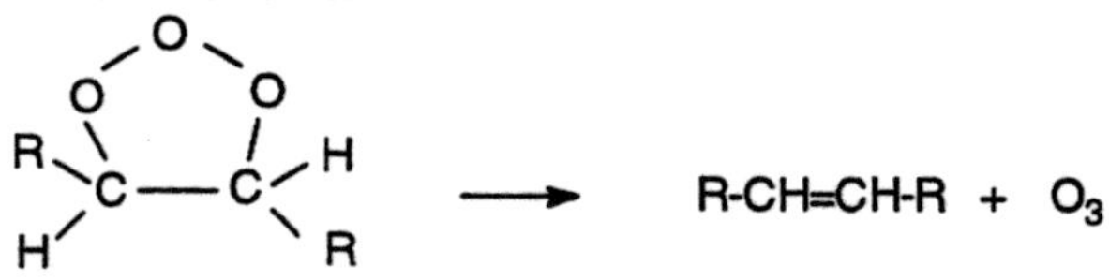

at blowing through the equilibrium will be shifted to the starting compounds and the ozone concentration will be equal to the whole amount absorbed by the system. In this case an olefin rather than PO should be identified. Upon repeatedly ozonation the olefin will absorb again an equimolar amount of ozone. The experiments, however, show that only PO is identified in the reaction mixture, practically no olefin is found and ozone is not consumed which evidences the absence of a reverse reaction.

The rate constants were measured: 1) through ozonation of a mixture composed of unknown olefin and a referense, for example cylcohexene [154] as discussed above; 2) through ozone bubbling through the olefin solution in an inert solvent (in regard to ozone) and registration of the change of outlet ozone concentration [155] and 3) through fast mixing of ozone and olefin solutions by means of stopped-flow techniques [156].

All these methods provide similar results for the *k* values but the bubbling methods appears to be more convenient and experimentally available one. We have applied successfully this method throughout our studies in all cases when it is possible.

The rate constants of ozonolysis of some olefins in liquid phase are presented in Table 3.4.

Table 3.4

Rate constants ($k.10^{-5}.M^{-1}.s^{-1}$) of olefins ozonolysis in CCl_4 at 20°C

No.	Olefin	[153]*	[155]	[156]	No.	Olefin	[153]	[154]	[155]
1.	Ethene	-	-	0.4	8.	Cyclopentene	4.5	2.0	4.0
2.	Butene-1	-	-	1.3	9.	Hexene-1	0.27	0.76	1.4
3.	*trans*-Butene-2	0.4-1.6	-	-	10.	Hexene-2	-	1.48	5.0
4.	*cis*-Butene-2	0.3-1.0	-	-	11.	Tertamethylethene	-	2.0	-
5.	Isobutene	0.2	0.97	-	12.	Styrene	0.30	1.0	3.0
6.	Pentene-1	2.4	-	5.0	13.	*p*-Methylstyrene	0.31	-	-
7.	2-Methyl-butene-2				14.	*m*-Methylstyrene			

Note - measured in $CHCl_3$ with respect to cyclohexene

The comparison of the data in Tables 3.3 and 3.4 reveals that, mostly, the rate constants measured in liquid phase are several times of order higher than those found in the gas phase which is in accordance with the collision theory [156].

The study of the rate dependence on the concentration of the reagents shows that in most cases it is first order with respect to ozone and slightly greater than 1 in regard to olefin. However, the kinetics approximates first order with respect to olefin at low or sufficiently low olefin concentrations. The increase of the concentration results in an increase of the reaction rate more than that predicted by the second order kinetics. This could be explained by the fact that the PO formation is preceded by the formation of ozone-olefin complex (Scheme 3.10):

Scheme 3.10.

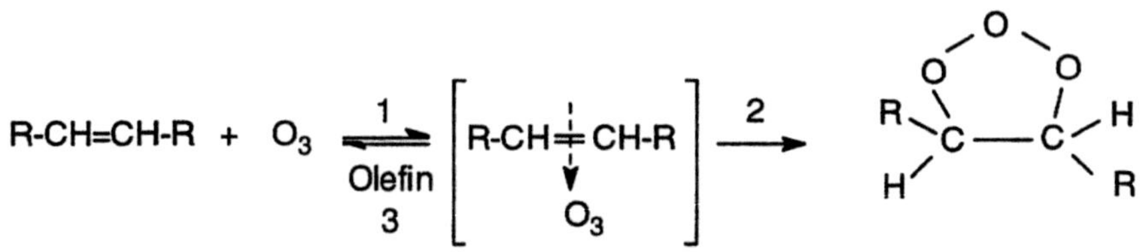

This complex is thermodynamically stable and at collision with a new olefin molecule is destroyed to the initial products. Then the dependence of the observed rate constant (k_{obs}) on the olefin will be described as follows:

$$k_{obs} = k_1.k_2/(k_2 + k_3.[\text{Olefin}]) \quad \text{(Eq.3.7)}$$

$$1/k_{obs} = 1/k_1 + k_3/(k_1.k_2).[\text{Olefin}]) \quad \text{(Eq.3.8)}$$

Upon ozonolysis of acrylic acid the values of k_1 and k_3/k_2 amount to $4.2.10^3$ $M^{-1}.s^{-1}$ and 8.10^6 $M^{-1}.s^{-1}$, respectively. It should be pointed out that these values are higher than those reported in Ref.154 as the olefin concentrations are lower at the points of rate constants determination (denoted by arrows). In this case the rate constant was calculated by applying the real olefin concentration which is equal to the difference between the initial and the consumed one up to the moment of calculation. Generally, this difference is within the limits of the experimental error and thus the reported constants could be used for kinetic characteristics of the process. However, when possible we have used further the values of k_1.

The temperature dependencies of the rate of hexene-1 and maleic anhydride (MA) ozonolysis, obtained by us, are presented in Table 3.5.

Table 3.5

Dependence of the rate constant k_1 of the ozone reaction with some olefins on temperature

Hexene-1	17^0C	0^0C	-20^0C	-40^0C	-60^0C	-78^0C	E, kcal
k_1, $M^{-1}.s^{-1}$	$1.4.10^5$	$1.5.10^5$	$1.2.10^5$	$0.7.10^5$	$1.0.10^5$	$0.6.10^5$	0.5±0.5
Maleic anhydride		43^0C	35^0C	26^0C	17^0C	9^0C	E, kcal
k_1, $M^{-1}.s^{-1}$		68.3	61.5	36.1	25.8	18.4	6.1±0.5

As seen the rate constant of hexene-1 does not change with temperature while the activation energy of maleic anhydride reaches the value of 6.1 kcal/mol. The dependence of the rate constant on the temperature for maleic anhydride, according to our data [32, 33], is described by the following expression:

$$k_1 = (1.1\pm0.2).10^6.\exp(-6100 \pm 500/RT) \quad \text{(Eq.3.9)}$$

The approximate value of A for hexene-2 is 10^6-10^7 $M^{-1}.s^{-1}$.

Generally the olefin ozonolysis is characterized by very low values of E_a= 1-10 kcal/mol. However the inaccurate determinations of the rate constants often does not allow the precise definition of the Arrhenius parameters which in its turn could lead to misleading interpretation of the experimental results. The values of E_a are commensurable with a number of other energy factors, influencing the reactivity, which impedes the evaluation of their contribution [157]. Among them are the reorientation of the solvent surrounding, the thermal energy of the molecules, the temperature dependence of the diffusion coefficients, etc. Consequently, the dependence of the rate constant on temperature can not be outlined since according to definition:

$$E_a = E_0 + 0.5.RT \quad \text{(Eq.3.10)},$$

where: E_o is the height of the potential surface at 0^oK.

The rate constants for olefins with various structures are given in Table 3.6.

Table 3.6

Rate constants of olefin ozonolysis in CCl_4 at 20°C.

No.	Olefin	$k_1.10^{-5}$, $M^{-1}.s^{-1}$	No.	Olefin	$k_1.10^{-5}$, $M^{-1}.s^{-1}$
1.	Ethene	0.4	24.	Metacrylic acid	0.078
2.	Butene-1	1.3	25	Methylmetacrylate	0.078
3.	cis-Butene-2	1.6	26.	Maleic acid	$1.42.10^{-3}$
4.	Isobutene	1.0	27.	Fumaric acid	0.084
5.	1,3-Butadiene	0.74	28.	Maleic anhydride	$3.6.10^{-4}$
6.	Pentene-1	1.4	29.	2,6-Dimethylhept-2-en-2-carbonic acid	0.026
7.	2-Me-Butene	5.0	30.	Oleic acid	10.0
8.	Hexene-1	1.4	31.	Methyloleate	10.0
9.	Hexene-2	5.0	32.	Vinylchloride*	0.022
10.	2-Me-Pentene-1	5.0	33.	Allylchloride*	0.085
11.	Octene-1	1.3	34.	cis-1,2-Dichloethane*	$3.57.10^{-4}$
12.	Octadecene-1	1.8	35.	Vinylidene chloride*	$2.11.10^{-4}$
13.	Cyclopentene	4.0	36.	Trichloroethene*	$3.6.10^{-5}$
				Tetrachloroethene*	$1.0.10^{-5}$
14.	Cyclohexene	4.0	37.	2-Bromopropene	0.028
15.	Cyclodecene	4.0	38.	Acrylonitryle	$1.0.10^{-3}$
16.	Tri-trans-cyclo-dodecatriene-1,5,9	3.5	39.	Tetracyanoethene	$0.4.10^{-5}$
17.	Norbornene	5.0	40.	1,1,2-Triphenyl-2,2-dinitroethene	$3.6.10^{-5}$
18.	Styrene	3.0	41.	1,1-Diphenyl-2,2-dinitroethene	$0.7.10^{-5}$
19.	Stylbene	1.8	42.	Polybutadiene	0.6
20.	1,4-Diphenybutadiene	0.9	43.	Squalene	7.5
21.	1,1,4,4,-Tetraphenyl butadiene	0.8	44.	Natural rubber	4.4
22.	Acenaphthylene	2.2	45.	Gutta-percha	1.7
23.	Acrylic acid	0.042	46.	Polychloroprene	0.042

Note- the values are preprinted from [154, 158].

The rate constants of the reaction of ozone with olefins with terminal double bonds (hexene-1, octene-1 and decene-1) are high and similar. The isomeric olefins with more internal double bonds (hexene-2 and methyl oleate) react much more readily that those pointed above. The rate of the ozone attack increases as the number of alkyl substituents on the double bond increases (hexene-1 and 2-methylpentene-2, acrylic acid and methacrylic acid, polybutadiene and polyisoprene). Cycloolefins C_5-C_{12} react with relatively similar rate constants but slower than those with open chain (cyclohexene $k_1=4.10^5$ and k_1 for methyl oleate $=1.10^6$ $M^{-1}.s^{-1}$). The presence of phenyl groups results in acceleration of the reaction (styrene) while the introduction of second, third and fourth ones (stilbene, 1,1,4,4-tetraphenylbutadiene) has no further influence on the reaction rate. The olefins ozonolysis can be regarded as an eletrophilic 1.3-addition and similarly to the reactions of Prilezhaev and Diels-Alder which has close electronic multiplicity, one can expect that electron-accepting substituents in the molecule should decrease the reaction rate. Actually, Br, Cl, -COOH and NO_2 substituted olefins react much more slowly with ozone [32-35, 158-160].

The comparison of the rate constants allows to disclose effects depending on the structure of the olefins under study. Thus, the values of *k* for highmolecular compounds (polybutadiene, polyisoprene and squalene) are lower than those found for their low molecular analogs (methyl oleate and tri-*trans*-cyclododecatriene and 2-methylpentene-2) [161]. It has been also found that the reactivity of *cis*- and *trans*-isomers (maleic and fumaric acids, cylcohexene and tri-*trans*-cyclododecatriene-1,5,9, gutta-percha and natural rubber) is various as the *trans*-isomers are more active in their reaction with ozone [132, 158, 159]. The rate constant of *cis*-diphenylethene is $8.9.10^4.M^{-1}.s^{-1}$ while for *trans*-diphenylethene it is $1.8.10^5$ $M^{-1}.s^{-1}$. Usually these facts are attributed to the occurrence of compensation effect (CEF). Huisgen [162], however, explained this fact by the repulsion of the *cis*-substituents in the transition state. The hybridization of the carbon atoms of the double bond changes from sp^2 to sp^3 in the acivated complex which results in overlapping of the van der Waals radii of the *cis*-substituents. Thus the rate of the ozone attack is retarded whereas the overlapping is impossible in the case of *trans*-isomers and thus the reaction is faster. Similar differences in the reactivity of *cis*- and *trans*-isomers have been also observed in Prilezhaev reactions.

The small sizes of the ozone molecule predetermine the insignificant role of steric effects in the reaction. Only one of the examples in Table 3.7 shows an anomaly in the rate which could be ascribed to steric influence.(*k* of acrylic, metacrylic and 2,6-dimethyl-hept-2-ene-2-carbocxylic acids are $2.3.10^3$, $7.8.10^3$ and $2.8.10^3$ $M^{-1}.s^{-1}$, respectively). If one takes into account only the inductive effects of the substituents, then the rate constant of 2,6-dimethyl-hept-2-2-carboxylic acid would be $7.8.10^3$ $M^{-1}.s^{-1}$. In fact the latter is 3 times greater than the experimental one which evidences the effect of steric factors on the reactivity.

The rate of olefin ozonolysis depends significantly on the olefin structure (*k* for oleic acid is sixfold higher than that for tetracyanoethylene) (Table 3.6). The quantitative relation between the electronic and geometric structure of olefins and their reactivity is a very important stage in studying the ozone chemistry. The first attempts in this respect were the studies of Cvetanovic et al. [154] who found a good linear relationship between lg of the rate constant and the ionization potential of chlorinated ethenes.

It has been found that the linear dependence of the reactivity on the free energy change in the system is valid for many organic reactions. The Hammett relationship is well known and widely applied [160]:

$$\lg k_i / k_0 = \rho.\sigma \qquad \text{(Eq.3.11)},$$

where: k_j are the rate constants of the compounds studied; k_0 - the rate constant of the reference; ρ - coefficient accounting for the reaction character; σ - inductive constants of the substituents.

Some other relationships, like that of Taft, have been also used for evaluation the contributions of inductive, steric, and mesomeric effects of various substituents on the rate constant.

A dependence of Hammett type has been found upon decomposition of PO and ozonides [161, 163]. It is quite natural that similar relationship might be observed during olefins ozonolysis.

The rate constant of the reaction of ozone with ethene, equal to $4.10^4.M^{-1}.s^{-1}$, is accepted as a standard in studying the kinetics of various olefins ozonolysis. It is calculated on the basis of the *k* value in gas phase multiplied by a coefficient of 25 which accounts for the heat of dissolution, 1.9 kcal/mol.

Table 3.7

Logarithms of the relative values of the rate constants and the Hammett inductive constants (σ_{para}) of the substituents on the double bonds

No.	Olefin	Substituent	$\Sigma\sigma_{para}$	lgk_i/k_0
1.	Methyloleate	2 C_nH_{2n+2}	-0.302	1.40
2.	2-Me-pentene-2	2 CH_3, C_2H_5	-0.491	1.097
3.	Hexene-2	CH_3, C_3H_7	-0.321	1.097
4.	Styrene	Ph	-0.01	0.875
5.	Stylbene	2Ph	-0.02	0.65
6.	Hexene-1	C_4H_9	-0.151	0.544
7.	Octadecene-1	$C_{16}H_{33}$	-0.151	0.65
8.	Ethene	H	0	0
9.	Methylmethacrylate	CH_3, $COOCH_3$	0.22	-0.7
10.	Methacrylate acid	CH_3, COOH	0.28	-0.71
11.	3-Chloropropene-1	CH_2Cl	0.18	-0.67
12.	2-Bromopropene	CH_3, Br	0.062	-1.15
13.	Acrylic acid	COOH	0.45	-0.98
14.	Acrylonitryle	CN	0.66	-2.6
15.	Maleic acid	2 COOH	0.9	-2.45
16.	Fumaric acid	2 COOH	0.9	-1.68
17.	1,1,2-Triphenyl-2-nitroethene	3 Ph, NO_2	0.748	-4.04

The Hammett relationship in Fig.3.2 is plotted on the bases of the data of Table 3.7. As seen the data fall on a straight line (Fig.3.2) thus indicating a linear dependence.

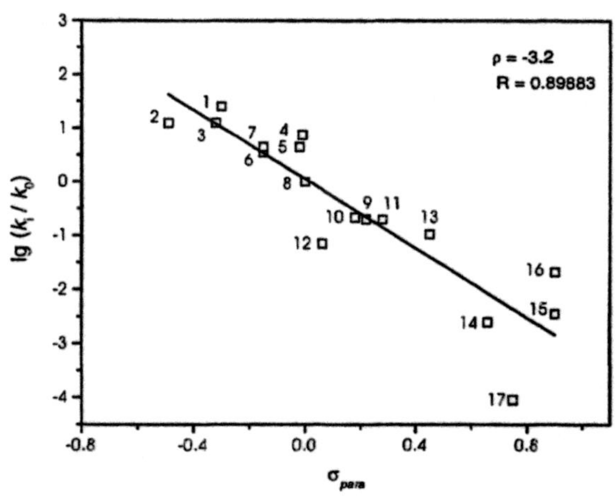

Fig.3.2 Hammett relationship of ozone reaction with olefins

The negative value of $\rho = -3.2$ indicates the eletrophilic nature of the ozone reaction with olefins. The low value of the correlation coefficient R=0.89883 could attain much more higher values if the steric effects of the substituents and the electrostatic interactions are considered. Regardless of this the dependence in Fig.3.2 is quite acceptable for evaluating the rate constants of reactions with new olefins.

Another very important question related to the olefins reactivity, which is not enough studied, is the relationship between the reactivity of C=C bonds and the deformation and steric factors responsible for the olefin conformation. These factors depend on the inter, intra, supra, micro- and macro effects. Thus the intramolecular strains arising in the cyclic olefins are due to the deformation of valent, dihedral angles, interatomic distances and transanular interactions [163-171]. The values of the different deformational and steric components responsible for cyclohexene conformation, as calculated by us, are presented in Fig.3.3.

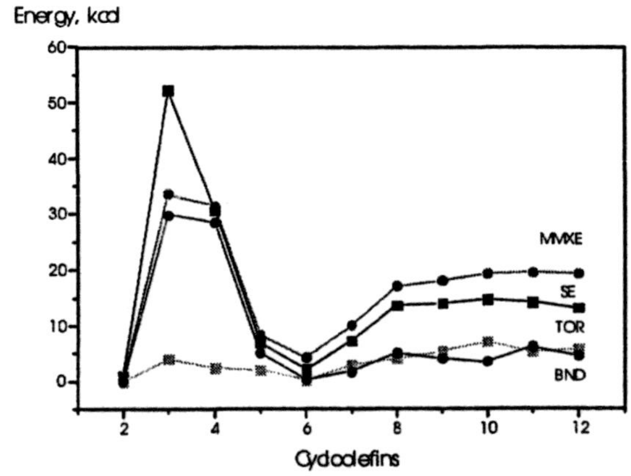

Fig.3.3 The values of C_2-C_{12} cylcoolefins energies calculated according to the molecular mechanics theory: MMXE - the Alinger energy; SE - the strain energy; TOR - the torsion energy and BND - the bond energy

It is seen that the dependencies have specific character with minimum in case of cyclohexene.

Among the olefins, given in Table 3.6, there are a number of olefins with highly strained molecules such as cyclododecene, cyclododecatriene, norbornene, ethylidenenorbornene, acenaphthylene and 2,6-dimethylhept-2-2-carbonic acid. It has been found that these olefins react with diazomethane in contrast to those with strain-free molecules. The reduction of strain in the sequence of C_4, C_5 and C_6 cycloolefins correlates with the reduction of the corresponding pyrazolines yield [171].

The reactivity and cycloolefins size relation is demonstrated in Fig.3.4. It is seen that cyclohexene, the olefin with the lowest strain energy of the initial molecule, exherts the lowest reactivity towards ozone. Simultaneously, its activated complex has the highest conformation energy. Thus E_a - the difference between the energy of the starting olefin and that of the AC, would be the highest which accounts for the lower reactivity of cyclohexene with respect to the other cyclolefins [1].

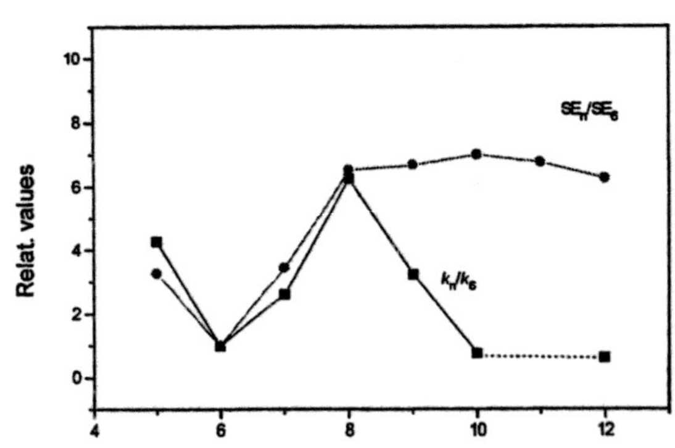

Fig.3.4 Dependencies of the relative rate constants and strain-energies of cyclolefins on the cycle size

Olefins may contain several double bonds separated by three simple C-C bonds. In these cases the C=C bonds react as independent kinetic units. The ozone addition to one of them practically does not affect the neighbouring ones. The substituent inductive effect is reduced by a factor about 2.7 through each C-C bond and thus its influence on the other carbon atoms is significantly reduced [171, 172].

Upon ozonolysis of compounds with conjugated bonds (1,3-diene, p-divinylbenzenes, etc. the different reactivity is determined by the structure of the starting diene, the nature of the olefin after the first ozone addition and the effect of the ozonide inductive factor on the second C=C bond. This is illustrated by the ozonolysis of divinylbenzene (Scheme 3.11).

Scheme 3.11

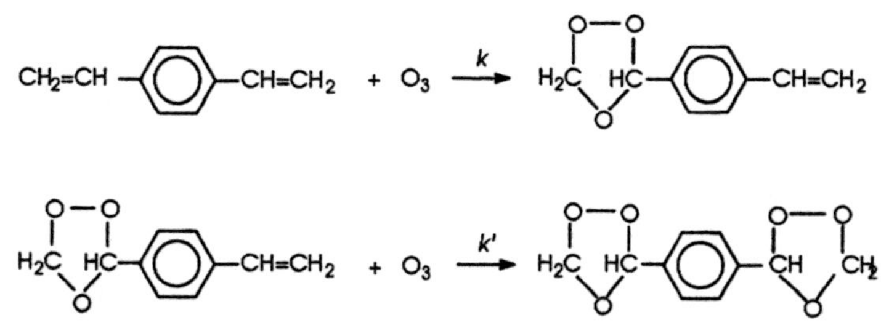

The proceeding of the reaction in two steps results in appearance of two stationary parts on the curve $[O_3]_g=f(\tau)$ thus allowing the determination of the two rate constants *k* and *k'* (Table 3.8).

Table 3.8

Rate constants of ozone reaction with dienes and trienes in CCl_4 at 20°C

Compounds	$k.10^{-5}$, $M^{-1}.s^{-1}$	$k'.10^{-5}$, $M^{-1}.s^{-1}$
Divinylbenzene	2.0	0.2
Ethylidenenorbornene	3.6	2.0
Vinylnorbornene	3.3	0.8
Vinylcyclohexene	1.4	0.8
Tetrahydroindene	1.4	1.4
Cyclododecatriene-1,5,9*	3.5	3.5 (3.5)

Note: * - the third double bond reacts with the same rate as the second one.

On the example of divinylbenzene the change in the rate constants after ozone addition could be followed. Its correlation with the general principles of the theory can be also evaluated. The rate constant of the ozone reaction with the first C=C bond should be slightly smaller than that of styrene because of the higher conjugation in divinylbenzene. After the ozone attack on the first C=C bond the conjugation degree is reduced and ozone interaction with the second C=C bond should take place with a rate similar to that of styrene ($3.10^5 M^{-1}.s^{-1}$). However, taking into account the electroacceptor effect of the ozonide cycle, the rate of the of the second step should be smaller than that of the first one, which has been also confirmed by the experimental data.

The comparison of *k* for ethylidene- and vinylnorbornene, allows to assume that ozone attacks first the double bond in the side chain snce the changes in its structure affect the value of *k* while the values of *k'* are equal for the two isomers. The C=C bonds in cyclododecatriene, 1,5,9 are equivalent due to the long distance between them and the equal configuration of the neighbouring environment. This is demonstrated by the equal rate constants for the three bonds and by the value of the hydrogenation rate of cyclododecatriene-1,5,9 on palladium [173].

3.1.2.2.1. Effect of Solvent

It is known that the rate of many reactions depends strongly on the solvent polarity [169, 170]. There are a number of examples whereby the reaction mechanism changes with solvent variation and the olefins ozonolysis is among them. It has been found that alkoxy- and aryloxy-hydroperoxides [174, 175] instead of ozonides, are the major ozonolysis products of olefin ozonolysis in solvents such as alcohols, acids and other compounds.

In order to assess the influence of the solvent on the rate constant we have followed the kinetics of acrylic acid ozonolysis in various solvents. The choice of acrylic acid was done mainly for two reasons: 1) its good solubility in the solvents used and 2) its relatively low rate of interaction with ozone which simplifies the registration of the reaction progress. The following solvents which were used for this study - CCl_4, decane, acetic acid, and water, differ essentially in regard to their dielectric constants (ε) and their ability to dissolve ozone. Decane and CCl_4 are inert towards the intermediate and end products but acetic acid and water react with the dipolar ion. We have found that the reaction obeys first order kinetics with respect to each reagent.

Table 3.10

Rate constants of acrylic acid ozonolysis in various solvents at 20°C

Solvent	Dielectric constant (ε)	Henry's constant (α)	$k.10^{-3}, M^{-1}.s^{-1}$
CCl_4	2.24	1.8	2.9
n-Decane	2.05	1.5	2.8
Acetic acid/H_2O, %			
99	6.2	1.70	2.6
84	18.2	1.26	2.6
67	30.9	0.95	2.6
50	43.6	0.77	2.6
16	69.0	0.65	2.8
100	80.2	0.42	2.7

The data in Table 3.10 show that the value of *k* does not depend significantly on the solvent. Thus two basic conclusions emerge: 1) the solvent does not affect the first step of the reaction or, if participates, it is in fact during some of the next reaction steps, as :

$$RHC^{+}OO^{-} + R'COXOH\ (RH(HOO)C\text{-}O(O)CR')$$
$$RHC^{+}OC(OO^{-})CHR + R'C(O)OH\ (RHC[O(O)CR']\text{-}O\text{-}C(OOH)HR$$

2) formation of polar compounds or polar transition states [179, 180] is not observed:

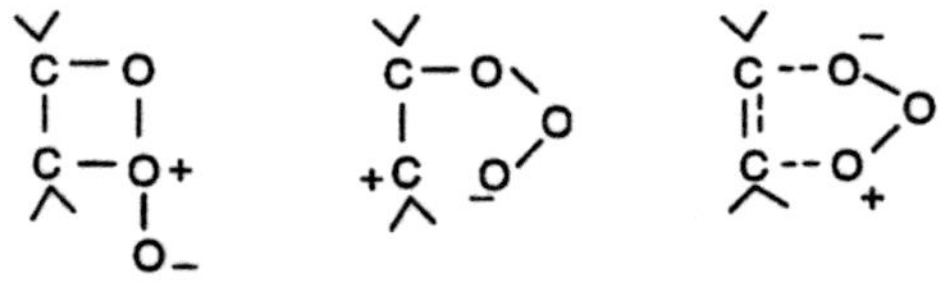

The polar intermediate compounds should solvate or react with "the active solvents". This should change the *k* value with the increase of the solvents polarity, and lead to the formation of reaction products with C-C bonds instead of C=C ones. However similar compounds have not been experimentally identified.

The constancy of *k* values in various solvents testifies that the reaction occurs without formation of kinetically active polar compounds due to the high rate of electron delocalization in the AC. The latter is much more slower than the relaxation time of the solvent shell (~10^{-13} s).

3.1.2.2.2. Effect of Configuration

We have studied the ozonolysis of structural derivatives of one and the same chemical compound: butenedicarbonic acid and its anhydride - maleic anhydride (I)-cyclic form, maleic acid (II) *cis*-form and fumaric acid (III) *trans*-form:

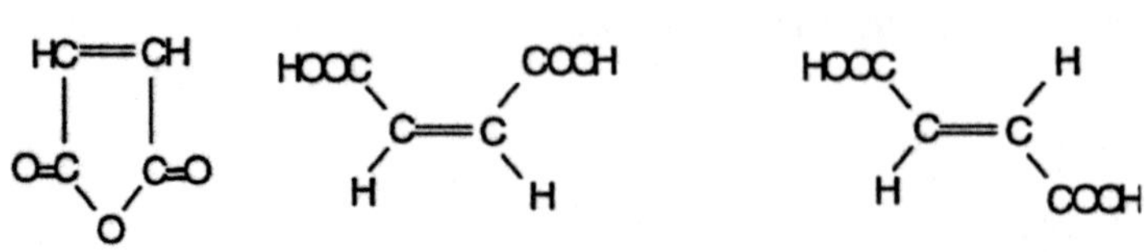

I II III

These compounds, similarly to acrylic acid, react relatively slow with ozone which makes them suitable objects for investigating the reaction kinetics [32, 33]. The concentrations of the reagents vary from 6.10^{-5} to 0.9 M. Water, glacial acetic acid and chloroform were used as solvent media. The products of ozone reaction with the double bond in the ethers of maleic and fumaric acids have been identified using chemical methods and IR spectroscopy . The low values of the rate constants suggest that at definite conditions ozone may also attack the C-H bond (1b) in addition to the basic interaction with the C=C bond (1a):

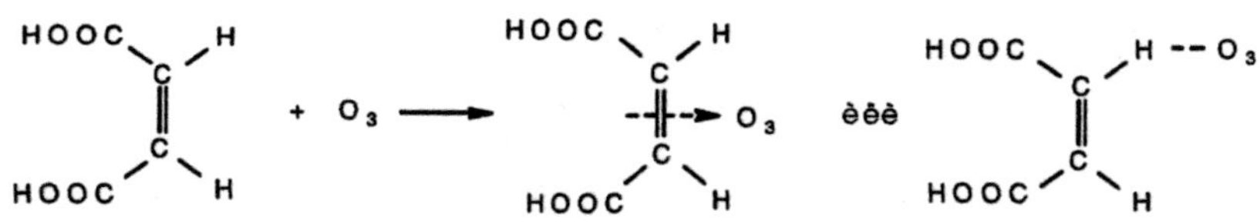

1a 1b

In this connection we have determined the reaction stoichiometry and analyzed the IR spectra of the solutions during the reaction course. The stoichiometry we found to be 1:1. The iodometrically estimated content of active oxygen in the samples corresponds to the amount of absorbed ozone. The progress of the reaction was accompanied by an increase in the intensity of the C-O bands in the range of 110-1200 cm^{-1} and narrowness of the C=O bands in the range of 1710-1850 cm^{-1}. It is seen (Fig.3.5) that the band at 1125 cm^{-1} that is assigned to the presence of ozonides appears even at low conversion degrees.

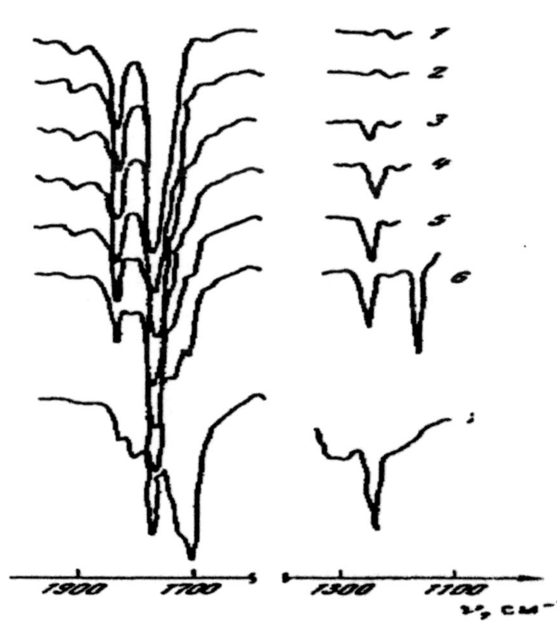

Fig.3.5 IR spectra of 0.3 M ozonized maleic anhydride in $CHCl_3$ versus the conversion degree in %. 1 - 0; 2 - 6; 3 - 12; 4 - 21; 5 - 27; 6 - 31; 7 - 100%.

However, at high conversion degrees the C=O bands are moved from their initial range due to the inductive effect of the ozonide cycle. With increase of the ozonation duration the bands of the C=C valent vibrations in the 1610-1680 cm^{-1} range decrease their intensity and an abrupt decrease of the rate of ozone uptake is observed. At complete consumption of C=C bonds the ozone concentration at the reactor outlet becomes equal to that at the reactor inlet. This fact suggests that ozone interacts mainly with the C=C bonds, while its consumption in parallel and consecutive steps is very low. This occurs regardless of the aldehyde formed in ozonolysis, that may also react with ozone. On the other hand the band intensity of the C-H and H-O valent vibrations at 3020 cm^{-1} and 3550 cm^{-1}, respectively, do not change significantly in the course of the ozonation. These experimental observations confirm the assumption that ozone does not attack the C-H bonds at the C=C bonds. This is also evidenced by the low *k* value for butandicarbonic acid (1.10^{-3} $M^{-1}.s^{-1}$) at room temperature. Briener et al. [181-184] has also confirmed the absence of ozone interaction with the hydrogen of the OH group. Upon ozonation of the aforementioned acids and their esters he obtained a similar products composition. Consequently ozone reacts predominantly with the C=C bonds of I, II and III according to pathway 1a.

It has been found that the rate of the reaction and the products compositions do not vary substantially with the solvents used, i.e. water ((= 80.1), acetic acid ((= 6.2) or chloroform ((= 10). However it should be noted that because of the lower solubility of fumaric acid in chloroform most of the comparative kinetic studies were conducted in acetic acid medium (Table 3.11).

We have established that at relatively low reagents concentration the rate of ozone absorption is not proportional to the reagent concentrations though the amount of absorbed ozone is equal to that of the consumed olefin. The reaction exhibits first order with respect to ozone and the dependence on the olefin concentration is more complicated. The rate constant decreases with the grow of olefin concentration. The reciprocal magnitude of the observed rate constant depends linearly on the olefin concentration (Fig.3.6).

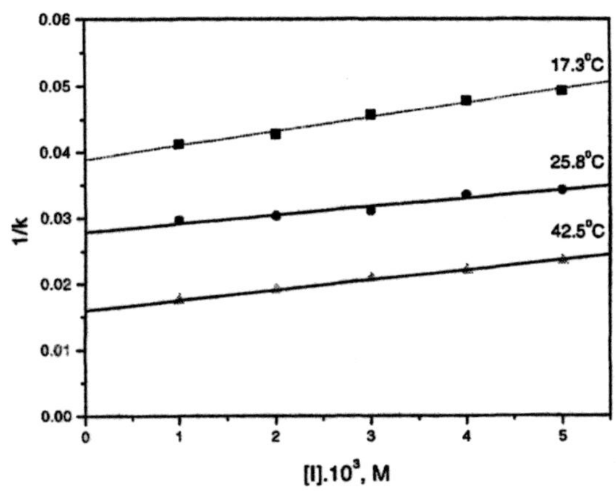

Fig.3.6. Dependence of the reciprocal value of the rate constant on the concentration of maleic anhydride at various temperatures in acetic acid and $\{O_3] = 5.10^{-6}$ M

These results indicate the occurrence of a side-nonefficient reaction, in which the olefin is consumed via a reaction with an intermediate product formed between the ozone and olefin:

$$\text{Olefin} + O_3 \rightarrow P_1 \quad (1)$$
$$P_1 \rightarrow \text{Products} \quad (2)$$
$$P_1 + \text{Olefin} \rightarrow 2\ \text{Olefin} + O_3 \quad (3)$$

On the basis of the kinetic analysis of this scheme an expression for the reciprocal value of the observed rate constant, similar to Eq.3.8, is derived:

$$1/k_{obs} = 1/k_1 + (k_1.k_3/k_2)\ [\text{Olefin}] \qquad \text{(Eq.3.13)}$$

Actually Fig.3.6 illustrates graphically equation 3.13.

The values of the rate constants as determined on the basis of Scheme 3.1.2 are given in Table 3.11.

Table 3.11

Kinetic parameters of the ozone reaction with maleic anhydride, maleic and fumaric acids in acetic acid at 2 ÷6 .10^{-6} M ozone concentration

Substrate	k_1, $M^{-1}.s^{-1}$	k_3/k_2, M^{-1}	lgA_1	E_1±0.5, kcal/mol
I	26(17.3^0); 36 (25.8^0); 51 (35.0^0); 64 (42.5^0)	63 (17.3^0)	6.30	6.5
II	114 (17.0^0); 142 (23.0^0); 162 (35.0^0); 232 (42.0^0)	4560 (17.0^0)	5.50	4.6
III	765 (17.0^0); 840 (22.0^0); 919 (28.5^0);1050 (42.5^0)	1000 (22.0^0)	5.13	3.0

The analysis of the data in Table 3.11 reveals two significant facts: 1) the high values of k_3/k_2 ratios and 2) the increase in the rate constant k_1 in the sequence: maleic anhydride (cyclic form)< maleic acid (*cis*-isomer)< fumaric acid (*trans*-isomer). The first observation discloses the high efficiency of reaction 3 and the second one accounts for the anomalous reactivity of the geometrical isomers. Thus it was expected that the *trans*-isomer would react faster that the *cis*-one. The high value of k_2 may be attributed to the specific character of P product which probably is an unstable π-complex and its decomposition to the initial products is facilitated by the olefin molecules.

We have also determined the kinetic parameters of the ozonolysis of maleic anhydride, maleic and fumaric acids at very low olefins concentrations whereby reaction 3 can be ignored. In these experiments the rate of the ozone reaction with I, II and II depends linearly on the concentration of the two reagents and the Arrhenius parameters differ from those determined at high olefins concentrations (Table 3.12). The reaction stoichiometry is 1:1 and the rate of ozonide formation is proportional to the rate of ozone uptake. The kinetics of ozonides formation was followed both iodometrically and spectrally by monitoring the IR band intensity at 1125 cm^{-1}.

The scheme of the reaction accounting for the observed experimental data is:

Scheme 3.13

$$>C=C< + O_3 \rightarrow \text{Products}$$

However the literature provides data that the olefins ozonolysis probably takes place via the formation of π-complex according to the following scheme:

Scheme 3.14

$$>C=C< + O_3 \rightarrow \pi\text{-Complex} \rightarrow \text{Products}$$

3.1.2.2.2.1. APPLICATION OF ERM

In order to propose the most probable mechanism of the ozone reaction with I, II and II we have applied the method based on comparison of the calculated values of the prerexponential A according to the activated complex method with different forms of the activated complex in the transition state with those obtained experimentally using two forms of TS:

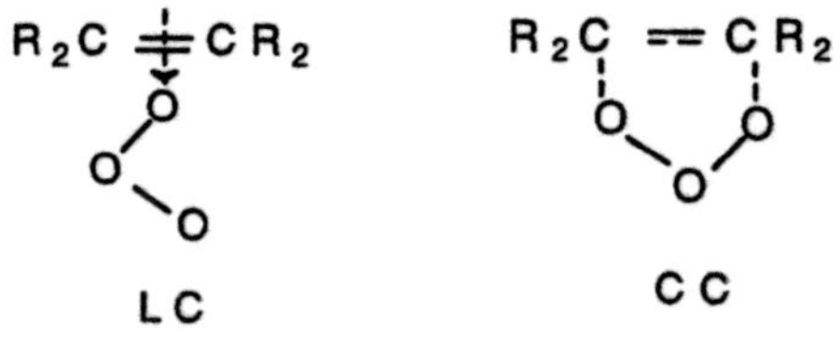

The calculated and experimental values for A are presented in Table 3.12.

Table 3.12

Arrhenius parameters for the ozonolysis of I, II and III

Compound	k_{obs} (20°C), $M^{-1}.s^{-1}$	lg A^{CC}	lg A^{LC} at E_{fr}= 0	lg A^{LC}, at real E_{fr}	lg A_{exp}	E_a, kcal/mol
Maleic anhydride	29.1	4.48	7.48	6.30 (1.60)	6.30	6.50
Maleic acid	128	4.42	7.42	5.50 (2.60)	5.50	4.60
Fumaric acid	810	4.33	7.33	5.13 (3.00)	5.13	3.00

It is seen that the values of A calculated with CC of the activated complex are lower that those with LC. The values of A calculated on the assumption that the energy of free rotation is zero are given in column 4 and those taking account of this energy are placed in column 5. Compared to $A_{exp.}$, the values of A-CC are lower by about 1-2 orders. The conclusion based on these facts is that the transition state can not have a cyclic form since these values of A are the highest possible predicted by the theory with this form of the activated complex. The values of A calculated with linear form of the activated complex are similar to those obtained experimentally. Therefore, it can be assumed that the reaction proceeds via the formation of a π-complex according to Scheme 3.14:

$$k_{obs} = k_1.k_2/(k_{-1} + k_2)$$

(Eq.3.14)

at $k_2 >> k_{-1}$ and $k_{obs} = k_1$

The assumption that $k_2 >> k_{-1}$ is supported by the good fit between the calculated value for A of the right reaction with that obtained experimentally for I, II and III ozonolysis. This is also confirmed by the absence of any additional bands in the IR-spectra of the ozonized solutions of I, II and II in methylene chloride registered for 48h at -80°C.

3.1.2.2.3. Effect of Structure

For detailed analysis of the effect of olefins chemical structure on their reactivity, we have synthesized a number of substituted cyanostilbenes:

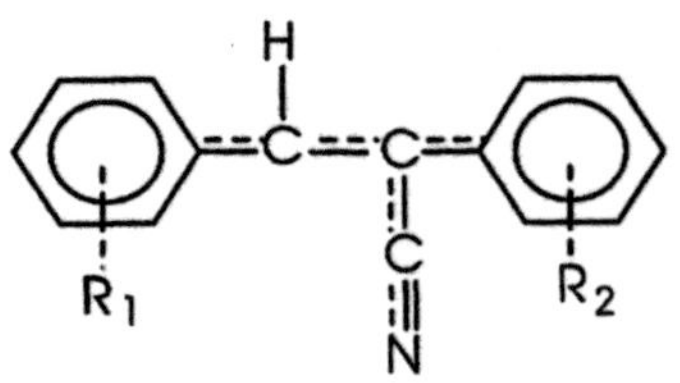

Earlier, these compounds have been investigated as phosphorescent and coloring agents [184, 185]. Their interaction with ozone, however, has not received considerable attention though their specific electronic structure provides unique possibility to assess the effect of such an important chemical factor on the reaction ability.

Some characteristics of the synthesized and investigated cyanostilbenes are presented in Table 3.13 [35, 36].

Table 3.13

Kinetic and physical constants of synthesized and studied cyanostilbenes in CCl_4 at 20°C

No.	Cyanostilbene	$k.10^{-5}$, $M^{-1}.s^{-1}$	M.p., ^0C	log k	Σσ
1.	2-Phenyl-3-phenyl	1.3		5.114	0
2.	2-Tolyl-3-(4-chlorophenyl)	1.3		5.114	0.687
3.	2-(4-Chlorophenyl)-3-tolyl	1.2		5.079	0.687
4.	2-(3,4-Dimethoxyphenyl)-3-nitrophenyl	-		-	0.625
5.	2-(4-Nitrophenyl)-3-(4-chlorophenyl)	0.6		4.778	1.005
6.	2-(4-Nitrophenyl)-3-(4-nitrophenyl)	-		-	1.556
7.	2-(4-Nitrophenyl)-3-(4-dimethylaminophenyl)	1.6		5.204	-0.052
8.	2-(4-Aminophenyl)-3-(4-methoxyphenyl)	3.0		5.477	-0.928
9.	2-(4-Nitrophenyl)-3-(4-methoxyphenyl)	1.1		5.041	0.510
10.	2-(4-Methoxyphenyl)-3-(4-methoxyphenyl)	2.6		5.415	-0.536
11.	2-(4-Aminophenyl)-3-(4-methoxyphenyl)	0.5		4.699	0.608
12.	2-(4-Nitrophenyl)-3-(4-ìåòèëphenyl)	2.4		5.380	-0.091
13.	2-(3,4-Dimethoxyphenyl)-3-(4-ôëóîðphenyl)	-		-	0.074
14.	2-(3,4-Dimethoxyphenyl)-3-(4-chlorophenyl)	1.7		5.230	-0.170
15.	2-(4-Cyanophenyl)-3-(4-dimethylaminophenyl)	5.9		5.771	-0.200
16.	2-Tolyl-3-(4-aminophenyl)	1.4	184	5.146	0.470
17.	2-(3,4-Dimethoxyphenyl)-3-(3,4-dibromphenyl)	-	107	-	-0.163
18.	2-(3,4-Dimethoxyphenyl)-3-(1-naphtyl)	-	-	-	0.768
19.	2-Phenyl-3-(4-dimethylaminophenyl)	-	-	-	-0.840
20.	2-(3,4-Dimethoxyphenyl)-3-piperonyl	-	145	-	-
21.	2-(3,4-Dimethoxyphenyl)-3-(3,4-dimethoxyphenyl)	-	-	-	-0.306
22.	2-(3,4-Dimethoxyphenyl) - 3- aminophenyl	-	85	-	-0.813
23.	2-(4-Nitrophenyl) - 3- phenyl	-	-	-	0.768
24.	2-(3,4-Dimethoxyphenyl) - 3- (2-methoxy-5-bromophenyl)	-	175	-	-0.156
25.	2-(3,4-Dimethoxyphenyl) - 3- (3-fluorophenyl)	-	-	-	0.184

Note: the constants are measured only for tetrachloromethane soluble compounds

The values of the melting points are given only for the newly synthesized compounds. The concentration of cyanostilbenes in the kinetic experiments amounts to 1.33 mM.

For determining the rate constants upon high reaction rate we have applied the absorption theory which is enhanced by the proceeding of a first order chemical reaction [34]:

$$R^0 = [O_3]^*(k_1.D)^{1/2},$$

(Eq.3.15)

where: R^o is the absorption rate at bubbling liquid in stationary conditions, mol/(cm^3); $[O_3]^*=\alpha.[O_3]_g$ - equilibrium ozone concentration on liquid-gas surface (mol/cm^3), where α=2 is Henry' s coefficient at 20°C; $D=7.8.10^{-8}(\kappa M_B)^{1/2}.T.(\mu_B.V_A^{0.6})^{-1}$ = $3.06.10^{-5}$ cm^2/s - the coefficient of ozone diffusion (A) in CCl_4 (B), calculated by the Wilkiey and Chang equation [186]; κ=1 - coefficient of association; M_B. is the molecular weight of CCl_4, g;; μ_B=1.2 cP - viscosity of the B; V_A=30 cm^3/g.mol- molecular volume and k_1 - the rate constant.

Equation 3.15 is true at high accelerations of the physical absorption resulting from the chemical reaction at $(D.k_1)^{1/2}/k_L>>1$. At known values for $D=3.06.10^{-5}$ cm^2/s, $k_L\approx 10^{-2}$ cm/s and $k_1=10^6.10^{-3}=10^3 s^{-1}$ the condition $(D.k_1)^{1/2}/k_L>>1$ is fulfilled since the left side of the inequality has a value of 15.

The ozone reaction with the cyanostilbenes is a bimolecular process as the rate of the reaction increases linearly with increasing the concentrations of the both reagents. Consequently ozonolysis is first order with respect to each of the reagents. The quantitative analysis of the band at 1152 cm^{-1}, which is responsible for the ozonide formation, reveals that its accumulation rate is almost equal to the rate of cyanostilbene consumption.

It has been found that 4,4'-di-nitro-cyanostilbene ozonolysis yields 4-nitro-benzenecyanate and 4-nitro-benzaldehyde as the major products, while the ozonide amount is very low. The scheme of the reaction is:

Scheme 3.15

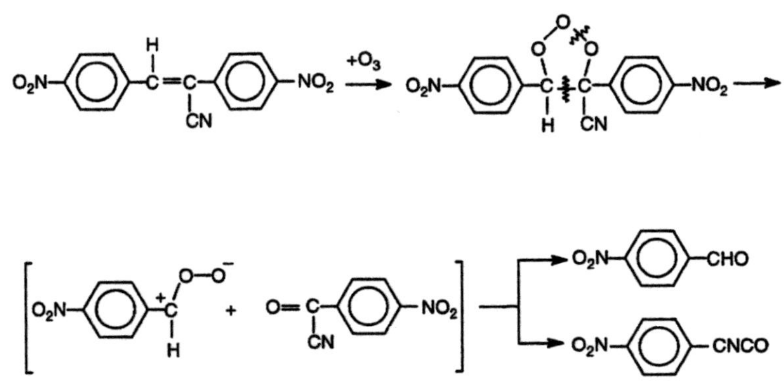

In view of the concepts on PO decomposition mechanism, with electron donating substituents, a zwitterion at the carbon atom with CN-group could not be formed due to its destabilizing effect. The conjugation between the carbonyl compound formed at this carbon atom and the CN and Ph-groups prevents the ozonides formation via zwitterion. For this reason the zwitterion leaves the cage and then it interacts with another zwitterion in the volume liberating oxygen. This reaction results from the strong electronaccepting action of the nitro groups in *para*-position and provides an example of anomalous ozonolysis.

Generally, the cyanostilbenes studied obey the Hammett's relationship and give a straight line with ρ= -0.4 (Fig.3.7). This value, though substantially lower than that found for substituents directly attached to the double bond, shows clearly the occurrence of conjugation leading to re-distribution of the substituents inductive effects.

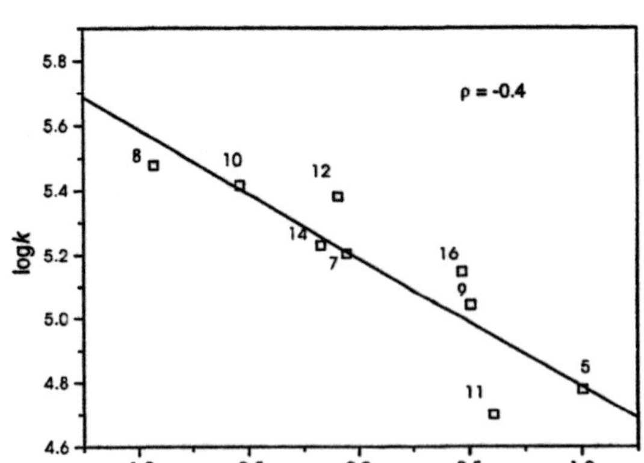

Fig.3.7 Dependence of the rate constants on the Hammett inductive constants

The small difference in the rates in dependence on the inductive constants can be attributed to the low A values, most probably, because of the low activation enthropy.

The cyanostilbenes studied have substituents in the both benzene rings (α and β). This could be regarded as the reason for the neutralization of the substituents effect, which contributes to the low value of ρ. The electronic structure of cyanostilbene is shown as:

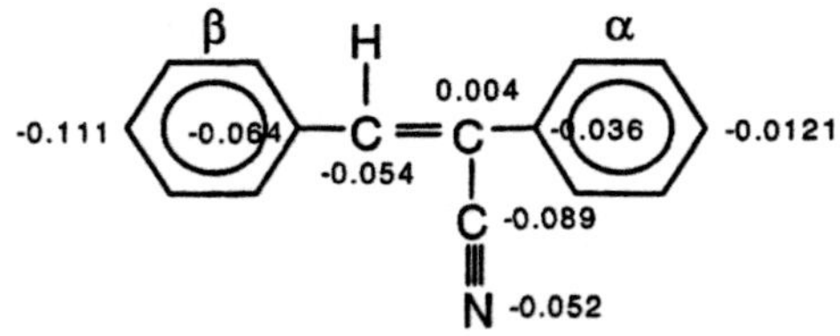

The values of the charges are determined by quantum-chemical calculations with AM1 Hamiltonian (*mopac6*).

Aiming at more accurate assessment of the substituents effect on the reaction ability we have synthesized 13 new compounds substituted only in α- and 14 in β-benzene ring The rate constants of the ozone reaction with all these compounds have been evaluated and are outlined in Table 3.14.

Table 3.14

Rate constants of cyanostilbene ozonolysis in CCl_4 at 20°C

No.	α-Substituents/σ	$k.10^{-5}$, $M^{-1}.s^{-1}$	No.	β-Substituents/σ	$k.10^{-5}$, $M^{-1}.s^{-1}$
1.	p-NH_2/-0.66	4.13	14.	p-Br/0.232	1.90
2.	m,p -CH_3O/-0.153	4.13	15.	p-Cl/0.227	2.09
3.	m-CH_3O/0.115	2.82	16.	p-NHCOXCH_3/0.00	2.14
4.	m-NO_2/0.710	4.36	17.	p-NO_2/0.778	1.39
5.	p-F/0.062	2.08	18.	m-NO_2/0.710	1.50
6.	p-CH_3/-0.170	3.06	19.	m-Cl/0.373	1.80
7.	p-Cl/0.227	2.08	20.	m-CH_3O/0.115	1.56
8.	p-CN/0.660	2.08	21.	m-Br/0.391	1.56
9.	Naphthyl/-0.01	1.49	22.	p-CH_3O/-0.268	2.12
10.	Pyrene/-0.01	0.56	23.	p-CH_3/-0170	2.60
11.	Biphenyl/-0.01	2.22	24.	Pyridine	2.01
12.	Stilbene	1.80	25.	Naphthyl	3.05
13.	Cyanostilbene	1.30	26.	m, p-CH_3O/-0.153	3.13
			27.	m-OC_2H_5,p-CH_3O/-0.258	3.08

The correlation of these constants with the Hammett constants (Fig.3.8 a, b) gives the following values of ρ = -0.266 and -0.286, for substitution in the α- and β-benzene ring, respectively. The correlation coefficients are not higher than 0.8.

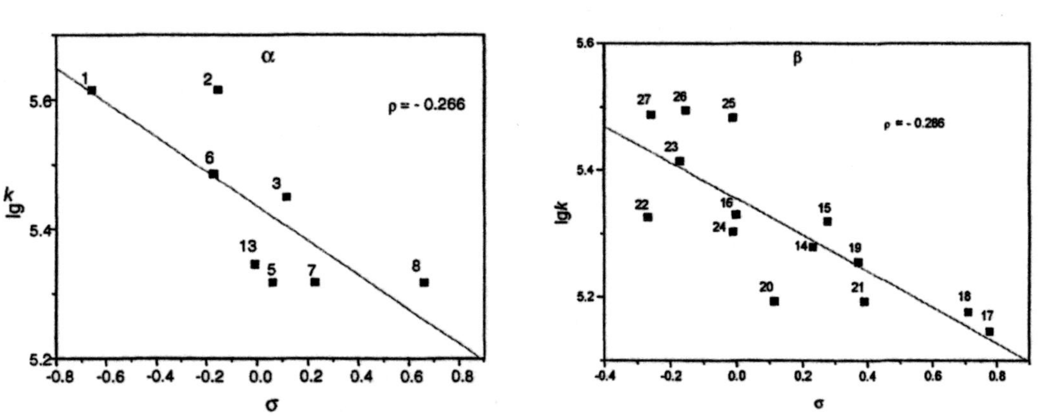

Fig.3.8 Dependence of the rate constants on the Hammett inductive constants

It is evident that the two coefficients (are almost equal since the difference of about 7 % is within the limits of the experimental error. Thus the reactivity actually does not depend on the substitution position. The similar slopes of the kinetic curves suggest that the role of the CN group in the re-distribution of the electron density and its effect on the reaction ability of these compounds can not be differentiated in this reaction.

3.2. POLYDIENES

The interest in the ozone reaction with polydienes is closely related the problem of ozone degradation of rubber products. The reaction kinetics is found to be strongly dependent on the polymeric state, a question which attracts a special interest in the recent years [187].

Florry [188] has shown that the reactivity of the functional groups in the polymer molecule does not depend on its length. It is also known that some reactions of the polymers proceed much slower as compared with their low molecular analogues (catalytic hydrogenation) At the same time many fermentative reactions with polymers take place rapidly while similar reactions with low molecular compounds do not occur at all [189]. The folded or unfolded form of the macromolecules provides various conditions for contact of the reagents with the reacting parts [190]. By using the modified version of this principle [191] it was possible to explain the proceeding of reactions without specific interactions between the adjacent C=C bonds and the absence of diffusion limitations. The study of the mass-molecular distribution (MMD) is in fact a very sensitive method for establishing the correlation between molecular weight (M.w.) and reactivity. The theory predicts that the properties of the system: polymer-solvent can be described by the parameter of globe swelling (γ) which defines the free energy (F) of the system and thus the rate constant of the reaction. For reversible reaction, i.e. polymerization - depolymerization, the dependence of the rate constant of the chain length grow on the molecular weight is expressed by the following equation:

$$\ln k_{pj}/k_{p\infty} = -\,\text{const}.(5\gamma - 3/\gamma).(d\gamma/dM).M_0 \qquad \text{(E.3.12)},$$

where: M_o is the molecular weight of the studied sample and $k_{p(}$is the rate constant for infinitely long macromolecules

A good correlation between the theoretical and experimental data for polystyrene solutions in benzene has been found in Ref. 192.

The study of the polymer degradation is complicated by their structural peculiarities on molecular and supramolecular level and diffusion reasons. It is difficult to find simple model reactions for clarification of particular properties and for the express examination of the proposed assuptions. An exception in this respect is the ozone reaction with C=C bonds whose mechanism has been intensively studied and could be successfully applied upon ozonolysis of polymeric materials [166]. This question will be discussed in detail at the end of this Chapter and in Chapter IV.

In Table 3.15 are summarized the rate constants of the ozone reaction with some conventional elastomers and polymers and their low molecular analogues, synthesized by us.

Table 3.15

Rate constants of ozone with polymers and low molecular analogues in CCl_4, 20°C

Compound	M.w.	$k.10^{-4}$, $M^{-1}.s^{-1}$
Polyvinylchloride	8.10^5	0.42±0.1
Vinylchloride	62.45	0.18
2-Bromopropene	121	0.28±0.05
Polybutadiene	$3.3.10^5$	6.0±1
Cyclododecatriene-1,5,9	162	35±10
Polybutadienestyrene	8.10^4	6±1
Gutta-percha	3.10^4	27±5

Natural rubber	1.10^6	44±10
2-Me-pentene-2	85	35±10
Squalene	410	74±15
Polystyrene	5.10^5	$0.3.10^{-4}$
Cumene	120	$0.6.10^{-4}$
Polyisobutylene	$1.7.10^5$	$0.02.10^{-4}$
Cyclohexane	84	$0.01.10^{-4}$

It is seen that the reactivity of elastomers and polymers and their corresponding low molecular analogues, as demonstrated by their rate constants, are quite similar, thus suggesting similar mechanisms of their reaction with ozone. This statement is also confirmed by: 1) the dependence of k on the inductive properties of substituents like as k of polychloroprene is higher than that of vinylchloride due to the presence of two donor substituents and 2) the dependence of k on the configuration of the C=C bond in *trans*-isomer (gutta-percha) and *cis*-isomer (natural rubber).

It has been found that the effects related to the change of the macromolecule length or the folding degree do not affect the ozonolysis in solution. Probably this is due to the fact that the reaction is carried out with thermodynamical elastomers solutions in which the macromolecules can do free intramolecular movements and do not react with adjacent macromolecules. Moreover, the rate of macromolecules reorganization is probably higher than the rate of their reaction with ozone since the experiment does not provide evidence for the effects of the change of the parameters pointed above [194, 195]

However, it should be noted that k of the elastomers are about 2-6 times lower that those of the low molecular analogues. The accuracy of E_a determination does not allow to to estimate on account of which of the two parameters A or E_a one should relate the decrease in k. If we assume that the mechanism of ozone reaction with monomers and elasomers is similar, i.e. the reactions are isokinetic, then $A_{mon}=A_{pol}$. At $k_{mon}/k_{pol} = 2 \div 6$ the difference in E_a at 20°C will be 0.5-1.0 kcal/mol. At the low experimental values of E_a, these differences will become commensurable with them and thus the determination of E_a is not enough accurate. In this case two assumptions could be made which can give a reasonable explanation for the lower values of k_{pol}.1) reorientation of the macromolecules is a slower process that that of olefins which would results in A_{pol} lower than A_{mon} and 2) the addition of ozone to C=C bonds is accompanied by the rehybridization of the C-atoms from sp^2-sp^3 and the movement of the polymer susbstituents during AC formation will be more restricted than those in olefins mainly because of their greater molecular mass and sizes. This will ultimately result in decrease of the rate constant.

In Table 3.15 are shown some examples of ozonolysis of saturated polymers - polystyrene and polyisobutylene. These reactions, as mentioned earlier in Chapter I, take place not via the mechanism of ozone reaction with the double bonds but through a hiden radical mechanism with rate constants of 4-5 orders lower.

3.2.1. Polybutadiene

Some industrial samples of rubbers have been the object of investigation (Table 3.16). The polymers were purified by tripple extraction with methanol in a Soxhlet extractor. The experiments were conducted in a bubbling reactor containing a polymer solution (0.06-1 %) in tetrachloromethane. The progress of the ozonation was followed viscosimetrically, gel-chromatographycally, IR and NMR spectrally, iodometrically and through triphenylphosphine reduction.

Table 3.16

Some characteristics of polydiene samples

Elastomer	Monomeric unit	Unsaturation degree, %	1,4-cis, %	1,4-trans, %	1,2-, %	3,4-, %	$\bar{I}_v.10^{-3}$	n
SCD	-CH=CH-	95-98	87-93	3-8	3-5	-	454	2.1
SKI-3S	$-C(CH_3)=CH-$	94-98	94-97	2-4	-	1-2	380	2.0
Denka M40	-C(Cl)=C-	94-98	5	94	-	-	180	1.8

Note: M_v is the average molecular weight, determined viscosimetrically after equation $[\eta]=k.M_v^{\alpha'}$, where $[\eta]=(\eta_1/C)(1+0.333\eta_1)$, $\eta_1=\eta_{rel}-1$, η_{rel} - is the intrinsic viscosity; C-solution concentration; k=1.4.10^{-4} - Staudinger constant and α' = 0.5.-1.5 - constant depending on the rubber type, being one for natural rubber; $M_v \approx M_w$; $n=M_w/M_n$, where M_w and M_n are the average weight and number average molecular mass, respectively.

Because of the high viscosity and large rate constants the reaction takes place in the diffusion and mixed region. In order to obtain correct kinetic data we have used the theory of boundary surface [34]:

$$[O_3] = \alpha[O_3]_0.\exp[-\delta(k.c.D)^{1/2}], \quad \text{(Eq.3.16)}$$

where: $[O_3]$ is ozone concentration at the boundary surface; α - Henry's coefficient; $[O_3]_0$ - ozone concentration in the gas phase at reactor inlet; δ - boundary surface depth; k - rate constant of the ozone reaction with diene; c - concentration of the monomeric units; D - diffusion coefficient of ozone in the liquid phase.

It was found that the relative viscosity decreases exponentially upon ozonation of SKD and SKI-3S solutions (Fig.3.9)

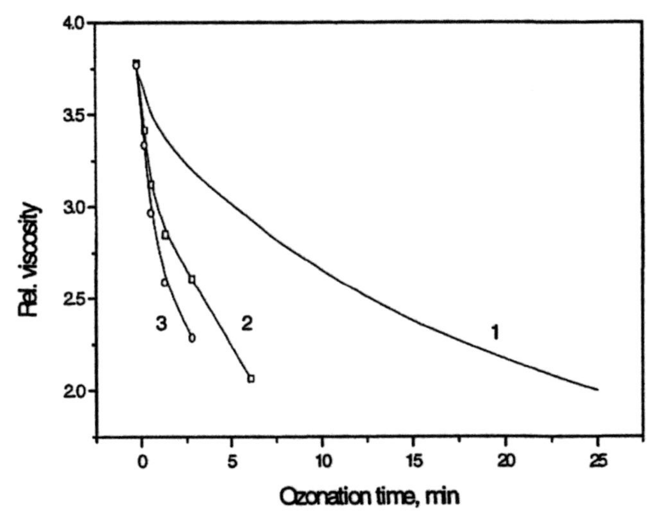

Fig.3.9. Dependence of the relative viscosity (η_{rel}) of SKD solutions (0.6 g in 100 ml CCl_4) on reaction time at ozone concentrations of : 1-1.10^{-5} M; 2 - 4.5.10^{-5} M; 3 - 8.25.10^{-5} M.

The dotted lines denote the theoretical exponential functions. As the viscosity is proportional to the molecular weight it follows that the polydiene consumption is described by first or pseudofirst order kinetics.

The value of ϕ, standing for the number of degraded polymeric molecules per one absorbed ozone molecule can be used to calculate the degradation intensity. The value of this parameter (ϕ) may be estimated using the following equation:

$$\phi = 0.5\,[(M_v^t)^{-1} - (M_v^0)^{-1}].P/G, \quad \text{(Eq.3.17)}$$

where: M_{vt} is the molecular weight at time T; M_{vo} - the initial molecular weight; P - the polymer amount; G - amount of consumed ozone.

The dependence between ϕ and G is a straight line for a given reactor and depends on the hydrodynamic conditions in the reactor.

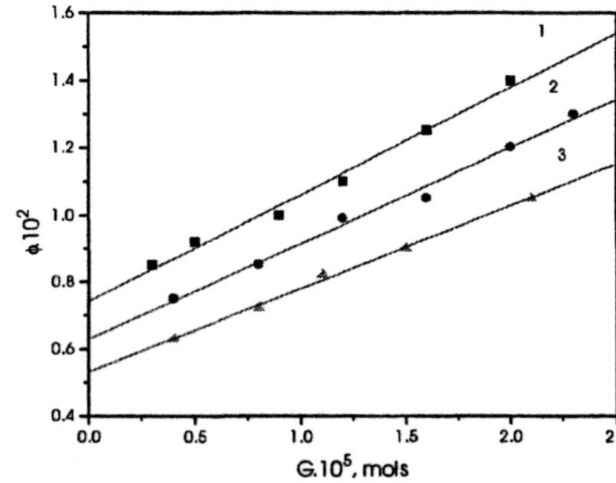

Fig.3.10 Dependence of (on G for SKD (0.6/100) at various ozone concentrations: 1-1.10^{-5} M; 2 - $4.5.10^{-5}$ M; 3 - $8.25.10^{-5}$ M.

It is seen from Fig.3.10 that the ϕ values increase linearly with the reaction time and decrease with increase in ozone concentration. The corresponding (values for SKI-3S and Denka M40 ozonolysis are similar. The ϕ values for G→0 were used to avoid the effect of hydrodynamic factors on them.

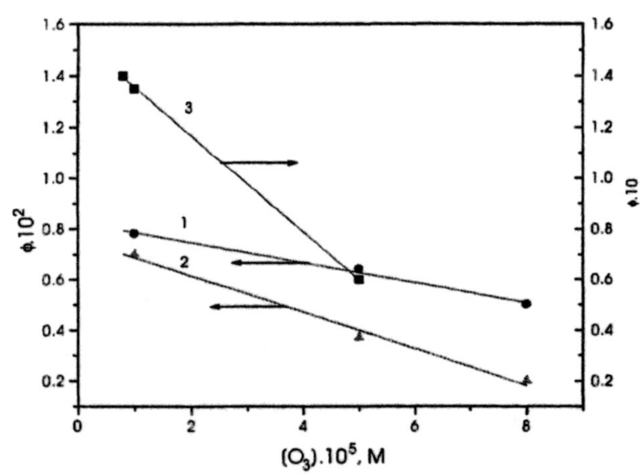

Fig.3.11. Dependence of ϕ on ozone concentration for elastomer solutions: 1 - SKD (0.6/100); 2 - SKI-3S (0.6/100); 3 - Denka M40 (1/100).

The values of (found for SKD, SKI-3S and Denka M40 at $[O_3] = 1.10^{-5}$ M amount to $0.7.10^{-2}$, $0.78.10^{-2}$ and 0.14, respectively and the slopes are : -40, -70 and 200 M^{-1} ,respectively.

Substituting with the known values for the parameters in equation 3.16 we have obtained δ within the range of 1.10^{-3} - 2.10^{-4} cm, which indicates that the reaction takes place in the volume around the bubbles, and hence in the diffusion region.

The ozonolysis of polydienes in solutions is described by the Criegee mechanism. The C=C bonds in the macromolecules are isolated since they are separated by three simple C-C bonds. According to the classical concepts, the C=C bonds configuration and the electronic properties of the groups bound to them affect also the polymer reactivity, similarly they do with the low molecular olefins. The only difference is that the polymer substituents at the C=C bonds are less mobile which influences the sp^2-sp^3 transition and the ozonides formation. In the first stage, when PO are formed, the lower mobility of the polymer substituents requires a higher transition energy, the rate being respectively lower as compared to that with low molecular olefins and the arising strain accelerates the PO decomposition to zwitterion and carbonyl compound. The lower mobility of the polymer parts impedes the further ozonide formation and brings about the zwitterion leaving the cage and going into the volume which in its turn accelerates the degradation process. The latter is associated with either its monomolecular decomposition or its interaction with low molecular components in the reaction mixture. The efficiency of degradation is determined by the C=C bonds location in the macromolecule, for example, at C=C bond location from the macromolecule center to its end, it is in the range from 2 to 1.

$$M_1 = (1/\gamma).M_0,$$

(Eq.3.18)

where: $M_2 = M_0 - M_1$; $1 \bullet \gamma \bullet 2$ - coefficient pointing the C=C-bond location; M_0, M_1 and M_2 - the molecular weights of the initial macromolecule and the two degradated polymer parts, respectively.

At γ=2, i.e. when the broken C=C bond is located in the macromolecule center, the values of M_1 and M_2 will be exactly equal to $M_o/2$, at $\gamma \rightarrow 1$, i.e at terminal C=C bond in the polymer chain, the value of M_1 will approximate to M_o and thus the value of M_2 will be practically insignificant. For example, M_2 may be 50-1000 which is of 3-4 order less than that of the macromolecule and in fact degradation process will not occur. The viscosimetric determination of the molecular weight which we have applied in our experiments is with accuracy of ±5 % and does not allow the differentiation of molecular weights of 22700, 19000 and 9000 for the corresponding rubbers. This suggests that the cleavage of C=C bonds located at a distance of 420, 280 and 100 units from the macromolecule end would not affect the measured molecular weight.

Since the reaction of elastomers ozonolysis proceeds in the diffusion and diffusion-kinetic region, at low conversions each new gas bubble in the reactor would react with a new volume of the solution. From the other hand, the reaction volume is a sum of the liquid layers surrounding each bubble. It is known that the depth of the penetration from the gas into the liquid phase is not proportional to the gas concentration and thus the rise of ozone concentration would increase the reaction volume to a considerably lower extent than the ozone concentration. This leads to the occurrence of the following process: intensive degradation processes take place in the microvolume around the bubble and one macromlecule can be degraded to many fragments while the macromolecules out of this volume, which is much greater, may be not changed at all. Consequently with increase in ozone concentration, one may expect a reduction of MMD and increase of the oligomeric phase content. This will result in apparent decrease of ϕ at the viscosimetric measurements. The discussion above make possible the correct interpretation of the data in Fig.3.11.

On ozonation of SKD solution (1/100) at 10 % conversion an intensive band of C-O in ozonides [195, 196] was observed at 1115 cm^{-1} in the IR spectrum. A less intensive band at 1735 cm^{-1}, characteristic of C=C vibrations in aldehydes and esters (Fig.3.12, 2), was also registered. Since the ozonides did not react quantitatively with HI their reaction with thriphenylphosphine (Ph_3P) was used for the determination [197]:

$$\text{Ozonide} + Ph_3P = Ph_3PO + 2\ \text{Aldehyde}$$

The amount of ozonide as estimated from the band intensity at 1735 cm^{-1} (Fig.3.12 3) was 70±5 % with respect to the amount of the absorbed ozone

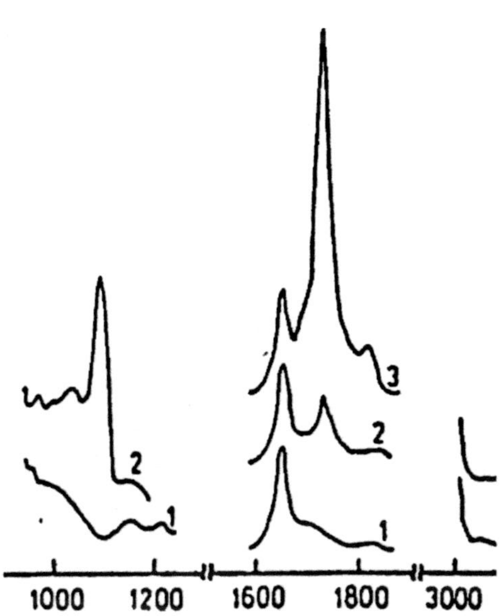

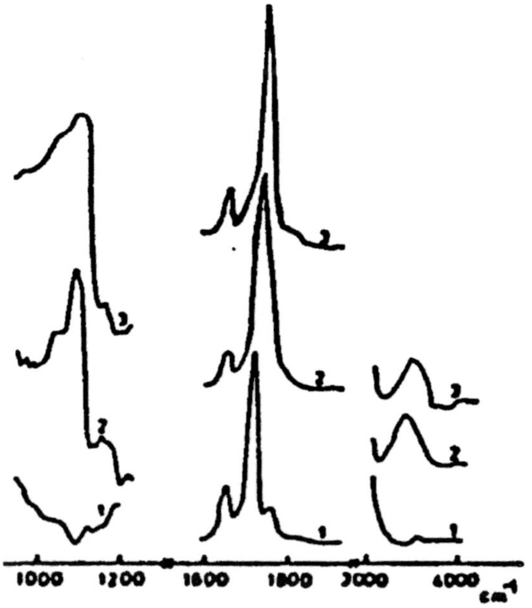

Fig.3.12. (left) Infra-red spectra of SKD solutions (1/100): 1 - non-ozonized; 2 - ozonized to 10 % conversion; 3 - solution after Ph_3P treatment.

Fig.3.13. (right) Infra-red spectra of SKD solutions (1/100): 1 - non-ozonized + butyric acid; 2 - ozonized to 20 % conversion; 3 - ozonized to 50 % conversion without butyric acid.

It has been found that the amount of C=O groups as estimated from the band intensity at 1735 cm^{-1} that appears prior the Ph_3P treatment, is to 11 % of the absorbed ozone (the extinction coefficient is taken from the calibration to >C=O in butyric aldehyde). These functional groups are formed as a result of the proceeding of reactions parallel with those of ozonide formation. Control experiments have demonstrated the relatively high thermal stability of SKD ozonide [198, 199]. Usually it is assumed that the zwitterion izomerization into acid is responsible for the degradation of macromolecules. However, absorption in the 1710-1716 cm^{-1} was not detected, although the extinction coefficient for the >C=O bonds in acids is two to three times higher than that of >C=O in aldehydes and ketones [196, 197]. The acid groups may react with the zwitterions, so that their concentration in solution remains below the threshold sensitivity of the spectroscopic method. This assumption is supported by the fact that butyric acid added to the SKD solution is consumed during the reaction (Fig.3.13).

The higher band intensity at 1735 cm^{-1}, obtained on ozonolysis in the presence of acid, as compared with that of the same band in its absence, reveals overlapping of the >C=O absorption in carbonyl and ester groups [197]. The interaction between the zwitterion and carboxylic group is also confirmed by the appearance of a band at 3480 cm^{-1} at 50 % conversion (Fig.3.13). The latter may be attributed to vibrations of the hydroperoxide OH-groups. Thus, it can be concluded that the decay of the excited acid molecule resulting from the zwitterion izomerization to free radicals [198] is not the only route for their consumption. Degradation may also result from the interactions between zwitterions, leading to di- and poly-peroxides which are very unstable and readily decompose into polymer carbonyl compounds with lower molecular weight and an oxygen molecule. It has been found that the most efficient way for reduction of the molecular weight is provided by the interaction of zwitterions formed in the chain center with low molecular compounds and oligomers.

The thermal decomposition data (DSC spectra) of ozonized rubber (SKD) in an inert atmosphere (Ar) in the temperature range of 323-453 K were processed using the equation suggested by Gorbachev et al. [199-201]:

$$\ln h/h_1/[\ln(1-\alpha)/(1-\alpha_s)] = \{(E_a/RT_s) . [(T-T_s)/T] . 1/[\ln(1-\alpha)/(1-\alpha_s)]\} + n \quad \text{(Eq.3.19)}$$

where: α is the conversion degree, H - the peak height, T - temperature, s - an index standing for the values of the DSC peak maximum, n - the reaction order,

Equations (Eq.2.2 -4) given in the previous Chapter and the expression of Julai and Grinhau [201]:

$$\alpha_s = 1 - 1.062n^{1/(1-n)} \qquad \text{(Eq.3.20)}$$

where: α_s -accounts for the transformation degree in the DSC maximum.

The results from the study on the thermal degradation of SKD and Diene 35 NFA rubbers are summarized in Table 3.17.

Table 3.17

DSC analysis of products obtained from ozonolysis of polybutadiene rubbers SKD and Diene 35 NFA

Heating rates, ^{0}C/min	1	2	5
ΔH_1, kJ/mol ozonide	274/216	338/260	355/276
ΔH_2, kJ/mol O_3	-/-	370/-	370/-
T, K	402.3/398.7	409.3/406.2	420/417
α_s	0.65/0.70	0.67/0.70	0.68/0.75
E, kJ/mol	99.9/91.2	105.5/93.7	121/109.6
n	0.7/0.73G60	0.73/0.73	0.79/0.73

3.2.2. *Cis*-1,4-Polyisoprene

The positive inductive effect of the methyl group in polyisoprene enhances the rate of ozone addition to the double bonds from 6.10^4 for SKD to 4.10^5 $M^{-1}.s^{-1}$ for SKI-3S. The asymmetry double bond allows the formation of two zwitterions: one as in butadiene (A) and the corresponding ketone and the other, in which one of the hydrogen atoms is substituted by a methyl group (B) and the corresponding aldehyde. Since the stabilization of the zwitterion depends on the electron donor properties of the substituents, the ratio of zwitterions A/B< 1 and is equal to -0.56. The interaction between zwitterions and aldehydes leads predominantly to ozonides formation while that with ketone is more difficult. though the conflicting literature opinions on the latter [202-207]. The band at 1725 cm^{-1} observed in the spectrum of the ozonized SKI-3S solutions (Fig.3.13) May be attributed to the absorption of the keto group in the ketone formed together with the zwitterion (A) [208]. By means of a calibration using methylethyl ketone, it was calculated that the amount of ketone is ca. 40 % of that of the absorbed ozone. Consequently, the higher degree of SKI-3S degradation compared with that of polybutadiene is due to impeded zwitterion A interaction with the corresponding ketone resulting in its insignificant conversions according to reactions leading to decrease of the molecular weight. The ozonides formation has been demonstrated by the occurrence of IR bands at 1110 and 1070 cm^{-1} (Fig.3.14, 2).

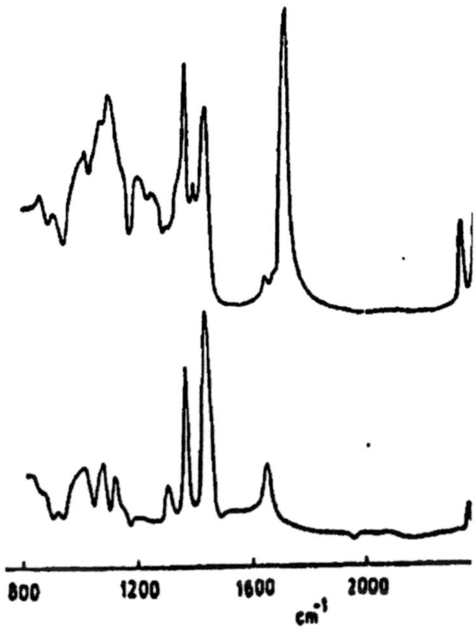

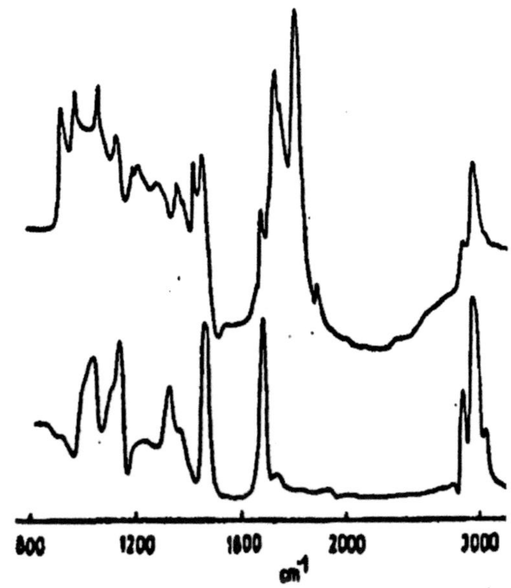

Fig.3.14 (left) Infra-red spectra of SKI-3S solutions: 1 - non-ozonized; 2 - ozonized to 50 % conversion.

Fig.3.15. (right) Infra-red spectra of Denka M40 solutions: 1 - non-ozonized; 2 - ozonized to 40 % conversion.

3.2.3. Polychloroprene

The electronaccepting properties of the chlorine atom at the polychloroprene double bond reduces the reactivity of Denka M40 as demonstrated by its relatively low rate constant, i.e. k=4.10^3 M^{-1}.s^{-1}. In this reaction the ratio between the zwitterion A and B, according to theoretical calculations, is in favour of A, the ratio being A/B = 4.55. The A formation is accompanied by choroanhydryde group formation and that of B with aldehyde one. In both cases the ozonides formation is insignificant and the zwitterions react predominantly in the volume resulting in enhancement of the degradation process. The intensive band detected at 1795 cm^{-1} in the IR spectrum of ozonized Denka M40 solutions (Fig.3.15) is characteristic of chloroanhydride group [208]. This fact correlated well with the conclusion about the direction of the primary ozonide decomposition. It also reveals insignificant conversion of chloroanhydride to monomer and polymer ozonides. It is difficult to ascribe the absorption maximum at 1735 cm^{-1} to a definite functional group because besides aldehydes and esters, chloroanhydrides also absorb in this region. However, the band at 955 cm^{-1} is typical of chloroanhydrides. Two other bands - at 1044 and 905 cm^{-1} may be attributed to the C-O vibrations. The valent vibrations characteristic of HO-groups are observed in the 3050-3500 cm^{-1}.

The iodometrical analysis of active oxygen in the ozonized Denka M40 solutions shows that the amount of ozonides is ca. 43 %. It is of interest to note that the HI reaction with ozonized polyisoprene solutions occurs quantitatively for 3-4 h, while in SKD the same reaction proceeds only to 20 % after 24 h. The above data, however, provide insufficient information for the preferable route of the ozonides formation (via dimerization, polymerization of zwitterions or secondary processes).

The DSC analysis of the products of Denka M40 ozonolyis together with those of SKD and cumeneperoxide as reference, are presented in Table 3.18.

Table 3.18

DSC analysis of Denka M40 ozonolysis

Parameter	Denka M40	SKD	Cumene peroxide
Convesion, %	28	18	-
Weight loss, %	7-13	6-11	-
ΔH_1,, kJ/mol O_3	239	337	-
ΔH_2,, kJ/mol	-	-	215
$E_à$, kJ/mol	70	128	140
T_i,, (^{0}C)	51	86	-
T_m, (^{0}C)	96	147	-
T_e, (^{0}C)	132	175	-

Note: ΔH_1' is the enthalpy related to one mol ozone; ΔH_2' - the enthalpy of cumeneperoxide, E_a - the decomposition activation energy; T_b, T_m and T_e - the temperatures at the beginning, maximum and end of DSC curves.

The data of Table 3.17 reveals that the chloroprene rubber ozonolyis yields polyperoxide as the enthalpy of its decomposition is found to be very close to that of dicumylperoxide. The higher value of E_a (ca. two times) testifies the possible formation of polymer peroxides. Moreover, the SKD ozonolysis leads predominantly to the formation of ozonide (89 % yield), its enthalpy being significantly higher than that of Denka M40.

3.2.4. Butadienenitrile Rubbers

The most important characteristics of the nitrile rubbers under study are shown in Table 3.19

Table 3.19.

Characteristics of the investigated butadiene-nitrile rubbers

Elastomer	Acrylonitrile, %	1,4 -trans, %	1,4-cis, %	1,2-, %	$M_v.10^{-3}$	M_w/M_n
SKN-18	20	64	21	15	380	3.0
SKN-28	28	65	24	11	300	3.0
SKN-40	39	72	22	6	230	2.7

The results from the NMR study of the nitrile rubbers structure are listed in Table 3.20.

Table 3.20

Distribution of the triad sequence of 1,4-butadiene (B) and acrylonitrile (A) using ^{1}H NMR data

Triads	SKN-18	SKN-26	SKN-40
BBB	73	56	33
BBA	9	14	18
ABA	18	30	49

It was shown that the viscosimetric measurements provide lower values for of the ozone degradation level as compared with those obtained by the IR-data. This observation may be attributted to the fact that the reaction between ozone and the rubber is localized in a very thin layer around the gas bubbles, while the major part of the macromolecules do not undergo degradation in the solution volume. The intensive progress of the reaction in a very small volume provides conditions for the formation of significant amounts of low molecular products without significant change of the molecular weight. Nevertheless, the viscosimetric determinations of a series of samples confirm some tendencies of the degradation process. The results from the viscosimetric studies of three nitrile rubbers are shown in Table 3.21.

Table 3.21

Number of chain scissions per molecule of reacted ozone (ϕ) for ozone degradation of nitrile rubbers at various ozone concentrations in $CCl_4/CHCl_3$ - solvents

Elastomer	Conc., g/100ml	$\phi.10^2$, 0.0025mM $[O_3]_0$	$\phi.10^2$, 0.005mM $[O_3]_0$	$\phi.10^2$, 0.01mM $[O_3]_0$
SKN-18	0.45	2.5/4.8	0.9/2.1	0.6/1.6
SKN-28	0.49	-/7.0	-/5.4	-/1.7
SKN-40	0.40	-/7.5	/6.1	-/2.5

From Table 3.21 one can see that ϕ depends on three factors: 1) the ozone concentration; 2) the nature of the solvent and 3) the rubber structure.

Under the experimental conditions of the bubbling method, the influence of ozone concentration is associated with the insignificant increase of the reaction volume and thus to the number of interactions of one macromolecule.

The role of the solvent is associated with::

- providing conditions for higher equilibrium ozone concentrations, due to the different values of Henry' s coefficients for different solvents (CCl_4 =2 and $CHCl_3$ = 2.8);
- creation of new conformation conditions for the macromolecules; effect on the electronic properties of the reaction center and the free access of ozone to it and thus on the degradation degree;
- participation in the ozonolysis process (which has not been experimentally registered during nitrile rubbers ozonolysis).

It is seen that the increase in the acrylonitrile content in the elastomers studied reduces the probability of the formation of low molecular fragments and consequently the degradation degree should increase. Indeed, such a dependence has been experimentally observed (columns 3-5).

It should be noted that during the ozonolysis of SKN-18 (for $CHCl_3$ solutions up to 25 %) gel formation is practically not obsedrved to 30 % conversion. Thus, it can be assumed that under the employed experimental conditions the contribution of crosslinking reactions (interaction between the zwitterions and aldehydes with highmolecular fragments or other structural reactions between macromolecules resulting in rise of the molecular weight) to the overall balance of reacted ozone is very small. It was shown that the degradation of macromolecules is a result of the zwitterions reaction with low molecular compounds in the bulk solution, monomolecular izomerization of the former to acids and di- and polyperoxides decomposition with oxygen evolution [209-211].

Several significant features were found in the ^1H NMR spectra of the ozonized CCl_4 solutions of SKN-18: a signal due to the aldehyde proton at 9.75 ppm; a few signals in the 4.97-5.20 ppm region]211, 212] ascribed to methine protons and a multiplet at 2.75 ppm with an integral intensity of about 40 % of that of the aldehyde peak. The ratio of the aldehyde to the ozonide groups calculated from the integral intensities is about 40:60.

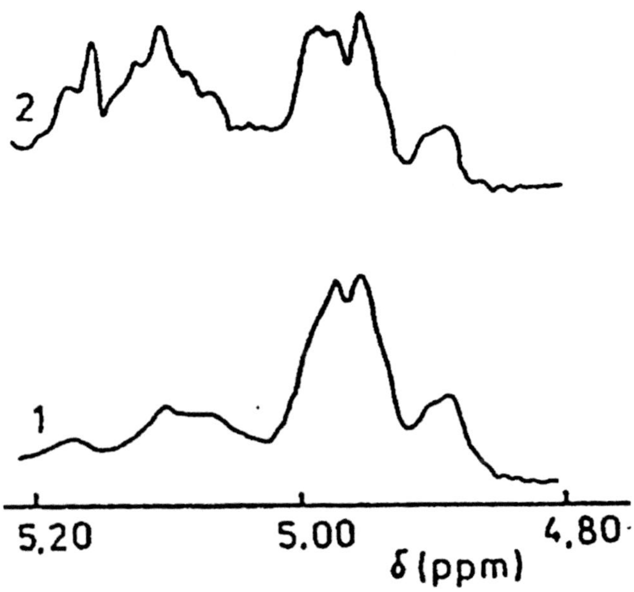

Fig.3.16. ^{1}H NMR spectra of SKN solutions (0.75/100); 1 - non-ozonized; 2 - ozonized to 20 % conversion.

As SKN-26 and SKN-40 are not dissolved in CCl_4, the ozonolysis and the ^{1}H NMR spectra were recorded in $CHCL_3$ and $CDCl_3$ solutions. The insoluble gel after ozonation was separated by filtration and $CDCl_3$ was added. The following peaks were detected in the ^{1}H NMR spectra: 9.76 ppm(s); 4.97-5.20 ppm (n.d.); 2.73 ppm (m) and a series of very weak peaks in the 3.12-4.20 ppm range.

Simultaneously the IR spectra (Fig.3.17) of nitrile rubbers before and after ozonation were registered.

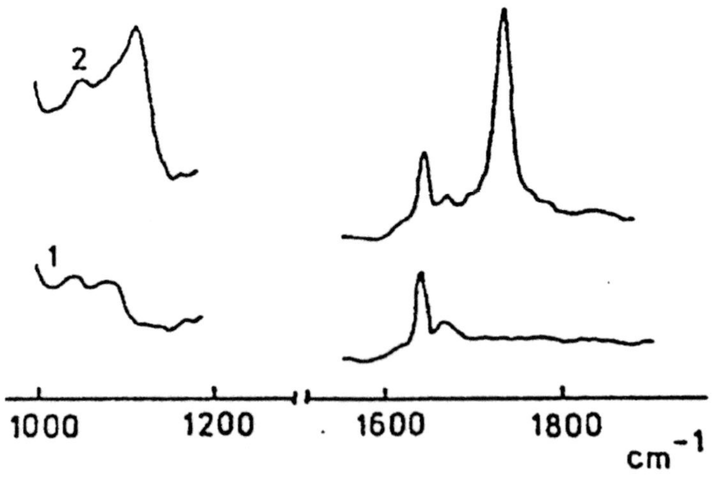

Fig.3.17. Infra-red spectra of SKN-18 solutions (0.75/100): 1 - non-ozonized solution; 2 - ozonized to 20 % conversion

The IR spectra of SKN-26 and SKN-40 are very similar to that of SKN-18 shown in Fig. 3.17. The kinetics of ozonides and aldehyde groups formation for the three rubbers was studied by monitoring the changes in the optical density at 1110 and 1731 cm^{-1}, respectively. The effect of the reacted ozone and the type of solvent, i.e. CCl_4 or $CHCl_3$ was investigated. It may be seen that in all cases the kinetic curves are linear up to 20 % conversion. The slopes of the dependence for the solution of DSN-18 depend on the solvent nature and it amounts to $2.4.10^{-2}$ M aldehyde/2.10^{-4} mol absorbed ozone in $CHCl_3$ and is 1.2 times higher than that in CCl_4. The registered optical density of the ozonides do not depend on the solvent but the slope of the curve is found to be 2 times lower than that of the aldehydes in $CHCl_3$. The slopes of the kinetic curve for aldehyde formation for the solutions of SKN-26 and SKN-40 ($CHCl_3$) are identical. However the accumulation of aldehydes takes place with 2.25 times higher rate. The corresponding slopes are about 8 % greater than those of SKN-18. Using literature values for the extinction coefficients for ozonides (((115 [211]) and aldehydes (ε(230 [211, 214] it was possible to evaluate the ratio aldehyde/ozonide for the three rubbers studied (SKN-18; SKN-26 and SKN-40) in $CHCl_3$ which is 1.03; 1.12 and 1.12 at ambient temperature and 20 % conversion. For SKN-18 in CCl_4 this ratio amounts to 0.85. The difference in the values of this ratio obtained in the two solvents is probably due to the stabilization of the zwitterion in the more polar solvent, i.e.chloroform, and the differences between the various rubbers is associated with the various amounts and size of the polymer fragments containing 1,2-double bonds in them. The role of nitrile groups is demonstrated by the electronic effects of the strong electroaccepting CN group and its inductive effect transferred via two σ-bonds.

The quantitative interpretation of the NMR spectra of SKN-18 ozonized in $CHCl_3$ gives the value of 0.82 for the aldehyde/ozonide ratio which is within the accuracy limits of the extinction coefficients determination and correlates with the IR data. It has been found that from one mol of absorbed ozone, 0.52 moles of aldehyde and 0.72 moles of ozonide are formed in SKN-18 ozonolysis in CCl_4. Thus the ratio amounts to 0.72.

The results from the two spectral studies support our suggestions for the important role of zwitterion reactions in the ozone degradation of nitrile rubbers. Their reactions out of the cage lead to the formation of di- and polyperoxides which are converted to aldehydes with evolution of oxygen.

The effect of temperature on aldehyde/ozonide ratio is illustrated by the kinetic parameters given in Table 3.22.

Table 3.22

Activation energy for the formation of aldehydes and ozonides in SKN-18 ozonolysis

Solvent	E_a, kJ/mol, ozonides	E_a, kJ/mol, aldehydes
CCl_4	-0.8	4
$CHCl_3$	-0.4	0.8

In the ozonolysis of nitrile rubbers the amount of aldehyde groups formed varies within the range of 50-70 %, while on ozonolysis of 1,4-cis-polyisoprene it is ca. 11 %. The significant difference could be explained by the presence of nitrile groups in the polymer molecules and the different in the configuration of the C=C bonds: 64-72 % *trans*- for the nitrile rubbers and 94-97 % *cis*- for the isoprene rubber. Literature data point out that the ozonide yield from *trans*-olefins is considerably lower than that from cis-olefins [211, 212].

3.2.5. Ethylenepropylene Rubbers

Ethylenepropylene rubbers (EPDM) are copolymers of 65-80 % ethylene with 35-20 % propylene and a third copolymer - diene 1-2 % (for example, cyclopentadiene, 1,4-hexadiene, 2-ethylene-norbornene-5). They are characterized by high resistance towards the action of temperature, oxygen and ozone, mainly because of the extremely low content of C=C bonds. The vulcanizing properties are due to the presence of the diene copolymer. In most cases the EPDM-rubbers are employed alone, and only in particular cases in combination with diene elastomers due to their low compatibility with them, for the preparation of efficient antizonant and antioxidant materials. Their application as antiozonant additives will be discussed further in detail.

The investigations were carried out on Keltan 778 samples (ethylene:propylene: ethylenenorbornene = 75.5:23:1.5 with M_v = 193 000 and M_v/M_n = 1.84) in CCl_4 (0.6/100).

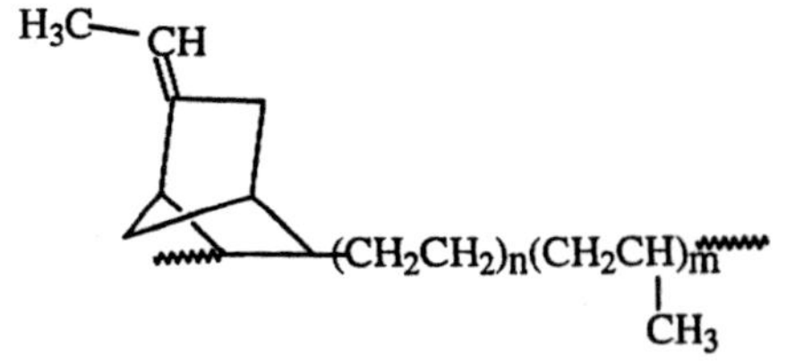

1

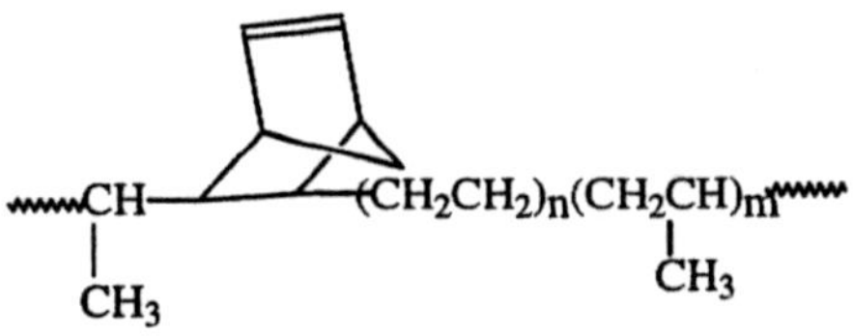

2

The diene copolymer undergoes polymerization both via the vinyl and norbornene double bond (1 and 2) in 1:4.66 ratio. These structures have been identified by means of ^1H NMR -250 MHz. The spectra analysis suggests the existence of two preferable conformations for the methyl group, during the polymerization via the norbornene C=C bond. The first one is when the methyl group is oriented towards the polymer chain and the second one when it is oriented in the opposite direction.

Upon ozonolysis of EPDM the reaction centers are the C=C bonds of the diene copolymer though their concentration is almost 70 times lower as compared with that of the C-H bonds, which is due to the substantially higher rate constants (up to 4-6 orders) [215]. Because of the relatively high viscosity of the solutions the rate constants of ozonolysis could not be determined accurately by using Eq. 1.4. The value of the activation energy estimated from the dependence of the viscosity change of the ozonized EPDM-rubber solutions on the temperature is found to be E_a=33 kJ/mol. Ozone attacks these double bonds resulting in the formation of two zwitterions in the volume. The interaction between the latter leads rather to structurization of the macromolecules and grow of the polymer molecular mass than to its reduction. At low temperatures this process is more efficient and thus the activation energy has a negative value. The presence of peroxide bridges is easily established. The treatment of the ozonized solution with PPh_3 gives PPh_3O and the molecular weight is restored.

The thermal decomposition of ozonized Keltan 778 solutions in an inert atmosphere makes possible the estimation of the Arrhenius parameters for the monomolecular decomposition of the cross-linked molecules. The activation energy was estimated as 19 kcal/mol which is in a very good agreement with other data on the decomposition of compounds containing peroxide bridges [216].

In order to determine the type of the bridges the ^1H NMR spectra of the initial rubber have been recorded. It reveals four doublets assigned to the four methine protons: 10 *cis*-vinylene- 5.08 ppm; 2) *trans*-vinylene - 5.30 ppm; 3) methine proton in norbornene, located in the direction of the polymer matrix - 5.58 ppm and 4) methine proton orientated in the opposite direction to the polymer chain - 5.70 ppm. The quantitative analysis show that *trans*-vinyl radicals (2) constitutes the major portion of the free radicals while the norbornene double bond is found to be the most reactive in the polymerization process due mainly to the highly strained hydrocarbon structure. The latter could not form ozonide at the double bond rupture but easily leads to polyperoxides formation.

REFERENCES

1. Razumovskii, SD, SK Rakovski, DM Shopov, GE Zaikov. *Ozone and Its Reactions with Organic Compounds (in Russian)*. Publ. House of Bulgarian Academy of Sciences, Sofia, 1983.
2. Bailey, PS. *Ozonattion in Organic Chemistry*. Acad. Press, New York, v.I, 1978, v.II, 1982, *Adv. Chem. Ser. - A*, (Ed. HH Wasserman), 39 -I, 39 - II, 1978, 1982.
3. Bailey, PS, *Chem. Rev*., 58, 925 (1958).
4. *Adv. Chem. Ser*., 21, (1959).
5. *Adv. Chem. Ser*., 77 (1968).
6. *Adv. Chem. Ser*., 112 (1972).
7. *Div. Petr. Chem., Preprint*., 16 (2) (1972).
8. Ishmuratov, GY, RY Kharisov, VN Odinokov, GA Tolstikov. The ozonolysis of unsaturated compounds in the synthesis of insect pheromones and juvenoids. *Uspekhi Khimii*, 64: 6, 580-608 (1995).
9. Menyailo, AT, MV Pospelov. *Uspehi Khimii*, 36, 662 (1967).
10. Bålew, J. *Oxidation*, 1, 259 (1969).

11. *Ozone Chemistry and Technology*. Murphy, J.S., J.R. Orr. (Ed.). The Franklin Inst. Press, 1975.
12. Manley, T., S. Niegowski. *Kirk-Othmer Encycl. Chem.Tech.*, 2nd ed., 14, 410 (1967).
13. Nebel, C. Ozone. *Kirk Othmer Encyclopedia of Chemical Technolog.* (3-rd ed.), 16, 650-713 (1981).
14. Horwath, M, L Vitzky, G Huttner. *Az Ozone*. Budapest, Academy Kiadoi, 1985.
15. Vol'nov, IM. *Peroxides, superperoxides and Ozonides of Alkali Metals.* Moscow, Nauka, 1964.
16. Gushchin, GR. *Investigation of Ozone in Atmosphere. Leningrad, Gidrometizdat,* 1963.
17. Andrews,.H.D. Barnard, A.N. Gent. *Ozone attack on rubbers.* L. Bateman (Ed.), J.W., N.Y., 1963.
18. Dormagen, DB, Ozonolyse von ungesattigten Kautschuken und Vilkanisaten ein Uberblick. *Kautschuk + Gummi Kunststoffe*, 42, 760-769 (1989).
19. Bently, K.W. *Technique of organic chemistry*, XI, part 2, N.Y., Int. Publ. 1963.
20. Logan, R., A. Maggiolo. *Ind. Fatty Acids and their Applications.* S. Pattison (Ed.). Reinhold Publ.Corp., 1959, 41.
21. Razumovskii, SD, GE Zaikov. *Uspehi Khimii*, 49, 2344 (1980).
22. Odinokov, VN, GA Tolstikov. *Uspehi Khimii*, 50,1207 (1981).
23. Aleksandrov, EL, YuS Sedunov. *Man and Sratospheric Ozone.* Leningrad, Gidrometeoizdat, 1979.
24. Benson, S. *Thermochemical kinetics. Methods for the estimation of thermochemical and rate parameters.* N.Y., J.W., 1968.
25. Greene, CR, R Atkinson. Rate Constants for the Gas-Phase Reactions of O_3 with a Series of Cycloalkenes and alpha,beta- Unsaturated Ketones at 296±2 K. *Int.J.Chem.Kinet.*, 26: 1, 37-44 (1994).
26. Atkinson, R, EC Tuazon, J Arey, SM Aschmann. *Atmospheric Environment*, 29, 3423-3432 (1995).
27. Atkinson, R. Gas-phase tropospheric chemistry of volatile organic compounds. 1. Alkanes and alkenes. *J Phys Chem Ref Data*, 26, 2, 215-290 (1997).
28. Horie, O, P Neeb, GK Moortgat. The reactions of the Criegee intermediate CH_3CHOO in the gas-phase ozonolysis of 2-butene isomers. *Int J Chem Kinet*, 29, 6, 461-468 (1997).
29. Zhang, XM, QY Xhu. Olefinic ozonation electron transfer mechanism. *J.Org.Chem*, 62: 17, 5934-5938 (1997).
30. Andrews,.H.D. Barnard, A.N. Gent. *Ozone attack on rubbers*. L. Bateman (Ed.), J.W., N.Y., 1963.
31. Dormagen, DB, Ozonolyse von ungesattigten Kautschuken und Vilkanisaten ein Uberblick. *Kautschuk + Gummi Kunststoffe*, 42, 760-769 (1989).
32. Rakovsky SK, SD Razumovskii, GE Zaikov. Kinetic Parameters of the Ozone Reaction with Maleic Anhydride, Maleic and Fumaric Acids. *Izv. Akad. Nauk SSSR, ser. khim.*, 38, (8), 1766 - 1769 (1974).
33. Rakovsky SK, DM Shopov. On the Mechanism of Ozone Reaction with Maleic Anhydride, Maleic and Fumaric Acids. *Izv. Khim. BAN*, XI, (1), 169 - 172 (1978).
34. Rakovsky SK, DR Cherneva, DM Shopov SD Razumovskii. Applying of Barbotage Method to Investigation the Kinetic of Ozone Reactions with Organic Compounds. *Izv. Khim. BAN*, XI, (4), 711 - 717 (1976).
35. Cherneva DR, SK Rakovsky, DM Shopov. Kinetics and Mechanism of the Ozone Reaction with Cyanstilbenes in Liquid Phase. *Izv. Khim. BAN*, X, (1), 84 - 87 (1977).
36. Rakovsky SK, DR Cherneva, DM Shopov. Kinetics of Interaction of Ozone with Cyanstilbenes. Communication II. *Izv. Khim. BAN*, X, (3), 359 - 397 (1977).
37. Rakovsky, SK, Diene Rubbers, Conventional (Ozone Degradation and Stabilization)., *Polym. Mater. Encyclopedia (ed. JC Solomone)*, 1878 - 1884 (1996).
38. Rakovsky, SK, VV Podmasterev, SD Razumovskii, GE Zaikov. The Mechanism of Crack Formation in Rubbers under the Effect of Ozone. *Intern. J. Polymeric Mater.*, 15, 123-133 (1991).
39. Anackov, MP, SK Rakovsky, DM Shopov. Studing of the Ozone Degradation of Some Elastomers in CCl_4 Solution. IUPAC, 1983 Prague Meetings on Macromolecules - 25-th Microsymposium "Processimg and Long-Term Stabilities of Hydrocarbon Polymers", Prague, July 18-21, 1983, pp. P34-1 - P34-3.

40. Anackov, MP, SK Rakovsky, SD Razumovskii. IR - Investigation of the Ozone Degradation of Some Elastomers in ClC_4. Intern. Rubber Conf. - Section A, 1-6.09. 1984 Moscow, pp.21-32.
41. Anachkov MP, SK Rakovski, SD Razumovskii, AA Kefely, GE Zaikov DM Shopov. Study of the Ozone Degradation of Polybutadiene, Polyisoprene and Polychloroprene in Solution. *Polymer Degradation and Stability*, 10, (1), 25 - 42 (1985).
42. Anachkov MP, SK Rakovski, DM Shopov. Ozone Degradation of Some Elastomers in Solution. *Commun.Dep.Chem.-Bulg.Acad.Sci.*, 18, (2), 187 - 194 (1985).
43. Anachkov M.P., S.K. Rakovski, S.D. Razumovskii, G.E. Zaikov, D.M. Shopov. A Spectroscopic Study of the Ozone Degradation of Some Elastomers in Solution. *Commun.Dep.Chem.-Bulg.Acad.Sci.*, 18, (2), 194 - 201 (1985).
44. Anachkov MP, SK Rakovsky, AK Stoyanov, RK Fotty. DSC Study of the Thermal Decomposition of 1-Decene Ozonide. *Thermochemical Acta*, 237, 213 - 217 (1994).
45. Anachkov MP, SK Rakovski, A Stoyanov. Thermal Analysis of Products Obtained Upon Ozonolysis of Polybutadiene Rubbers in Solutions. *Int.J.Polym.Mater.*, 15, (2), 12 (1991).
46. Anachkov, MP, SK Rakovsky, AK Stojanov. DSC-Study of Thermal Decomposition of Partialy Ozonized Diene Rubbers. *J. Appl. Polym. Sci.*, 61, 585-590 (1996).
47. Anachkov MP, SK Rakovski, RV Stefanova, DM Shopov. Kinetics and Mechanism of the Ozone Degradation of Nitrile Rubbers in Solution. *Polymer Degradation and Stability,* 19, (2), 293 - 305 (1987).
48. Anachkov, MP, RV Stefanova, SK Rakovski. 1H-NMR Study of Monomer Sequence Distribution in Some Comercial Acrylonitrile-Butadiene Copolymers *British Polymer Journal*, 21, (5), 429-432 (1989).
49. Anachkov MP, SK Rakovsky, RV Stefanova, AK Stoyanov. Ozone Degradation Of Polychloroprene Rubber in Solution. *Polymer Degradation and Stability*, 41, (2), 185 - 190 (1993).
50. Anachkov MP, SK Rakovski, DM Shopov, SD Razumovskii, AA Kefely, GE Zaikov. Study of the Kinetics and Mechanism of the Increase in the Molecular Mass of EPDM - Rubber During Ozonolysis in Solution. *Polymer Degradation and Stability*, 14, (1), 189 - 198 (1986).
51. Shonbein,. *J Prakt. Chem.*, 66, 282 (1855); 105, 232 (1866).
52. Berthelot, M. *Hebd.Seances Acad.Sci.*, 92, 895 (1881).
53. Fonrobert. Das Ozon. *Enke, Stuttgart*, 1916.
54. Harries, CD. *Ber.Dtsch.Chem.Ges.*, 36, 1933 (1905).
55. Harries, CD, P Reichard. *Ber.Dtsch.Chem.Ges.*, 37, 612 (1904).
56. Harries, *Justus Liebigs Ann.Chem.*, 343, 311 (1905).
57. Harries, *Justus Liebigs Ann.Chem.*, 390, 238 (1912).
58. Harries, C.D., *Untersuchungen uber die natrurlichen und kunstlichen Kautschukarten.* Berlin, 1919.
59. Staudinger, *Ber.Dtsch.Chem.Ges.*, 58, 1088 (1925).
60. Bailey, PS, TP Carter, CM Fisher, JA Tompson, *Can.J.Chem.*, 51, 1278 (1973).
61. Criegee, R., G. Wanner. *Ann.*, 564, 9 (1949).
62. Criegee, R. *Ann.*, 583, 1 (1953).
63. Criegee, R. *Rec. Chem. Progr.*,18, 111 (1957).
64. Criegee, R. *Angew. Chemie*, 21, 756 (1957).
65. Criegee, R., *Andew. Chem.*, 21, 765 (1975).
66. Criegee, R. *Peroxide Reaction Mechanisms*. J. Edwards (Ed.), N.Y., Int. Publ., 1960, 29.
67. Diaper, D.G.M. *J.Chem.Educ.*, 44, 354 (1967)
68. Rieche, A. *Magyar Tudomanyos Acad. Kem. Tudomanyok Oszlalynak Kîzlamenyei*, 12, 143 (1959).
69. Bischoff, C., A. Rieche. *Z.Chem.*, 5, 97 (1965).
70. Criegee, R. *Chem.Unserer Zeit*, 7, 75 (1973); *Chimia*, 22, 392 (1968).
71. Gunstone, F. *Educ.Chem.*, 5, 166 (1968).
72. Murray, R. *Acc.Chem.Res.*, 1, 313 (1968)
73. Pospisil, J. *Chemie* (Prague), 10, 1030 (1958).
74. Pryde, E., J. Cowan. *Top.Lipid Chem.*, 2, 1 (1971).
75. Visintin, B., S.Monteriolo. *Ind.Sanit.*,1965, 102.

76. Muray, R. *Trans.N.Y.Acad.Sci.*, 29, 854 (1967).
77. Muray, R. *Techn.Methods. Org.Organomet.Chem.*,1, 1 (1969).
78. Schulz, M., K. Kirchike. *Adv.Heterocycl.Chem.*, 8, 165 (1967).
79. Schroder, G. *Chem.Ber.*, 95, 733 (1962).
80. Criegee, R., H. Korber. *Chem.Ber.*, 104, 1807 (1971).
81. Loan, L., R. Murray, P. Story. *J.Am.Chem.Soc.*,87,737 (1965).
82. Lorenz, O., C. Parke. *J.Org.Chem.*, 30, 1976 (1965).
83. Murray, R., P. Story, L. Loan. *J.Am.Chem.Soc.*, 87, 3025 (1965).
84. Briner, E., E. Ballwigk, M. Picca. *Helv.Chem..Acta*, 41, 1390 (1958).
85. Briner, E., M. Picca. *Helv.Chem..Acta*, 41, 2578 (1958).
86. Briner, E., S. Fliszar. *Helv.Chem..Acta*, 42, 1310, 2063 (1959).
87. Criegee, R., S. Bath. B.von Bornhauph. *Chem.Ber.*, 93, 2891 (1960).
88. Kolsaker, P. *Acta Chem.Scand.*,19, 223 (1965).
89. Greenwood, F., B. Haske. *Tetrahedron Lett.*, 631 (1965); *J.Org.Chem.*, 30, 1276 (1965).
90. Murray, R., R. Youssefyeh, P. Story. *J.Am.Chem.Soc.*, 88, 3143 (1966).
91. Greenwood, F. *J.Am.Chem.Soc.*, 88, 3146 (1966).
92. Murray, R., R. Youssefyeh. *J.Am.Chem.Soc.*, 89, 2429 (1967).
93. Story, P., R. Murray, R. Youssefyeh. *J.Am.Chem.Soc.*, 88, 3144 (1966); 89, 4473 (1967).
94. Bauld, N.L., J.A. Thompson, C.E. Hudson, P.S. Bailey, *J.Am.Chem.Soc.*, 90, 1822 (1968).
95. Story, P., C. Bishop, J.R. Burgess, R.W. Murray, R.D. Youssefyeh. *J.Am.Chem.Soc.*, 90, 1907 (1968).
96. Muray, R.W., R. Hagen. *J.Org.Chem.*, 39,1103 (1971).
97. Story, P.R., J. Alford, J. Burges, W. Ray. *J.Am.Chem.Soc.*, 93, 3042 (1971).
98. Higley, D.P., R.W. Murray. *J.Am.Chem.Soc.*, 98, 4526 (1976).
99. Gallaher, K.L., R.L. Kuczkowski. *J.Org.Chem.*, 41, 892 (1976).
100. Fliszar, S., J. Carles. *J.Am.Chem.Soc.*, 91, 2637 (1969).
101. Castonguay, J., M. Bertrand, J. Carles, S. Fliszar, J. Rousseau. *Can.J.Chem.*, 47, 919 (1969).
102. Gillies, C., R. Kuczkowski. *J.Am.Chem.Soc.*, 94, 7609 (1972).
103. Gillies, C., R. Kuczkowski. *J.Am.Chem.Soc.*, 94, 6337 (1972).
104. Lattimer, R., R. Kuczkowski, C. Gillies. *J.Am.Chem.Soc.*, 96, 348 (1974).
105. Renard, J., S. Fliszar. *J.Am.Chem.Soc.*, 92, 2628 (1970).
106. Fliszar, S., J. Renard, D. Simon. *J.Am.Chem.Soc.*, 93, 6953 (1951).
107. Fliszar, S. *J.Am.Chem.Soc.*, 94, 7386 (1972).
108. Lattimer, R., R. Kuczkowski, C. Gillies. *J.Am.Chem.Soc.*, 95, 1348 (1973).
109. Hiberty, P. *J.Am.Chem.Soc.*, 98, 6088 (1976).
110. Rouse, R. *J.Am.Chem.Soc.*, 95, 3460 (1973).
111. Fliszar, S., J. Renard. *Can.J.Chem.*, 48, 3002 (1970).
112. Fliszar, S., M. Gaunger. *J.Am.Chem.Soc.*, 92, 3361 (1970).
113. Fliszar, S., M. Gaunger. *J.Am.Chem.Soc.*, 91, 3330 (1969).
114. Fliszar, S., J. Renard. *Can.J.Chem.*, 45, 533 (1967).
115. Fliszar, S., J. Carles. *Can.J.Chem.*, 47, 3921 (1969).
116. Bailey, P. *Chem.Ind.*, 1148 (1957).
117. Criegee, R. *Chem.Ber.*, 108, 743 (1975).
118. Criegee, R., A. Rustaiyan. *Chem.Ber.*, 108, 749 (1975).
119. Criegee, R., A. Bancin, H. Reul. *Chem.Ber.*, 108, 1642 (1975).
120. Murray, R.W., R. Hagen. *J.Org.Chem.*, 36, 1098 (1971).
121. Bailey, P.S., T.M. Ferrell, A. Rustaiyan, S. Seyhan, L.E. Unruh. *J.Am.Chem.Soc.*, 100, 894 (1978).
122. Bailey, P.S., T.M. Ferrell. *J.Am.Chem.Soc.*, 100, 899 (1978).
123. Rouse, R.A. *J.Am.Chem.Soc.*, 95, 34600 (1973); 96, 5095 (1974).
124. Hiberty, P.C. *J.Am.Chem.Soc.*, 98, 6088 (1976).
125. Rouse, R.D. *Int.Quantum Chem.Symp.*, 7, 289 (1973).
126. Benson, S.W., F.R. Cruickshauk, D.M. Golden, G.R. Haugen, H.E. O'Neal, A.S. Budgers, R. Shaw, R. Walsh. *Chem.Rev.*, 68, 279 (1968).
127. Goddard III, W.A., T.H. Dunning, Jr., W.J. Hunt, P.J. Hay. *Acc.Chem.Res.*, 6, 368 (1973).
128. Harding, L.B., W.A. Goddard III, *J.Am.Chem.Soc.*, 97, 6293, 6300 (1975); 98, 6093 (1976).

129. Bodrowicz, F.M., W.A. Goddard III.-In:*Modern Theoretical Chemistry; Methods of Electronic Structure Theory*, 3. H.F. Schaefer (ed.), N.Y., Plenum Press, 1977, 79.
130. Harding, L.B., W.A. Goddard III. *J.Am.Chem.Soc.*, 100, 7180 (1978).
131. Bailey, P.S., T.M. Ferrell, A. Rustaiyan. *J.Am.Chem.Soc.*, 98, 638 (1976).
132. Cremer, D. Theoretical Determination of Molecular Structure and Conformation. 7. Stereoselectivity of the Ozonolysis Reaction. *J. Am. Chem. Soc.*, 103, 13, 3619-3626 (1981).
133. M Olzmann, E Kraka, D Cremer, R Gutbrod, S Andersson. Energetics, kinetics, and product distributions of the reactions of ozone with ethene and 2,3-dimethyl-2- butene. *J.Phys.Chem. A*, 101: 49, 9421-9429 (1997).
134. Razumovskii, SD, Doctor of Sciences Theses, Institute of Chemical Physics, 1973.
135. US pat. 2857410 (1956); C.A., 53, 9067a (1959).
136. Noller, C., J. Garson, H. Martin, K.S. Havkins. *J.Am.Chem.Soc.*, 58, 24 (1963).
137. Treacy, J., Curley, M., Wenger, J., H. Sidebottom. Determination of Arrhenius parameters for the reactions of ozone with cycloalkenes, *J Chem Soc Faraday Trans,* 93, 16, 2877-2881 (1997).
138. Cadle, R.D., C. Schadt. *J.Am.Chem.Soc.*, 74, 6002 (1952).
139. Buffalini, J.J., A.P. Altschuller. *Can.J.Chem.*, 43, 2243 (1965).
140. Hanst, P.S., E.R. Stepfens, W.E. Scott, R.C. Doerr. *Atmospheric ozone olafine reaction.* Philadelphia, Pa, The Franklin Institute, 1958.
141. Vrbaski, T., R.J. Cvetanovic. *Can.J.Chem.*, 38, 1053 (1960).
142. Heickelen, J. *J.Phys.Chem.*, 70, 477 (1966).
143. Herron, J., R. Huie. *J.Phys.Chem.*, 78, 2085 (1974).
144. O'Neil, H., C. Blumstein. *Int.J.Chem.Kinet.*, 5, 397 (1973).
145. Japan, S.M., C. Wu, H. Niki. *J.Phys.Chem.*, 78, 8 (1974).
146. Toby, S., F.S. Toby, B.A. Kaduk. *Int.J.Chem.Kinet.*, 8, 25 (1976).
147. Finlayson, J. Pitts, R. Atkinson. *J.Am.Chem.Soc.*, 96, 5356 (1974).
148. Hull, L., I. Hisatsune, J. Heinlen. *Can.J.Chem.*, 51, 1504 (1973).
149. Japan, S.M., C. Wu, H. Niki. *J.Phys.Chem.*, 80, 2057 (1976).
150. Toby, F.S., S.Toby. *J.Phys.Chem.*, 80, 2313 (1976).
151. Hansen, D.A., J.N. Pitts. *Chem.Phys.Lett.*, 35, 569 (1975).
152. Toby, F.S., S. Toby. *Int.J.Chem.Kinet.*, 6, 417 (1974).
153. Pritzkow, W., G. Schappe. *J.Prakt.Chem.*, 311, 689 (1969).
154. Williamson, D.G., R.J. Cvetanovic. *J.Am.Chem.Soc.*, 90, 3668, 4248 (1968).
155. Razumovskii, SD. *Izv. Akad. Nauk SSSR, ser. khim.*, 335, (1970).
156. Benson, SW. *The Foundations of Chemical Kinetics*. Moscow, Mir, 1964.
157. Caldin, E.F. *J.Chem.Soc.*, 3345, (1959).
158. Razumovskii, SD, GE Zaikov. *Zhurn. Org. Khimii*, 8, 464 (1972).
159. Razumovskii, SD, LM Reutowa, GA Niazashvili, IA Tutorskii, GE Zaikov. Dokl. AN SSSR, 194, 1127 (1970).
160. Hammett, L.P. *Physical Organic Chemistry*. N.Y., London, p.188 1940.
161. Razumovskii, SD, YUN Yure'ev. *Zhurn. Org. Khimii*, 4, 1716 (1968); *Tetrahedron Lett.*, 3939 (1967).
162. Huisgen, R. The Concerted Nature of 1,3 - Dipolar Cycloadditions and the Question of Diradical Intermediates. MJ. Org. Chem., 41, 403-419 (1976); Huisgen, R. *Angew.Chem.*, 75, 604, 742 (1963).
163. Bernatek, E., P. Kolsaker, T. Ledaal. *Tetrahedron Letters*, 1317 (1969).
164. Bartlett, P.D., M.R. Rice. *J.Org.Chem.*, 28, 3351 (1963).
165. Woods, W., R. Carboni, J. Roberis. *J.Am.Chem.Soc.*, 78, 5653 (1956).
166. Aleksanyan, VT, HE Sterlin, AA Mel'nikov. *Izv. Akad. Nauk SSSR, ser. phys.*, 22, 1073 (1958).
167. Wiberg, E., B. Nist. *J.Am.Chem.Soc.*, 83, 1226 (1961).
168. Allinger, N. *J.Am.Chem.Soc.*, 80, 1953 (1958).
169. Krieger, H. Suomen. *Khem.*, 38, 136 (1962).
170. Sverdlov, LM, EP Krasnov. *Optika i spektr.*, 6, 334 (1959).
171. Paub, H., J. Sange, A. Kausmann. *Chem.Ber.*, 98, 1789 (1965).
172. Popov, AA, SD Razumovskii, VM Parfenov, GE Zaikov. *Izv. Akad. Nauk SSSR, ser. Khim.*, 282 (1975).

173. Gil'chenok, ND, SD Razumovskii, VK Tziskovskii in: *Catakytic Reaction in Liquid Phase.* Alma-Ata, Nauka, p.189 (1967).
174. Amis, E. *Influence of the Solvent on the Mechanisms of Chemical Reactions.* Moscow, Mir,1968.
175. Buligin, MG, GE Zaikov. *Izv. Akad. Nauk SSSR, ser. khim.*, 491 (1968).
176. Razumovskii, SD, YuN Yur'rev. *Neftekhimiya*, 6, 737 (1966).
177. Kerur, D.R., D.L.M. Diaper. *Can.J.Chem.*, 51, 3110 (1973).
178. Diaper, D.L.M. *Oxidat.Cumbust.Revs.*, 6, 145 (1973).
179. Bailey, P.S., C. Abshire, S. Manthia. *J.Am.Chem.Soc.*, 82, 6136 (1960).
180. Bailey, P.S., J.A. Thompson, B.A. Shoulders. *J.Am.Chem.Soc.*, 88, 4098 (1966).
181. Briner, E., D. Frank, *Helv. Chim. Acta.*, 21, 1927 (1938).
182. Dallwigk, E., E. Briner. *Helv. Chim. Acta*, 41, 1030 (1958).
183. Briner, E , E. Fliszar, *Helv. Chim. Acta*, 47, 2041 (1964).
184. Bauer, Rh, M Senfarth. Ber., 63B, 2691 (1930).
185. Plotnikov, IS. Izv. *Akad. Nauk SSSR, ser. khim.*, 1563 (1916).
186. Wilke, R, P Chang. *Am. Inst. Chem. Eng. Sci.*, 21, 999 (1966).
187. Alfrai, TV in: *Chemical Reactions of Polymers.* Moscow, Mir, p.9 1967.
188. Flory, P. *J.Principles of Polymer Chemistry.* N.Y., p.41 1953.
189. Kotliar, A.M., N.J. Moravetz. *J.Am.Chem.Soc.*, 77, 3692 (1955).
190. 133.Giazer, A.N., E.L. Smith. *J.Biol.Chem.*, 236, 2948 (1961).
191. Berlin, AA, AA Sayadyan, NS Enikolopyan. *Visoko Molekul Soedin. - A*, 11, 1893 (1969).
192. Razumovskii, SD, GA Niazashvili, YuN Yur'ev, IA Tutorskii. *Visoko Molekul Soedin. - A*, 13, 195 (1971).
193. Zuev, YuS, VF Molofeevska. in: *Degradation and Stabilization of Rubbers.* Moscow, Gostkhimizdat, p.27 1960.
194. Razumovskii, SD, ON Karpuhin, AA Kefely, TV Pokholok, GE Zaikov. *ÂÌÑ-À*, 13, 782 (1971).
195. Kefely AA, SD Razumovskii, GE Zaikov. *Visoko Molekul Soedin. - A*, 13, 803 (1971).
196. K.Nakanisi, Infrared spectra and the structure of organic compounds, Mir, Moscow (1965).
197. CRC Atlas of Spectral Data and Physical Constants for Organic Compounds, J.G. Grassell Ed., CRC Press Inc. (1973).
198. R. Murray and P. Story. Ozonation. in *Chemical Reactions of Polymers*, Moscow., Mir, 1967. (In Russian).
199. L.W.Crane, P.J.Dynes, D.H.Kaelble, *J. Polym. Sci., Polym. Letters* **11**, 533 (1973).
200. H.Kissinger, *Anal. Chem.*, **21**, 1702 (1957).
201. G. Gyulai and E.J. Greenhow, *Thermochim. Acta*, **6**, 254 (1973).
202. K.W.Ho, J. Polym. Sci. Part A, vol. **24**, 2467 (1986).
203. Yu.S.Zuev, Elastomer cracking under peculiar operating conditions, Khimia, Moscow, (1980). (In Russian).
204. G.G.Egorova, V.S.Shagov. in *Synthesis and Chemical Transformation of Polymers*, Izd. LGU, Leningrad (1986) (In Russian).
205. K.W.Ho, J.E.Gutman, *J. Polym. Sci, Part A: Polym. Chem.*, 27, 2435 (1989).
206. Tikhomirov, O.P.Baraban, and A.Y. Yakubchik, *Vysokomol. Soed. , Ser.A*, 11, 306 (1969).
207. P. Dole Robbe, *Bull.Soc.Chim. Fr.*, 3160 (1980).
208. McCullough and M. Nojima, Peroxides from Ozonization, in *Organic Peroxides*, W. Ando Ed., John Wiley & Sons Ltd, 1992.
209. Brazier, *Rubber Chem. Technol.*, 53, 438 (1980).
210. L.F.R.Cafferata, G.N.Eyler, E.L.Svartman, A.I.Canizo, and E.J.Borkowski, *J. Org. Chem.* 55, No 3, 1058 (1990).
211. Jefford, A.Jaber, J. Boukouvalas and P. Tissot, *Thermochim. Acta*, 188, 337 (1991).
212. A.C.Baldin, Thermochemistry of peroxides. in *The Chemistry of Functional Groups, Peroxides*, Ed. S.Patai, John Wiley & Sons Ltd, 1983.
213. M.J.Hackathorn, M.J.Brock, *Rubber Chem. Technol.*,45, 1295 (1972).

214. M.J.Hackathorn, M.J.Brock, *J. Polym. Sci., (A-1)*, 13, No 4, 945 (1975).
215. Brazier and N.V. Schwartz, *Thermochim. Acta,* 39, 7 (1980).
216. The Sadtler Handbook of Proton NMR Spectra, Sadtler, Philadelphia (1978).
217. Ellerstein, SM in *Analytical Calorimetry,* R.S.Porter and J.O.Johnson, Eds, N.Y., Plenum Press, 1968.

Chapter 4

DEGRADATION AND STABILIZATION OF REAL RUBBER SYSTEMS

In the light of the contemporary search of high efficient technologies the studies in the field of degradation and stabilization of elastomers, vulcanizates and rubber products receive considerable attention. The demand for producing new articles with new properties, the growing production of elastomers and the expanding of the fields of their application, together with the higher technical requirements and the worsening of the exploitation conditions [1-3] make the research in this field important and necessary. The study of the kinetics and mechanism of polydienes degradation and stabilization in real conditions makes possible: 1) establishment of the definite steps responsible for the degradation processes that result in deterioration of the physicochemical and mechanical properties of elastomers and vulcanizates and the products from them; 2) development and application of efficient methods for stabilization that prevent or retard the degradation processes; 3) more complete utilization of raw materials and prolongation of the "life" of rubber goods; 4) creation of scientific-quantitative bases for ensuring and prediction of the storage and exploitation terms of rubber products [4-6].

As ozone is one of the basic factors resulting in polydienes degradation we focus our attention on investigating its effect on rubber goods both alone and in combination with oxygen, temperature and mechanical stresses. We have studied the degradation processes in pure systems like nonfilled radiation vulcanized elastomers, in sulphur vulcanized model mixtures and real mixtures and products. The term degradation stands for the loss of physicochemical, mechanical and exploitation properties of elastomers and rubber products associated with molecular mass reduction under the action of various degradation agents. The contributions by various researchers and companies on this problemt so far have served as a base of our studies.

In regard to the processes of stabilization we have directed our studies to the synthesis of antiozonants, elaboration of their protective mechanisms, development of highly efficient antiozonants and antioxidative compositions and highly stabilized elastomers systems. We have also followed the role of the various components of rubber mixtures such as photostabilizers, antifatigue, etc. [7]. Stabilization is a general term used to describe the process of preventing the various materials from the harmful effect of degradation agents.

Aiming at the development of appropriate scientific bases for prediction of the behavior of elastomers and rubber products during storage and exploitation we have successfully applied the Arrhenius equation.

4.1. OZONE DEGRADATION

As we have already shown the ozonolysis of olefins and elastomers in liquid phase is an electrophilic process in regard to ozone. The rate of ozonolysis depends slightly on temperature and is decreased in the presence of electron-withdrawing substituents at the double bonds. It is also dependent on the olefin configuration, the *trans*-isomers being more reactive by about 10 fold than the *cis*-isomers. The ozonolysis of elastomers in solution is governed by molecule mobility and macromolecules conformation. In general, the values of the rate constant for the latter are lower as compared with those of olefins and the yield of the C=C cleavage products is also higher for polymers. The main reasons for these observations are: 1) the low mobility of polymer chains hinders the ozone access to the C=C bonds; 2) the formation of PO is also hindered since the sp^2-sp^3 rehybridization in the transition state is accompanied by reorganization of large polymer units; 3) the latter makes the PO decomposition slower than the olefins PO; 4) after the PO decomposition to zwitterion (carbonyl oxide - CO) and carbonyl compound (CC), the low mobility of the polymer fragments prevents the appropriate orientation of CO and CC that leads to ozonide formation which although being unstable does not change the molecular weight of the elastomer and 5) the low mobility of the polymer residual fragments increases the possibility for the CO interaction outside of the kinetic cage which in its turn enhances the degradation efficiency.

The physicochemical and mechanical properties of rubbers in solid-elastic state are determined mainly by their chemical structure, macromolecules configuration and conformation, the segmental mobility, the type and the properties of the supramolecular structures, defects in the mass and on the surface, etc.[8-17].

The segmental mobility, supramolecular structures and the ratio between the amorphous and crystalline phase play an essential role in the degradation process of elastomers in solid state. It is known that the diffusion of various gases including ozone in the amorphous phase is more facilitated than in the crystalline one. The greater access of the C=C bonds to ozone attack in this phase together with the higher ozone solubility provide suitable concentration conditions for acceleration of the process [18]. The degradation reaction that takes place on the boundary layer is facilitated by the concentration of stresses in the linking phases units. There should be mentioned the reactions of monomolecular decomposition of the various intermediate polymer compounds, including polymer zwitterions. The surface defects as well as those in the solution are actually the reaction centers from which the degradation process starts. They are also the entry for the ozone degradation

In the course of their exploitation the rubber materials are subjected to considerable mechanical stresses resulting from elongation, load, tearing, shrinking in static and dynamic conditions. These stresses play an important role in accelerating the degradation process acting as mechanical and thermal promoters of polymer molecules decomposition. Usually greater part of the mechanical stresses are converted into heat without deteriorating the rubber articles. They are also accumulated and concentrated, to a smaller extent, as stresses in net defects, interphase surfaces, in surface defects, thus changing the supramolecular structure, configuration and conformation of the macromolecule through retardation of the relaxation processes.

The rise of the temperature to 70-90°C has a dual effect: 1) it facilitates the molecules orientation and contributes to the removal of defects and spots of energy concentration and 2) accelerates the process of oxidative degradation.

We have made an attempt to evaluate the contribution of the main factors responsible for the degradation of elastomers and their vulcanizates. In this connection we have compared the rate of oxidative and ozone degradation. It has been established that the first one is proportional to the rate of the chain oxidation process as described by:

$$W_{O2}\ (k_p.(W_i/k_t)^{1/2}.[RH] \qquad \text{(Eq.4.1)}$$

where; k_p is chain propagation rate constant = $1.10^{-1}\ M^{-1}.s^{-1}$; $k_t = 1.10^{7}\ M^{-1}.s^{-1}$ rate constant of chain termination; W_i - initiation rate; [RH] = 14 M - rubber concentration.

Substituting the known values for the constants and concentration in Eq. 4.1 the following expression for the rate of oxidative degradation is obtained:

$$W_{ox}\ (0.014.(W_i)^{1/2} \qquad \text{Eq.4.2}$$

The second rate is proportional to the rate of ozone reaction with the elastomer :

$$W_{oz} = k_{oz}.[O_3].[RH] \qquad \text{(Eq.4.3)}$$

where: $k_{oz} = 10^5 M^{-1}.s^{-1}$ - is the rate constants of ozone interaction with C=C bonds; $[O_3] = 1.10^{-8}$ M - the real ozone concentration in the atmosphere.

Again by using the known values we get:

$$W_{O3} = 0.014 \qquad \text{Eq.4.4}$$

The comparison of Eqs 4.2 and 4.4 reveals that the two rates differ by the multiplier ($W_i^{1/2}$) which is always smaller than unity. This fact can be regarded as the main reason that makes the study and fight against ozone degradation so important. It is true that ozone degradation takes place in the presence of other degradation agents like oxygen, heat, light, mechanical stresses, etc. which complicates further the research in this field. Our experiments show that this problem can be satisfactorily solved as linear combination of its one-dimensional components:

$$\varphi = c_1\varphi_1 + c_2\varphi_2 + ... + c_X\varphi_X = \sum_{X=1}^{n} c_X\varphi_X$$

where: c- constant; φ- parameter related to the degradation process.

What is the main result arising from the ozone effect on rubber samples? Rubber samples pre-stretched (ε) more than 10 % exposed for a definite time in ozone atmosphere are covered by a network of cracks that grow with time and finally break down [20-23]. The longitudinal axis of the cracks depending on the deformation force (up to elongations ≤200 %) ant the rubber type is rectangular (orthogonal) to the extension direction while at higher deformations the cracks orientation begins to change and may reach parallel orientation. Another interesting phenomenon which has been observed by many authors [25-27] is the occurrence of critical deformation. At ε=10-50 % depending on the type of the elastomer a maximum in the degradation rate is observed with a simultaneous increase of the total area and cracks depth (Figs 4.1 and 4.2)[24]:

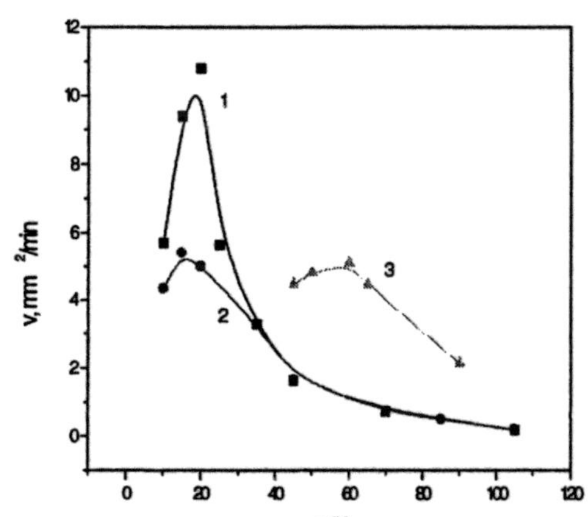

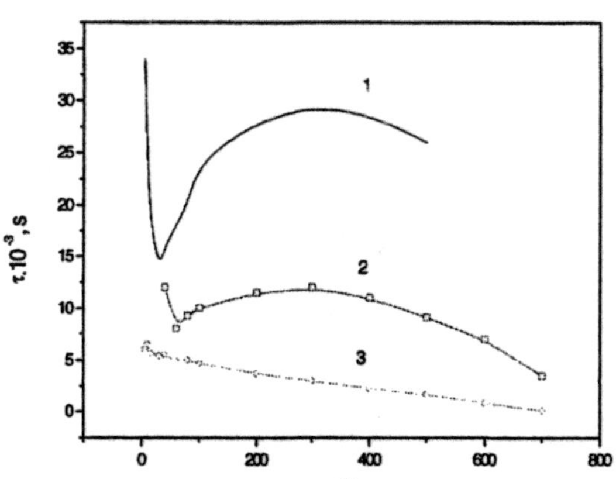

Fig.4.1. Dependence of the rate of cracks growth (v) on deformation during ozonation: 1 - Naturale rubber - NR, $[O_3] = 1.5.10^{-3}$ mol/m^3; 2 - Butadiene styrene ruber - SKS-30 with dominant *trans*-configuration of the butadiene units, $[O_3] = 0.9.\ 10^{-3}$ mol/m^3; 3 - Polychloroprene rubber (PCh) - with predominant *trans*-configuration, $[O_3] = 4.5.\ 10^{-3}$ mol/m^3

Fig.4.2. Dependence of the time break down on the deformation: 1 - NR, 3 - SKS-30, $[O_3] = 4.5.10^{-4}$ mol/m^3; 2 - PCh, $[O_3] = 2.2.10^{-3}$ mol/m^3

Figs 4.1 and 4.2 show that the maximum rate of ozone degradation is registered in the range of critical deformations, this effect being more pronounced in the case of natural rubber. Different points of view are put forward by the different authors with respect to the nature of this phenomenon. For that reason we have made an attempt to summarize and analyze the basic and widespread opinions related to the mechanism of this observation.

Newton [28] tried to explain this anomalous phenomenon by the mutual effect of cracks and relates the extremum character to the cracks number and sizes. He assumed that at small deformations a few cracks are formed in energetically rich surface defects, and with an increase in (the deformation energy begins to concentrate on energetically poorer defects thus transforming them in reaction cores for cracks formation. At ε equal to the critical deformations the number of the reaction centers reaches it is maximum and respectively the degradation rate its maximum values. At higher (the cracks begin to influence each other, thereby slowing their mutual growth and retarding the degradation. The further increase in ε results in the deformation energy accomulation of in the cracks lips - they increase their own depth and sizes leading to rise of the degradation rate. Bradden and Gent [29, 30] considered that the acks formation and growth is closely related to the appearance of critical deformation whereby a sufficient amount of energy is accumulated for the formation of new surfaces. Solomon and Tellamon [31, 32] put forward the opinion that the ozone cracking is an unique phenomenon that is characteristic only for elastomers with double bonds. Based on the similarity between static wear and corrosion decay Zuev [33, 34] assumed that elastomers degradation is related to the static wear of the vulcanizates, accelerated by the strong corrosive action of ozone. The basic argument in favor of this explanation is the observed similarity in the change of the rubber durability on exposure to ozone action with the change of hardness and resistance to shareness at deformation. He considered, however, that the nature of the corrosive agent is of no importance since the specificity of the chemical reaction is of minor importance. He attributed the observed changes only to the extent of change of the polymer orientation and its acceleration or relaxation resulting from the applied deformations.

The analysis of the mechanisms discussed above reveals principal differences between them with respect to the nature of the critical deformation. Some of them explain the cracks formation only by physical or only by chemical reasons. The simultaneous action of these two factors has not been considered in explaining the existence of critical deformations.

In our opinion the mechanism of ozone effect on rubber specimen could be much better understood if the study is concentrated on the initial steps of the reaction prior the appearance of the first cracks [35-37]. The change of the C=C bonds amount and the accumulation of C=O groups in the course of radiation vulcanizate samples from 1,4-cis-polyisoprene rubber (SKI-3 product of Russia) in dependence on ε are given in Figs 4.3 and 4.4

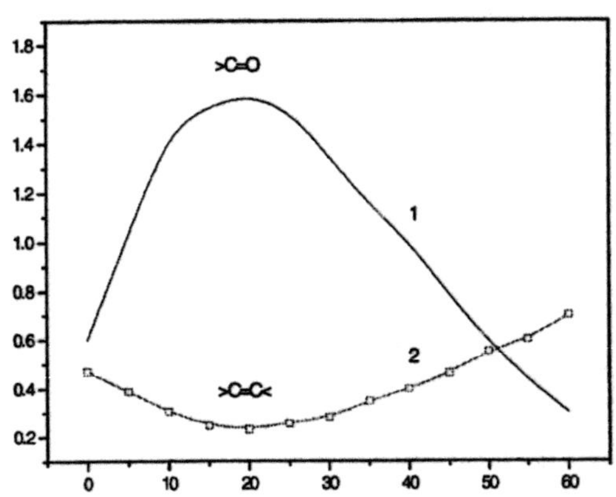

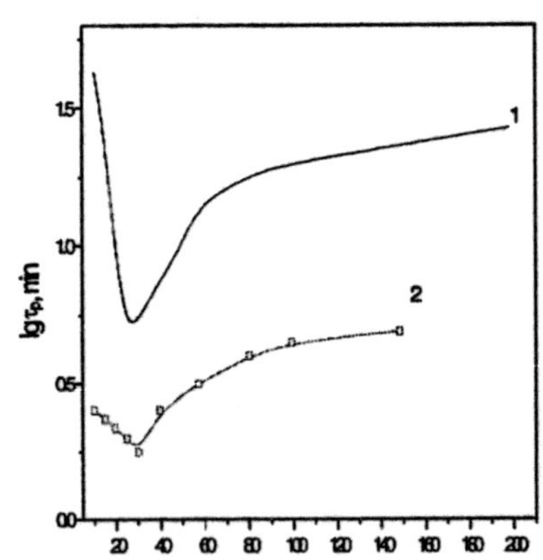

Fig.4.3. Dependence of the optical densities ratio on the deformation ε.
1 - D_{1720}/D_{1640} -C=O-groups; 2 - D_{1640}/D_{1375} - C=C-groups; ATR - KRS-5 - 45^0; $[O_3] = 7.10^{-5}$ mol/m^3, r.t.

Fig.4.4. Dependence of the chemical relaxation time on (of radiation vulcanizate samples of SKI-3 in ozone atmosphere at 20°C, 1- $[O_3] = 2.\ 10^{-3}$ mol/m^3 and 2 - $[O_3] = 1.\ 10^{-2}$ mol/m^3

The maximum rate of oxidation corresponds to ε = 20 % after which it goes down markedly. This testifies the effect of the strain on the rate of the chemical process. Comparing these data with the results from the chemical relaxation (Fig. 4.4) it becomes obvious that the maximum rate is achieved at ε = 20-30 % after which it begins to decrease with the further rise of ε. This effect is most pronounced at small ozone concentrations. The results in Figs 4.3 and 4.4. are in a good correlation with the data for the cracks formation and break down (Figs 4.1 and 4.1) This observation gives us ground to conclude that the initial rate of the ozone degradation predetermines the course of the degradation process. On the other hand we should note that the amorphous/crystalline phase ratios in crystallizing elastomers like NR do not change substantially up to the critical values of ε.

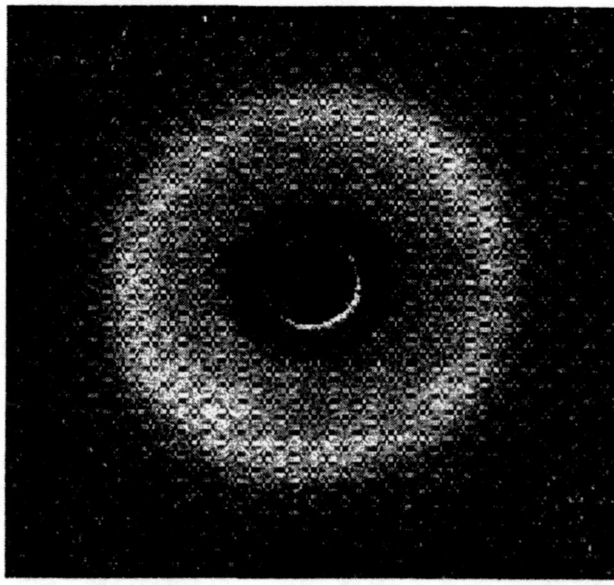

a.

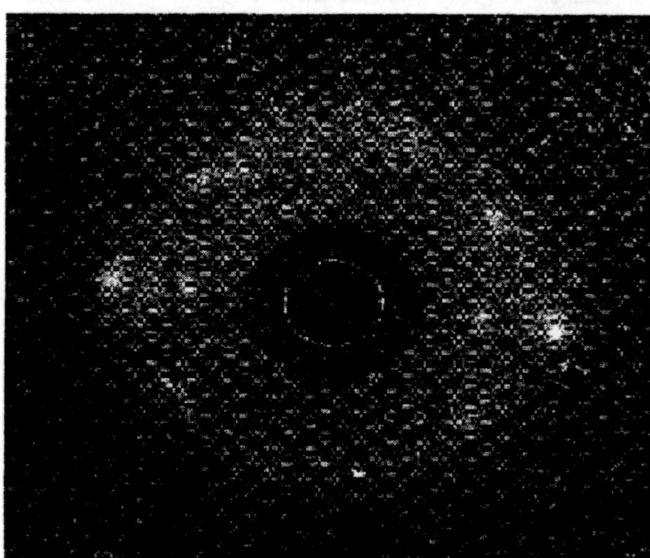

b.

Fig.4.5. Rentgenograms of *cis*-1,4-polyisoprene at:
a - ε = 100 % and b - ε = 400 %

It is clearly seen that while at deformations up to 100 % crystallization processes do not occur at ε = 400 % well shaped crystallites are formed.

The pattern of sample cracking is primary dependent on rubber type (Table 4.1) [38, 39].

Table 4.1

View and shape of rubber specimen cracks in ozone atmosphere - 6 ppm, 20°C and $\varepsilon = 20\%$ up to their break down

Rubber	Form of cracks	(a/b)**	Picture of cracks
NR*	Oval	2-4	A_4T_1
SKI-3	Oval with sharp ends	>4	A_4T_1
SKD, (1,4-cis-Polybutadiene - Russia)	Rectangular, with furrow tight side	>2	A_1T_3
SKI-3/SKD	Oval, with some cross fibers	>3	A_3T_2
Bulex 1500, (Polybutadienestyrene 70:30 - product of Bulgaria)	Cyrillic letter (ZH)	>1.7	A_2T_3
Bulex M27, (Oil filling BDS - Bulgaria)	Poor definite rectangular	3	A_2T_2

Note: * - also for filled sample; ** at oval - the ratio of ellipse axes, at rectangular - the ratio of faces; A - cracks number: A_1=1÷5, A_2=5÷40, A_3=40÷120, and A_4>120; T - cracks depth; T_1 - surface cracks; T_2≤1mm, T_3>1mm, and T_4 - cracks through the total depth of the specimen studied.

The process of cracks formation was registered by means of reflected-light optical spectroscopy methods using metalographic-MMR-4 LOMO microscope (Russia) with magnification x600[39]. Shape of the cracks depends on the type of rubber are prezented On Fig.4.6.

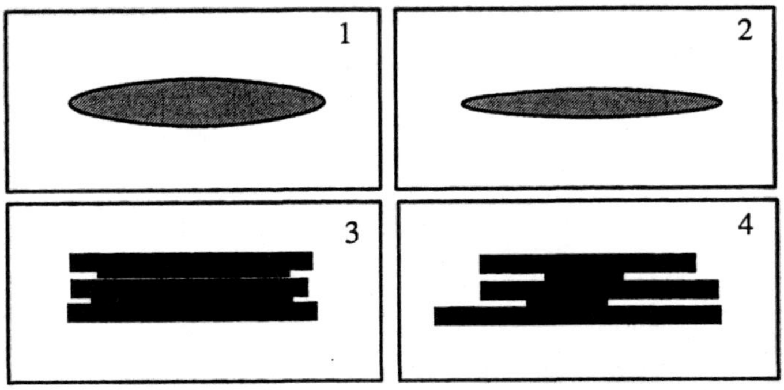

Fig.4.6. Shape of the cracks depends on the type of rubber at ε = 20%, 20°C, 10 ppm ozone. 1 - NR; 2 - SKI- 3; 3 - SKD; 4- Bulex 1500

In this case cracks are formed when the molecular interactions result in visible supramolecular changes.

The parameters of cracking: time of first appearance (t_o), shape and sizes of cracks and time of break down (t_c) of the samples depend on: 1) the elongation level; 2) static or dynamic deformations and their frequencies; 3) rubber type; 4) ozone concentration; 5) temperature; 6) for vulcanizates - on the reception of their preparation [40]. The Scheme pointed below visualizes the influence of external deformations on ozone interaction with rubbers:

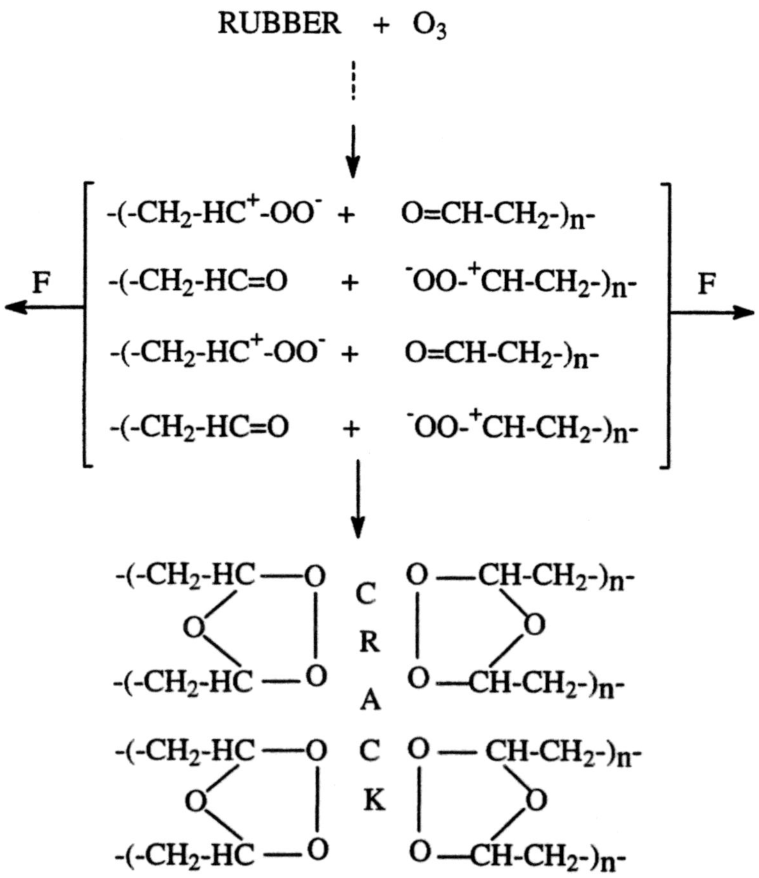

For case F the force acts along the longitudinal axis of the macromolecules. The first step which is not shown in the Scheme, includes the PO formation which decomposes further to zwitterion and aldehyde. The applied force facilitates the O-O and C-C bonds cleavage as part of the stretching mechanical energy is converted into bond energy thus forming zwitterion and aldehyde. The next step, as suggested by the Criegee mechanism, includes the reaction of the zwitterion in or outside the cage. The force application impedes the ozonides formation since it hinders the orientation of the broken ends of the macromolecules. Thus the macromolecules is cleaved at the place of the double bonds and a crack is formed. This Scheme examplifies the relation between the molecular changes and their influence on macro level.

The pattern of cracks formation in dependence on (indicates that the number of cracks for NR increases linearly with (grow up to 82% and their depth has a minimum at 32%.

We believe that the mechanism of degradation under definite conditions will be governed by a definite degradation factor and will depend on its statistical weight in the overall degradation process, but in each case the interaction of ozone with the double bonds will play a substantial role.

A number of Russian authors, Popov, Razumovskii and Zaikov [41-45] who have great contributions to the study of ozone reactions with organic substances have put forward similar opinions.

4.2. ANTIOZONANTS

The diene elastomers: natural and synthetic butadiene, isoprene, butadiene-styrene rubbers which compose the major part (over 80%) of all rubber products are mostly exposed to ozone degradation.

Synthetic rubbers such as polychloroprene, butadiene-nitrile, polyisobutylene, ethylene-propylene, fluoro, silicon, polyurethane, etc. are relatively more stable towards the ozone action.

In our studies we have focus our attention on ozone degradation of conventional rubbers.

As we have already pointed the ozone concentration in the atmosphere varies in the range of 0.2-0.6 ppm. However it can reach values up to 60-100 ppm in some areas arising from the photochemical oxidation of organic compounds in the presence of nitrogen oxides in the polluted city atmosphere (smog), photochemical decay of algae and sea organisms on the seashore, the UV and microwave radiation, electric charges and other radiation on the atmospheric oxygen. As the degradation rate is proportional to the ozone concentration (Eq. 4.3) it is obvious that in these areas the degradation will increase almost twofold.

The basic protection against ozone involves the application of small amounts up to 10 php (php-parts per 100 rubber parts) of organic, inorganic, polymeric and other additives, called antiozonants (AO) as components of the rubber compositions.

The creation and selection of AO is a complex problem and irrespective of the numerous empirical approaches there is no so far generally accepted quantitative theory on this problem. There are, however, several basic requirements and rules which direct the synthesis and selection of the compounds suitable for application as AO: They should:

-ensure prolonged ozone resistance of the rubber materials under static and dynamic conditions;

-be thermally stable, have an antiscorching effect and be fire-safe;

-be compatible with the rubber mixtures, discoloring, diffused easily in the polymer matrix and not be easily washed away;

- be colorless, not coloring and stable at storage;

- be healthy-safe;

- the raw materials and technologies for their preparation should be easily available.

In addition, according to the assumed mechanism of ozone interaction with rubber solutions, discussed in the foregoing chapter, the following specific requirements can be formulated:

1. AO should deactivate ozone and C=C bonds as:

-react with ozone in orders faster than with the elastomers molecules;

-form complexes with ozone or decompose it catalytically;

- form complexes with C=C bonds;

- the products of their interaction with ozone and the AO themselves should possess structurizing functions as AO are polyfunctional compounds.

2. AO and their ozonation products should physically prevent the ozone access to the C=C bonds :

-being inert compounds towards the ozone attack;

- facilitate the relaxation of the strains formed, i.e. act as antifatigue additives.

It is obvious that practically all these requirements can not be satisfied by a single compound. That is why the stabilization of elastomers is carried out by using stabilizing systems - mixtures of various AO which could ensure a complex protection, both chemical and physical one.

The rate constants of ozone reaction with amines, hydroquinolines, thiosemicarbazides, some aminophosphorous compounds, thiophenols, aminophenols, etc. are up to 3-4 orders higher than with elastomers [48]. Among them the derivatives of N,N'-substituted-*p*-phenylenediamines (PPHDA), the salts of dialkyldithiocarbamates, dihydroquinolines, etc.are the most powerful antiozonants. Some inorganic compounds like dithiocarbamate and dithiophospate complexes of some alkali-earth and transition metals exhibit also good antiozonant properties. It has been also found that the oxides of the latter are good catalysts of ozone decomposition.

However, the rate constants of ozone reaction with saturated organic compounds such as paraffins and their oxygen-, halogeno-, nitrile, nitro-, thio-, sulpho-, phospho-containing derivatives, waxes, i.e. paraffin mixtures with definite molecular-mass distribution, polyolefins, saturated fluoro-, chloro- and silicon containing elastomers [48] appear to be much lower (up to 4-6 orders) than with elastomers. These compounds prevent the degradation process by reducing the ozone access to the elastomer, form surface layers, and at the simultaneous application with lowmolecular AO guarantee favorable conditions for their diffusion to the reaction zone. Saturated polymer additives operate also through a kinetic factor associated with the decrease of the surface C=C bonds.

Many organic and inorganic metal-containing compounds and complexes, the active and nonactive fillers in the rubbers compositions act as deactivators of the double bond. The formation of complex and surface compounds with participation of the C=C bonds decrease substantially their reactivity (manifold) towards the ozone action. There should be mentioned the complex compounds of zinc, magnesium, nickel, copper and other metals formed in the course of vulcanization.

The derivatives of PPHDA and dihydroquinolines exert also cross-linking functions and the latter are the best known so far antifatigue additives.

4.2.1. Mechanism of Action

The use of stabilizing additives as AO depends primarily on the type of rubbers and the field of their application for producing rubber goods (RG) and car and trunk pneumatic tyres [50].

The selection of a particular AO is based on considering the probability of its interaction with the various ingredients of the rubber composition and its compatibility with the vulcanizing agents [51].

PPHDA being one of the most efficient and widely applied antiozonants in the rubber industry has been intensively studied with a view of elaborating its mechanism of action.

The most widespread mechanism for the AO action is the kinetic approach based on the concurrent interaction of ozone with AO and polydiene [52-56].A characteristic peculiarity of these AO is that they alone or their ozonation products can form protective surface film [57] thus preventing the ozone access to rubber [58].

Lorenz and Parks [59] considered that the AO react mainly via three reaction routes: 1) directly with ozone; 2) with the peroxide compounds reducing them and 3) with two polymer molecules containing aldehyde groups.

Other authors [60] related the AO efficiency to the inhibition of free radical reactions initiated by ozone.

Braden and Gent [62] distinguished two basic classes of AO: 1) AO that prevent the cracks growth and 2) AO that increase the critical energy of cracks growth.

Storey et al., [62] attributed the high antiozonant activity of PPHDA to the presence of two conjugated nitrogen atoms, one of which can act as a nucleophile due to the presence of the second one. The different efficiency of AO is due to the various nucleophility of the nitrogen atoms, which in its turn affect the decomposition rate of the peroxide compounds on the rubber surface. The stronger the nucleophilic effect is, the higher the rate of nitroxyl radicals formation. The latter initiate reactions with short kinetic chains and form cross bonds thus reducing the number of adjacent double bonds. Alkyl substituents that increase the nucleophility of the nitrogen atoms increase the AO activity in comparison with those without aryl substituents. This action of PPHDA is confirmed by the fact that ozone penetration in rubber is at depth of about $1.3.10^{-5}$ m.

Other authors stated that the bifunctional AO prevent the growth and formation of cracks in elastomers through combining the two zwitterions, obtained from the decomposition of two PO, thus forming a new bond between the broken ends of the macromolecules [63, 64].

The protective action of AO is also associated with the relative stability of the free radicals generated from them during the ozone degradation of elastomers [65].

Michaelis [66] has shown that the more effective PPHDA lead to the formation of stable semiquinone structures while the less active ones form unstable ones. On the bases of these results Barnhard and Newby [67] assumed that the stable semiquinone structure prevents the macromolecules degradation through destroying the ozonides and peroxides. In our studies on SKI-3 degradation in the presence of various PPHDA it has been demonstrated that some of them exhibit initiating properties in dependence on the applied deformations [68, 69].

The foregoing survey shows that the antiozonant action of PPHDA is a rather complex and ambiguitous process. We believe that the study on the kinetics and mechanism of AO action and the establishment of relationships for prediction of their efficiency or for the synthesis of new antiozonants with definite properties is of great practical importance [70-73].

Bailey suggested the following mechanistic pathways for ozone interaction with di-*tert*-butylamine as a lowmolecular model of PPHDA-AO:

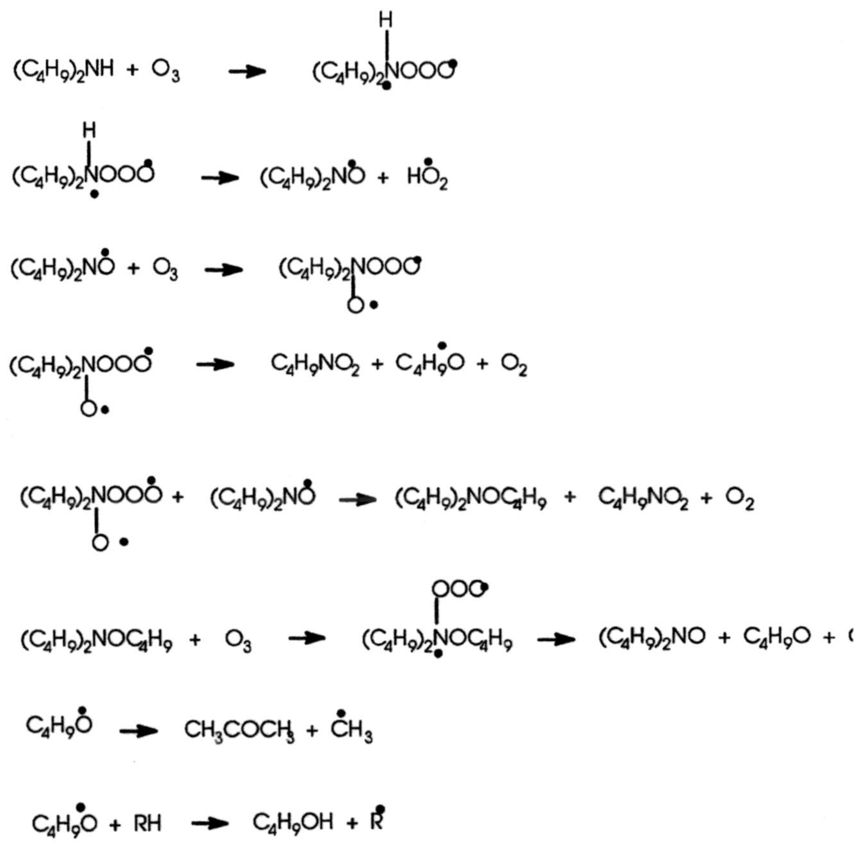

The ESR spectra characteristic for the nitroxyl radical, as well as the obtained 2-methyl-di-nitro-methane, tert-butyl alcohol, acetone and oxygen confirm this mechanism. One molecule amine decomposes up to 4 ozone molecules. If one takes into account the possibility of ozone interaction with radicals, without nitrogen, then the number of decomposed ozone molecules should be greater which additionally enhances the amine efficiency.

The influence of ozone on pure rubber and AO solutions in inert solvents has been the subject of numerous investigations. These studies have contributed to the elaboration of some of the basic steps of AO transformations and rubber decomposition products which can undergo mutual interactions and affect the elastomers degradation process [74-76].

Upon the ozone reaction with elastomers with low C=C bonds content or with electron-accepting substituents at the C=C bonds such as butylrubber, polychloroprene, etc., the basic reactions of degradation are related either to chain-radical route or molecular reactions of ozone addition to the C-H bonds and izomerization of RO_2 and RO-radicals. We have evaluated the role of ozone as an initiator of chain-radical oxidation processes in Chapter I. The protection of such rubbers is accomplished via the AO reaction with ozone thus converting the active oxy- and peroxy-radicals into non active molecular products.

Thiosemicarbamates and thiosemicarbazides [77, 78] have been used as good non coloring antiozonants instead of PPHDA.

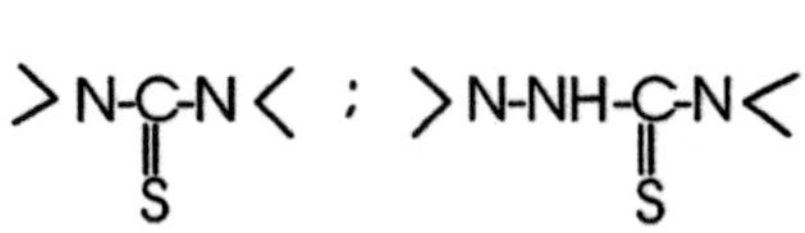

These compounds easily donate electrons and the values of *k* for their reaction with ozone amount to 10^6-10^7 $M^{-1}.s^{-1}$ (by about 2-3 orders higher than *k* found for the ozone reaction with C=C bonds in elastomers) and E_a = 1.3 kcal/mol [79, 80]. It is known that the value of E_a of diffuison is of the same order while that for solid state may be higher. This suggests that the degradation reaction will be limited rather by the mass transfer than by the change of temperature. For this reason the antiozonants should be compatible with the rubber mixture and should easily diffuse in the polymer matrix.

As we have mentioned above the stabilizing compositions are composed of various in nature AO or of AO blends with polyolefins and saturated elastomers such as butyl rubber or triple copolymers of ethylene with propylene and diene, with ozone resistant elastomers like chorine, fluorine or nitrile rubber, etc. The AO action can be enhanced by the use of synergetic mixtures and the ozone resistance of the rubber products can be increased by using AO mixtures spread on the surface. The latter are AO solutions in gasoline, chloroform, tetrachloromethane, turpentine, etc.

4.2.2. Synthesis

The organic compounds used as AO and antioxidants usually contain various number of heteroatoms: N_1, N_2, N_3, N_4, N_1O_2, O_2S_1, O_1, P_2S_4, etc. The synthetic methods for their preparation depend on the type of the AO and the initial products which are mostly obtained from petroleum.

4.2.2.1. Paraphenylenediamine

In our [81-83] and other studies [84-84] it has been convincingly established that among the great variety of chemical compounds used as antiozonants the substituted N,N-dialkyl, diaryl- or alkylaryl-PPHDA are the most powerful antiozonants exhibiting a broad spectrum of useful properties.

The modern tendencies in PPHDA-AO preparation is closely associated with the creation of new high technologies. The latter require that the whole process of additives preparation including the separate reactions of amination, alkylation (arylation), reduction and oligomerization up to the commercial state of the AO should be carried out in one reactor by using a highly selective and efficient catalytic system. The semi-products for the PPHDA preparation: *p*-chloronitrobenzene, *p*-dinitrobenzene, *p*-chlorophenol, *p*-dichlorobenzene, phenyl-*p*-phenylenediamine, naphthyl-*p*-phenyldiamine, at these technologies in fact become semi-products for producing the commercial AO. The search basically is directed to expanding the range of raw materials by including low-cost and available products from petroleum processing, coke preparation, etc.

4.2.2.1.1. Alkylation in the Presence of Hydrogen

Catalytic alkylation in the presence of hydrogen is widely applied to synthesis and industrial preparation of PPHDA-AO. Catalysts based on Pt, Pd, Ni, Co, Ce and Zr are most suitable for reducing alkylation process.

This is a process of introducing an alkyl group into the molecule by the interaction of amino, nitro or nitroso compounds with ketones, aldehydes or alcohols in the presence of hydrogen. The most efficient AO are obtained upon alkylation with ketones, mainly, unsymmetrical one like methylethylketones with alkyl substituent varying from C_3 to C_5-C_7. The most used ketones are methyl ethylketone (MEK), methyl isobutylketone, methyl isoamylketone, methyl hexylketone, etc. Alcohols should be preliminary dehydrogenated, which makes the process more expensive.

N,N'-dialkyl-PPHDA is obtained as the mixture of *p*-nitroaniline or *p*-phenylenediamine with at least 2 moles of aldehydes or ketones is hydrogenated at a temperature of 100-200°C and a hydrogen pressure of 10-200 at in the presence of CuO, Cr_2O_3 and BaO or their ratios [87]. Other catalysts used for the same purpose are: CaO, Ce_2O_3 and Bao, Ba, $CuCrO_2$ or Cu/ZnO [88], with a ratio between the ketone and the *p*-nitroaniline in the range from 2:1 to 10:1 under hydrogen pressure of 140 at temperature over 100°C.

The interaction of paranitroaniline with methyl ethylketone in the presence of copper chromite at P_{H2} - 50 at and temperature 110-160°C leads to N,N'-di-*sec*-butyl-PPHDA. The reaction proceeds according to the scheme:

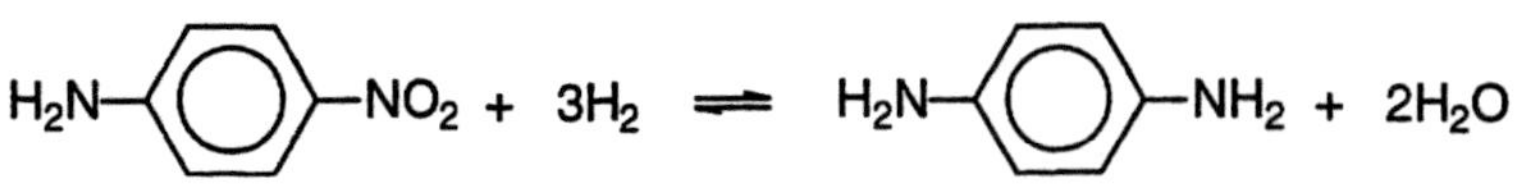

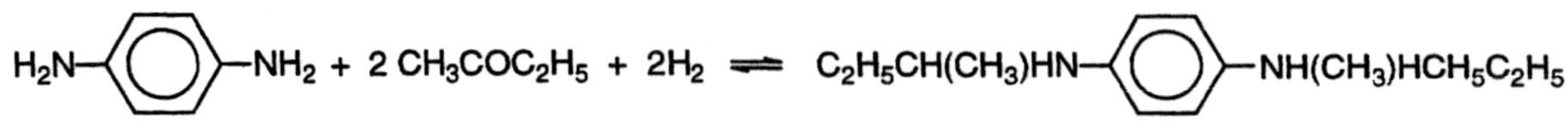

On the basis of Refs 91, 92 concerning the conversion of paranitroaniline into N,N'-di-*sec*-butyl PPHDA, the most suitable conditions for this reaction are a temperature of 180-200°C, P_{H2} = 30-60 at and a molar MEK:PPHDA ratio = 10. Other conditions for the alkylation of p-nitroaniline with MEK are presented in Refs 91-99.

Copper chromite has been used as a catalyst not only in the above reaction but also in the preparation of 4-$C_6H_5NCH_6H_4NHCH(CH)_3C_5H_{11}$ from 4-aminodiphenyl amine and methyl hexyl ketone in a ratio of 1:2 [100].Substitution of 4-CH_3O-4-CH_3-pentene-2-one for methyl hexyl ketone leads to the formation of 4(4-CH_3O-4-CH_3-pentyl)-amino diphenyl amine [101, 102]. Alkylation with a sec butyl alcohol proceeds in the presence of 15% Ni, 5% Cu and 1% Mn on SiO_2 and P_{H2} = 14 at [103]. The yield of the final product increases after re-distillation of aniline or *p*-nitroaniline [104, 105].

When PPHDA or N-aryl-N-(4-aminophenyl)amine is alkylated with furfurol or tetrahydrofurfuryl alcohol, N, N'-difurfuryl-PPHDA is obtained on the Pd/C $LiAlH_4$ or Raney-nickel catalyst [106, 107].

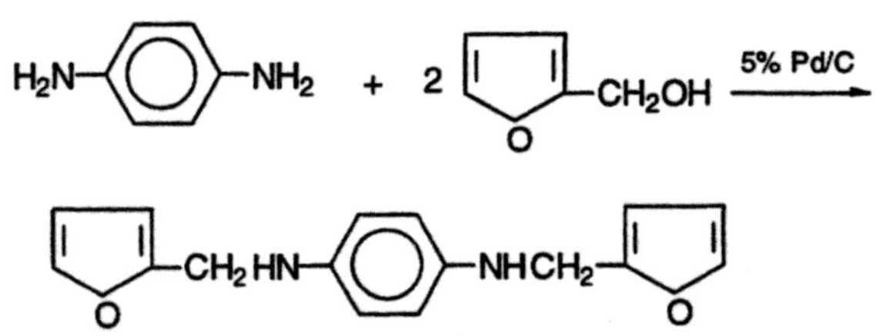

If a mixture of diphenylamine and furfuryl alcohol is allowed to stand for 80 days at room temperature, 4,5-di(diphenylamino) cyclopentene-2-one with antiozonant properties is obtained [107].

Diatomite-supported Ni (63 %) is used for the preparation of N(1,3-dimethyl butyl)-N'-phenyl-PPHDA from 4-amino-diphenylamine and methyl isobutyl ketone in a ratio of 1:2 [108, 109].

$$C_6H_5\text{-}C_6H_4\text{-}NH_2 + CH_3COCH_2CH(CH_3)_2 \xrightarrow{\text{50-60 atm 165°C}} (CH_3)CH_2CH(CH_3)CH_2NH\text{-}C_6H_4\text{-}NH\text{-}C_6H_5$$

Modification of Ni catalysts with cobalt leads to an increase of its selectivity [110]. This catalyst is very active and can cause hydrogenation of the phenol ring. That is why it is recommended to add negligible amounts of organic sulphur compounds such as sulphates, mercaptans, thio acids and sulphides to the reaction mixture [111, 112].

Selenium and tellerium as catalysts preserve their activity for a long time. They are not poisoned by sulphur compounds (no purification of hydrogen is needed) and lead practically to no side reactions [113, 114].

Nickel catalysts are also used during the conversion of N-alkyl-4-nitrosoaniline into N-alkyl-N-phenylenediamine, while iron -copper-chromium catalysts are utilized in the reaction of the N-alkyl-N-phenylenediamine with ketones. The yield (74 %) can be enhanced up to 96.3% by the addition of 0.34% triphenyl methane[115, 116].

One of the most effective antiozonants, N-(1,3-di-methyl-butyl)-N'-phenyl-PPHDA is obtained from nitro-, nitroso- or *p*-aminophenylenediamine and *iso*-butylmethyl ketone in the presence of Pd/C catalyst under hydrogen atmosphere [117-121]. The reaction proceeds according to the following general scheme:

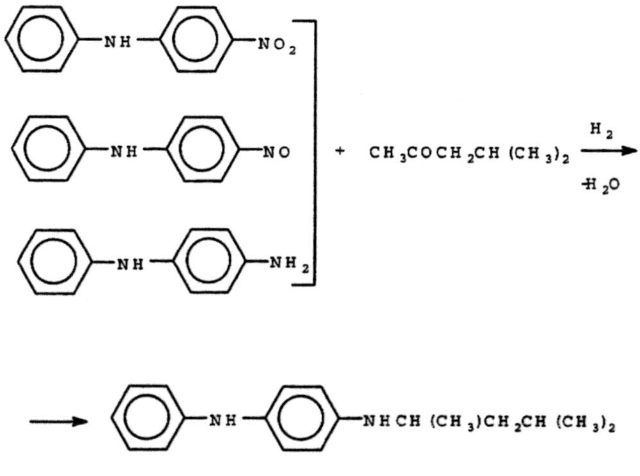

If alkylation is conducted with acetone at a hydrogen pressure ranging from 5 to 300 at and a temperature of 100-250°C, the final products will be N-*iso*-propyl-N'-phenyl-PPHDA [122, 123].

$$C_6H_5NHC_6H_4NO + CH_3COCH_3 = C_6H_5NHC_6H_4NHCH(CH_3)_2$$

The yield can be enhanced by modification of the catalyst with thiocarbamide [125].

Another catalyst used is Pt (0.2-2%) on an appropriate support. This catalyst is reduced at a temperature of 95°C and its activity is enhanced by addition of $(CH_3)_2S$ or hexyl mercaptan [126, 127]. When platinium catalysts are used, the initial *p*-nitrosodiphenylamine should contain neither sulphur nor chlorine since these elements deactivate the catalyst [128].

N,N'-di-(1-ethyl-3-methylphenyl)-PPHDA, N,N'-di-(1-methyl-heptyl)-PPHDA and N,N'-(1,3-dimethyl-pentyl)-PPHDA obtained on 1% Pt/C belong to the widely applied antiozonants [129,130]. N,N'-di-(1,3,3-trimethyl)-PPHDA with a m.p. of 98°C is more appropriate from a technological point of view than are liquid PPHDA. It is obtained from *p*-nitroaniline and methyl-neopentylketone according to the scheme:

$$O_2NC_6H_4NH_2 + 2CH_3COCH_2C(CH_3)_3 \xrightarrow[P_{H_2}]{CH_3COOH,\ Pt/C} (CH_3)_3CCH_2CH(CH_3)NHC_6H_4NHCH(CH_3)CH_2C(CH_3)_3$$

The preparation of lower-toxicity AO such as N-(1,3-dimethylbutyl)-N'-(1,4-dimethylpentyl)-PPHDA and N-isopropyl-N"-methylpentyl-PPHDA is described in Refs 131-133.

When halogens are present in the alkyl chain of substituted PPHDA, their antiozonant activity increases. For instance, N-(2-chloroallyl)-N-phenyl-PPHDA has a 75% higher activity than N-alkyl-N'-phenyl-PPHDA [134].

A mixture of differently substituted PPHDA is prepared by reducing alkylation of 4-nitrodiphenyl or (N-isopropyl-*p*-nitroaniline, *p*-nitroaniline or PPHDA) with two or more ketones at a temperature of 25-120°C, a hydrogen pressure of 28 at and 1% Pd/C catalyst [135, 136].

Hydrogenation of a mixture of $O_2NC_6H_4NH_2$ and the ketones $CH_3COCH_2CH(CH_3)_2$ and $CH_3COCH_2CH_2CH(CH_3)_2$ in a ratio of 1:2 leads to 4.2% *p*-$(CH_3)_2CHCH_2CHCH_3NH_2C_6H_4$ and 41.8% $(CH_3)_2CHNC_6H_4NHCH(CH_3)CH_2CH(CH_3)_2$ [137]. When the ratio between the ketones is 1:1, about 20% of symmetric di-*sec*-alkyl-PPHDA is also obtained [137]. The products of the reaction between *p*-nitroaniline and a mixture of methyl-*iso*-butylketone and methyl-*iso*-amylketone (1:1) are 18.6% N,N'-di(1,3-dimethylbutyl)-PPHDA, 30.6% N,N'-di(1,4-dimethylpentyl)-PPHDA and 50.2% N-1,3-dimethylbutyl-N'-(1,4-dimethylpentyl)-PPHDA.

We have synthesized a number of complexes -catalysts on the bases of Pt(II) and Pd(II) with phenyl-azo-2-naphthol (1), 1,2,4-trianthraquinone (purpurin) (2), 8-oxyquinoline (3), anthranilic acid (4), chloroanilic acid (5) and dimethylglyoxime (6) on γ-Al_2O_3 [138].The complexes are obtained by the treatment of γ-Al_2O_3 supported ligands (mentioned above) with aqueous solutions of K_2PtCl_4, K_2PdCl_4, $Pd(OAc)_2$ at a metal ion: ligand ratio of 1:2.5.

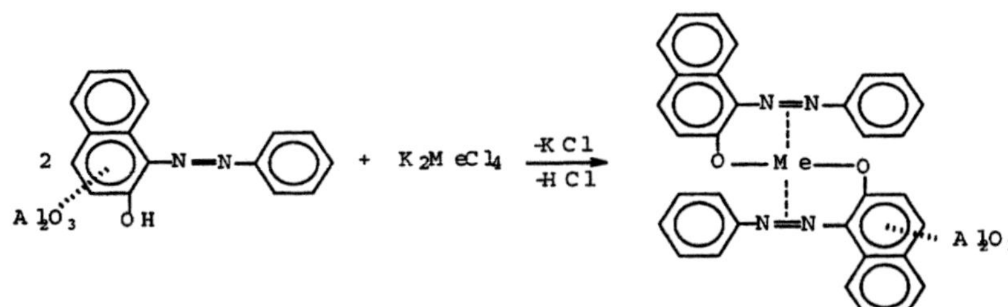

1

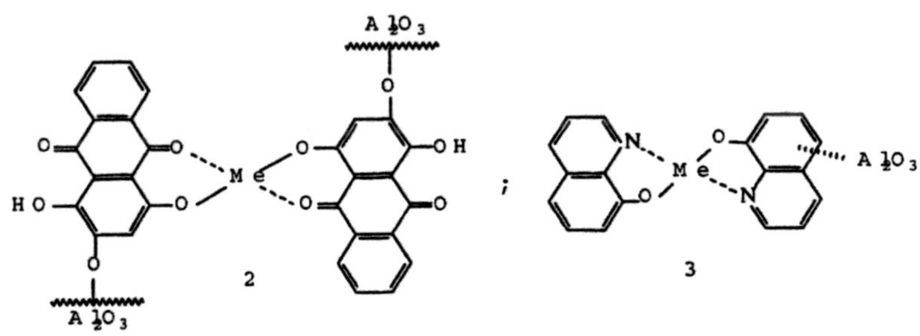

2 ; 3

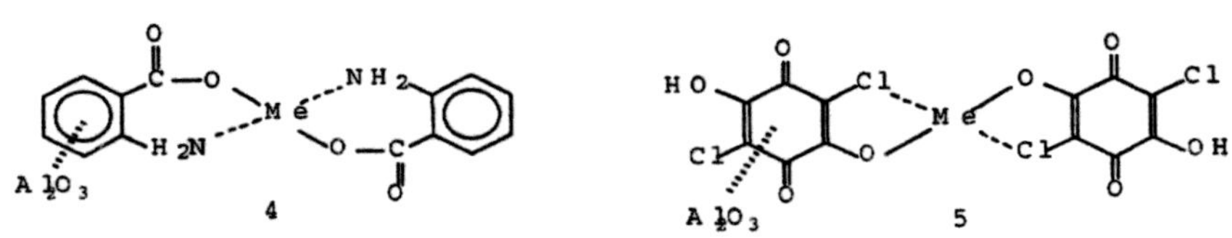

4 5

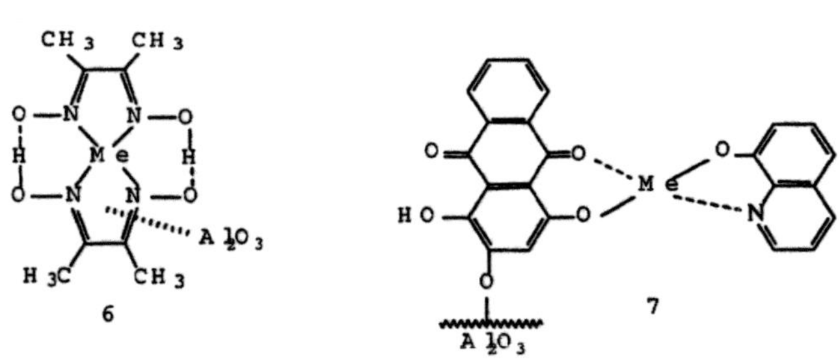

6 7

The supported metal content is 0.5-3% wt. In a separate tests with the homogeneous and immobilized complex (1) we have established that the immobilization does not affect the catalyst activity. This parameter has been evaluated in a model reaction of nitrobenzene hydrogenation in various solvents (Table 4.2).

Table 4.2

Catalytic activity of Pd(II)-complexes in nitrobenzene hydrogenation to aniline. Catalysts weight -0.1 g, Pd=0.62%, nitrobenzene - 1Mm, 67°C, P_{H2} = 1 at.

Catalysts	Solvent	Rate of Hydrogenation, mol H_2/(min.g.atom Pd)
$Pd(purpurin)_2$	Octane	62.3
$Pd(purpurin)_2$	Heptane	47.2
$Pd(purpurin)_2$	Decane	32.0
$Pd(purpurin)_2$	Cycloheptane	22.0
$Pd(purpurin)_2$	DMFA	21.2

Pd(purpurin)$_2$	Ethanole	22.4
Pd(purpurin)$_2$ + 8-oxyquinoline*	Octane/DMFA	59.5/27.1
Pd(purpurin)$_2$ + phenylazonaphthol	Octane/DMFA	59.5/27.1
Pd(purpurin)$_2$ + dimethylglyoxime	Octane/DMFA	47.7/30.9
Pd(purpurin)$_2$ + chloroanile	Octane/DMFA	18.5/16.9

The catalysts preserve their activity for a long time, even after 14-fold application. These catalysts hydrogenate nitrobenzene to phenylhydroxylamine in DMFA and to aniline in alcohols. The reaction kinetics was followed by registration the ozone uptake, chromatographically (HPLC, gas and thin-layer chromatography), the nitrobenzene consumption and phenylhydroxylamine and aniline formation. The strong complexing functions of DMFA stabilize phenylhydroxylamine and the reaction kinetically stops at the absorption of 2 moles hydrogen, while in alcohols (ethanol, *iso*-propanol) the reaction goes on up to 3 mol ozone uptake forming aniline. The hydrogenation of *p*-nitroaniline in DMFA in the presence of Pd(purpurin) gives a quanitative yield of *p*-phenyldiamine.

4.2.2.1.2. Alkylation in the Absence of Hydrogen

Substituted paraphenyldiamines can be obtained without using hydrogen and expensive noble metals as catalysts [139, 140]. When N-alkyl-4-nitroaniline is mixed with *n*-hexanol and 59% NaOH, and heated for 1 h at a temperature of 100°C, it yields monosubsituted PPHDA:

$$RHNC_6H_3(R_1)NO_2 + 2R_2CH_2OH \rightarrow RHNC_6H_3(R_1)NH_2,$$

where: R is C_1-C_{12} alkyl or naphthyl radicals, and R_1 is C_1-C_{12} alkyl radical.

Monosubstituted PPHDA can also be obtained by interaction of N-substituted aminophenols with ammonium in the presence of an acid catalyst (Lewis acid):

$$RHNC_6H_4OH + HNH_2 \rightarrow RHNC_6H_4NH_2 + H_2O$$

Other catalysts used in this reaction are $FeCl_3$ or a mixture of free iodine and powdery iron in a ratio of 6:1 [141, 142].

$$RNH_2 + HOC_6H_4NH_2\ (RNHC_6H_4NH_2 + H_2O,$$

where: R = C_6H_5-, $C_6H_4CH_3$ or C_6H_{11}-.

Heating at 215-220°C in a nitrogen atmosphere of a mixture of aniline, cyclohexylamine and N-(1,3-dimethylbutyl)-*p*-aminophenol in the presence of iodine, powdery iron and NH_4Cl leads to the formation of 10% N,N'-(1,3-dimethylbutyl)-PPHDA, 5.5% N-cyclohexyl-N'-(1,3-dimethylbutyl)PPHDA, 30% N-(1,3-dimethylbutyl)-N'-phenyl-PPHDA, 8% N-cyclohexyl-N'-phenyl-PPHDA and 9% N,N'-diphenyl-PPHDA as final products [143].

Substituted N,N'-diphenyl-PPHDA with low melting points appear after a reaction of hydroquinone with a mixture of amines (35% aniline, 25% toluidine and 40% xylidines) during heating in the presence of a catalyst ($FeCl_3$) of condensation [144].

Reaction of substituted PPHDA with thiols or formaldehyde in an acid medium results in their conversion into active stabilizers having a complex behavior against the action of oxygen, ozone, light and heat [145-147]. Their general formula is:

$$R_1\text{-}S\text{-}CH_2\text{-}C_6H_2\text{-}(R_2)(R_3)\text{-}NH\text{-}C_6H_4\text{-}NH\text{-}R_4,$$

where: R_1 is alkyl C_1 - C_{20}, aryl C_6 - C_{14}, alkylaryl; R_2 and R_3 are H, alkyl C_1-C_7, aryl C_6-C_{14} and R_4 is [-C_6H_4-CH_2-S-R_1].

Alkylated 2,4-toluenediamines are good antiozonants and antioxidants, which interrupt the chain-radical process taking place during the oxidation of polymeric materials [148]. A mixture of 2,4-toluenediamine, dicyclopentadiene and pentene is stirred in the presence of a catalyst (12% Li and 87% Si) under elevated pressure in a nitrogen atmosphere. A mixture of 3-(cyclo-pent-2-enyl)-2,4-toluenediamine and 5-(cyclopent-2-enyl)-2,4-toluene is obtained. The conversion of 2,4-toluenediamine amounts to 64%.

4.2.2.2. Hydroquinolines

Hydroquinolines are prepared by the classical method of Scraup using a reaction of aromatic amines with aldehydes or ketones in the presence of oxidants and acid catalysts according to the following scheme:

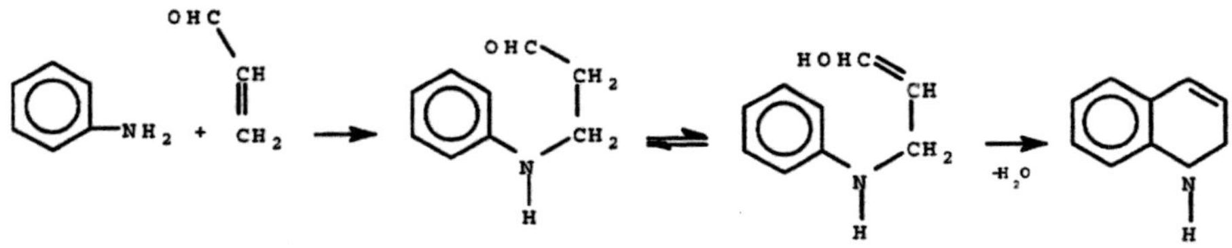

A mixture of aniline and C_8-C_{15}-alkylated amines (45% C_{12}, 15% C_{11} and 12% C_{10}) in an acid medium is heated up to a temperature of 140°C for 12h, after which acetone is added. The final product is an oil containing 58% 6-dodecyl-1,2-dixydro-2,2,4-trimethylquinoline, 6% dodecylaniline and oligomer of 1,2-dihydro-2,2,4-trimethylquinoline [149, 150].

Addition of acetone to 4,4'-(ethylenedioxyl)-dianiline for 12 h in the presence of benzenesulphonic acid at 150-160°C yields 6,6'(ethylenedioxyl)-di-1,2-dihydro-2,2,4-trimethylquinoline [151].

When 2,2,4-trimethyl-1,2-dihydroquinoline is allowed to polymerize in the presence of HCl, an active polymeric antiozonant is obtained. The monomer is formed during reaction of aniline with acetone at a temperature of 120°C in the presence of *p*-toluenesulphonic acid as a catalyst [152]. The polymer is isolated by dissolution of the reaction mixture in benzene, followed by neutralization and evaporation of the solvent. The reaction mixture can be dispersed in water and then granulated. The obtained antiozonants are usually applied in a combination with N-phenyl-N'-isopropyl-PPHDA [153, 154].

4.2.2.3. N,N'-Disubstituted Hexahydropyrimidines

These are nonstaining antioxidants [155], which are prepared by condensation of cyclohexanone or its derivatives with ammonia. Tetrahydropyramidine is first formed and then reduced to hexahydropyrimidine whose pyramidine ring is open to form diamine. The two amino groups form bonds with aldehydes and yield N,N'-substituted hexahydropyrimidine. The reaction proceeds as follows:

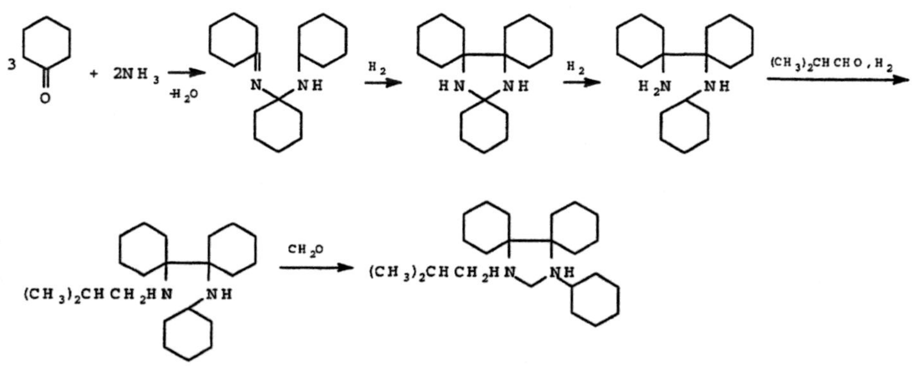

4.2.2.4. N-Substituted Dimethylpyrols

These compounds are also nonstaining antiozonants [156]. They are obtained from 1,4-dicarbonyl compounds passing through an enol intermediate which reacts with amines giving:

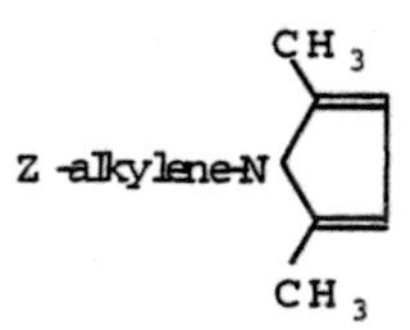

where Z is substituted NH_2; RNH-; R_2N-; R-NH-C(S)-N=N- or -(R)N-C(S)-NH_2

4.2.2.5. Enamines

The enamines are nonstaining antiozonants with the general formula:

$$(R_1)(R)N\text{-}HC{=}C(R_2)(R_3),$$

where: R_1, R_2 and R_3 are alkyl groups with C_1 - C_6.

These compounds are prepared by condensation of amines with ketones. The amines used are pyrolidine, pyperidine, substituted morpholines, etc.

$$RR_1NH + (R_2)(R_3)CHC(R_4)O \rightarrow (R)(R_1)N\text{-}(R_4)C{=}C(R_2)(R_3) + H_2O$$

If the ketones are unsymmetric, a mixture of products is obtained. Condensation of 2,6-dimethylmorpholine with 2-ethylhexanal yields 1-(2,6-dimethylmorpholino)-2-ethyl-hexene-1 which is usually applied for natural rubber and butadienestyrene rubber [157].

$$RCH_2CHO + HN(R')_2 \rightarrow RCH_2CH(OH)N(R')_2 \longleftarrow (RCH{=}CHN(R')_2 + H_2O$$

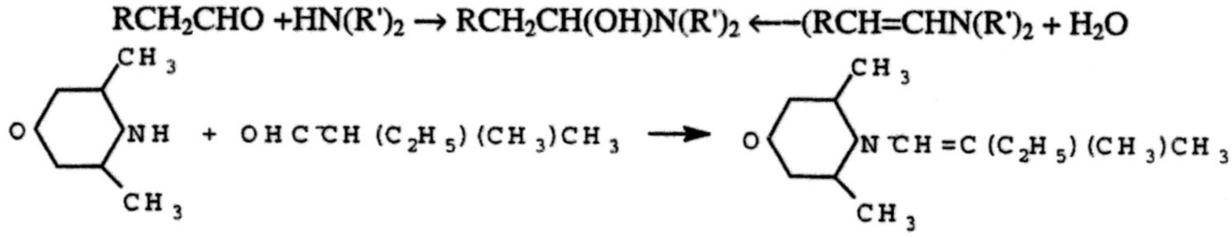

4.2.2.6. Nitrone Compounds

The C-substituted or unsubstituted aryl-N-substituted nitrones are appropriate not only as antiozonants but also as antifatigue inhibitors [158]. Particularly effective in this respect are aldonitrones of the type:

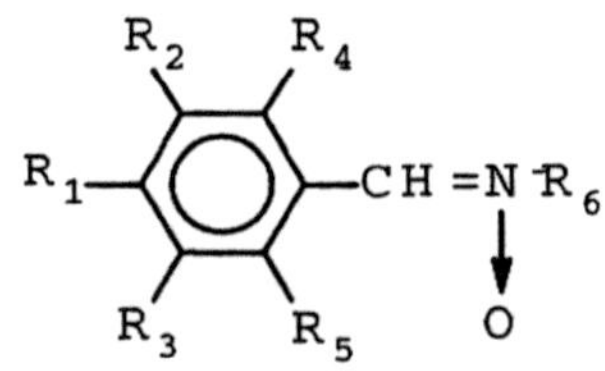

where: R_1 is mono- or disubstituted aminogroup; R_2 and R_3- alkyl or alkoxy groups; R_4, R_5= R_2 and R_6 - *iso*-propyl, *sec*-butyl or *tert*-butyl group.

The nitrone compounds may be prepared by a variety of ways, for example:

a) by oxidation of the corresponding aryl-N-substituted or unsubstituted (branched alkyl or cycloalkyl) hydroxylamines;

b) by reaction of aryl ketones with primary hydroxylamine;

c) by N-alkylation of the corresponding oxime;

d) by reaction of the corresponding ketimine with primary hydroxylamine ;

e) by oxidation of the corresponding N-substituted imine.

The most used nitrone is 4-hydroxy-3,5-dimethyl-N-*tert*-butylnitrone obtained when N-*tert*- butyl-hydroxylamine and 3,5-dimethyl-4-hydroxybenzaldehyde in a ratio of 1:1 are dissolved in absolute alcohol and the mixture is allowed to stand for 5 days at room temperature.

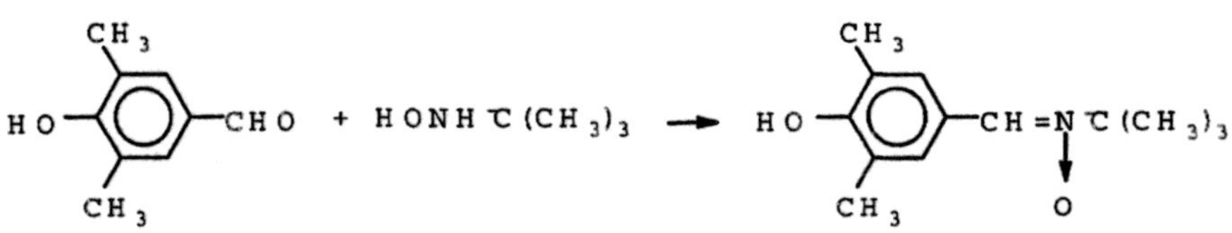

4.2.2.7. Derivatives Of 3(5)-Methylpyrazone

The general formula of these compounds is:

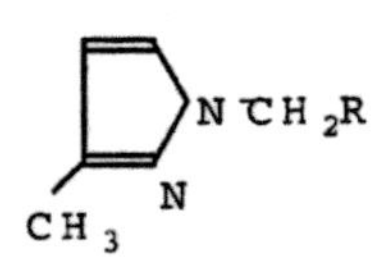

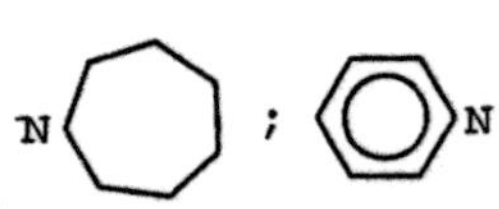

where: R -

The ozone resistance of rubber mixtures based on natural rubber with Ozonine-16 (pyperidinomethyl-3,5-methylpyrazole) and Ozonine-51 (hexamethyleneimino-3,5-methylpyrazole) exceeds 1.5 to 2 times that of mixtures containing 2246 (2,2'-methylene-*bis*-4-methyl-6-*tert*- butylphenol) and is close to that reached by tributylcarbamide. A mixture of Ozonine-16 or Ozonine-51 with 2246 in a ratio of 1:1 displays the highest activity [159].

4.2.2.8. Enolethers

These compounds are used as antiozonants for natural and synthetic rubber. They are slightly staining. The following enolethers are synthesized: 2H-pyran-2-methyl-(3,4-dihydro-2H-pyran-2-carboxylate); 3,4-

dihydro-2,5(dimethyl or diisobutyl, didecyl)-2H-pyran-2-methyl-(3,4-dihydro-2,5-dimethyl-2H-pyran-2-carboxylate) with the general formula:

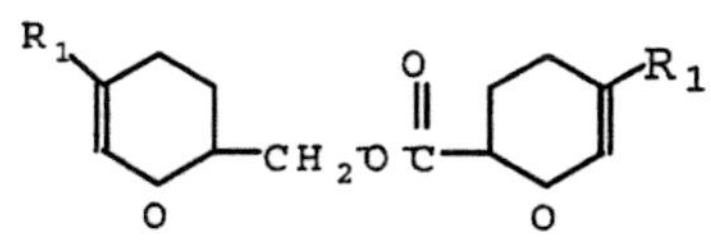

where: R is alkyl, aryl or alkylaryl radical.

By condensation of 3,4-dihydro-2H-pyran-2-carboxyaldehyde and aluminium isopropylate with stirring for 4 h at 40°C, 75% of 3,4-dihydro-2H-pyran-2-methyl-(3,4-dihydro-2H-pyran-2-carboxylate) is obtained [160]. Mixing of different enoethers results in a better antiozonant activity. For instance, a mixture of hexanediol-1,6-di-(2-methyl-prop-1-ethyl)ether and 1,1'-bis-(2-methyl-prop-1-epoxy)-dietheylether is prepared by aldol condensation of *iso*-butyl aldehyde and 1,2-hexanediol in the presence of toluenesulphonic acid, quinoline and cyclohexane as a solvent [161].

4.2.2.9. Ethers

Ethers with the general formula:

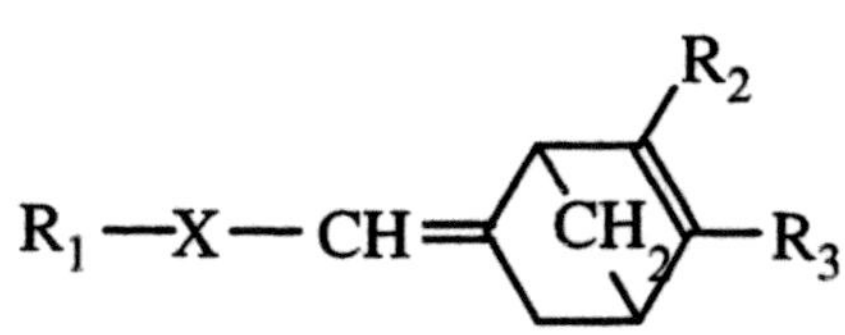

are odorless antiozonants where R_1 is a hydrocarbon group which may contain a heteroatoms; X is S or O; R_2 and R_3 are H, $-CH_3$.

These ethers are used together with R-CO-NH-NH_2 (R-alkyl group with C_3- C_{10}) and exhert good antiozonant properties [162].

4.2.2.10. Cyclic and Acyclic Acetals and Ketals

Cyclic and acyclic acetals and ketals are nonstaining antiozonants with general formula [163-166]:

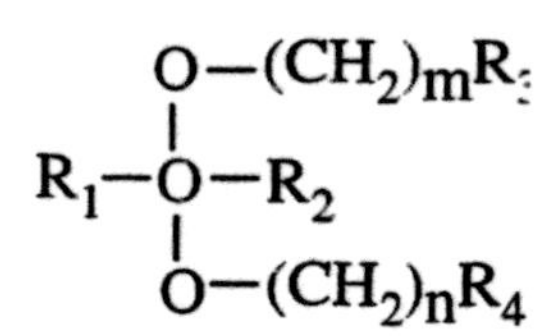

where: n and m range from 0 to 12; R_1, R_2=H, alkyl group with C_1-C_3, alkylene group containing halogens, cyclo or bicycloalkenes with C_5 - C_7, aryl groups with C_6 -C_{10} or alkylaryl groups.

They are prepared by the reaction of alcohols with aldehydes or ketones in the presence of catalysts:

$$RCHO + 2RCH_2OH \rightarrow RCH(OCH_2R)_2$$
$$(R)_2CO + 2RCH_2OH \rightarrow (R)_2C(OCH_2R)_2$$

The catalysts used are: H_2SO_4, H_3PO_4, *p*-toluenesulphonic acid, boron trifluoride, etc. These cyclic acetals are usually applied with microcrystalline waxes [167]. Acetals and ketals with unsaturated cyclic groups are less volatile, have no odor and are compatible with rubbers [168].

4.2.2.11. Other Classes of Compounds

4.2.2.11.1. *Bis*-Alkylaminophenoxy Alkanes

Bis- alkylaminophenoxy alkanes with the general formula:

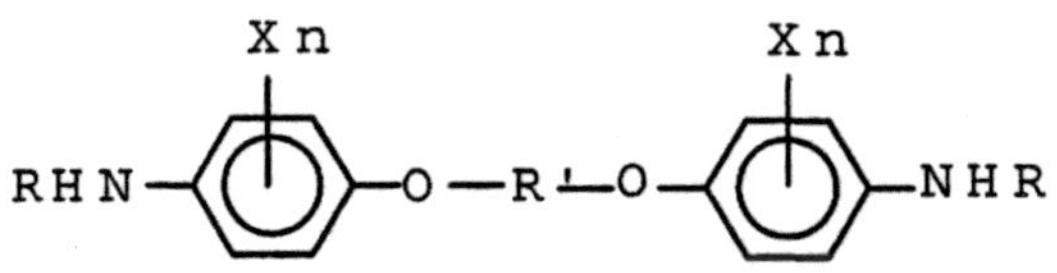

(where: X=halogen, R= *sec*-alkyl radical with C_3 - C_8, R' is $-CH_2-$, $-CH_2-CH_2-$) are obtained from nitrophenol by reducing alkylation in the presence of Pt/C as a catalyst [169].

4.2.2.11.2. Derivatives Of 2,2,7,7,-Tetramethyl-1,4-Diazocyclopentane

The derivatives of 2,2,7,7-tetramethyl-1,4-diazocyclopentane with the formula

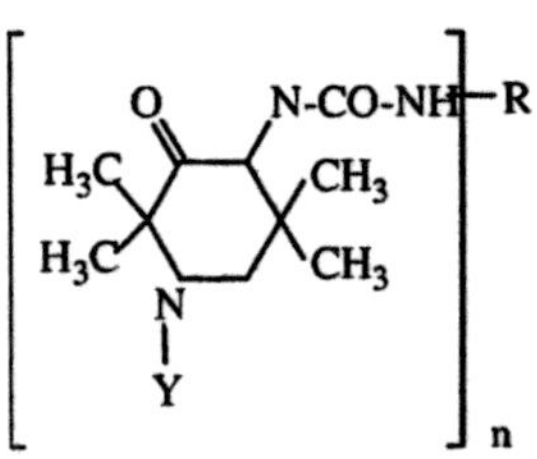

where: n is 1, Y is H, CH_3, O and -NH, R denotes alkyl groups with $C_1 = C_{20}$; where n is 2, R is alkylene groups with C_1 -C_{20}, cycloalkene groups with C_7 -C_{18} or alkylaryl groups with C_5 - C_{12}.are also applied as antiozonants for polymers. Their preparation is accomplished in a reaction of 1,4-diazocyclopentane-5-one with isocyanate [170].

4.2.2.11.3. *Bis*-Cyclopentadienyl Compounds

Bis-cyclopentadienyl compounds with the general formula:

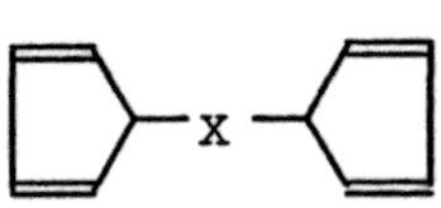

where: X is α,α'-*p*-xylilene and 4,4'-oxydiphenyl-dimethylene.

are good antiozonants [171].

4.2.2.11.4. Alkylnaphthenes

Alkylnaphthenes are applied as stabilizers of cables and insulating materials. They are obtained by alkylation of naphthaline with dodecane chloride in the presence of $AlCl_3$ at 80°C or 1-hexadecanol in the presence of BF_3 at 130°C.

$$C_{10}H_8 + C_{12}H_{25}Cl \leftrightarrow \beta\text{-}C_{12}H_{25}\text{-}C_{10}H_7 + HCl$$

$$C_{10}H_8 + C_{16}H_{33}OH \rightarrow \beta\text{-}C_{15}H_{30}\text{-}C_{10}H_7 + H_2O$$

The yield is 60% [172].

4.2.2.11.5. Aminomethylene Derivatives of Furane

Aminomethylene derivatives of furane are cheaper and nonstaining antiozonants [173].

4.2.2.11.6. Lactams

Strong antiozonants are the lactams with the general formula:

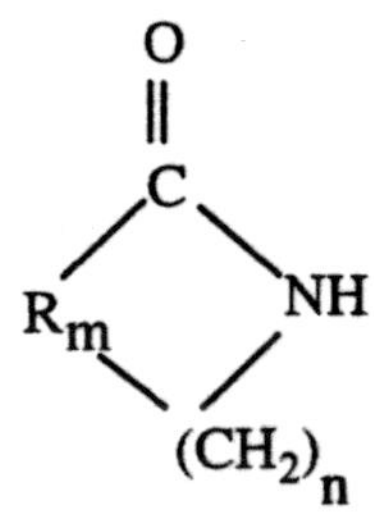

where n=3 - 20; m=1 - 2; R - alkyl or phenyl radical.

Comparison based on antiozonants activities indicates ε-caprolactam and capryllactam to rank first [174].

4.2.2.12. S-Containing Compounds

The derivatives of isothiocarbamate used as antiozonants for natural rubber, butadiene rubber, isoprene rubber and butadienestyrene rubber are noncolouring and nonstaining, and have general formula:

$$R_4\text{-}N{=}(SR_1)\text{-}N\text{-}R_2R_3$$

where: R_1, R_2, R_3 and R_4 are alkyl C_1 - C_{30}, phenylalkyl C_1 - C_6 or cycloalkyl radicals [175].

A high antiozonant activity is exhibited by sulphides R-S-R [176], where R is $[C_6H_5NHC_6H_4NHCO]_nX$, n is 1 or 2, X denotes methylene, ethylene or 1,2-propylene radicals. When R=H, the sulphides are obtained by a reaction of 4-aminodiphenylamine with acids having the general formula:

$$H\text{-}S\text{-}X\text{-}(\text{-}SO_2H)_n$$

If R' = -S-R, the acids should have the formula $(HOOC)_n$-X-S-S-X-$(COOH)_n$, (dithioacetic, β-dithiopropionic). The sulphides obtained are *bis*-(4-anilinophenyl aminocarbonyl methyl) disulphide and *bis*-(4-aminophenyl aminocarbonyl ethyl) disulphide. These antiozonants are easily soluble in rubbers compositions.

Sulphonate antiozonants with the general formula $R_1R_2SO_3N(R_3OH)$ (R is alkyl radical with C_1 - C_{13}, R_2 - aryl radical with C_1 - C_6, R_3 -alkylene with C_4 - C_6) are nonstaining, nontoxic and, above all, cheap to prepare [177]. The initial substances used are alkylated products, resulting from the preparation of dodecyl benzene (hydrocarbon radicals with C_{24} - C_{30}) which is sulphonated with oleum and then reacted with alkanolamines at pH=7.5.

The sulphides of 3,4-dehydropyperidine are efficient antiozonants. They have the general formula:

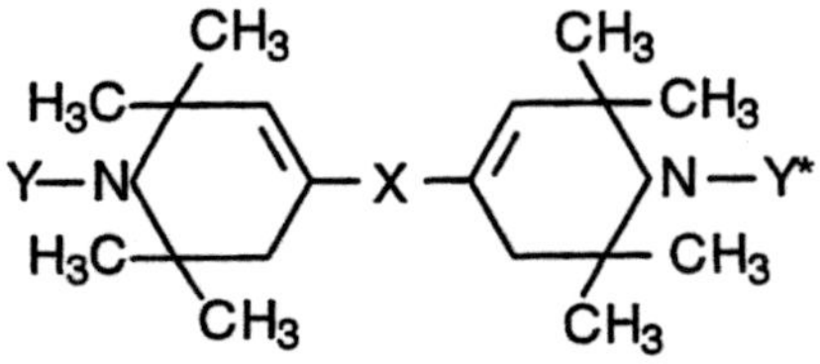

where: X = S, SO, SO_2; Y, Y' = methylene, isopropyl, *sec*-butyl radicals.

Di-4-(3,4-dihydro-2,2,6,6-tetramethyl-pyperydinite)-sulphide, sulphoxide or sulphone are prepared by interaction of:

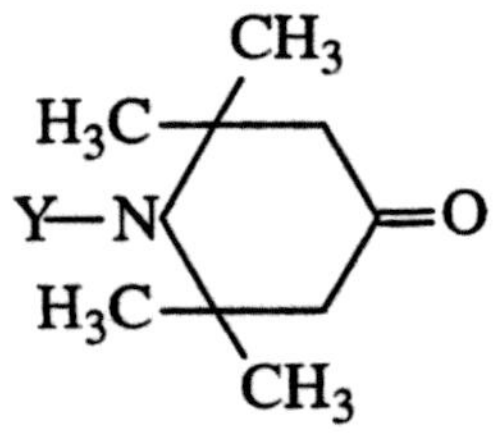

with a gaseous mixture of HCl and H_2S [178, 179].

Ozone resistance is observed with the products of mercaptoaldehyde condensation with phenols [180-183].

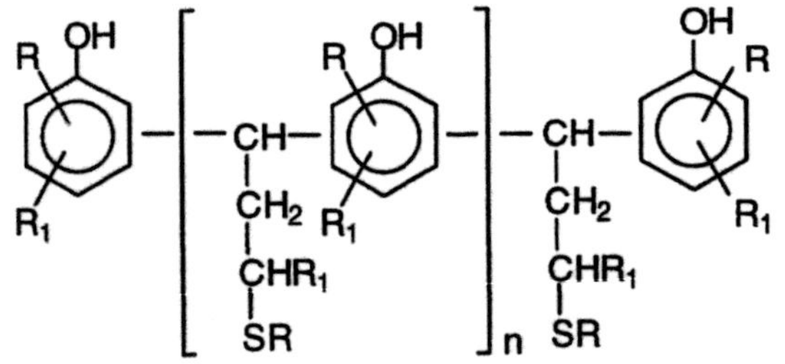

where: R, $R_1$are H, CH_3, *tert*-butyl radicals.

Triazinedisulphide derivatives to be used as antiozonants are relatively easily prepared by a reaction of benzthiazoldisulphide with 2(4-morpholinyl di-thio)benzthiazole and 1,3,5-triazine dithiol in the presence of a neutralizing agent (NaOH) [184, 185].

In comparison with Zn-dialkyldithiophosphates, dithiolderivatives exhibit a higher antiozonant activity. They have the general formula:

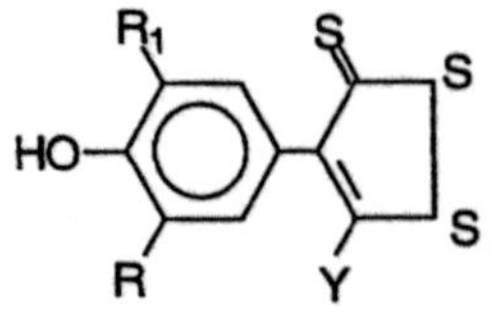

where: R_1 and R_2 are alkyl radicals with C_1 -C_{12}; T is H or SR_2; R_2 denotes alkyl radicals with C_1 -C_{20}.

The reaction of alkylated phenols with sulphur in the presence of a basic catalyst yields 4-(3,5-di-iso-prpyl-4-hydroxyphenyl_-1,2-dithio-3-thione [186, 187].

2-Vinylphenothiazine enhances the stability of rubber products towards ozone, temperature and repeated deformation [188]. Butoxyethyl N,N'-dimethyldithiocarbamate is used as antiozonant of light rubbers [189].

Nitrile antiozonants with the general formula

$$R_1-S-\underset{\displaystyle R_2}{\underset{|}{(CH)_n}}-C\equiv N\ .$$

where: R - alkyl radicals with C_1 - C_{18} or $(CH_2)_n$-CN and R_2 - H, CH_3 or C_2H_5

are nonnstaining AO and they are characterised with low toxicity (LD_{50}=5300 mg/kg) and low production expenses [190].

4.2.2.13. Si-Containing Compounds

Aminoorganosilanes with general formula:

$$R_nSi(CH_2NXY)_{4-n},$$

where: R - alkyl radical; X - aryl, alkoxyaryl, aminoaryl, phenyl, and naphthyl radical; Y - alkylaril, aminoaryl radical

show a good AO activity.

Dimethyl-*bis*-tert butylhydroxysilane used for the stabilizing of butyl rubber is prepared by interaction of 3,5-di-*tert*-butyl-4-hydroxy-benzyl alcohol with dimethylchlorosilane [191, 192].

The oligodiphenylsiloxyquinonezarylene, which is a good thermostabilizer, is isolated by heterophase condensation of diphenylchlorosilane with aromatic diols in the presence of suitable catalysts [193].

4.2.2.14. P-Containing Compounds

Phosphites and phosphates of 5-norbornene-2-methanol are used as antiozonants of polychloroprene rubber and have the following general formula:

$$HP(O)(OR)_2,\ (RO)_3P,\ (RO_3)PO,$$

where: R -5-norbornene-2-yl methyl.

The interaction of PCl_3 with 5-norbornene-2-methanol in a pyridine medium results in a mixture of *bis*-(5-norbornene-2-methyl) phosphite (17%), *tris*-(5-norbornene-2-methyl) phosphate (22%) and *tris*-(5-norbornene-2-methyl) phosphate (15%). The reaction with $POCl_3$ leads to the formation of phosphates only [194]. Aryl phosphites are applied as stabilizers of synthetic rubber [195, 196].

Organic polymers are stabilized by amide esters of phosphonic acid having the general formula:

$$P(ZR)(Z'R')(NR_2R_3),$$

where: Z, Z' - O, NH; R_2, R_3, R and R' - linear or cyclic, substituted or unsubstituted aliphatic or aromatic radicals [197].

4.2.3. Application

The stabilizing systems on the basis of well-known inhibitors and their mixtures developed by various authors are presented in Table 4.3.

Table 4.3

Antiozonant systems for rubbers and specific protective properties

No.	Stabilizing system	Elastomer	Protecting ability	Ref.
1.	a) PPHDA type AO b) Imidazole	EPDM	Increasing the O_3 resistance at elongation	198
2.	a) N-Phenyl, N'-iso-propyl-PPHDA (4010 NA) b) N-Pheny-β-naphthylamine (Neozone D)	Diene rubbers (DR's)	O_3	199
3.	a) Reaction products of acetone with diphenylamine b) Zn-2-mercaptotoluylimidazole	EPDM	O_2, O_3, t^o	200
4.	a) 2,6-Di-tert-butyl-cresole b) 4,4'-thio-bis (6-tert-butyl)-m-cresole (SC)	Butadienstyrene rubbers (BSR's)	O_3, t^o	201
5.	Cd salts of I. $ArNRArNR_7$-R_1-COOH II. ArNHAr-R_4-COOH III. $OH(R_5)(R_6)ArR_2$-COOH R - H, halogen atom, C_1-C_8 alkyl; R_1 - C_1-C_{20} alkylene, -CO-R_2; R_2 - C_1-C_{20} alkylene; R_3 - C_1-C_8 alkylene; R_4 - -OCO-R_2, R_5 è R_6 - H or C_1-C_8 alkyl, R_7 - H or C_1-C_4 alkyl	Nitrile rubbers (NiR's)	O_3	202
6.	a) N, N'-Bis-(2-hydroxy-4,6-dimethylbenzyl)-piperazine b) 2, 2' - Methylene-bis (4-methyl, 6-tert-butylphenol) -(2246)	Natural rubber (NR)	t^o	203
7.	a) Na-Aluminiummethylsilicate	Carboxylate synthetic latecs	O_3, nonstaining, nontoxic	204
8.	a) PPHDA type AO - 1-4 php b) Imidazole-0.2-4 php	Cl - sulfonated polyethylene	t^o, O_2	205
9.	a) PPHDA type AO - 0.1-5 php b) p-Di-nitroso-arene - 0.1 php	NR	O_3, O_2	206
10.	a) PPHDA type AO b) Bis-maleimide a : b=80-20 : 20-80	Vulkanizates (V's)	O_3	207
11.	a) N, N'-Diphenyl-PPHDA, 0.1-5 php b) Petroleum type waxes 0.1-10 php	Isoprene rubber (IR)	O_3, nonstaining	208
12.	a) Polymerized 2,2,4-trimethyl-1,2-dihydroquinoline (Flectol H) b) 4010 NA c) N - Phenyl - N' - (1,3-dimethylbutyl) - PPHDA (S13) a:b:c=50:25:25	NR	O_3, inhibitor of flex cracking, antifatiguing	209
13.	a) N, N'- Diphenyl-PPHDA b) N, N'- ditoluyl - PPHDA a : b = 10-50 : 90-50	NR	O_3, non-staining	210
14.	a) 4010 NA b) Montmorilonite	NR	O_3	211
15.	a) Phenothiazine b) N - Alkyl - N' - phenyl - PPHDA a : b = 20-60 : 80-40	NR	O_3 non-staining	212

16.	a) Di-iso-propyl-di-thiophosphate b) Di-hydro-di-phenylpropane c) 4010 NA a : b : c = 1 : 0.5 : 0.5	BSR's	O_3, ligth	213
17.	a) S13 b) Flectol H c) 6-anilino-2,2,4-trimethylquinoline	DR's	O_3, non-staininig	214
18.	a) 4010 NA b) S13	DR's	O_2, O_3, t^o	215
19.	a) Neozone D - 0.5-2 php b) 9 - Phenyl - 10 - p - vinyl-phenyl-anthracene - 2-7 php	V's	O_3, t^o	216
20.	a) Deca-cis-(nonylphenyl)-hepta-cis-(di-propylenglycole- octaphosphite) b) N, N' - Di - (2 - octyl) - PPHDA a - 0.75-1.75 php b - 1.25-0.25 php	NR	O_2	217
21.	a) N - iso - propyl - amino-di-phenyl-amine b) 2, 6 - Di - tert - butyl - phenol	DR's	t^o	218
22.	a) Phenothiazine - 20-80% b) Flectol H - 80-20%	BSR's	thermal, antifatiguing	219
23.	a) 2246 - 0.3-0.75% b) 2 - Mercaptobenzimidazole - 0.15-0.5%	NR	O_2, O_3	220
24.	a) 1, 1, 3 - Tris - (2 - methyl - 4 - hydroxy - 5 - tert - butyl - phenyl) butane (Topanol CA) b) 4010NA a - 5-50; b - 95-50 w.%	Synthetic rubbers (SR's)	O_2	221
25.	a) 1, 3 - Bis - pyperidino - methylene - imidazoline - 2 - thion b) 2246	DR's	O_3	222
26.	a) S13 b) Flectol H	NR	O_3	223
27.	a) N - Aminophenyl-phosphoramidate ester b) Zn - di - hydro - carbyl- dihydro phosphate	NR and SR	O_3, non-staining	224
28.	a) S - Containing sulfur ester b) Amine or phenolic AO	NR è SR	t^o, O_2	225
29.	a) Topanol CA, 5-50 wt.% b) Tributoxyethylphosphate 55-50 wt.%	SR	O_2, increase dispersity and penetration ability	226
30.	a) Diethyldithiocarbamate b) Co -diethyldithiocarbamate c) 4010NA	DR's	O_3	227
31.	a) Phenylhydrazone b) Amine or S-containing AO	NR	O_2	228
32.	a) 2 - Mercaptomethylbenzimidazole b) Flectol H	EPDM and DR's	O_3,O_2	229
33.	a) 4010 NA b) Flectol H	DR's	O_3, O_2, antifatiguing	230
34.	a) 4010 NA, 1.7 php b) Flectol H, 1.3 php or c) 4010NA 1.9 php d) SC 1.1 php	IR, BSR's	t^o, O_3	231

35.	a) 4010 NA b) Flectol H a : b = 1-2.5 : 0.5-2 php	NR, IR, and BSR's	t^o, O_3	232
36.	a) 4010 NA b) SC a : b = 1-2.5 : 0.5-2 php	NR, IR, BSR's	O_3	233
37.	a) N, N' - Bis (1,5- dimethylpentan) - PPHDA (S77) b) SC a : b = 1-2.5 : 0.5-1 php	NR, IR, BSR's	O_3	234
38.	a) S77 b) Flectol H a : b = 1 : 0.5 php	NR, IR, BSR's	O_3	235
39.	a) S77 b) 2 - Methylmercaptobenzimidazole a : b = 1-2.5 : 0.5:2 php	IR and BSR's	O_3	236
40.	a) 4010 NA b) Flectol H a : b = 1.7 : 1.3 php	BSR's	O_3	237
41.	a) 2 - Methylmercaptobenzimidazole b) N - Phenyl N' - octyl - PPHDA	NR	antifatiguing, O_3, t^o	238
42.	a) 4010 NA b) Neozone D	V's	t^o	239
43.	a) Unsaturated cycloaliphatic carboxylated ester, 1 php b) Paraffin - 0.25 php	All rubbers	O_3, non-odoring	240
44.	a) 2 - Mercaptobenzimidazole b) Castor oil	Chloroprene rubbers	O_3	241
45.	a) 2246 - 0.5 php b) Tetrakis (2, 4 - di - tert - butylphenyl - 4, 4' - diphenyl) phosphite - 0.25 php	NR, BSR's	O_3, t^o	242
46.	a) PPHDA type AO - 1-3 php b) Paraffins - 1-5 php	V's	O_3, t^o	243
47.	a) S13 - 1 php b) Naphthenes - 0.5-10 php	DR's	O_3, t^o	244
48.	a) Di-tert-butyl phenols - 1-5 php b) Zn - salt of mercaptoimidazole - 2-8 php	EPDM	t^o,O_3	245
49.	a) Amines b) Imidazoles	EPDM	t^o,O_3	246

Note: AO - antiozonant

We have developed methods for evaluating the antizonants efficiency, a series of synergetic stabilizing systems for rubber mixtures, highly stabilizing elastomers compositions for sidewalls and protectors of light and heavy PT and highly effective stabilizing compositions for hose faces and chambers [247-257].

By applying Eq.1.4 the rate constants of six industrial antiozonants of PPHDA type, were determined and their values are listed in Table 4.4 [258].

Table 4.4

Rate constants of PPHDA-type antiozonants at 20°C in CCl_4

No.	Antiozonant	$k.10^{-7}$, $M^{-1}.s^{-1}$
1.	N, N' - Dinitroso-di-(C_7-C_9) - alkyl - PPHDA	0.5
2.	N, N' - Di - (2 - ethylhexyl) - PPHDA	0.7
3.	N - Phenyl - N' - iso-propyl - PPHDA	1.2
4.	N, N' - Di - (1,4 - dimethylpentyl) - PPHDA	1.3
5.	N - iso-propyl - PHDA	1.4
5.	Polymerized 2, 2, 4 - trimethyl - 1, 2 - dihydroquinoline	2.2
6.	N - Phenyl - N' - cyclohexyl - PPHDA	2.4

An express method for determination of the AO efficiency in dependence on the rubber type is proposed. The AO activity is evaluated by comparison the values of the rate constant in chlorinated hydrocarbon. This method includes the separate determination of the molecular masses of rubbers in solution with and without antiozonants and of the corresponding amounts of absorbed ozone. The antiozonant efficiency is defined as a ratio between the amount of consumed ozone in an elastomer solution containing antiozonant and the amount of reacted ozone in the antiozonant-free solution (Fig.4.7) through which an equal change of the molecular mass of both the solutions is attained. The registration of the changes begins at initially equal molecular masses, further goes on with maintaining the ratio between the amount of absorbed ozone in solution containing antiozonant and the antiozonant amount in the range from 0.1 to 1 and ends at values of the antiozonants activity vary from 1 to 3.

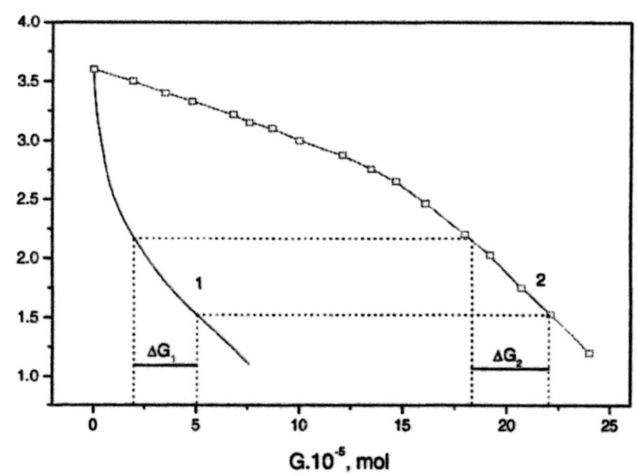

Fig.4.7. Variation of the relative viscosity of SKD (0.6/100)- 1 and the same solution containing 3.7 mM 4010NA (2) at 20°C.

The quantitative determination of the AO efficiency was estimated from the relationship [259]:

$$E_f = \Delta G_2/\Delta G_1 \quad \text{Eq.4.5}$$

where: ΔG_1 - the amount of consumed ozone changing the molecular mass of sample 1 from M_1 to M_2; ΔG_2 - the amount of absorbed ozone changing the molecular mass of sample 2 from M_1 to M_2, ΔG_2 values were selected in the region whereby the ozone uptake varying from 0.1 to 1 mol corresponds to 1 mol AO.

On the basis of the synergistic stabilizing system including N,N'-*bis*-1,4-dimethylpentyl-*p*-phenylenediamine (S77) and polymerized 2,2,4-trimethyl-1,2-dihydroquinoline (Flectol H) in a ratio from 0.5 to 2 and in amounts ranging from 1.5 to 4.5 ppm we have prepared a highly stabilizing rubber mixture (Table 4.5).

Table 4.5

Rubber compositions under vulcanization conditions: 153°C, 25 min.

No.	Ingradients, php	Base sample No.1	Base sample, No.2	New system, No.3
1.	NR	10.0	10.0	10.0
2.	Bulex M-27	20.0	20.0	20.0
3.	Bulex 1500	30.0	30.0	30.0
4.	SKI-3	40.0	40.0	40.0
5.	Pyrolen -highly aromatic resin	7.0	7.0	7.0
6.	Carbon black - PM - 50	25.0	25.0	25.0
7.	Carbon black - PM - 75	25.0	25.0	25.0
8.	ZnO	3.0	3.0	3.0
9.	Stearic acid	2.0	2.0	2.0
10.	PN-6SH - highly aromatic oil	10.0	10.0	10.0
11.	S77	3.0	-	1.7
12.	Flectol H	-	3.0	1.3
13.	2 - Benzthiazolylsulfenmorpholid - Santocur MOR	1.2	1.2	1.2
14.	Sulfur	1.9	1.9	1.9

The behavior of model samples was evaluated under conditions of ozone aging (30°C, ozone - 6 ppm and elongation deformation of 20%) and atmospheric aging (20 % elongation deformation on a roof in Sofia during the summer in 1982) and the results obtained are demonstrated in Table 4.6.

Table 4.6

Results from ozone and atmospheric aging tests

Sample	Ozone degradation, first cracks, min	Atmospheric degradation, first cracks, months	Atmospheric degradation, breaking down, months
No.1	80	9	11
No.2	20	5	6
No.3	>120	>12	>12

Among the investigated combinations, only system No 3 exhibits synergistic effect. Applying the same proportions and conditions we have also evaluated the synergistic properties of the following systems: No 4 - S77/Sc [4,4-thio-*bis*-(2-methyl-6-*tert*-butylphenol)] - 1.8/1.1 php; No 5 - 4010NA/SC = 1.9/1.1 php and No 6 - 4010NA/Flectol H = 1.7/1.3 php (Table 4.7).

Table 4.7

Results form ozone and atmospheric aging tests

Sample	Ozone degradation, first cracks, min	Atmospheric degradation, first cracks, months	Atmospheric degradation, breaking down, months
No.4	280	-	>12
No.5	120	9	>12
No.6	230	9	11

The stabilizing system S77/Vulcanox MB-2A (2-methyl-mercaptobenzimidazole) = 2.9/1,0 php was studied in two compositions: 1) SKI-3 - 50.9; Bulex M-27 - 50.0; carbon black - PM-50 - 45.0; highly aromatic oil - PN-6SH 5.0; ZnO - 5.0; stearic acid - 2.0; Vulkazit CZ (2-cyclohexyl-mercaptobenzthiazole)-1,2 and sulphur -2.2 and 2 and 2) SKI-3 -100.0; PM-50 - 45.0; PN-6Sh? - 5.0; ZnO - 5.0; stearic acid - 2.0; Vulcazit CZ - 1.2 and sulphur - 2.3 (php), vulcanized at 150°C for 15 and 10 min, respectively.

These samples are superior to those proposed and mentioned above with respect to their efficiency demonstrated by the values of the following parameters: relative change of physicomechanical properties (PMP); relative elongation and tensile strength under conditions of accelerated thermal aging of the standard specimen at ε = 20%, 70°C and 95°C by 16 % for the two parameters after 21 days aging at the both temperatures for the first composition and by 20% for the second one.

The proposed synergetic stabilizing systems were applied for the development of universal reception for PT sidewalls and protectors with high antiozonant and antioxidant stability, containing complex elastomer compositions with reduced content of NR and high PMP:

1) tyre's sidewall: NR - 10.0; SKI-3 - 40.0; Bulex M-27 - 20; PM-50 - 25.0: Carbon black PM-75 - 25.0; Highly aromatic resin - Pyrolen - 7.0; PN-6SH - 10.0; ZnO - 3.0; Stearic acid - 2.0; Microcrystalline wax (Ozonshuzvax 111 - BASF) - 2.0, Santocur MOR - 0.9 and Sulphur - 1.9 (php) vulcanized at 153°C for 25 min.

The rubber specimen resistance towards ozone and atmospheric aging is found to be superior by 29-40 % and 50% during exploitation and storage conditions, respectively as compared with the usual rubber compositions

2) tyre's protector: SKI-3 - 25.0; Bulex 1500 - 50.0; SKD-25 - 25.0: Carbon black PM-100 - 65.0.; Pyrolen - 2.0; ZnO - 3.0; Stearic acid - 2.0; Petroleum bitumen (Rubrax) - 5.0; Ozonshuzvax 111 - 2.0, Santokur MOR - 1.2 and Sulphur - 1.9 (php) vulcanized at 153°C for 25 min.

In dependence on the stabilizing system, the ozone and atmospheric aging resistance of the proposed compositions. is by 15 (exploitation) and 40%,(at storage) higher than the commonly used rubber systems.

Applying a similar approach we have proposed some highly stabilized elastomer compositions for hose sidewalls and chambers possessing high resistance to water and air action. On the basis of the developed synergetic stabilizing systems we put forward the following rubber compositions:

1) hose cover: Bulex 1500 - 100.0; PM-6SH - 9.0; ZnO - 3.0; Stearic acid - 1.5; Microcrystalline wax - 2.0; Vulcazit CZ- 1.0 and Sulphur - 1.9, vulcanized at 160°C for 25 min.

2) hose chamber: Oilfilled butadiene-styrene rubber (SKMK-30ARKM-15-Russia) - 100.0; PM-50 - 90.0; PM-75 - 5.0; PN-6SH - 8.0; ZnO - 3.0; Asphalt - 7.0; Stearic acid - 2.0; Vulkazit CZ - 1.3 and Sulphur - 2.0, vulcanized at 160°C for 15 min.

3) hose chamber: SKMK-30ARKM-15 - 92.0; Highmolecular styrene resin (KER 1904) - 8.0; PM-94 - 94.0; PN-6SH/ - 1.4; Asphalt - 11.0; ZnO - 2.5; Stearic acid - 2.8; Vulkazit CZ - 2.1; N-cyclohexyl-thio-phthalimide - 0.45 and Sulphur - 1.8., vulcanization conditions - 160°C and 15 min.

The resistance of the proposed mixtures for sidewalls and chambers during accelerated thermal (70°C) and ozone aging is improved if compared to the commonly used stabilizing compositions. Thus depending on the stabilizing system they are superior by 1.7 and 1.7 times for the faces mixtures and from 2.0- to 4.0 times for the chambers ones.

The degradation process in real conditions is closely related to two main reactions:oxidation and ozonolysis. The first one is initiated by: temperature, static and dynamic deformations, light, radiation, electrons action, defects in the solid body, initiators, catalysts, oxy-reduction agents, bases, acids, etc. The monolecular conversions of $R^{\bullet}$, $RO^{(}$and $RO_2^{(}$- radicals or in some cases of the ion intermediate are among the degradation factors which should be taken into account.

The ozonolysis degradation is influenced, mainly, by the ozone concentration, deformations and defects and to a smaller extent, by temperature due to the relatively low values of E_a of the ozone reaction of the C=C bonds and the zwitterions reactions.In this case the mono- and bimolecular reactions with zwitterions participation and the peroxide bonds decomposition which replace the strong C=C bonds formed in ozonolysis are responsible for degradation occurrence.

4.2.4. High Molecular Antiozonants

Due to their poor compatibility with polymer, high volatility, fast diffusion to the materials surface and possibility for wiping and easy susceptibility to hydrolysis, a number of highly efficient antiozonants can not be widely applied. In view of this, the problem of proposing and application of highly effective highmolecular antioxidants (HMAO) which are able to overcome the forgoing disadvantages is quite modern and provokes an intensive research.

The elastomers stabilization by HMAO can be accomplished via two basic routes: by addition of preliminary prepared polymer stabilizer to the parent polymer or by chemical bonding of the active additive in the polymer matrix.

The first approach for stability improvement is more and more often recommended. The ozone resistant polymer component may be PVC [260], ethylene-vinylacetate-copolymer [261], polychloroprene rubber copolymer of butadiene and butene *n*-butene, without C=C bonds [263], chlorosulphopolyethylene [264], etc. However recently the ethylenepropylene terpolymers (EPDM) are attracting much attention, particularly, their application as protective polymers. This is associated with two main reasons: 1) highly protective action; 2) negligible effect on important vulcanizate parameters [265, 266]. Among them EPDM with ethylidenenorbornene or cyclopentadiene [267] as a third copolymer are the most affective antiozonants additives. The ratio between the EPDM concentration and the basic elastomer group varies in broad ranges from 1:9 to 30:70, mostly from 10:90 to 25:75 and the addition of up to 5 php of PFDA to the rubber composition is highly recommended [268-271].

Andrews proposed that the HMAO particles prevent mechanically the growth of microcracks which are formed under ozone action [272]. This suggests that HMAO should form a continuos disperse medium which can block the microcracks formation and development in the phase of the second polymer. However this factor is necessarily but not sufficient to ensure high ozone protective properties of HMAO. Another mechanism for protection includes the blocking of the C=C bonds by the HMAO molecular segments thus reducing the ozone assess to them. This can be realized only in case of high degree of mutual polymers dispersion, i.e. at the formation of monophase systems. A mixture of butadienenitrile rubber (SKN 40) and PVC [273] is such an example. In the general case of two phase system the ozone-protective action of HMAO is determined by the dispersion level and their ability to cover the sample surface [273, 274].

Polymers containing antioxidant groups can be prepared by the following reactions:

1. Interaction of polydienes with nitrosocompounds, i.e. *p*-nitrosoamines [275], giving PPHDA bound to the rubber [276, 277]:

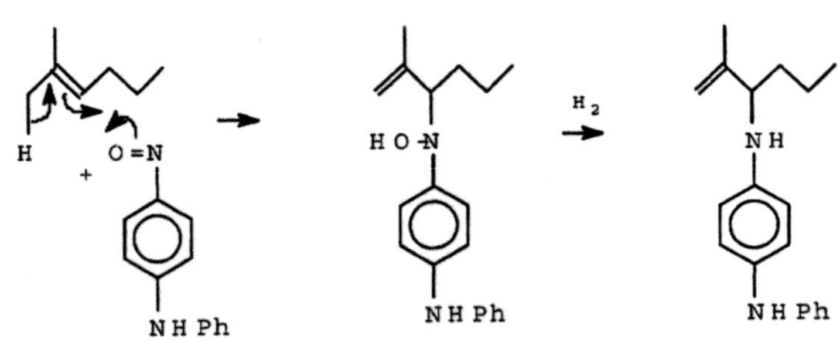

The nitrosocompound can be easily added to rubber simultaneously with the other ingredients

2. Interaction of unsaturated polymers containing epoxy-groups with primary aromatic amines [278]:

$$\text{-CH}_2\text{-CH-CH-CH}_2\text{-} \ (\text{epoxy, -O-}) + \text{H}_2\text{N-Ar} \rightarrow \text{-CH}_2\text{-CH(OH)-CH(NHAr)-CH}_2$$

or with 2,6-di-*tert*-butyl-4-alkylphenols derivatives [279]:

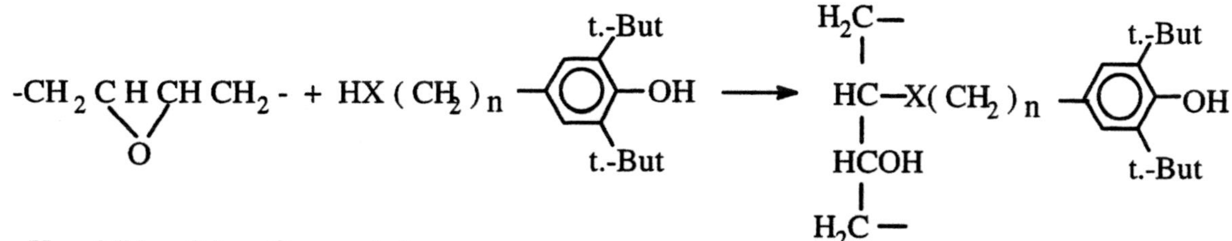

X = -NH-, -OO-, -O-; n = 2-3

Mostly, the HMAO obtained by modification via reaction (1) are equal or superior to N-phenyl-β-naphthylamine (Neozone D) and practically do not differ and even in some cases are superior to N,N'-diphenyl-*p*-phenylenediamine (DF), 4010NA, etc., while those prepared by reaction (2) are similar in their AO action to 2,6-di-*tert*-butyl-4-methylphenol (Ionol).

There are also other polymersimilar conversions for introduction of stabilizing groups, for instance, during the treatment of butadiene and metacrolein copolymer with aromatic amines and phenols [280]; hydrochlorinated *cis*-1,4-polyisoprene with 2-*tert*-butyl-resorsinol, whereas the efficiency of the obtained HMAO is higher than that of Neozone D and 2,2-methylene-*bis*-(4-methyl-6-*tert*-butylphenol)-2246; diene elastomer with unsaturated aliphatic antioxidant, i.e. N-(4-anilinophenyl)-metacrylamide CH_2=CH-CO-NH-Ph-Ph. The reaction can be conducted in solution, emulsion or in solid polymer. The stabilizer is added through stirring in a mill or closed mixer for 5-10 min at a temperature ranging from 20 to 160°C in the presence of peroxides initiator [281]; phosphorization of diene rubber with P_2S_5 and further treatment with aromatic amines [282] resulting in functionalizing of the polymer with functional groups of the type:

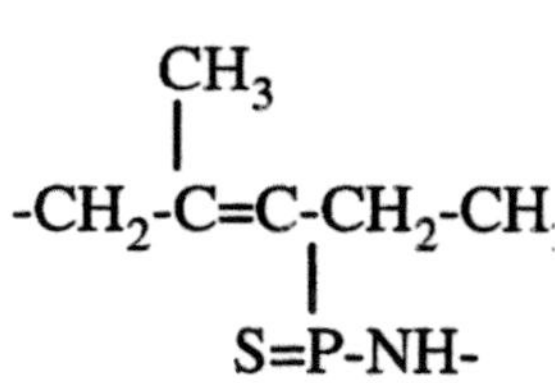

containing P-S and N-P bonds which are energetically weaker than the C-C and C-H bonds.

3. Copolymerization of diene monomers (butadiene with styrene, acrylonitrile, isoprene, 2-chlorobutadiene) in the presence of a third monomer with groups capable of inhibiting the oxidation processes [283-285] such as compounds with secondary aminogroups, i.e. Ph-NH-Ph-CO-C(CH_3)=CH_2 or phenol type antioxidants [286, 287], i.e. HO-Ph-CH_2-CH_2-O-CO-C(CH_3)=CH_2.

The mechanism of the protective action of HMAO is determined by: 1) the more efficient termination of the chain in oxidative degradation and ozone scavenging as a result of the more uniform distribution of HMAO; 2) structurizing of the polymer during recombination of HMAO molecules and elastomer radicals or its zwitterions [289]; 3) the itramolecular synergetic action in polyfunctional HMAO.

Some other factors, like π-conjugation, configuration, conformation and supramolecular structures in elastomers [288] affects also the HMAO efficiency. The kinetics of HMAO stabilization is controled rather by the simultaneous presence of the stabilizer and elastomer radical/zwitterion in the kinetic cage [289] than by the HMAO diffusion in the polymer matrix which is the case with lowmolecular AO.

The protective action of HMAO on the basis of nitroso compounds is due to the deactivation of the impurities catalyzing the oxidation, for example iron, as a result of the formation of complex compounds [275].

HMAO are most widely applied for stabilizing vulcanizates which are continuously exposed to the avtion of higher temperatures, various solvents, oils, water and static load under operating conditions.

4.2.4.1. Ethylene-Propylene Rubber with Diene Monomer (EPDM)

Recently the EPDM are considered as the most promising HMAO for vulcanizates [290-293]. However, EPDM addition can result in deterioration of PMP of the vulcanizates which is mainly due to:

different compatibility of the vulcanizing agents in EPDM and the diene elastomer and the insufficient compatibility of the elastomers used. This problem can be successfully solved by the proper choice of the vulcanizing groups which can ensure most similar kinetics of elastomers vulcanization.

In our studies we have used Keltan 312 (dicyclopentadiene as a third monomer) as a HMAO since it has a higher rate of vulcanization at temperatures over 150°C. Based on literature data three vulcanization groups have been selected: sulphur/2-(morpholinothio)benzthiazole (MBS) (I); sulphur/MBS/dibenzthiazoldisulphide (Altax) (II) and sulphur/diphenylguanidine (DFG) (III). Their efficiency was evaluated on rubber mixtures based on: Bulex, SKD and SKI-3 and their combinations [290. 294-298].

The kinetic curves obtained by the standard method with Moncanto TM-100 rheometer at 155°C show that similarly to other elastomers Keltan 312 undergoes an effective structurization under the influence of the vulcanizing systems used (Table 4.8)

Table 4.8

Rate constants of rubber mixtures vulcanization (k) with various vulcanizing groups (I-III) at 155°C

No.	Elastomer	I, k, min^{-1}	Ratio	II, k, min^{-1}	Ratio	III, k, min^{-1}	Ratio
1.	Bulex - M27	0.19	1.46	0.22	1.50	2.30	8.80
2.	Bulex - 1500	0.14	1.10	0.14	1.00	1.97	7.60
3.	SKD	0.31	2.40	0.38	2.70	4.57	13.30
4.	SKI-3	0.75	5.80	0.81	5.80	3.00	11.50
5.	Keltan 312	0.13	1	0.14	1	0.26	1

The addition of Altax to (I) forming a system (II) increases the rate constants of vulcanization by more that 7% and the structurizing effect for all rubbers including Keltan 312 is more than 25% (Table 4.9, deformation at 300% elongation).

Table 4.9

PMP of vulcanizates on the basis of Bulex 1500 (1), SKD (2), SKI-3 (3) and Keltan 312 (4) in dependence on the vulcanizing system (I and II).

Indicator	Dimension	1^{I}	2^{I}	3^{I}	4^{I}	1^{II}	2^{II}	3^{II}	4^{II}
Force of strain	kg/cm^2	186	130	222	167	202	140	237	170
Tension at 300% elongation	kg/cm^2	75	52	73	91	90	58	87	117
Relative elongation	%	555	460	565	455	540	455	545	445

It has been found that systems I and II are very suitable for covulcanization of Keltan 312 with Bulex 1500 and SKD and to a smaller extent with SKI-3, namely, due to the different vulcanization rates. The large differences in the vulcanization rates at using system III renders its application. The small differences in the vulcanization rate of Keltan 312 with diene elastomers favors the formation of homogenous vulcanization structure which preserve or improve their PMP. The optimum ratio of Keltan 312/dienes elastomers, according to our studies, should not exceed 20/80 as the best ratios are found in the range Of 5/95 - 15:85.

On the basis of the results obtained we have proposed rubber compositions for PT protectors in diene rubber/Keltan 312 ratio of 95:5 thus reducing the SKI-3 content. PMP of the vulcanizates with I and II are listed in Table 4.10.

Table 4.10

PMP of protective type vulcanizates of elastomer compositions Bulex 1500:SKD:SKI-3 = 50:27:23 (1) and (1):Keltan 312 = 90:10 (2) with vulcanizing systems: I - sulphur/MBS = 1.9/1 php and II - sulphur/MBS/Altax = 1.9/0.7/0.5 php

Indicator	1^{I}	2^{I}	2^{II}
Time of prevulcanization at 130^{0}C, min	48	51	40
Force of strain, kg/cm^2	175	150	162
Tension at 300% elongation, kg/cm^2	102	70	85
Relative elongation, %	515	490	500
Residual prolongation, %	8	14	10
Shor hardness, ^{0}Å	62-63	60	60
Elasticity, %	25	23	23
Strength of tear, kg/cm	5.2	4.9	4.9
Gurvich residual deformation, %	14.6	8.9	10.2
Heat formation, ^{o}C (according to Gurvich)	40	40	41

The ozone resistance of the studied protector compositions increases in the presence of Keltan 312 in the elastomer composition (Table 4.11).

Table 4.11

Ozone resistance at 6 ppm of ozone, 30°C, of elastomer compositions from Table 4.10 stabilized by 4010NA/Flectol H = 1.7+ 1.3

Time, min	1^{I}	2^{I}	2^{II}
20	A_1T_1		
45	A_1T_2	A_1T_1	A_1T_1
130	A_1T_3	A_1T_2	A_1T_2
365	T_4 - break down	A_1T_3	A_1T_3
660		T_4 - break down	T_4 - break down

It is seen that the type of the vulcanization group does not affect the ozone resistance of the vulcanizate.

The analysis of the results reveals that the amount of Keltan 312 in the PT protector compositions should be maximum up to 5 php. This content ensures high PMP and enhances substantially their ozone resistance. Among the vulcanizing groups used, Sulphur/MBS/Altax = 1.9/0.7/0.5 turns to be the most appropriate one.

4.3. Apparatus for Ozone Resistance Determination

The ozone resistance of vulcanizates is tested by various accelerated methods both under laboratory and real conditions [299, 300]. The laboratory methods are divided into two groups. The first one, using up to 6 ppm ozone, is intended for testing of natural rubber, polybutadiene, polyisoprene, butadiene-styrene rubbers which are usually more susceptible to ozone action and the second one, with 0.01-0.15% ozone concentration, is applied for testing of ozone resistant rubbers such as polychloroprene rubber, butylrubber, etc.

The time of cracks appearance (t_0) and breakdown (t_C) depends on ε and [O_3] and in the different standards they are evaluated at various deformations and ozone concentrations. This is the reason why the results from the different tests can not be properly compared [301-303]. The apparatus for testing are with fixed parameters and do not allow the tests to be conducted according to the different standards. This has stimulated us to design a new more suitable equipment which could make possible the application of different standards along with selection of appropriate test conditions [304].

The scheme of the apparatus developed by us for evaluation the vulcanizates ozone resistance is depicted in Fig.4.8.

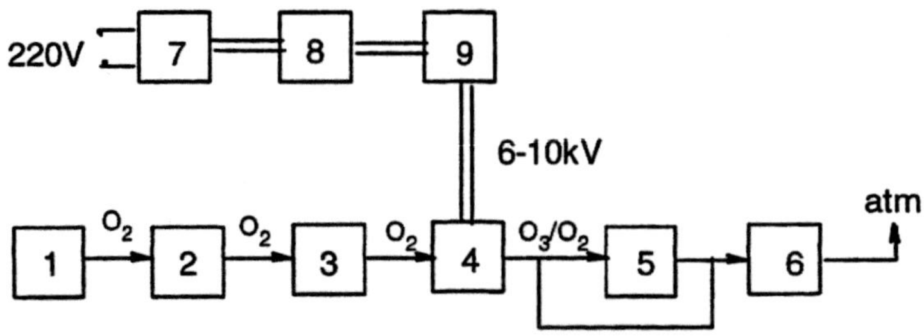

Fig.4.8 Block scheme for ozone resistance determination of vulcanizates. 1 - oxygen source; 2- reducing valve; 3-gas dryer; 4 - ozonator; 5- test chamber; 6 -UV analyzer of ozone; 7 -stabilizer; 8 - autotransfomator; 9 - highvoltage transformer

The ozone chamber with capacity of 20 l is equipped with an ozone inlet and outlet system and ventilator for equalization of ozone concentration.

The frames for deformation of the samples were designed according to our original schemes. In one frame can be placed up to 8 samples. Three frames can be placed in the test chamber, or totally 24 samples can be simultaneously tested.

Ozone was prepared by passing oxygen or air (preliminary dried) through a pipe generator at electrodes voltage of 5-8 kV.

The ozone analysis was carried out spectrophotometrically at 254 nm by means of (ADS-3) (double bonds analyzer, produced at the Institute of Chemical Physics Russian Academy of Sciences). The device sensitivity is about 10% vol. The samples are rubber strips with size 25x5x1 mm.

The ozone concentration may vary from 4 to 10^{-5} % vol, and (can reach 200%.

We have investigated the ozone resistance of Bulex-1500-based vulcanizates by means of the designed test unit. The sample compositions were as follows: Bulex-1500-100, Carbon black (PM-75) - 50, ZnO - 3, P-6SH - 3, N-cyclohexyl-2-benzothiazolsulphenamide (CZ) - 1.2, Sulphur - 1.8 and 4010NA - 0; 1.5; 2; 2.5; 3 and 3.5 php, vulcanization at 160°C for 10 and 15 min.

The test results are demonstrated in Fig. 4.9.

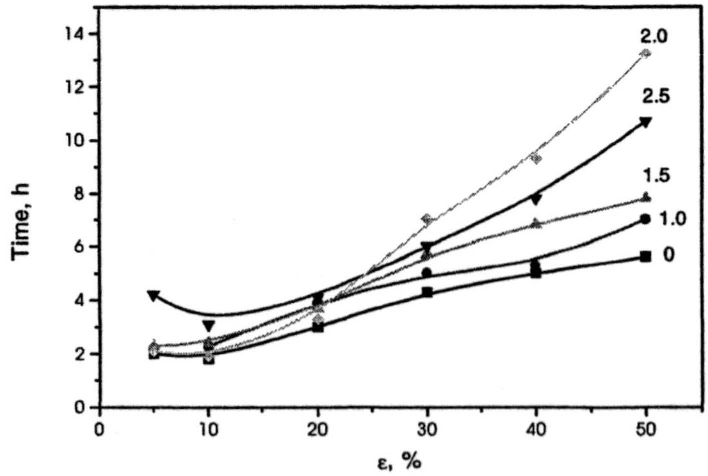

Fig.4.9. Dependence of the time of cracks breakdown on elongation at 42 ppm ozone, 20°C and different concentrations of 4010NA.

As the Figure shows the fastest breakdown of the samples takes place at ε = 10-15% - the range of critical deformations. The samples are more stable at lower (mainly because of the lower number of active degradation centers while with increasing (over its critical value orientation effects resulting in enhancement of vulcanizate resistance begin to operate [299]. The stabilization of vulcanizates at higher ε is also related to the decrease of the degradation rate as a result of the increase of the system potential energy due to the hindered sp^2-sp^3 transition [305]. At ε << 10-15% the microcracks on the samples surface do not grow not only because of the low (values but also because of the accumulation of oxygen containing compounds - a result of the ozone reaction with elastomer which together with the AO impedes the ozone assess [306, 307].

Upon ozone aging it was found that swellings and microcracks are first formed in the surface defects, then being gradually converted into cracks rectangular to the direction of the applied deformation. These changes can be observed first by optical means and after that with unaided eye. The cracks formation and growth is due to: 1) C=C bonds cleavage under the ozone action; 2) initiation of ozone oxidative degradation and 3) cracks development resulting from the deformations concentration in their lips. Other authors explain this step of elastomers degradation process in a similar way [308, 309].

With degradation proceeding the growth of some microcracks leads to the appearance of more and more deep cracks. The greater portion of the microcracks remain unchanged in longitude and depth up to the moment of breakdown. The cracks occurred at the defects are usually those which grow since these are the places of higher deformations concentration. Similar observations have been registered by other authors in investigating metal corrosion [310,] and degradation of some polymers [311].

The optimum amount of 4010NA is 2 - 2.5 ppm (Fig.4.10) which is in a good agreement with the data on the ozone resistance during rubber products exploitation in real conditions [299, 301, 312]. In this sense the apparatus designed by us proves to be quite suitable for accelerated studying the ozone resistance of vulcanizates.

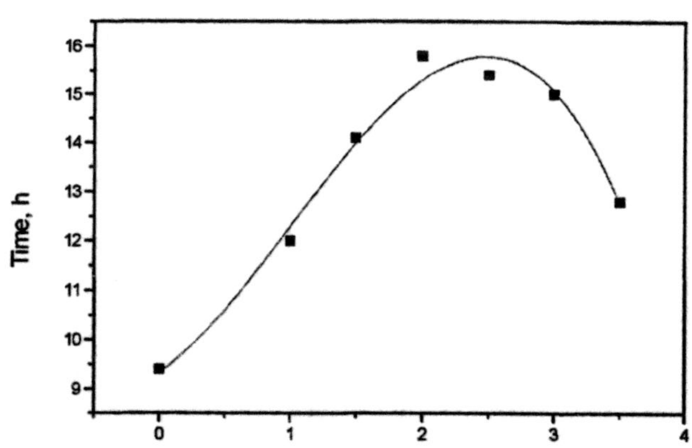

Fig.4.10. Dependence of breakdown time on the amount of 4010NA

The apparatus has the following advantages: it is small-sized, easy operable and allows: 1) the variation of ozone concentration in wide ranges; 2) the continuos registration of the change in ozone concentration during exposure time; 3) continuos monitoring of the changes occurring on the samples surface and their photographing; 4) to apply static deformations in a wide range of elongations. Thus this equipment can be used for studying the ozone resistance of any type of elastomers, their mixtures and vulcanizates on their base.

4.4. Evaluation of Industrial Stabilizers

The antioxidant and antiozonant efficiency of some industrial stabilizers and their mixtures were studied [313-320] with the purpose of proposing new highly efficient stabilizing systems for rubbers. The estimation of their efficiency was carried out by applying model compounds, rubbers solutions and real samples upon conditions of oxidative, ozone and atmospheric degradation.

Analysing the antiozonants used in rubber industry we select the following stabilizers: 2,2'-methylene-*bis*-(6-*tert*-4-methylphenol) (I) -2246; Cyanamid Co. product (CY); 2,2'-methylene-*bis*-(6-*tert*-4-ethylphenol) (II) - 425 - CY; 2,6-di-*tert*-butyl-4-methylphenol (III) - Ionol - Shell; N,N'-*bis*-(1,4-dimethyl-pentyl)-*p*-phenylenediamine (IV) - S77 - Monsanto (MO); N,-phenyl-N'-*iso*-propyl-*p*-phenylenediamine (V) - 4010NA - BASF; N-phenyl-β-naphthylamine (VI) - Neozone D - Du Pont; 6-ethoxy-2,2,4-trimethyl-1,2,-dihydroquinoline (VII) - AW - MO; 4,4-thio-*bis*-(6-*tert*-butyl-m-cresol) (VIII) - SC - MO and polymerized 1,2-dihydro-2,2,4-trimethylquinoline (IX) - Flectol H - MO.

The stabilizers were investigated without further purification.

The reactions in liquid phase were conducted in tetrachloromethane and cumene using SKD, NR, SKI-3 and Bulex-1500 rubbers. The latter were used either without purification (technical grade) or after triple precipitation with ethanol from carbon tetrachloride (p.a.).

The cumene oxidation was carried out in a volumetric device [321, 322], the ozonation of AO and their mixtures in CCl_4 medium with or without rubber in a bubbling reactor [323], the degradation of the solid specimen - upon atmospheric conditions and in an ozone chamber - 3-6 ppm ozone and (= 20% according to the Bulgarian and German standards BSS-8851-77 and TGL-16890, respectively.

4.4.1. Antioxidant Action

The antioxidant efficiency of antiozonants (I-IX) and their mixtures was studied in an initiated oxidation of cumene. The initiation was accomplished by the addition of Azo-*bis*-(isobutyronitrile) (AIBN) ensuring an initiating rate (W) of 2.10^{-6} $M.s^{-1}$ at 70°C. According to the classical concept [322] the antioxidant efficiency of the stabilizers was estimated by: 1) the induction time - τ and 2) the coefficient of inhibition (μ). Both parameters are defined by the following equation:

$$\tau = \ln\{1 - (\mu/2e).[InH]_0/[AIBN]_0\}.1/k_i \qquad \text{(Eq.4.6)}$$

where: e is the probability for the initiation radicals to leave the kinetic cage, for AIBN it is equal to 0.6 at 70°C; $[InH]_0$ = 5 and 2.5 mM and $[AIBN]_o$ = 50 mM are the initial concentrations of AO and AIBN; k_i - the rate constant of the AIBN decay at 70°C equal to $3.36.10^{-5}$ s^{-1}.

If we substitute the above-mentioned values for e, $[InH]_o$ and $[AIBN]_o$ and accept the values of 1, 2, 3 and 4 for μ, the following respective values for (are obtained: 44, 90, 142 and 200 min. for $[InH]_o$= 5 mM and 22; 43; 66 and 91 min. for $[InH]_o$= 2.5 mM. The experimentally observed values for (of the AO ranging from I to IX were: 194; 123; 85; 97; 123; 41; 54; 126 and 66 min. The experimentally found values of μ for I-IX (within unity) were 4; 3; 2; 2; 3; 1; 1; 3 and 2.

On substituting the experimentally found values of (in Eq. 4.6 the following values of (were obtained: 4; 6.6; 2.96; 1.99; 2.33; 2.95; 0.98; 1.30; 3.0 and 1.6. Bearing in mind the structure of the stabilizers, it can be concluded that these values of μ are in compete concent with the AO structure and properties.

Moreover, the reaction of cumene oxidation was used in examining mixtures of two stabilizers. Comparing the experimental values of (with those estimated by means of Eq. 4.6., one get that systems Nos 1, 2, 3, 4, 5, 6, 8, 9, 10, 11, 13 and 14 exhibit a synergistic effect (Table 4.12).

Table 4.12

Induction periods of stabilizers mixtures in the reaction of cumene oxidation

No.	Stabilizing system	τ_{adit}, min	τ_{exp}, min	$\Delta\tau$, min
1.	IV / I	134	242	108
2.	IV / II	109	225	116
3.	IV / III	86	119	33
4.	IV / VIII	109	205	96
5.	IV / IX	86	138	52
6.	V / I	157	93	-64
7.	V / II	132	94	-38
8.	V / III	109	154	45
9.	V / VIII	132	172	40
10.	V / IX	109	135	26
11.	VI / I	113	198	85
12.	VI / II	88	76	-12
13.	VI / III	63	130	65
14.	VII / V	88	123	35
15.	VII / II	88	88	0

Further on, we selected systems Nos 1, 4, 5, 6, 9 and 10 for experiments with vulcanizates. The choice was justified by the strong synergism and the reference data on their antiozonant and antifatigue properties.

4.4.2. Atmospheric Aging

Standard samples (TGL-16890), from mixtures based on Bulex-1500, NR and SKI-3 at a ratio of 50:20:30 were mixed with stabilizing systems Nos 1, 4, 5, 6, and 9 in 3 php concentration. We have introduced the AO in slight excess (by about 1-0.5 php more than the optimum) (Fig.4.8) in order to ensure a maximum activity. The samples were fixed in frames at ε = 20% and were placed on a building roof in spring - summer period in Sofia. A great numbers of cracks of 1-mm depth were observed only with system No 5 (after 5 months exposure). All other samples remain unchanged during the exposure time.

4.4.3. Antiozonant Action

Samples (10 each) of the investigated mixtures for atmospheric degradation were also exposed to ozone (ozone concentration of 3 ppm, 30°C and ε = 20%). Figure 4.11 shows the behavior of the samples from the formation of the first cracks (t_o) to the complete breakdown (t_b) and the full lines demonstrate the degradation time.

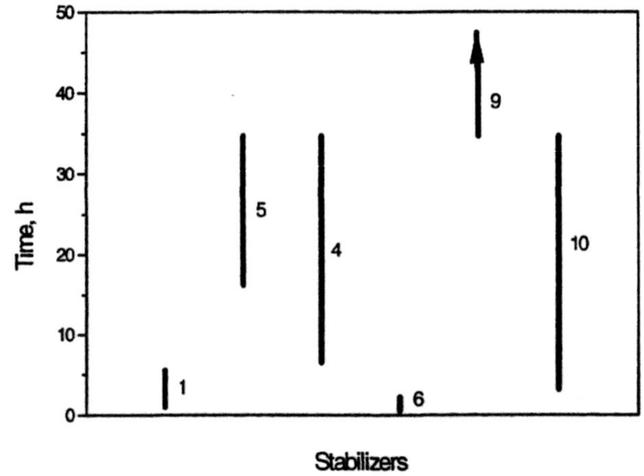

Fig.4.11. Dependence of t_o and t_b of the stabilizers systems Nos 1, 4, 5, 6, 9 and 10 at 3 ppm ozone, 30°C, ε = 20%.

As seen systems Nos 4, 9, 5, and 10 show higher protective properties as compared to systems Nos 1, 6. For system No 9 - S77/SC the first cracks appear when the other samples have already broke down completely and breakdown of its cracks is not observed during the test time.

The effect of the stabilizing systems was studied by using rubber solutions containing the stabilizing system 10 and reference solutions without a stabilizer. A bubbling reactor (10 ml capacity) was loaded with a 0.5%-rubber solution (CCl_4) mixed with an amount of the stabilizing system corresponding to 3 php AO in the rubber mixture. The ozone concentration was 10^{-4} M and the rate 0.1 l/min. The stoichiometric coefficients (the ratio of the amount of absorbed ozone per rubber monomer unit) are presented in Table 4.13

Table 4.13

Stoichiometric coefficients (SCo) of the ozone reaction with technical and pure rubbers in the presence and absence of a system No.10.

No.	Rubber	Purity	No. 10	SCo
1.	Bulex-1500	technical grade	-	4
2.	Bulex-1500	pure	-	2
3.	Bulex-1500	technical grade	added	3
4.	Bulex-1500	pure	added	1.8
5.	NR	technical grade	-	2
6.	NR	pure	-	1
7.	NR	technical grade	added	1.7
8.	NR	pure	added	1
9.	SKI-3	technical grade	-	2
10.	SKI-3	pure	-	1
11.	SKI-3	technical grade	added	1
12.	SKI-3	pure	added	0.8
13.	Bulex-1500 : NR : SKI-3	technical grade	-	3
14.	Bulex-1500 : NR : SKI-3	pure	-	1
15.	Bulex-1500 : NR : SKI-3	technical grade	added	1
16.	Bulex-1500 : NR : SKI-3	pure	added	0.9

The concentration of rubbers was 5.10^{-4} M, of AO = 1.10^{-5} M and the rate constants of ozone interaction with the double bonds were equal to 10^5 $M^{-1}.s^{-1}$ and that with AO = 10^7 $M^{-1}.s^{-1}$ [323]. It is seen that the stabilizing system decreases the stoichiometric coefficient both with the technical grade rubbers and the purified ones. In all cases it is related to lowering of the concentration of the reactive double bonds. Considering the pure rubbers, the change of concentration is also associated with the removal of unsaturated olygomers during purification. Bearing in mind the structure of stabilizers V and IX, we can assume that the AO screen the C=C bonds thus becoming unaccessible to ozone action. Most probably, the stoichiometry decrease is consistent with deactivation of the double bonds by the stabilizer molecules, noting that one molecule of stabilizer deactivates from 2 to 4 double bonds. This model appears to be qualitative but it is in conformity with the reported mechanism which explain the antiozonant effect [324].

Figure 4.12 shows the behavior of these vulcanized samples in the presence of antiozonants IV and V.

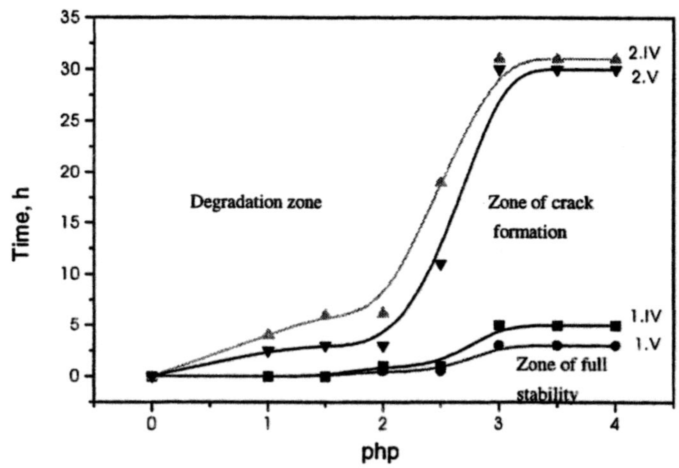

Fig.4.12. AO - IV and Y efficiency in dependence on their concentrations. 1.IV, 1.V - time for the first crack appearance; 2.IV, 2.V - time of breakdown

It is seen that, depending on the concentration, three zones appear on the kinetic curves which can be defined as follows: period of breakdown, of relative stability, and of complete stability, respectively. Comparing these two stabilizers, it can be concluded that the both are highly efficient, however, the former one is slightly superior. This fact can be explained by the strong electrophilic properties of the ozone molecule and the stronger electron-donor ability of the two alkyl substituents than that of the arylalkyl substituents.

We have studied the behavior of various individual stabilizers and stabilizing mixtures (Table 4.14).

Table 4.14

Ozone resistance of vulcanizates based on Bulex 1500:NR:SKI-3 = 50:20:30 for tyre sidewalls (SW) and Bulex 1500:SKI-3:SKD = 50:23:27 for tyre protectors (P) at ozone concentration of 3 ppm, 30°C and 20% elongation in dependence on the stabilizing system (3php). In double mixture the ratio is 1:1.

No.	Stabilizing system	First crack, min	Character of cracks	Break down, h
1.	S77/2246 (SW)	70	A_1T_1	5.5
2.	4010NA/2246 (SW)	30	A_3T_2	2.3
3.	S77/Flectol H (SW)	370	A_1T_2	35
4.	4010NA/Flectol H (SW)	220	A_1T_1	35
5.	S77/SC (SW)	970	A_1T_2	35
6.	4010NA/SC (SW)	>35 h	-	-

7.	S77/Flectol H (P)	155	A_1T_1	32
8.	4010NA/Flectol H (P)	140	A_1T_1	32
9.	S77/SC (P)	23	A_1T_2	32
10.	4010NA/SC (P)	>32 h	-	-

It can be summarized that a detailed research is necessary to select the most efficient stabilizing systems including : invention of model conditions for a primary selection and presence of synergism, further selection by model conditions close to the real ones and final selection of stabilizers examined under operating conditions. Simultaneous check of stabilizers efficiency towards the effect of ozone, oxygen, heat and light is appropriate at each stage of research. It was established that among the stabilizers studied systems Nos 4, 5, 9 and 10 are the most effective.

4.5. PREDICTION

The aging and deterioration of PMP of manufactured rubber articles during their storage and exploitation is related to the change of the molecular mass of the polymer molecules. Among the numerous chemical and physical factors resulting in polymers degradation, as it has been already mentioned, two basic factors should be outlined:

a. Oxidation

That proceeds via two reaction routes: mainly radical pathway, which includes the formation of alkyl (R), alkoxy (RO) and peroxy (RO2) radicals and in some particular cases ionic one giving cation, anion, metal and metalloid species at different oxidation states. The reactions of monomolecular decomposition, cyclization and disproportionation are the basic reactions which lead to molecular mass reducing. It is known that any oxidation process begins with the reaction of chain initiation, i.e. the reaction of free valence generation. Some of the initiators of oxidation degradation are listed below:

- oxygen, ozone
- chemical initiators
- temperature
- light
- radiation
- mechanical action
- catalysts
- defects
- simultaneous action of the initiators pointed above

Temperature, defects and mechanical effects are the main initiators for rubber articles oxidation - PT and rubber goods under exploitation conditions.

b. Ozonolysis

That takes place upon the ozone attack on the the elastomer double bonds. The reaction is accelerated by the presence of defects and application of mechanical deformations while temperature has an insignificant effect. The main reactions responsible for occurrence of degradation process are the reactions with participation of polymer zwitterions. They replace the strong C=C bonds by the relatively weak peroxide bonds. The latter are rapidly decomposed thus decreasing the molecular mass. Polymeric zwitterions can also interact with lowmolecular compounds or undergo monomolecular decomposition which also result in reduction of the molecular mass.

On predicting the variations in the service properties of manufactured articles one should bear in mind the proceeding of these two reactions. Considering the prolonged duration (years) of storage and exploitation of rubber goods, it is very important to look for and propose accelerated tests for estimation of their behavior thus providing reliable information for prediction of the storage and exploitation terms of rubber products.

In this connection some standards for accelerated tests of rubbers are proposed:

-determination of the physicomechamical parameters (PMP) of standard rubber specimen (BSS 9014-77): wear resistance (BSS 8871-71), dynamic Godrich test (BSS 9013-71);

-atmospheric aging at temperatures of 70°C, 85°C and 100°C for 56 days (BSS 8800-77). The estimation of PMP involves the determination of the change of two parameters: tensile strength and relative elongation in % at a definite temperature:

$$a_\tau = (1 - A_\tau/A_0).100, \% \quad \text{(Eq.4.7)}$$

where: A_o and $A_\tau$are the values of the parameter before and after aging time τ.

The experimental results from the accelerated test according to this standard are shown in Fig.4.13 a and b.

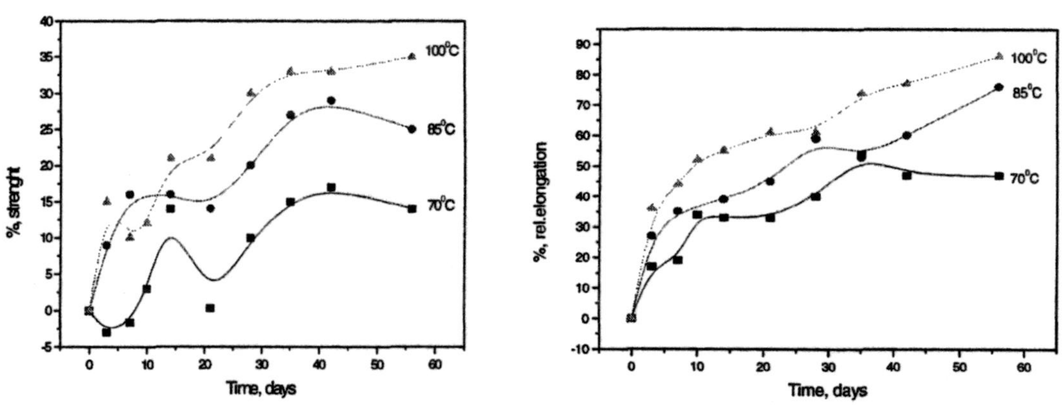

Fig.4.13. Variation of tensile strength (a) and relative elongation (b) in time depending on the temperature at $\varepsilon = 20\%$ for Bulex-1500-based vulcanizate with 3 php Santoflex 77.

It is obvious that the stabilizing system that preserves the PMP of the samples for a more prolonged time at the three temperatures studied is more efficient in its protective action.

We have analyzed the results obtained and established that the change of A can be quite reliably described by the exponential law

$$A_\tau = A_0.\exp(-k\tau) \quad \text{(Eq.4.8)}$$

where: k is the rate constant of A change

and then:

$$a = 1 - \exp(-k\tau) \quad \text{(Eq.4.9)}$$

The experimental points of Fig.4.13 are linearized when ploting in coordinates ln(1-a)/((Fig.4.14 a and b).

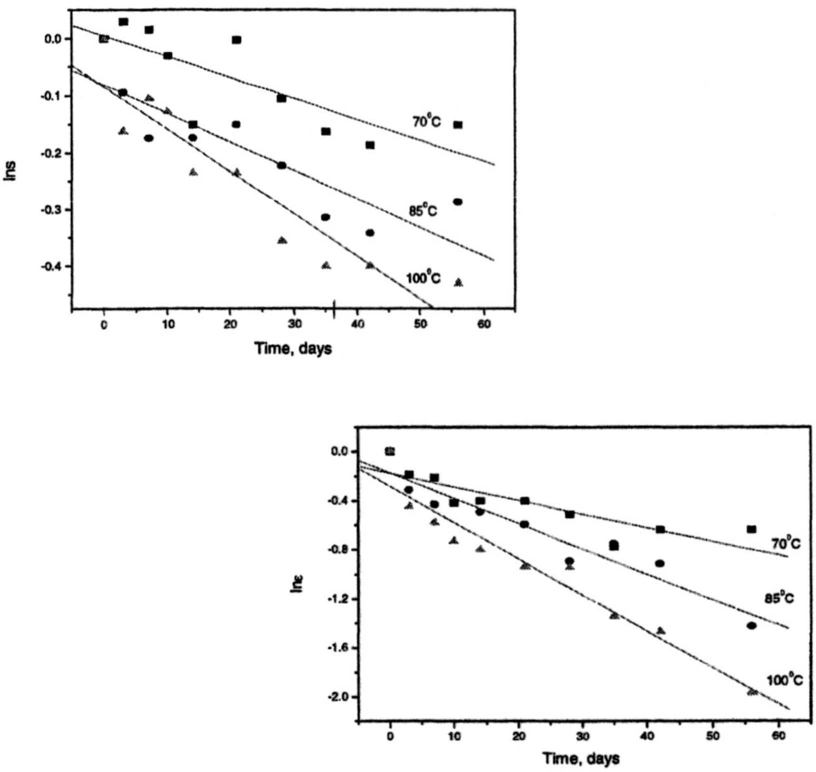

Fig.4.14. Semilogarithmic anamorphosis of experimental points of Fig.4.13, where ln(s) - strenght and ln(e) relative elongation.

The value of k estimated from the curve tangent is (3, 5 and 7).10^{-3} $days^{-1}$ and (11, 20 and 20).10^{-3} $days^{-1}$ for the three temperatures and the two parameters, respectively. The dependence of k on temperature is described by the Arrhenius equation:

$$k = k_0.\exp(-E_a/RT)$$

(Eq.4.10)

The activation energy of the parameter - tensile strength and that of the relative elongation amounts to 7.25 and 8.3 kcal/mol, respectively (Fig.4.15).

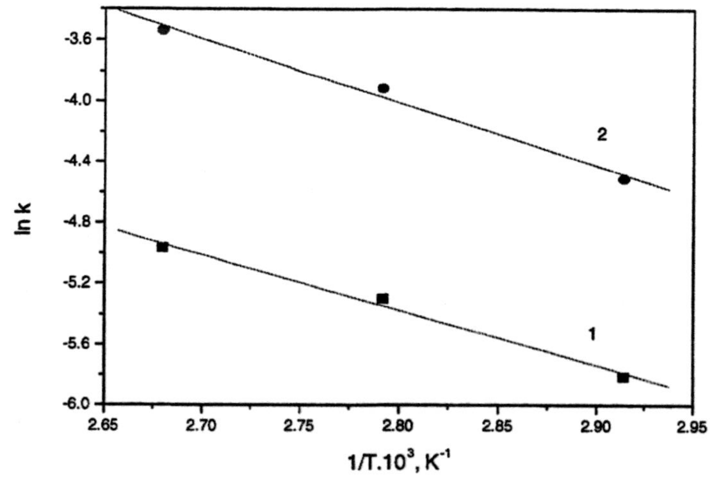

Fig.4.15 Arrhenius relationship of the rate constants on temperature

The calculated values of k at 20°C for the parameters are 5.10^{-4} and $1.4.10^{-3}$ $days^{-1}$, respectively.

According to the standard requirements, any product whose PMP properties has not been changed by more than 20% during storage, i.e. up to values of ln $(A_o/A_\tau) = 2.23$, can be used in practice. Then the time of its suitability upon storage, the temperature of which does not usually exceed 20°C, can be determined by the following expression:

$$\tau^{20} = 2.23/k$$

(Eq.4.11)

where: k is a constant at 20°C.

Applying the values of k at τ^{20} for vulcanizates in Fig.4.13 it has been found out that the variation in the tensile strength and relative elongation values by 20 % will be reached after 12 and 5 years. These values are in good correlation with those observed by us during the longterm storage of these unforced vulcanizates, (stored from 1982 at temperature of 20-25°C). We have registered 18 and 34 % change of these two parameters after 12 years storage. This observation gives us ground to assume that this accelerated test can be successfully used for predicting the service properties of rubber products in real storage conditions.

-aging in atmospheric medium with increased ozone content (BSS 8851-77) under the following conditions: temperature of 30±2°C, 3 ppm ozone and tensile deformation $\varepsilon = 20$ %, registering the time of the first crack appearance, the breakdown time and the change of the crack pattern.

-aging in air at relatively high moisture levels and elevated temperature (BSS 4975-71).

-aging in air at cyclic temperature, ultraviolet and infra-red regime (BSS 4976-71).

-aging in the open under Sofia climatic conditions, according to the conditions of TGL-16890 at $\varepsilon =$ 20%.

-determination of pneumatic tyre storage term according to the method of accelerated thermal aging (GOST 9.033-74).

At present the exploitation and storage terms are determined on the basis of the results observed during the real terms of PT storage and exploitation. For instance, according to BSS 10612-74, these terms are 3 and 5 years for car and truck tyres, respectively, that corresponds to 60 000 and 60 000 km. According to the Russian standard these values amount to 45 000 and 70 000 km for car and truck tyres, respectively.

The corresponding terms are guaranteed by the standard PMP values of vulcanizates and PT element specimen, the stend tests of random selected PT from the serial production and the observations related to the real life of PT.

The analysis of PMP of the protected mixtures reveals that the modern compositions and materials properties can ensure a working guarantee period of 80 000 and 100 000 km for car and truck tyres, respectively. The Bulgarian standard guarantees 45 and 60% of them leaving a considerable reserve. The main factor for such a high reserve is due to the insufficient level of the tyres stabilization.

Thus the stend test with serial specimen:

-car tyres 165SR/13 containing stabilizing system 4010NA -2.0/Ozonschutzvaxe 111-2.0 php on the side wall and 4010NA - 1.3/Santoflex AW -1.0/Ozonschutzvaxe 111 - 2.0 on the protector, preserve their PMP up to 4500±400 km at a speed of 80 km/h, i.e. up to complete breakdown.

-truck tyres 9.00R/20 containing a stabilizing system for the face - 4010NA - 1.0/Santoflex AW - 2.0/Ozonschutzvaxe 111 - 2.0 php and a stabilizing composition for the protector 4010NA -1.3/Santoflex AW -1.0/Ozonschutzvax 111 - 2.0 php guarantee service life of 2500±400 km up to complete breakdown.

However the rubber mixtures for sidewalls and protectors on the bases of the highly effective stabilizing systems as proposed by us give the following results;

- tyres 165SR/13 with a stabilizing system 4010NA - 1.9/Santowhite crystals - 1.1/Ozonschutzvaxe 111 - 2.0 php at the standart stand test at a speed of 80 km/h reach 5200±300 km run.

- tyres 9.00R/20 with the same stabilizing system reach - 3100±200 km run.

These data clearly manifest the great significance of the stabilization of rubbers and rubber materials and the effect of the new stabilizing systems offered by us.

The plant-manufacturer guarantees the necessary quality of rubber mixtures and their finished articles in conformity with standard requirements by control of their PMP during their production. Further on the variation of the parameters with time under the effect of temperature, static and dynamic deformation proceeds in consumer's possession and it is defined as a time for reaching a 1 mm depth of protectors design and below 0.5 mm crack depth of regenerated tyre sidewalls.

In our opinion the conformity with PMP is a necessary but not sufficient condition for evaluation of the future behavior of rubber-based articles due to the unobligatory application of accelerated aging tests. The standard values of PMP guarantee the protection of the rubber products against harmful factors arising from the different road conditions. However, the prediction of the useful life of rubber products could not be laid on scientific bases without knowledge of the tendency and namely of the rate of PMP variations under the effect of various degradation agents. The physicomechnical properties are closely related to the elastomers molecular mass and its preservation with time. As it has been already mentioned the longer the molecular mass remains unaltered the longer the PMP of the elastomer-based articles will be preserved reduction [325, 326]. Based on general considerations the rate of degradation of unstabilized diene elastomer will result in 100-fold loss of molecular mass for 1 year, i.e. from 10^{5-6} to 10^{3-4}, which will lead to its complete breakup of the article. For these reasons the adoption of accelerated methods for evaluation of degradation and for predicting purposes as obligatory is of great practical importance. For instance, the method for accelerated thermal aging, standardized by GOST 9033-71 similar to BSS 8800-77, is very appropriate and widely applied.

Actually it represents a simplified version of the theoretically developed by us modification of BSS 8800-77. For that purpose the vulcanizates are subjected to thermal aging at three temperatures T_1, T_2 and T_3, 50, 70 and 90°C(100°C). The aging duration is determined by the variation of a defined PMP by 80% at the highest temperature applied. First the curves describing the change of PMP with time at the three temperatures are plotted afterwhich they are replotted in coordinates $\lg(1/\tau)(1/T)$ where (is the time of 40% change of PMP for the three temperatures. Thus, across the points corresponding to 10, 30 and 50 % change of PMP at 70°C are drawn lines parallel to the line with coordinate $(1/\tau_{40})(1/T_2)$. Thus the nomogram obtained allows the extrapolation at various temperatures and PMP values.

The application possibility of this nomogram is demonstrated by the data in Tables 4.15 and 4.16.

Table 4.15

Experimental (τ_{80}^{90}, τ_{80}^{50}) and calculated (τ_{20}^{25}) storage terms of vulcanizates for protectors and sidewalls containing different stabilizing systems

Vulcanizate	τ_{80}^{90}, days	τ_{80}^{90}, days	τ_{80}^{90}, days
Side wall - car tyre*	162	1300	2250
Protector - car tyre	53	600	1125
Side wall - truck tyre*	126	1010	1900
Protector - truck tyre*	41	410	770
2-5-02A	220	1800	3100
2-5-02B	260	2100	4000
2-5-02C	196	1570	2910
2-5-02D	197	1580	2960
1-5-02A	156	1600	3000
1-5-02B	135	1400	2625
1-5-02C	120	1200	2250
1-5-02D	127	1270	2380

Note: the first four vulcanizates are Bulgarian tyre plant "Vida"-Vidin products after an original reception of the French Company Ronk Polenk. Symbols 1-5-02 (A-D) are protector compositions proposed by us for car and truck tyres after an unified reception for car and truck tyres protectors were prepared containing: Bulex 1500 -50; SKI-3 - 25; SKD - 25; Carbon black PM-100 - 65; ZnO - 3; Stearic acid - 2; Softener PM-6I - 13; Rubrax - 5; Pyrolen -2; MBS - 1.2; Sulphur - 1.9; Stabilizer - 3php: A - 4010NA /Flectol H = 1.7:1.3; B - 4010NA/SC = 1.7:1.3; C - S77/Flectol H = 1.7:1.3 and D - S77/SC = 1.7:1.3. Symbols 2-5-02 (A-D) denote the vulcanizates composition developed by us for sidewalls of car and truck tyres: NR - plasticate - 10; SKI-3 - 40; Bulex 1500 - 30; Bulex M27 - 20, Carbon black PM-75 - 25; Carbon black PM-50 - 50; ZnO - 3; Stearic acid - 2; PM-6Sh - 13; Pyrolen - 7; MBS - 0.9; Sulphur - 1.9; Stabilizer - 3 php: A - 4010AN/Flectol H = 1.7:1.3; B - 4010NA/SC = 1.7:1.3; B - S77/Flectol H = 1.7:1.3 and D - S77/SC = 1.7:1.3).

Table 4.16

Experimental (τ_{80}^{90}, τ_{80}^{50}) and calculated (τ_{20}^{25}) storage terms of proguced tyres (their potectors and sidewalls) containing different stabilizing systems

Tyres	τ_{80}^{90}, days	τ_{80}^{90}, days	τ_{80}^{90}, days	Test run,km
VIDA - base				
165-SR-13 (SW)	134	1071	2000	4500
165-SR-13 (P)	82	820	1540	
9.00-R-20 (SW)	127	960	1800	2500
9.00-R-20 (P)	67	670	1260	
NEW				
165-SR-13 (SW) tick off A	188	1500	2810	5100
165-SR-13 (P) - A	128	1150	2400	
9.00-R-20 (SW) - A	191	1530	2870	3100
9.00-R-20 (P) - A	132	1320	2480	
165-SR-13 (SW) - B	198	1590	2980	5300
165-SR-13 (P) - B	154	1540	2890	
9.00-R-20 (SW) - B	182	1460	2740	3000
9.00-R-20 (P) - B	135	1350	2530	
165-SR-13 (SW) - C	192	1530	2880	5300
165-SR-13 (P) - C	113	1130	1950	
9.00-R-20 (SW) - C	300	2400	4500	-
9.00-R-20 (P) - C	125	1250	2340	

165-SR-13 (SW) - D	190	1520	2850	-
165-SR-13 (P) - D	119	1190	2230	
9.00-R-20 (SW) - D	300	2400	4500	2000
9.00-R-20 (P) - D	115	1150	2160	

The simultaneous application of this method and BSS 8800-77 for evaluation the behavior of static dynamic strained specimen coupled with the data from ozone and atmospheric aging can be regarded as a reliable base for prediction the storage and exploitation terms of tyres.

4.6. Efficiency of Antiozonants under the Conditions of Various Deformations

There are literature data that the efficiency of AO is decreased at application of large deformations [327]. However, systematic studies in this respect has not been carried out and there is no generally accepted view on this problem.

We have investigated the behavior of the following elastomers: SKI-3 with *cis*-1,4-units content of 95%, unsaturation level - 94-98%; SKS-30, *trans*-1,4-units - 33%, *cis*-1,4-units - 43.4%, 1-2-units - 23.3%, ash - 0.2% and residual monomers content of 0.15 % wt in a broad range of stretch deformations using some of the most widely used AO: PFHDA, hydroquinoline, dithiocarbamates, etc. and antioxidants of mono, bis and tris-phenol type.

The rubbers were pre-purified as follows: dissolution in CCl_4 (1% solution) and precipitation by adding a three-fold volume of methyl alcohol followed by drying. The stabilizer was added to the concentrated rubber solution before preparing the films for its uniformly distribution over the whole sample volume. Then the rubber specimen without sulphur and promoters were subjected to radiation vulcanization with a 20 MRad dose (Co source) at intensity of 0.9 MRad/h, while the sulphur-containing samples were vulcanized at 150-170°C for 10-20 min.

The cross-linking degree of the samples was determined by the swelling method [328]. The IR-spectra (the method of multiple attenuated total reflection - ATR with KRS-5 crystal (reflections number - 20, 45^0, penetration 10-40μ) were monitored during the exposure duration of the specimen to ozone atmosphere - 10 ppm and 20°C. The spectra of butadiene-styrene rubber in a transmittance (1) and reflectance (2) mode is given in Fig 4.16.

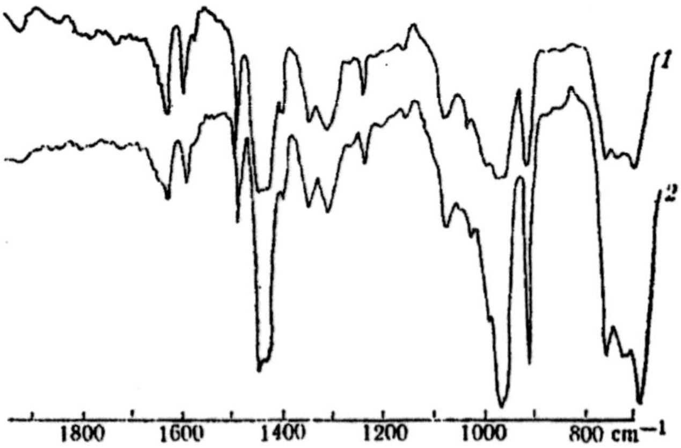

Fig.4.16 IR spectra of SKS-30. 1 - transmittance; 2 - reflectance

As seen the IR spectrum monitored by the ATR method has a good resolution and can be used for qualitative studies.

The correlation time (τ_c) that stands for the rotation mobility of the nitroxyl radical was estimated on the basis of the ESR spectra of the paramagnetic marker - 2,2,6,6-tetramethyl-pyperidine-1-oxyl - ≤1017 spin/cm^3, introduced through evaporation [329].

$$\tau_c = 6.65.\Delta H.[(I^+/I^-)^{1/2}-1].10^{-10},$$

(Eq.4.12)

where: ΔH - the width of the lowfield spectrum components; I^+, I^- - intensity of the final spectrum components, in high and lowfield, respectively.

The stress relaxation in the rubber specimen in ozone atmosphere - 50 ppm was measured at a temperature of 17°C in the 5-160% deformation range. Based on the experimental curves the time of chemical relaxation (τ), which is inversely proportional to the chemical relaxation rate v_r, can be estimated using the following expression [330]:

$$\sigma/\sigma_0=\exp(-t/\tau)\ ((1-t/\tau),$$

(Eq.4.13)

where: σ is the equilibrium deformation obtained by extrapolation to zero time in the linear part on the relaxation curve; t - time.

The initial parts of the relaxation curves allows the determination of the physical relaxation.

In order to study the effect of AO in strained speciment it is necessary to know the structural changes occurring in the rubbers under the influence of applied deformations (ratio between the amorphous and the crystalline phase).

Though the crystallization processes in *cis*-1,4-polyisoprene rubbers are well studied [328], due to the specific cross-linking degree at the doses, applied by us, we have to define the deformation range for this particular case. Visible reflections of the crystalline phase appear at deformations over 300% (Fig.3.5 a, b). All tests were conducted in 0-100 % deformation range whereby the rubber crystallization is of minor significance [328].

It is well known [331] that the rate of many chemical reactions proceeding in solid state polymer depend on the segmental mobility of the macromolecules. For studying the influence of the strething deformations on the segmental mobility some experiments with rubber samples containing nitroxyl radical were carried out. The dependence of the correlation time of the marker - radical on the elongation degree ε is presented in Fig.4.17.

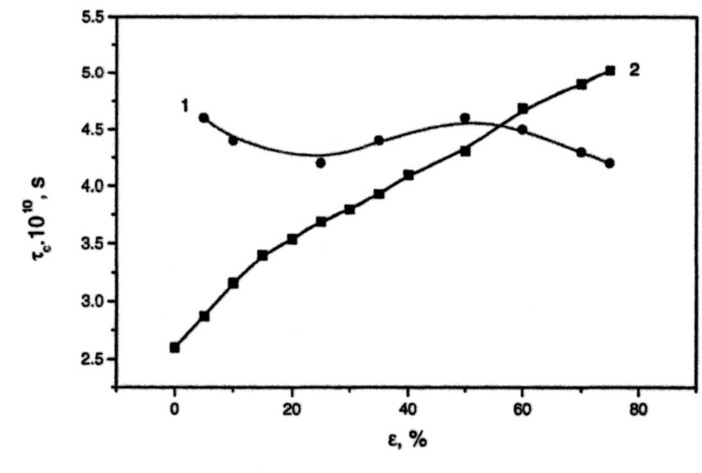

Fig.4.17 Dependence of the marker - radical correlation time of (τ_c) on elongation level (ε). 1 - SKI-3; 2 - SKS-30.

It is seen from the figure that the value of τ_c for SKI-3 samples does not change with ε which testifies the lack of change in the rotational diffusion of the marker - radical in the used deformation range. However with SKS-30 (Russia), a decrease of the rotational diffusion of the nitroxyl radical is observed as the value of τ_c rises with the growing ε. These results suggest that tensile deformation in SKI-3 up to 100% does not change substantially the value of τ_c, and consequently the rotational mobility of the AO, whose molecules are very similar in structure and composition to that of the probe-radical. Moreover the translational mobility of the latter does not change too since the rotation and translational diffusion are interconected [332].

The literature data [333] clearly manifest the close relationship between the AO efficiency and its rate constant in its reaction with ozone. In Table 4.7 are listed the rate constants (*k*) which we have measured by using the bubbling method [334].

Table 4.17

Rate constants of ozone with AO in CCl_4 at 25°C

No.	Antiozonant	$k.10^{-6},M^{-1}.s^{-1}$
1.	4-Methyl, N, N'-bis (iso-propyl)-m-PHDA	24
2.	N-Phenyl, N'-isopropyl-PPHDA	12
3.	N-isopropyl-p-anisidine	14
4.	N, N'-bis (1, 4-di-methylpentyl)-PPHDA	13
5.	N, N'-bis-(1, 3-di-methylbutyl)-PPHDA	100
6.	4-Methyl, N, N'-bis-(2-ethylhexyl)-m-PHDA	7.3
7.	N-Phenyl, N'-1, 3-methylbutyl-PPHDA	25
8.	N, N'-(di-nitroso, di-sec-octyl)-PPHDA	5
9.	N-Phenyl-β-naphthylamine	20
10.	N-Phenyl, N'-cyclohexyl-PPHDA	24
11.	N, N'-Bis (phenyl)-PPHDA	27
12.	Diphenylamine	40
13.	2, 2'-Methylen-bis-(6-tert-butyl-4 ethylphenol)	5.10^{-3}
14.	2, 2'-Methylene-bis-(6-tert-butyl-4-methylphenol)	4.10^{-3}
15.	Ni-bis-(N-dibutyldithiocarbamate)	1
16.	2, 2, 4-Trimethyl-1, 2-dihydroquinoline	0.8
17.	2, 2'-Thio-bis-(6-tert-butyl-4-methylphenol)	$4.5.10^{-3}$

The rate constants of ozone reaction with amine stabilizers are about 2-3 order higher than those with olefins and about 3-4 orders as compared with phenol stabilizers (Nos 13 and 14). Among all stabilizers studied only N-dithiocarbamate exhibits a similar antiozonant efficiency (No 15). The differences in their reactivity towards ozone is related to their different structure and various donor properties of the substituents. However it can be accepted that they exhibit similar protective ability with respect to ozone, particularly if one considers that in real conditions the diffusion of AO plays a very important role. Upon radiation vulcanization the antiozonants are stable and only in the cases of of amines with long alkyl groups, the alkyl chain is cleaved (Nos 4 and 6) with preservation of the PPHDA structure while at 200°C (the maximum vulcanization temperature) in an inert atmosphere the amine does not undergo any changes.

The efficiency of AO was evaluated by: 1) the kinetics of C=O bonds formation and 2) the deformation reducing.

The rate of change of 1720/1640 cm^{-1} ratio, $\Delta D = (D_{1720}/D_{1640})_{t} - (D_{1720}/D_{1640})$ depending on the stabilizing system used is depicted in Fig.4.18.

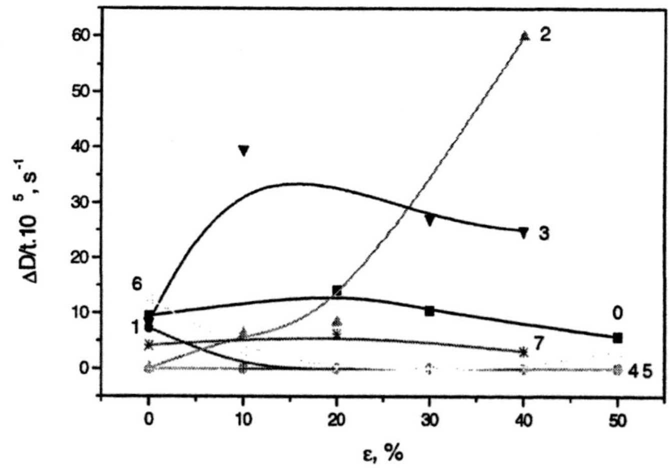

Fig.4.18 Initial rates of C=O accumulation in SKI-3 in dependence on AO (1-7) or without AO (0) at $[O_3] = 1.10^{-6}$ M, 20°C

Upon ozonolysis of the SKI-3 samples without antiozonant (0) the accumulation of the reaction products was linear in the whole range of deformations applied. At the same time the constancy in the value of ΔD for AO Nos 1, 4, 5, 6 (the designation from Table 4.17) indicates the absence of any visible chemical changes on the sample surface. Antiozonants Nos 2, 3 7 are also efficient in the absence of deformation.

It should be noted that ΔD rises at relatively small values of (while further increase in stress deformation results in a decrease of the rubber oxidation level (Fig.4.19). The existence of critical deformation was also observed for the pure rubber samples and the specimen containing antiozonants Nos 2, 3 and 7.

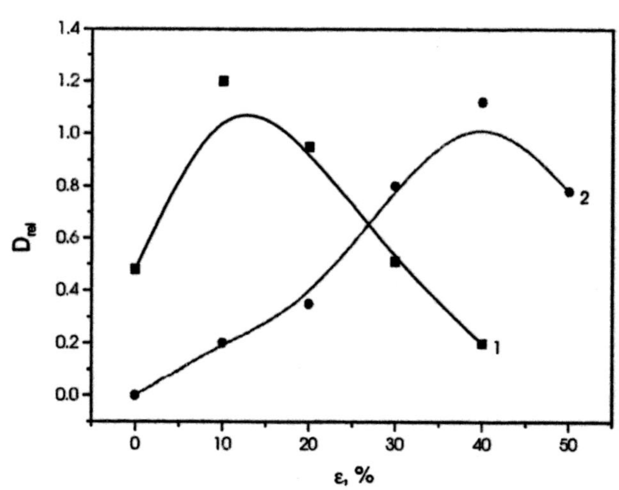

Fig.4.19. Dependence of the oxidation level of SKI-3 ($D_{1720}/_{1640}$) on ε: 1 - without AO; 2 - in the presence of 4010NA

As it has been already mentioned this phenomenon is characterised by a maximum crack propagation rate and a minimum longevity being observed in a certain range of tensile deformations [327, 328]. The critical elongations were investigated by means of relaxometer. The relaxation curves of ozone-induced SKI-3 samples are demonstrated in Fig.4.20.

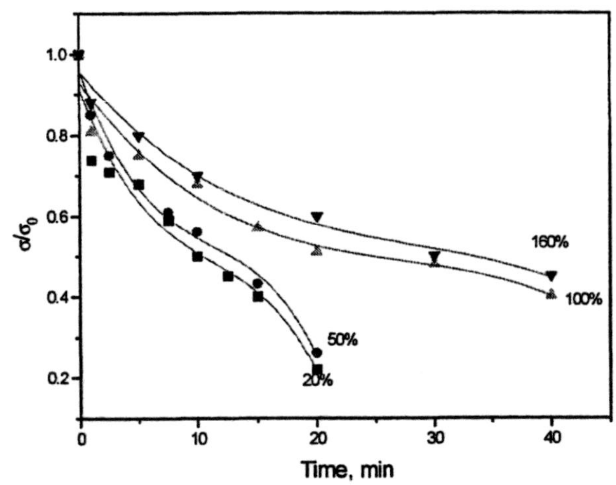

Fig.4.20. Stress relaxation (σ/σ_0) in SKI-3 at different ε on ozonation duration. AO - No 1 = 2.10^{-6} M.

Highest rate of stress decay is observed in unstabilized rubber sample at ε = 20% (the curve is not shown in Fig.4.20) while in the presence of AO under the same conditions the rate of stress decay declines slightly with time. It has been established that at higher deformations (50, 100 and 160%) the presence of AO in the rubber specimen reduces their ozone resistance as manifested by the relatively higher relaxation rate than the expected one.

This supposition is exemplified by the dependence of logarithm of the chemical relaxation time on the tensile deformation without AO (1) and with AO addition (2) shown in Fig.4.21. It is clearly seen that the presence of AO exerts a substantial stabilizing effect in the initial range of small deformations.

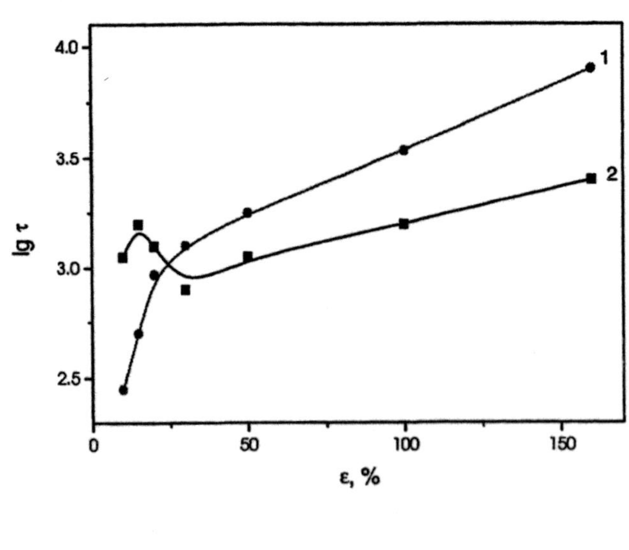

Fig.4.21. Dependence of the chemical relaxation on (in samples without AO (1) and with AO addition (2)

4.7. EFFECT OF VULCANIZATES STRUCTURE

The antiozonant efficiency depends also on the chemical structure of rubbers [335, 336]. Different points of view are put forward with respect to this relationship. For instance, one of them supposes that the stabilizer activity is related antibatic to its ability to form hydrogen bonds with polymers [337].

For evaluating the role displayed by the chemical nature of elastomers on AO efficiency we have investigated several rubbers specimen - NR, SKI-3, SKD and SKS-30. Mixtures with the following composition: elastomer - 100 php, ZnO - 3.0, Stearin - 1.0, Vulcazit CZ - 1.0, Sulphur - 1.8, AO - 3.0 were vulcanized at 160°C for 15 min. We have studied the following antiozonant systems: No 2, 4, 15, 16 and 17 (Table 4.17).

We have measured the relaxation time which (τ) is inversely proportional to the crack growth rate v_p [337]:

$$v_p = -\partial(S/S_0)/\partial\tau = -(\sigma_0/\sigma_\infty).\partial(F/F_0)/\partial\tau, \qquad \text{(Eq.4.14)}$$

where: F -the stress applied to the sample cross section S ; σ_0 - initial stress; σ_∞- equilibrium stress resulting from the fast physical relaxation.

Fig.4.22 shows typical curves of stress decay in the presence of ozone.

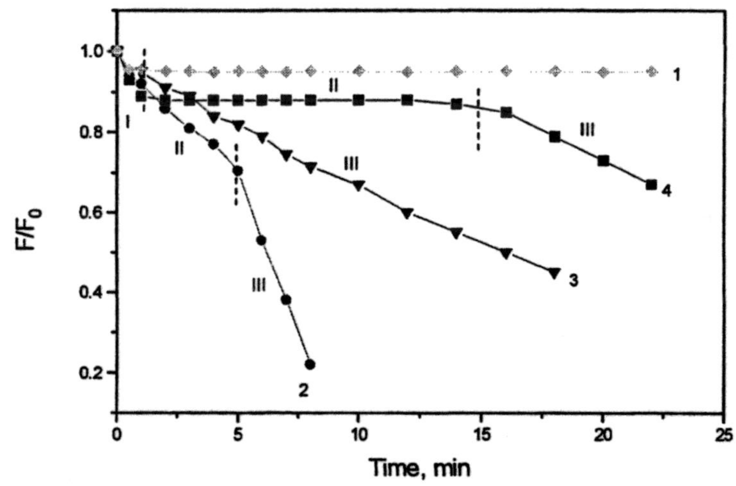

Fig.4.22. Relaxation curves of SKI-3 samples without stabilizer and with stabilizer addition in air and ozone atmosphere - 9.10^{-4} mol/m^3 at 293K. 1 - without AO, in air, ε = 10%; 2 - without AO, in ozone atmosphere, ε=10%; 3 - without AO, in ozone atmosphere, ε = 100%; 4 - in the presence of 5 php 4010NA, in ozone atmosphere, ε = 10%.

Curve 1 is assigned to the occurrence of fast physical relaxation taking place within several seconds, up to one min and does not change with time. It is characterized by a slight decline at the beginning which is related to the physical re-orientation of the macromolecules and the constancy of the parameter F/F_0 with the time after that. This region designated in the figure as (1) is observed in all curves. In the presence of ozone (curves 2 and 3) the pattern changes radically. Moreover, on curve 2 one clearly observes three regions: I - fast physical relaxation up to 1 min, II - chemical relaxation up to 5 min, a linear stress decay is observed accompanied by visible changes on the sample surface, and consequently with cracks origination on them, III - after 5min - chemical relaxation which accelerates the decay of stress and is associated with cracks appearance and propagation. If one assumes the rate of region II as unity, then the acceleration standing for region III will be 3.2 times. Curve 3 is characterized by an absence of region II since immediately after region I (physical relaxation) portions of cracks grow at a constant rate is observed. The slow down of the stress decay is about 1.7 times as compared with the rate for curve 2, region II and remains constant up to 18 min. The greater elongations change the surface rheology making it energetically more isotropic and thus the risk of appearance and growth of separate cracks to sizes resulting in sharp decrease of the sample resistance is decreased (region II in curve 2). The addition of AO (curve 4) leads to marked expansion of region II with almost zero value of change rate, up to 15 min, and in region III (after 15 min) the rate of relaxation decay is by a factor of 2 smaller than that of region II in curve 2.

Another parameter characterizing the process of ozone interaction with rubber samples is the change in cracks growth rate (v_p) with ε (Fig.4.23).

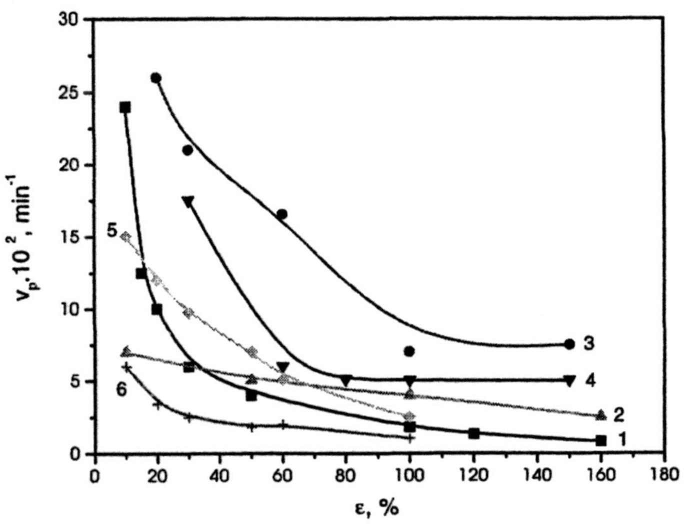

Fig.4.23. Dependence of v_r on ε for SKI-3 without AO (curves 1, 3, 5) and in the presence of AO - 4010NA (curves 2 and 4), S77 (curve 2) and N-*iso*-propyl-*p*-anizidine (curves 2, 4, 6) in amounts of 1, 3 and 5 php, respectively, at $[O_3] = 5.10^{-4}$ mol/cm^3 - (curves 1 and 2), $6.4.10^{-2}$ (curves 3 and 4) and 9.10^{-4} mol/cm^3 (curves 5 and 6).

Note: the experimental points for AO - 4010AN and S77 on curves 2 and 4 could not be seen for their values are very close to those of the third AO.

The rate of cracks growth depends on the ozone concentrations as it is shown in Fig.4.24.

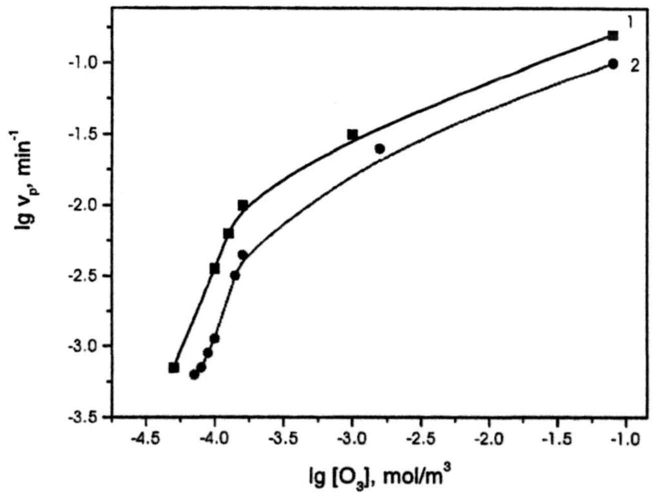

Fig.4.24. Dependence of lg v_p for SKI-3 samples, $\varepsilon = 20\%$, at 20°C on lg [O_3]; 1 - without AO; 2 - in the presence of 3 php 4010AN.

It is seen that the value of τ' is constant in the stationary zone of stress decay and has the same meaning as the relaxation time τ' in Tobolski's formula [338].

Fig.4.25 shows the dependence of the value of ln (on the degree of tensile deformation in the vulcanizates SKI-3 (a), NR (b), SKD (c) and SKS-30 (d).

A similar pattern is observed only in two cases. SKI-3 and NR. This similarity is not only qualitative but also quantitative both for the control (curve 1) and the protected samples. The shape of all the curves, irrespective of the presence of a stabilizer, is also quite similar; the lowest stability (the highest rate of cracks growth rate of cracks) is observed in the range of critical deformations $\varepsilon = 10\text{-}20\%$. However with an increase in the degree of strething the stability increases (the parameter τ_c rises).

The stability increase can be explained by the decrease in the reactivity during the extension of *cis*-fragment -C-C=C-C- in all the reactions accompanied by the rearrangement of two carbon atoms at the double bond from sp^2 into sp^3 state [339] and it is well known that ozone attacks the double bond with the formation of a primary ozonide.

There also exists another interpretation of the "critical deformation" phenomenon, based on the mechanism of orientation strengthening [335]. However, as shown by ESR spectral studies performed with the use of a marker radical [340, 342], at least up to 200% the tensile deformation of such elastomers of SKI-3, NR and SKD does not change the polymer matrix rigidity (segmental mobility of macrochains remains constant). That is why in these cases orientation strengthening cannot, in our opinion, play any significant role.

A different situation is registered in the case of SKS-30. The strething of its vulcanizate raises substantially the rigidity of its polymer matrix [340]. It is with the structural-physical changes that the rise in (value with the growing ε, observed in Fig.4.25, c, seems to be associated.

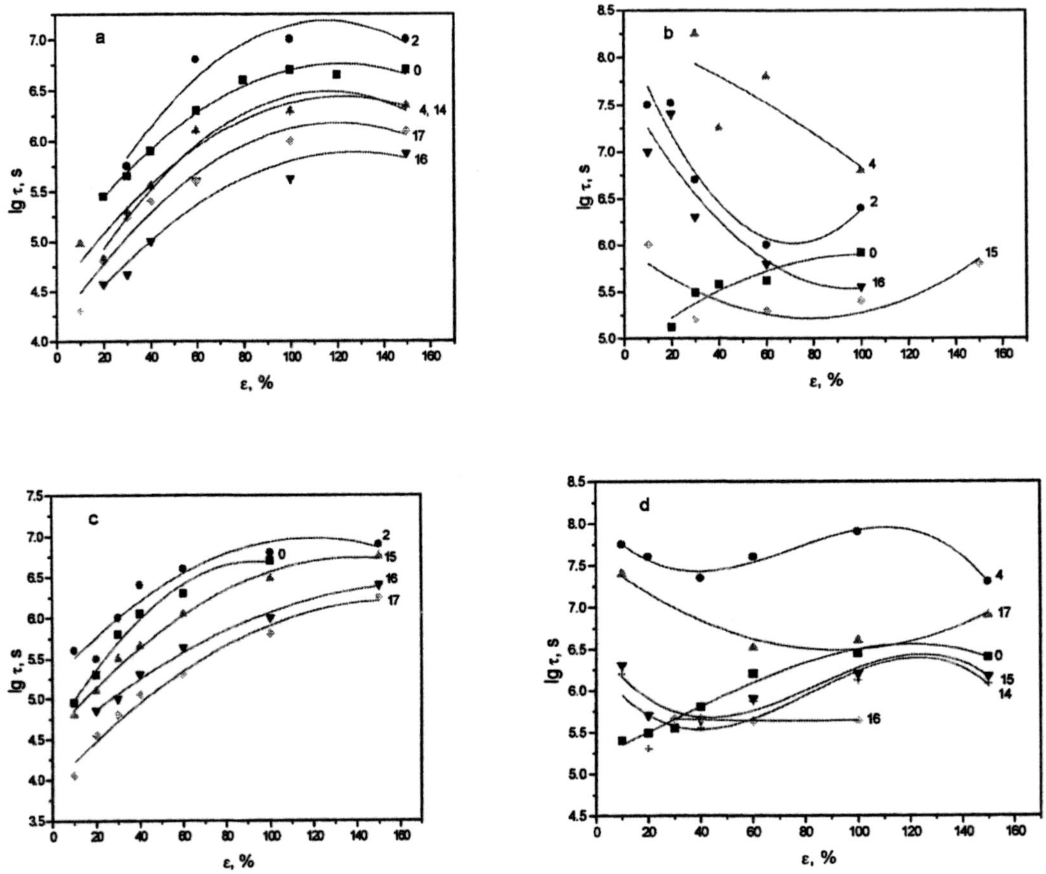

Fig.4.25 Relationship between lnτ and the degree of deformation ε in specimen from SKI-3 (a), NR (b), SKD (c) and SKS-30 (d) at ozone concentration of $6.4.10^{-8}$ M, (the number of the curves correspond to the number of the stabilizer in Table 4.17).

An efficient stabilizer for SKI-3 and NR under the chosen testing conditions is the widely applied antiozonant - 4010NA. In the whole range of deformations this stabilizer slows down the growth of cracks, which follows from the comparison of curves 2 and 1 in Fig.4.26 (curves 2 lie above curves 1). This slowing down, however, is not very great: in SKI-3 the v_r values decreases by a factor of approximately 1.6, in NR - 1.3. Compositions containing other stabilizers are less resistant to ozone effect under these conditions, as compared with the control samples.

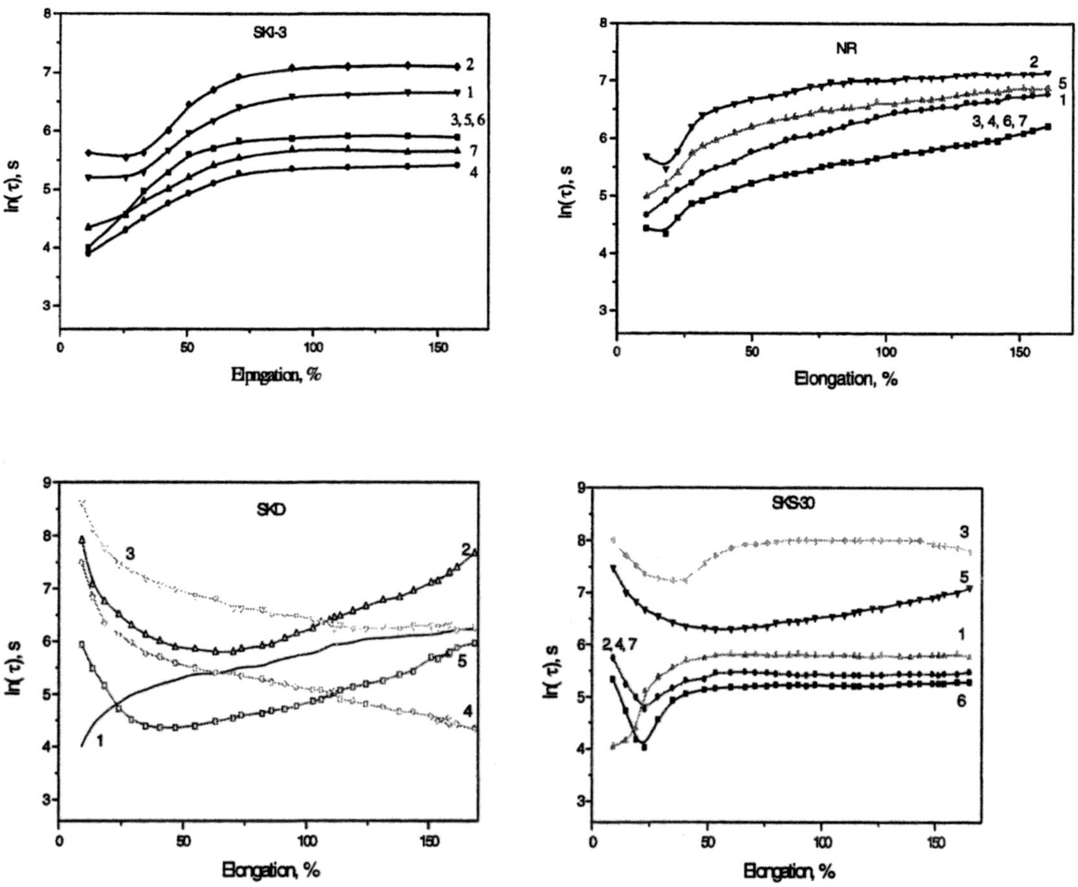

Fig.4.26. Dependence of lnτ on the degree of deformation ε in: a) SKI-3; b) NR; c) SKD; d) SKS-30., $[O_3] = 6.4.10^{-2}$ mol/cm^3, 20°C. The numbers of the curves correspond to the numbers of the samples in Table 4.17.

The similarity in the results obtained for SKI-3 and NR is caused by the identity of their monomer unit. The significant role of the macrochain chemical nature is confirmed by the difference in the results obtained for elastomers with other monomer units. A relatively small difference in the structure of *cis*-1,4-polyisoprene (the presence of CH_3-substituent) leads to a quite significant difference in the stabilizer efficiency. Thus, it is seen from Fig.4.18c that SKD vulcanizate is stabilized much better than SKI-3 and NR. Apart from 4010AN, other stabilizers prove also to be efficient in the whole deformation range: No 4 (Fig.4.26 b, curve 3) and No 15 (the curve is not given because it is identical to curve 1 and lies above it), in the region of small deformations the stabilizing effect is manifested by No 16 (up to 60%, curve 4) and No 17 (up to 20%, curve 5).

It should be emphasized that for SKD samples especially characteristic is a decrease in the efficiency of stabilizers with the growing of ε. The only exception is AO No 15. Table 4.18 illustrates the stabilizing effect, expressed as a degree of a decrease in the crack growth rate at various deformations of samples, as compared with samples that contain no stabilizers.

Table 4.18

Degree of a decrease in the crack growth rate in SKD samples in the presence of a stabilizer, compared with samples without AO, at different deformations

Antiozonant	ε=20%	ε=30%	ε=60%	ε=100%
No. 4	-	14.7	5	2.4
No. 2	10	3.6	1.4	1.5
No. 16	8.1	2.2	1.1	0.7
No. 15	1.4	1.6	1.6	1.6

We have observed a decrease in the AO efficiency with an increase in the degree of *cis*-1,4-polyisoprene deformation. Thus, in Fig.4.25 b the highest stability effect of 4010NA (curve 2) is displayed for $\varepsilon = 10\%$ (two-fold decrease in v_p; and at larger deformations the effect decreases. But, as it follows from the above data, this effect is much more pronounced in the case of SKD. Therefore, one can speak of a different nature of $\ln\tau_c = f(\varepsilon)$ dependencies obtained for the protected vulcanizates of butadiene and isoprene rubbers (Fig.4.25, a-c).

For SKS-30 vulcanizate one can note the following peculiarities. This elastomer, as well as SKD, is stabilized by No 4 in the whole range of ε (curve 3). Inhibiting effect is also exerted by No 17 at all deformations (curve 5), although to a smaller extent than No 4. All other stabilizers slow down ozone degradation only at small ε (up to 20-30%), and at large deformations their presence has an opposite effect - accelerate the cracks growth.

Another similarity of SKS-30 with SKD manifests itself in a decrease in the efficiency of stabilizers with the growth of ε, although it is not so substantial as in SKD. This is demonstrated by the data on the decrease in v_p in strained SKS-30 samples in the presence of AO (Table 4.19).

Table 4.19

Degree of a decrease in the crack growth rate in SKS-30 samples in the presence of a stabilizer, compared with samples without AO, at different deformations.

Antiozonant	ε=10%	ε=20%	ε=60%	ε=100%	ε=150%
No. 4	10	8	5	4	3
No. 17	6.7	-	1.4	1.3	1.7
No. 14	2.3	0.8	0.8	0.8	0.8
No. 15	2.3	1.2	0.8	0.8	0.8

In our opinion, the results presented in Fig.4.20 cannot be explained from the standpoint of the quenching of ozone molecules by the stabilizer, i.e. with the help of a mechanism based on a higher AO reactivity with respect to ozone in comparison with the polymer C=C bonds. This follows from the obtained data related to the influence of the macromolecule chemical structure on the stabilizers efficiency. Most clearly this is seen for Nos 2 and 4. The rate constants of their reactions with ozone are practically the same and equal to $1.2.10^7$ and $1.3.10^7$ $M^{-1}.s^{-1}$, respectively [341, 342]. Their protective effect, however, differs considerably and depends on the type of the polymer. Indeed, stabilizer No 2 raises and No 4 lowers the stability of SKI-3 and NR samples, whereas for SKD and SKS-30 stabilizer No 4 is a very efficient one.

The facts of such kind can only be explained by taking into account the active role played by the polymer matrix itself. It should be noted that this role is hardly caused by the intermolecular interaction of the polymer with the stabilizer, since the elastomers in question are by their nature of low polarity. From the standpoint of intermolecular interaction it is difficult to explain the considerable difference between SKI-3 and NR, on the one hand, and SKD and SKS-30, on the other. Most probably, the stabilizers enters into a chemical reaction with the products of the polymer oxidative degradation (cross-linking of the cut-off fragments). In this case the nature of the monomer unit can predetermine the course of the rate of such a reaction, thereby influencing the AO efficiency. From this point of view it proves possible to explain the No 4 efficiency in the two cases of SKD and SKS-30 because they have similar chemical fragments resulting from butadiene polymerization. And, possibly, it is with the products of ozone oxidation of these very fragments, that a given stabilizer interacts effectively. In polyisoprene chains a similar chemical process most probably does not take place at all or occurs with appreciably lower rate.

It thus seems that, as noted above, two factors (in the case of polyisoprene samples) are being superimposed [343]: the interaction of AO with ozone and the formation of a protective film on the polymer surface, whereas in the presence of butadiene units another mean of protection appears - cross-linking of the broken-down AO fragments. The latter, though more effective, are quite sensitive to deformation. Indeed, strething must facilitate the drawing apart of the molecule cut-off ends and thereby impede their cross-linking.

The conclusion put forward for the active role of the chemical nature of the polydiene monomer unit is in agreement with the data from the literature data. It is known that the two end aldehyde groups resulting from the ozonation of the polybutadiene monomer unit are capable of reaction with diamines, thus restoring the broken polymer chain. The ozonation of the polyisoprene monomer unit leads to aldehyde and ketone groups formation. That is the reason why diamines cannot restore the length of the polyisoprene macromolecule [345].

In conclusion it should be noted that the ozone degradation of diene polymers is a result of ozone reaction with the C=C bonds in the polymer macromolecule whereby the C=C bond is broken or is replaced by the weak peroxide bonds in ozonides and in di- and polyperoxides. The most susceptible areas to ozone attack are the defects on surface, amorphous and interphase regions in the elastomer structure. The application of external deformations in all cases accelerates the ozone degradation transforming the strain energy during C=C bonds cleavage. The latter proceeds during the zwitterion and carbonyl formation thus causing the appearance of cracks. The critical deformations, in our opinion, are determined by the initial rates of ozone interaction with the elastomer and the restoration of the degraded surfaces in addition to the mechanisms cited in literature.

The ozone stabilization of elastomers is accomplished by using AO-systems which can impede the harmful ozone effect by exerting simultaneous protective action, both chemically and physically. They include ozone quenching, C=C bonds and ozone deactivators, stucturizing bi- and polyfunctional compounds, which connect the cut-off ends of the macromolecules and resulting in fast stress relaxation, compounds that facilitate the AO transport from the volume to the reaction center, physical deactivators and concentration diluters.

REFERENCES

1. Razumovskii, SD, VV Podmasterev. Conformationally-nonequilibrium Macromolecules, and their reaction ability: Kinetics of ozone addition to deformed and undeformed polydienes *Vysokomol Soedin*, 38, 12, 2049-2052 (1996).
2. Reyx, D, I Campistron. Controlled degradation in tailor-made Macromolecules, elaboration Controlled chain-cleavages of polydienes by oxidation and by metathesis. *Angew Makromol Chem*, 247, 197-211 (1997).
3. Kang, ET, KG Neoh, X Zhang, KL Tan, DJ Liaw. Surface modification of electroactive polymer films by ozone treatment. *Surface and Interface Analysis*, 24: 1, 51-58 (1996).
4. Varghese, S, B Kuriakose, S Thomas. Short Sisal Fibre Reinforced Natural Rubber Composites - High-Energy Radiation, Thermal and Ozone Degradation. *Polymer Degradation and Stability*, 44: 1, 55-61 (1994).
5. Boochathum, P, M Shimizu, K Mita, Y Tanaka. Structure of Solution-Grown trans-1,4-Polyisoprene Crystals 1 Determination of Stem Length and Fold Surface Structure by Ozonolysis gpc Measurement. *Polymer*, 34: 12, 2564-2568 (1993).
6. Tanaka, Y, S Kawahara, T Ikeda, H Tamai. Analysis of the Sequence Distribution of 1,2 Units in Polybutadiene by an Ozonolysis-GPC Method. *Macromolecules*, 26: 19, 5253-5255 (1993).
7. Razumovskii, S D, S K Rakovski, D M Shopov, G E Zaikov. *Ozone and Its Reactions with Organic Compounds (in Russian)*. Publ House of Bulgarian Academy of Sciences, Sofia, 1983.
8. Dormagen, DB. Ozonolyse von ungesattigten Kautschuken und Vilkanisaten ein Uberblick. *Kautschuk + Gummi Kunststoffe*, 42, 760-769 (1989).

9. Tager AA. *Physical Chemistry of Polymers*. Moscow, Khimiya, 1978.
10. Dogatkin, BA. *Chemistry and Physics of Elastomers*. GOSKHIMIZDAT, 1947.
11. Dogatkin, BA, AA Dontzov, VA Shershnev. *Chemistry of Elastomers*. Moscow, Khimiya, 1981.
12. Kuzminskii, AS, SM Kavun, VP Kirpichev. *Physico-Chemical Bases of Preparing, Manufacturing and Applying of Elastomers*. Moscow, Khimiya, 1976.
13. Zuev, YuS. *Degradation of Polymers by Acction of Aggressive Medias*. Moscow, Khimiya, 1972.
14. Zuev, YuS. *Degradation of Elastomers at the Conditions Close to Real Exploitation*. Moscow, Khimiya, 1980.
15. Zuev, YuS, TG Degtereva. *Stabilitiy of Elastomers at the Conditions of Exploitation*. Moscow, Khimiya, 1986.
16. Razumovskii, SD, GE Zaikov. *Degradation and Protection of Polymeric Materials in Ozone. Dev Polym Stabil (Ed G Scott)*, 6, 239-294 (1983).
17. Jellinek, HHG. *Reaction of Polymers with Pollutanat Gases Chapter 9 in Aspects of Degradation and Stabilization of Polymers (Ed). HHG Jellinek*, Ellsevier, Amsterdam, 1978.
18. Denisov, ET. *The Oxidation of Alcohols, Ketones, Ethers, Esters and Acids in Solution Chapter 3 in Copr Chem Kinet (Ed CH Bamford and CFH Tipper, Elsevier, Amsterdam*, 16, 195-203 (1980).
19. Benson, SW. *The Termochemistry and Kinetics of Gas Phase Reactions In Frontiers of Free Radical Chemistry (Ed) WA Pryor*. Acad Press, New York, 1980.
20. Stewart, R. *Oxidation Mechanisms: Applications to Organic Chemistry*. WA Benjamin, Inc, 1964.
21. Emanuel', NM, GE Zaikov, ZK Majzus. *Role of the Media in the Radical-Chain Oxidation Reactions of Organic Compounds*. Nauka, Moscow, 1973.
22. Emanuel' NM, ET Denisov, ZK Majzus. *Chaine Oxidation Reactions of Hydrocarbons in Liquid Phase*. Nauka, Moscow, 1965.
23. Emanuel', NM (Ed.). *Theory and Practice of Liquid Phase Oxidation*. Nauka, Moscow, 1974.
24. Shopov, DM, SK Ivanov. *Reaction Mechanism of Inhibitors Peroxide Decomposers*. Publ House Bulg Acad Sci, Sofia, 1988.
25. Harries, CD. *Ber Dtsch Chem*, 36, 1933 (1903).
26. Harries, CD. *Untersuchungen Uber das Ozon und Seine Einwirkung auf Organische Verbindungen*. Springer-Verlag, Berlin und New York, 1916.
27. Harries, CD. *Untersuchungen Uber die Naturlichen und Kunstlichen Kautschukarten, Springer-Verlag*. Berlin und New York, 1919.
28. Criegee, R. *Rec Chem Progr*, 18, 111 (1957).
29. Criegee, R. *Chem Ber*, 108, 749 (1975).
30. Criegee, R. *Angew Chemie*, 87, 765 (1975).
31. Bailey, PS. *Chem Rev*, 58, 925 (1957).
32. Bailey, PS. *Ozonation in Organic Chemistry*, Vol I Acad Press, New York, 1978.
33. Bailey, PS. *Ozonation in Organic Chemistry*, Vol II Acad Press, New York, 1982.
34. Bernatek, E, P Kolsaker, T Ledaal. *Tetrahedron Lett*, 1317 (1963).
35. Lorenz, O, CR Parks. *Rubber Chem Technol*, 36, 194, 201 (1963).
36. Briner, GP, Rossety, S Fliszar. *Helv Chem Acta*, 48 1076 (1965).
37. Nangia, PS, SW Benson. *J Am Chem Soc*, 102, 3105 (1980).
38. Benson, SW. *Thermochemical Kinetics*, J Wiley, New York, 1976.
39. Kuczkowski, R. *Chem Soc Rev*, 79 (1992).
40. Lattimer, R, RL Kuczkowski, C W Gillies. *J Am Chem Soc*, 96, 348 (1979).
41. Murray R, PR Story, in "*Cemical Reactions of Polymers*" (E M Fettes, ed)., J Wiley (Interscience)., New York, 1964.
42. Murray, R. *Acc Chem Res*, 1, 313 (1968).
43. Williamson, DG, RJ Cvetanovic. *J Am Chem Soc*, 92, 2949 (1970).
44. Williamson DG, RJ Cvetanovic. *J Am Chem Soc*, 90, 3668 (1968).
45. Williamson, DG, RJ Cvetanovic. *J Am Chem Soc*, 90, 3248 (1968).
46. Pryor, WA, DH Giamalva, DF Church. *J Am Chem Soc*, 105, 6858 (1983).; 107, 2793 (1985).
47. Griesbaum K, G Zeick. *Chem Ber*, 119, 229 (1986).
48. Razumovskii, SD, VV Podmasterev, GE Zaikov *Phys Chem* (Russia), 129 (1983).
49. Popov AA, NYa Rapoport, GE Zaikov. Oxidation of Oriented and Strained Polymers. Khimiya, Moscow, 1987.

50. Razumovskii, SD, VV Podmasterev, GE Zaikov. *Intern J Polymeric Mater,* 13, 81 (1990).
51. Razumovskii SD, GE Zaikov, *Development in Polymer Stabilization,* (G Scott ed)., Appl Sci Publ, 6, 239 (1983).
52. Razumovskii SD, GE Zaikov, *Ozone and its Reactions with Organic Compounds*, Elsevier, Amsterdam, 1984.
53. Parfenov, VM, SK Rakovski, DM Shopov, AA Popov, GE Zaikov. *Commun Dep Chem, Bulg Acad Sci,* 9, 180 (1978).
54. Parfenov, VM, SK Rakovski, DM Shopov, AA Popov, GE Zaikov. *Commun Dep Chem, Bulg Acad Sci,* 12, 51 (1979).
55. Popov, AA, VM Parfenov, DR Cherneva, SK Rakovski, DM Shopov, AN Neverov, YaA Gurvich, G E Zaikov. *Commun Dep Chem, Bulg Acad Sci,* 12, 308 (1979).
56. Dobreva, RG, KS Rakovski, SK Rakovski, AJ Vajarova. *Coll Sci Rep Chem Ind, Polymer Ser, Central Inst Chem Ind,* No 5, 275 (1979).
57. Rakovski, SK, DR Cherneva, AJ Vajarova, RG Dobreva, KS Rakovski. *Coll Sci Rep Chem Ind, Polymer Ser, Central Inst Chem Ind,* No 5, 283 (1979).
58. Nenchev, LK, SK Rakovski, DM Shopov, KS Rakovski. *Coll Sci Rep Chem Ind, Polymer Ser, Central Inst Chem Ind,* No 5, 299 (1979).
59. Rakovski, SK, DM Shopov, KS Rakovski, and RG Dobreva. *Chem Ind* (Bulgaria)., 52, 33 (1980).
60. Rakovski, SK, AJ Vajarova, RG Dobreva, LK Nenchev, KS Rakovski, T Tzarianska. *Coll Sci Rep Chem Ind, Polymer Ser, Central Inst Chem Ind,* No 15, 9, (1982).
61. Rakovski, SK, AJ Vajarova, RG Dobreva, DR Cherneva, KS Rakovski, T Tzarjanska. *Coll Sci Rep Chem Ind, Polymer Ser, Central Inst Chem Ind,* No 15, 17 (1982).
62. Rakovski, SK, KS Rakovski, T Tzarianska, AJ Vajarova, LK Nenchev. *Coll Sci Rep Chem Ind, Polymer Ser, Central Inst Chem Ind,* No 15, 33 (1982).
63. Dobreva, RG, KS Rakovski, T Tzarjanska, AJ Vajarova, SK Rakovski. *Coll Sci Rep Chem Ind, Polymer Ser, Central Inst Chem Ind,* No 15, 83 (1982).
64. Dobreva, RG, KS Rakovski, T Tzarjanska, AJ Vajarova, SK Rakovski. *Coll Sci Rep Chem Ind, Polymer Ser, Central Inst Chem Ind,* No 15, 97 (1982).
65. Razumovskii, SD, G E Zaikov. *Ozone Reaction with Organic Compounds*, Monograph, Nauka, Moscowa, 1974, (in Russian).
66. Anachkov, MP, SK Rakovski, SD Razumovskii, AA Kefely, GE Zaikov, DM Shopov. *Polym Deg Stabil,* 10, 25 (1985).
67. Dobreva, RG, KS Rakovski, SK Rakovski. *Chem Ind* (Bulgaria)., 57, 448 (1985).
68. Anachkov, MP, SK Rakovski, DM Shopov, SD Razumovskii, AA Kefely, GE Zaikov. *Polym Deg Stabil,* 14, 189 (1986).
69. Rakovski, SK, KS Rakovski, LK Nenchev, RG Dobreva, AJ Vajarova. *Commun Dep Chem, Bulg Acad Sci,* 18, 181 (1985).
70. Anachkov, MP, SK Rakovski, DM Shopov. *Commun Dep Chem, Bulg Acad Sci,* 18, 187 (1985).
71. Anachkov, MP, SK Rakovski, SD Razumovskii, GE Zaikov, DM Shopov. *Commun Dep Chem, Bulg Acad Sci,* 18, 194 (1985).
72. Rakovski, SK, DR Cherneva, K S Rakovski, and D M Shopov. *Commun Dep Chem, Bulg Acad Sci,* 18, 202 (1985).
73. Rakovski, V Ivanova, V Georgieva. *Commun Dep Chem, Bulg Acad Sci,* 18, 220 (1985).
74. Anachkov, MP, SK Rakovsky, R V Stefanova, and D M Shopov. *Polym Deg Stabil,* 19, 293 (1987).
75. Parfenov, VM, AA Popov, SK Rakovsky, and GE Zaikov. *Visoko Mol.Soed.-Ser A,* (Russia), 31, 1250 (1989).
76. Anachkov, MP, RV Stefanova, and SK Rakovsky. *Brit Polym J.*, 21, 429 (1989).
77. Rakovski, SK, DM Shopov, GE Zaikov. *Int J Polym Materials,* 14, 1 (1990).
78. Rakovski, SK, DR Cherneva, GE Zaikov. *Int J Polym Materials,* 14, 21 (1990).
79. Rakovski, SK, V Kasparov, V Parfenov, and A Popov. *Int J Polym Materials,* 13, 223 (1990).
80. Anachkov, MP, SK Rakovsky, and SD Razumovski. *Int J Polym Materials,* 14, 79, (1990).
81. Popov, AA, VM Parfenov, SK Rakovski, DM Shopov. *Int J Polym Materials,* 13, 123 (1990).
82. Rakovski, SK, V Podmasterev, SD Razumovskii, GE Zaikov. *Int J Polym Materials,* 15, 123 (1991).

83. Anachkov, MP, SK Rakovsky, A Stoyanov. *Int J Polym Materials*, 15, 12 (1991).
84. Nenchev, LK, SK Rakovski, P Ivanov. *Oxid Commun.*, 16, 104 (1993).
85. Anachkov, MP, SK Rakovsky, RK Fotty, and AK Stoyanov. *Polym Deg Stabil*, 41, 185 (1993).
86. Garmonova IV (Ed.). *Synthetic Rubbers*, (in Russian)., Khimiya, Leningrad, 1983.
87. US Pat 2969394, 1961.
88. Fr Pat 1553993, 1969.
89. Rakovski S, D Cherneva, D Shopov. *Commun Depart Chem, Bul Acad Sci*, 18, 202 (1985).
90. US Pat 2323948, 1944.
91. US Pat 2381015, 1945.
92. Brit Pat 568230, 1945.
93. Brit Pat 712100, 1954.
94. Brit Pat 716239, 1954.
95. Prausnitz, JM, M Srausnitz, TK Sherwood. *The properties of Gases and Liquid.* 2-nd Edn., McGraw Hill, N Y, 1966.
96. Rossini, FD et. al. *Selective values of physical and thermodynamic Properties of Hydrocarbons and Related Compounds.* Carnegie Press, London, 1953.
97. Brit Pat 902768, 1963.
98. US Pat 2779789, 1957.
99. US Pat 3135796, 1961.
100. US Pat 3384664, 1968.
101. BRD Pat 1179947, 1966.
102. US Pat 3209030.
103. US Pat 3538162, 1970.
104. Brit Pat 1475628, 1977.
105. US Pat 3238177, 1966.
106. Jpn Pat 19732, 1976.
107. Brit Pat 961898, 1964.
108. BRD Pat 1077667, 1961.
109. Fr Pat 1324155, 1963.
110. Jpn Pat 82 123148, 1982.
111. Jpn Pat 57156466, 1982.
112. US Pat 3634313.
113. BRD Pat 2605166.
114. Fr Pat 1553993, 1969.
115. US Pat 373026.
116. Fr Pat 1463529, 1965.
117. US Pat 3384664, 1968.
118. Brit Pat 1098236, 1968.
119. US Pat 3538161, 1970.
120. US Pat 3538162, 1970.
121. USSR Pat 899532, 1982.
122. USSR Pat 891639, 1981.
123. Pol Pat 63933, 1971.
124. Brit Pat 1241569, 1971.
125. US Pat 3538161, 1970.
126. US Pat 3739026, 1973.
127. US Pat 3336386, 1963.
128. Brit Pat 1098236, 1968.
129. BRD Pat 2736228, 1976.
130. USSR Pat 183736.
131. US Pat 336684, 1968.
132. Brit Pat 753740, 1954.
133. US Pat 3384664, 1968.
134. Brit Pat 817142, 1961.
135. US Pat 2902466.

136. Brit Pat 1096487.
137. Brit Pat 1475620, 1978.
138. US Pat 3929855.
139. US Pat 3933739; BRD Pat 2550421; US Pat 4297269.
140. US Pat 3668254.
141. Eur Pat 84527.
142. Brit Pat 1475620.
143. Can Pat 1064691.
144. US Pat 3968080, 1976.
145. Eur Pat 0208456, 1987.
146. Jpn Pat 56125343,1981.
147. Brit Pat 1067987,1966.
148. Brit Pat 1486691.
149. Brit Pat 142676, 1972; Brit Pat 1486641,1977.
150. US Pat 4269763.
151. US Pat 4124565, 1978.
152. Jpn Pat 58074735, 1983.
153. US Pat 4021474.
154. USSR Pat 1027177.
155. Eur Pat 0147378, 1984.
156. Jpn Pat 56014516, 1981.
157. US Pat 4244864,1981.
158. NL Pat 8004296, 1981.
159. US Pat 4247664, 1981.
160. US Pat 732311, 1980.
161. US Pat 4354025, 1982
162. US Pat 3580885, 1971.
163. US Pat 4082706.
164. Brit Pat 2137619, 1983.
165. USSR Pat 529185, 1977.
166. BRD Pat 2146263, 1971.
167. BRD Pat 3228861, 1981.
168. BRD Pat 3228864, 1984.
169. BRD Pat 3114639.
170. BRD Pat 3431694.
171. BRD Pat 2548911.
172. USSR Pat 802321.
173. BRD Pat 3133567
174. EP Pat 62809, 1982.
175. BRD Pat 2850995.
176. Brit Pat 1424430, 1976.
177. BRD Pat 1954165, 1969.
178. BRD Pat 251864,1975.
179. USSR Pat 430108,1974.
180. EP Pat 139602, 1984.
181. Fr Pat 2491077.
182. Brit Pat 2491077.
183. US Pat 4186126
184. Brit Pat 1407168.
185. BRD Pat 2425425, 1974.
186. USSR Pat 193051.
187. USSR Pat 519407.
188. Brit Pat 1450205.
189. EP Pat 174277,1986.
190. Jpn Pat 58083037, 1983.

191. Jpn Pat 660186549, 1985.
192. Brit Pat 1419481, 1971.
193. EP Pat 351850, 1981.
194. USSR Pat 1027177.
195. USSR Pat 1085975,1984.
196. BRD Pat 2146263,1971.
197. USSR Pat 1196366, 1985.
198. USSR Pat 472962, 1976.
199. USSR Pat 1016329.
200. BRD Pat 2727584.
201. RD Pat 90565.
202. DDR Pat 216005, 1981.
203. DDR Pat 146464,1981.
204. US Pat 4419475, 1983.
205. US Pat 4303574, 1981.
206. USSR Pat 922117,1982.
207. Jpn Pat 58053924.
208. US Pat 4296007, 1981.
209. USSR Pat 761508, 1980.
210. USSR Pat 759556, 1981.
211. Fr Pat 2459266, 1981
212. US Pat 4568711, 1986.
213. Jpn Pat 60245652, 1985.
214. Jpn Pat 61168640, 1986.
215. Jpn Pat 60250051, 1985.
216. Jpn Pat 79153851, 1983.
217. Jpn Pat 58108236, 1983.
218. Jpn Pat 57117545, 1981.
219. USSR Pat 594138.
220. Jpn Pat 8644004,1986.
221. Jpn Pat 32528.
222. USSR Pat 954404,1982.
223. US Pat 3969315, 1976.
224. Jpn Pat 57057733, 1982.
225. Jpn Pat 57117544, 1982.
226. USSR Pat 235871, 1986.
227. Jpn Pat 57109840, 1982.
228. USSR Pat 834013, 1981.
229. Jpn Pat 56125440, 1981.
230. EP Pat 70596, 1983.
231. Jpn Pat 57147531, 1982.
232. Jpn Pat 57109840, 1982.
233. USSR Pat 859396, 1981.
234. Jpn Pat 58138775, 1983.
235. Jpn Pat 58122943, 1983.
236. Bulg Pat 36860, 1985.
237. Bulg Pat 36559, 1984.
238. Bulg Pat 34675, 1983.
239. Bulg Pat 34674, 1983.
240. Bulg Pat 34673, 1983.
241. Bulg Pat 34383, 1983.
242. Bulg Pat 36559, 1983.
243. Bulg Pat 36859, 1983.
244. Jpn Pat 58122946 1983.
245. USSR PaT 933117, 1982.

246. EP Pat 69908, 1983.
247. Jpn Pat 62141044, 1987.
248. Jpn Pat 62109838, 1987.
249. USSR Pat 1260376, 1986.
250. Jpn Pat 61168640, 1986.
251. US Pat 4459380, 1981.
252. Fr Pat 2459266, 1981.
253. Chtourou, H, B Riedl, BV Kokta. Surface Modification of Polyethylene Pulp Fiber by Ozone Treatment - An Analytical and Thermal Characterization. *Polymer Degradation and Stability*, 43: 1, 149-156 (1994).
254. Ishizu, K, F Naruse, R Saito. Preparation of Core Shell Type Polymer Microspheres from Anionic Block Copolymers. *Polymer*, 34: 18, 3929-3933 (1993).
255. Fargere, T, M Abdennadher, M Delmas, B Boutevin. Synthesis of Graft Polymers from an Ozonised Ethylene- Vinyl Acetate Copolymer 2 Determination of the Efficiency and of the Decomposition Rate of Macroinitiators. *Macromolecular Chemistry and Physics*, 195: 5, 1673-1685 (1994).
256. Fargere, T, M Abdennadher, M Delmas, B Boutevin. Synthesis of Graft Polymers from an Ozonized Ethylene Vinyl Acetate Copolymer, (EVA). 1 Study of the Radical Polymerization of Styrene Initiated by an Ozonized EVA. *Journal of Polymer Science, Part A - Polymer Chemistry*, 32: 7, 1377-1384 (1994).
257. Keshavaraj, R, RW Tock. Oxidative Effects of Ozone on the Aging of Structural Silicone Elastomers. *Advances in Polymer Technology*, 13: 2, 149-156 (1994).
258. Pickett JE. Effect of Stratospheric Ozone Depletion on Terrestrial Ultraviolet Radiation - A Review and Analysis in Relation to Polymer Photodegradation. *Polymer Degradation and Stability*, 43: 3, 353-362 (1994).
259. Callen, BW, ML Ridge, S Lahooti, AW Neumann, RNS Sodhi. Remote plasma and ultraviolet-ozone modification of polystyrene. *Journal of Vacuum Science, & Technology A - Vacuum Surfaces and Films*, 13: 4, 2023-2029 (1995).
260. Deng, YX, CP Peng, P Liu, RJ Zhang, LM Zeng. Studies on the microozonization-TLC method for the separation and analysis of copolymerization products of alpha-pinene styrene. *Chemical Journal of Chinese Universities - Chinese*, 16: 8, 1306-1310 (1995).
261. Hong, SG, CM Liao. The surface oxidation of a thermoplastic olefin elastomer under ozone exposure: ATR analysis. *Polymer Degradation and Stability*, 49: 3, 437-447 (1995).
262. Varghese, S, B Kuriakose, S Thomas. Short Sisal Fibre Reinforced Natural Rubber Composites - High-Energy Radiation, Thermal and Ozone Degradation. *Polymer Degradation and Stability*, 44: 1, 55-61 (1994).
263. Kulik, EA, MI Ivanchenko, K Kato, S Sano, Y Ikada. Peroxide generation and decomposition on polymer surface. *Journal of Polymer Science, Part A - Polymer Chemistry*, 33: 2, 323-330 (1995).
264. Basfar, AA, J Silverman Improved Ozone Resistance of Styrene Butadiene Rubber Cured by a Combination of Sulfur and Ionizing Radiation. *Polymer Degradation and Stability*, 46: 1, 1-8 (1994).
265. Krisyuk, BE, AA Popov, VV Kotov. Ozone Resistance, Relaxation Properties, and Molecular Mobility of Butadiene Styrene Thermoplastic Elastomers. *Vysokomolekulyarnye Soedineniya Seriya A & Seriya B*, 37: 1, 94-100 (1995).
266. Greenwood, OD, S Tasker, JPS Badyal. A comparative study of the silent discharge treatment of saturated and unsaturated hydrocarbon polymers. *Journal of Polymer Science, Part A - Polymer Chemistry*, 32: 13, 2479-2486 (1994).
267. Tanaka, Y, S Kawahara, T Ikeda, H Tamai. Analysis of the Sequence Distribution of 1,2 Units in Polybutadiene by an Ozonolysis-GPC Method. *Macromolecules*, 26: 19, 5253-5255 (1993).
268. Zuev, YuS. Degradation of Polymers by Aggressive Media. Moscow, Khimiya, 1972.
269. Kuzminskii, AS, SM Kawun, VP Kirpichev. Physico-chemical bases of Synthesis, Processing, and Application of Elastomers.Moscow, Khimiya, 1976
270. Zuev, YuS. *Zh. Vsesoyuz. Obstest. Im. D. Mendeleeva*, 1966, 3, 228
271. ASTM D 1149-64 and ASTM D 471-64
272. BR À23, 1963

273. Parfenov, VM, SK Rakovski, DM Shopov, AA Popov, GE Zaikov. *Commun.Dep.Chem., Bulg.Acad.Sci.*, 12, 1, 254 (1979).
274. Fettes E (Ed.) Chemical Reactions of Polymers. v.2, Moscow, Mir, 1976.
275. Foigt I. *Stabilization of Sinthetic Polymers againts Lithg and Temperature.* Leningrad, Khimiya, 1972
276. Bartenev, GM, YuS Zuev. *Strength and Degradation of Highelastic Materials.* Moscow, Khimiya, 1964.
277. Kuzminskii AS (Ed.). *Degradation and Stabilization of Polymers.* Moscow, Khimiya, 1966
278. Ratner SB. *Strength and Plasticity of Metals.* Moscow, Oborongis, 1950
279. Gul', VE. Structure and Strenght of Polymers. Moscow, Khimiya, 1978.
280. Avous, F. *Rubber Plast Weekly*, 147, 17, 671 (1961).
281. Emanuel', NM, ET Denissov, ZK Majzus. *Chaine Oxidation Reactions of Hydrocarbons in Liquid Phase.* Moscow, Nauka, 1965.
282. Emanul', NM (Ed.) *Theory and Practice of Liquid Phase Oxidation.* Moscow, Nauka, 1974
283. Karpukhina, GV, ZK Majzus, TV Zolotova, LI Mazaletzka, M Meskina. Mechanism of Synergestic Antioxidative Action of Mixtures of Different Types Inhibitors. Neftekhimiya, 708 - 715 (1978).
284. Emanuel', NM. Structure-kinetic aspects of the Degradation and Stabilization of Polymers. *Visoko Molek. Soed. Ser. A*, 2653 - 2661 (1978).
285. Karpukhina, GV, ZK Majzus, M Meskina. *Kinet. Kataliz*, 9, 245 (1968).
286. Denisov, ET, VV Kharitonov. *Zhurn. Fiz. Khim.*, 38, 699 (1964).
287. Denisov, ET, AL Aleksandrov. *Zhurn. Fiz. Khim.*, 38, 491 (1964).
288. Zolotova, TV, GV Karpukhina, ZK Majzus, NM Emanuel'. Dokl. Akad. Nauk SSSR, 223, 120 (1975).
289. Karpukhina, GV, ZK Majzus, NM Emanuel'. *Dokl. Akad. Nauk SSSR*, 182, 870 (1968).
290. Fedotov, BP, AF Lukovnikov, PM Mamedov, VV Edemska, VA Sukhov. *Izv. Akad. Nauk. SSSR, ser. khim.* 268 (1966).
291. Rakovski SK, DR Cherneva, VM Parfenov. *Commun.Dep.Chem., Bulg.Acad.Sci.*, 153 (1978).
292. Ambelang, J C, et al. *Rubb Chem Techn*, 42, 1186 (1969).; KV Boguslovskaya et al. *Kauchuk i Rezina.*, 1, 3 (1970).
293. Jpn Pat 48-18568, 1973.
294. Jpn Pat 6895, 1972
295. Ustinova TA et al. *Kauchuk I Rezina*, 4, 16 (1972).
296. Khamin SE, LG Angert et al. *Kolloid J.*, 37, 99 (1975).
297. Zacheev VS, AI Zvonkova. *Production of Tyres, Rabbur and Asbest Goods*, 1, 11 (1974).
298. Troitskaya NI et al., *Kauchuk i Rezina* 6, 18 (1974).
299. US Pat. 3817884, 1974; 3704741, 1972.
300. Jpn Pat. 7500030; Fr Pat.15555; UK Pat.1303068, 1972.
301. Italian Pat. 871056, 1970.
302. Mix, MY. *Eur Rubb J*, 6, 27 (1975).
303. Andrews, EH. *Rubb Chem Techn*, 40, 633 (1967)
304. Khaninya, SE, LG Angert et al., *Kauchuk i Rezina*, 1, 30 (1974).
305. Maloshuk, US, VN Kuleznev, SE Khanin. *Colloid J.*, 35, 408 (1973).
306. Ashworth, BT, BN Leisland, P M Quan Rubb Conf '72, *Proc Int Rubb Conf, Brighton*, 1972, F6/1 London, 1972
307. Cain, ME, et al *J Rubb Res Inst Malaya*, 22, 289 (1969).
308. Cuneen, JI. *Rev Gen Caut Plast*, 49, 521 (1972).
309. Kirpichov, VP et al., *Kauchuk i rezina*, 7, 20 (1972).
310. Kirpichov, VP, NS Domnina. *Vest. Penz.Gosud. Univ.*, 3, 16, 126 (1978).
311. Gordon, TYa. Stabization of Synthetic Polymers. Moscow, Khim. Lit.,1963.
312. UK Pat. 1498707, 1978.
313. Bebikh, GF, VP Saraeva. *Visok. Molek. Soed.*, 18, 461 (1976); Saraeva, VP et al., *Kauchuk i Rezina*, 5, 29;(1976); USSR Pat. 348581, 1969.
314. Kline, RH, YR Miller. *Rubb Chem Technol*, 46, 98 (1973).
315. Mayer, GE, RW Kavchok, FJ Naples. *Rubb Chem Technol*, 46, 196 (1973).

316. Levy, R. *Rev Gen Caout Plast*, 51, 243 (1974).
317. US Pat. 3714122, 973.
318. US Pat. 3962187,1976.
319. Plate, NA (Ed.) *Kinetics and Mechanism of the Formation and Transformation of Molecules.* Moscow, Nauka, 1968.
320. Askarov, MA., AS Bank. *Chemical Stabilization of Polymers*. Tashkent, Uzb. Akad. Nauk., Uzb. SSR (1974).
321. Gorbunov, BN, YaA Gurvich, IP Maslova. *Chemistry and Technology of Stabilizators for Polymeric Materials.* Moscow, Khimiya, 1981.
322. Emanuel', NM, AL Buchachenko. *Chemical Physics of the Degradation and Stabilization of Polymers*. Moscow, Nauka, 1983.
323. Emanuel', NM., D Gal. *Oxidation of Ethylbenzene*. Moscow, Nauka, 1984.
324. Scott, G, E Setoudeh. Mechanism of Antioxidant Action: Effects of Rubber-bound Antioxidants in Rubber Modified Polypropilene. *Polym Degr Stab*, 5, 11-22 (1983).
325. Rotschova, J, J Pospishil. Antioxidant Action of 1,4 - Phenylenediamine. *Coll Czech Chem Commun*, 47, 2501-2513 (1981).
326. Taimr, L, J Pospishil. Polym Degr Stabil, 8, 23-35 (1984).
327. Furukawa, J, S Yamashita, T Kotani. Mechanizm of Antioxidation and Antiozonation of Amines for Rubber Chapter 9 in Stabilization of Polymers and Stabilizer Processes, *Adv Chem Ser*, 85, 113-125 (1968).
328. Razumovskii, SD, VV Podmasterev, GE Zaikov Refresh of the Elastomer Surfice during the Ozonation. *Dokl Akad Nauk SSSR*, 129-132 (1982).
329. Popov, AA, GE Zaikov. The Effect of Stresses on Polymer Oxidation Quantitative Aspects 1 Elastomers. *JMS-Rev Macromol Chem Phys*, C23(1)., 1-32 (1983).
330. Amsden, CS. The Annulus Ozone Test Lecture befor AGM of Inst Rubber Tech -Sweden, 15-16 June, 1967.
331. Rapoport, NYa, GE Zaikov in Dev.PolymDegr., 6, 207-258 (1985).
332. Chtourou, H, B Riedl, BV Kokta, A Adnot, S Kaliaguine Synthetic Pulp Fiber Ozonation - An ESCA and FTIR Study. *Journal of Applied Polymer Science*, 49: 2, 361-373, (1993).
333. Ebdon, JR, S Rimmer. A one-pot method for the preparation of latices of telechelic oligomers, by ozonolysis of latices of polymers containing main-chain unsaturation. *J Polym Sci A-Polym Chem,* 35, 15, 3255-3262 (1997).
334. Couve, J, Dobrevaschue, R, Sledz, J, Schue, F, Nicol, P. Characterization of star-branched styrene-butadiene copolymers by ozonolysis. *Eur Polym J*, 33, 2, 137-141 (1997).
335. Gatenholm, P, Ashida, T, Hoffman, AS. Hybrid biomaterials prepared by ozone-induced polymerization 1 Ozonation of microporous polypropylene. *J Polym Sci A-Polym Chem,* 35, 8, 1461-1467 (1997).
336. Kim, WS, Huh, TS. Ozonolysis of 2-methylthio-3-methyl-2-butene on polyethylene: A new access to tetramethyl-1,2,4-trioxolane. *Bull Kor Chem Soc,* 18, 5, 543-544 (1997).
337. Rimmer, JR Ebdon, MJ Shepherd. Polymer-supported reagents as aids in controlled degradation of polymers containing ozonides in the main chain to give telechelic oligomers. *Reactive & Functional Polymers,* 26: 1-3, 145-155 (1995).
338. Kang, ET, KG Neoh, X Zhang, KL Tan, DJ Liaw. Surface modification of electroactive polymer films by ozone treatment. *Surface and Interface Analysis,* 24: 1, 51-58 (1996).
339. Fargere, T, M Abdennadher, M Delmas, B Boutevin. Synthesis of Graft Polymers from an Ozonised Ethylene- Vinyl Acetate Copolymer 2 Determination of the Efficiency and of the Decomposition Rate of Macroinitiators. *Macromolecular Chemistry and Physics,* 195: 5, 1673-1685 (1994).
340. Ebdon, EJ, NJ Flint, S Rimmer. Degradative routes to telechelic oligomers and macromonomers. *Journal of Macromolecular Science, - Pure and Applied Chemistry A* 32: Suppl 5-6, 603-611 (1995).
341. Varghese, S, B Kuriakose, S Thomas. Short Sisal Fibre Reinforced Natural Rubber Composites - High-Energy Radiation, Thermal and Ozone Degradation. *Polymer Degradation and Stability,* 44: 1, 55-61 (1994).

342. Giurginca, M, T Zaharescu, A Meghea. Degradation of ethylene-propylene elastomers in the presence of ozone. *Polymer Degradation and Stability,* 50: 1, 45-48 (1995).

343. Hong, SG, CM Liao. The surface oxidation of a thermoplastic olefin elastomer under ozone exposure: ATR analysis. *Polymer Degradation and Stability,* 49: 3, 437-447 (1995).

344. Evtushenko, AM, IP Chikhacheva, GV Timofeeva, SD Stavrova, VP Zubov. Graft Polymerization of Acrylic Acid into Ozonated Siloxane Matrixes. *Vysokomolekulyarnye Soedineniya Seriya A & Seriya B,* 35: 6, B312-B315 (1993).

345. Bulanin, KM, Lavalley, JC, Tsyganenko, AA. Infrared study of ozone adsorption on CaO. *J Phys Chem B,* 101, 15, 2917-2922 (1997).

Chapter 5

QUANTUM CHEMICAL CALCULATIONS OF OZONOLYSIS OF ORGANIC COMPOUNDS

Quantum chemical calculations of organic compounds ozonolysis are continuously expanding and cover more and more new compounds and separate stages in already accepted mechanisms of interaction [1-10]. The reaction of ozone with C=C bonds [11-20] is one of the reactions that has attracted much attention. Here should be mentioned the fundamental contributions of Godart [20-22], Cremer [6, 7, 24], Rouz [17, 19], etc. who have applied various *ab initio* methods. A number of important and interesting ozone reactions, however, have remained out of the scope of research interest [21-25].

In this connection we devoted our studies on quantum chemical calculations of the ozonolysis of organic compounds containing H, O, S, N and P atoms. The transition states (TS) depending on the atom to which the ozone attack is directed, are denoted in the following manner: hydrogen bound to carbon atom - (TS1), hydrogen bound to heteroatom - (TS2) and heteroatom - (TS3).

We have carried out the calculations by using a *mopac6* program, whose considerations, parameterization and practical use are described in Refs 26-83.

5.1. ALKANES

The calculations were performed according to the method of the reaction coordinate, whereby all coordinates of the reagents are optimized with the exception of the fixed reaction coordinate. The latter was varied from 3-4Å to 5Å with a gradient from of 0.2 to 0.05 Å.

The transition states for the reactions investigated are described as follows:

CH_3-(H_2)C-H...O_3 $\quad$ (CH_3)$_2$(H)C-H...O_3 $\quad$ (CH_3)$_3$C-H...O

1) linear transition states that correspond to hydrogen atom abstraction mechanism. The C-H bond, being the bond of lowest bond energy, is the reaction center of the molecule:

CH_3-H_2C-H...O_3 $\quad$ (CH_3)$_2$-HC-H...O_3 $\quad$ (CH_3)$_3$C-H..O_3

2) cyclic transition state, which accounts for 1,3-cycloaddition mechanism including a transitory state with 5-coordinated carbon and resulting in hydrotrioxide formation in one step:

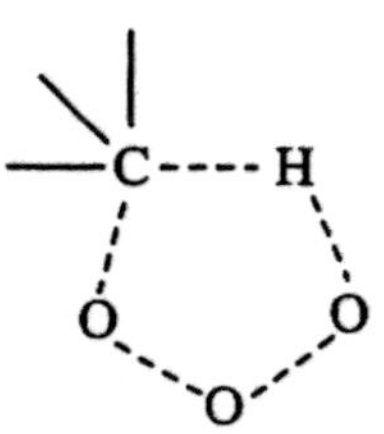

In Fig.5.1 are shown the summarized results from calculating the kinetics of ozone reaction with methane, ethane, propane and *iso*-butane with linear transitions states of TS1 type. The calculations with cyclic transition states are not given because we failed to observe formation of any structure resulting from C-H bond cleavage and formation of two new bonds C-O and O-H.

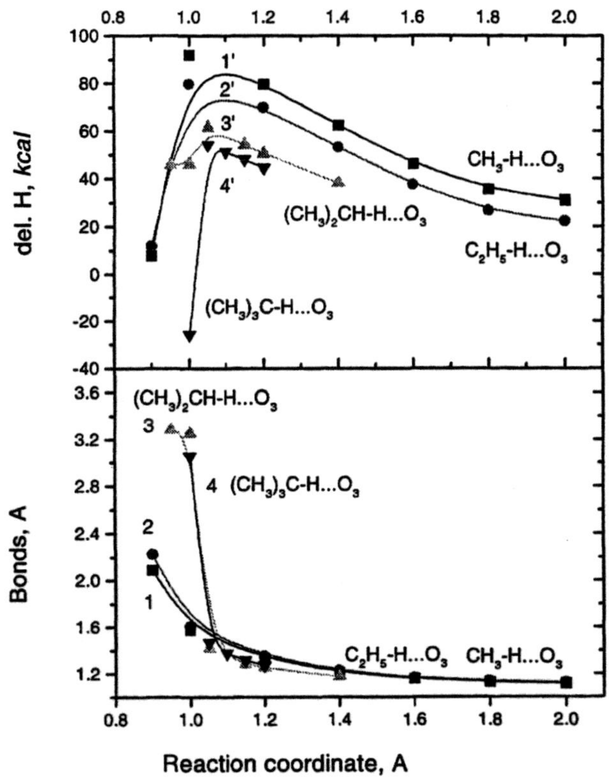

Fig.5.1. Interaction of ozone with methane, ethane, propane and *iso*-butane. 1, 2, 3, 4 - dependence of the C-H bond length on the reaction coordinate, respectively and 1', 2', 3'and 4' - dependence of ΔH on the reaction coordinate, respectively.

The data in Fig.5.1 supports the ozone reaction with C-H bonds. For methane the C-H bonds cleavage occurs at a reaction coordinate within 0.95 Å and 0.9 Å with simultaneous drop of ΔH from 96.3 kcal to 75.3 kcal. In case of ethane this takes place at RC from 1 Å to 0.95 Å and change of ΔH from 79.8 kcal to 11.9 kcal. For the ozone reaction with propane the C-H bond breaks occurs at RC from 1.05 Å to 1 Å and ΔH changes from 61.7 kcal to 46 kcal. The RC for *iso*-butane C-H bond fission lies between 1.05 Å and 1 Å and ΔH changes from 54.2 to -25.9 kcal. The analysis of the foregoing results reveals the following sequence of increasing reactivity of the studied alkanes: methane < ethane < propane < *iso*-butane. This conclusion is in a very good agreement with the experimental results presented in Chapter I.

Based on the results from the reaction coordinate method we proceed further to determination of the transition state. This procedure includes the determination of the first and second derivative of the energy in all coordinates. It is characterized by the fact that the first derivatives become zero at the saddle point, and the second ones are positive except one, the negative value of which stands for vibrations with pseudo frequency along the reaction coordinate. The results from utilization of this procedure are outlined in Table 5.1.

Table 5.1

Reaction of methane with ozone - localization of TS

GRADIENT NORM MINIMIZED USING NS01A
SCF FIELD WAS ACHIEVED

HEAT OF FORMATION	=	28.597233 KCAL/MOL
ELECTRONIC ENERGY	=	-2761.007473 EV
CORE-CORE REPULSION	=	1623.351869 EV
GRADIENT NORM	=	0.079615
RMS FORCE	=	0.018765
DIPOLE	=	1.78872 DEBAY
NO. OF FILLED LEVELS	=	13
IONIZATION POTENTIAL	=	9.738345 EV
MOLECULAR SYMMETRY:		C1
MOLECULAR WEIGHT	=	64.041
SCF CALCULATIONS	=	149
COMPUTATION TIME	=	1845.20 SECONDS

FINAL GEOMETRY OPTIMIZATION
AM1 PRECISE NS01A T=60M PULAY ITRY=1200
METHANE + OZONE
T.S. LOCALIZATION

										CHARGE
C	0.000000	0	0.000000	0	0.000000	0	0	0	0	-0.2678
H	1.111572	1	0.000000	0	0.000000	0	1	0	0	0.0649
H	1.111566	1	109.469708	1	0.000000	0	1	2	0	0.0634
H	1.111589	1	109.449504	1	119.975690	1	1	2	3	0.0648
H	1.112658	1	109.474622	1	-120.050073	1	1	2	3	0.0749
XX	1.000000	0	90.000000	0	0.000000	0	5	1	2	
O	2.409391	1	89.586115	1	179.856964	1	5	6	1	-0.2786
O	1.160081	1	168.570623	1	114.303161	1	7	5	6	0.5535
O	1.160081	1	120.954422	1	-168.405569	1	8	7	5	-0.2752
0	0.000000	0	0.000000	0	0.000000	0	0	0	0	

In the transition state any substantial change in the length of the C-H bond to which ozone attack is directed, is not observed as evidenced by its negligible change (by 1 %) and the reaction coordinate is 2.4 Å. Practically it can be assumed that TS is not localized. This compelled us to use other key words for localization. Looking for TS as a biradical the following results are obtained (Table 5.2):

Table 5.2

Reaction of methane with ozone - localization of TS

GEOMETRY OPTIMISED USING EIGENVECTOR FOLLOWING (EF).
SCF FIELD WAS ACHIEVED

HEAT OF FORMATION	=	68.776672 KCAL
ELECTRONIC ENERGY	=	-2913.845415 EV STATE: SINGLET A
CORE-CORE REPULSION	=	1770.900535 EV
DIPOLE	=	2.90916 DEBYE SYMMETRY: C1
NO. OF FILLED LEVELS	=	12
AND NO. OF OPEN LEVELS	=	2
IONIZATION POTENTIAL	=	5.330939 EV
MOLECULAR WEIGHT	=	64.041
SCF CALCULATIONS	=	49
COMPUTATION TIME	=	8 MINUTES AND 51.680 SECONDS

FINAL GEOMETRY OBTAINED
MNDO PULAY T=500M TS BIRADICAL
Methane + Ozone
R.C. calculation- looking for maximum; not found!

								CHARGE
C	0.00000000	0	0.0000000	0	0.0000000	0	0 0 0	0.0011
H	1.09600787	1	0.0000000	0	0.0000000	0	1 0 0	0.0257
H	1.09630043	1	113.5850685	1	0.0000000	0	1 2 0	0.0217
H	1.09587419	1	113.8612890	1	132.4350059	1	1 2 3	0.0230
H	1.29083650	1	104.5292213	1	-113.6886841	1	1 2 3	0.1295
XX	1.00000000	0	90.0000000	0	0.0000000	0	5 1 2	
O	1.25191432	1	85.9170670	1	179.9731171	1	5 6 1	-0.2060
O	1.26258040	1	112.7809188	1	147.2814073	1	7 5 6	0.2789
O	1.20288730	1	114.7207682	1	-93.6461817	1	8 7 5	-0.2740

It follows from this matrix Z that the transition state is more probably a biradical producing CH_3·-radical and ·O_3H - radical with 18 % extension of the C-H bond. The difficulties on TS localization with paraffins can be related to two reasons: parameterization of the semiempirical calculations and the low probability for proceeding of these reactions. In spite of this the observed tendency shows that the TS may be related to the hydrogen atom abstraction.

5.2. Oxygen Containing Compounds

5.2.1. Water

It can be assumed that ozone reaction with water may take place through the following transition states TS2 and TS3:

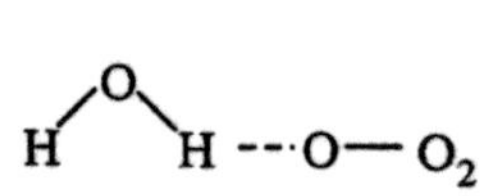

2

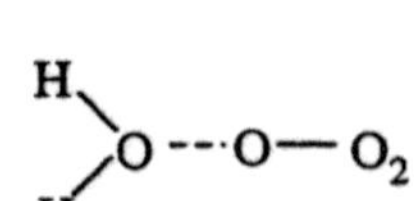

3

At TS2 the reaction coordinate is the H...O bond, while at TS3 - the O...O bond. In Fig.5.2 are shown the changes in the bond lengths and heats of formation (ΔH) in dependence on the reaction coordinate.

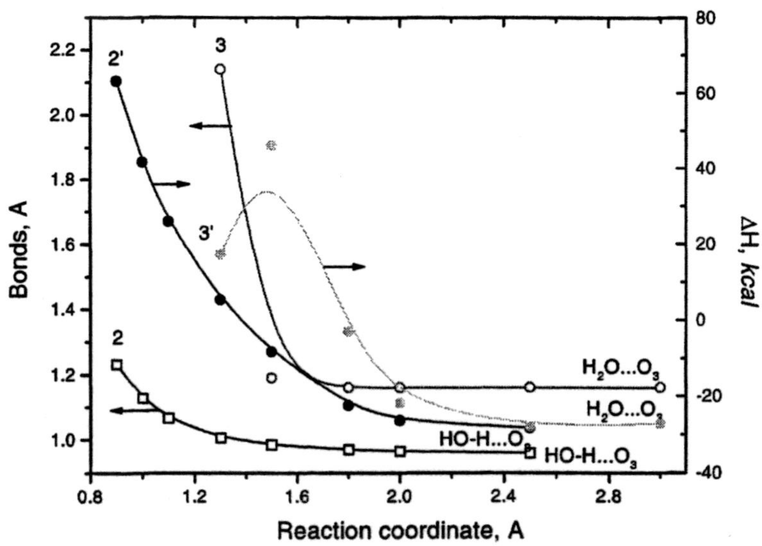

Fig.5.2. Ozone reaction with water: 2 - dependence of the O-H bond length on the reaction coordinate at TS2; 3 - dependence of the O-O_2 bond length on the reaction coordinate at TS3; 2' - dependence of ΔH (TS2) on the reaction coordinate and 3' - dependence of ΔH (TS3) on the reaction coordinate

The analysis of Fig.5.2 reveals that the ozone reaction with water through the O-H route is inefficient as the decrease in the reaction coordinate (RC) from 2.5Å to 0.9 Å leads to 27 % change of the O-H bond length (from 0.96 Å to 1,232 Å). The heats of formation of the transition states rise smoothly without any sharp leap. The interaction in oxygen atom direction is quite different. The change of RC from 3 Å to 1.3 Å does not change the O-O_2 bond up to 1.5 Å (1.16-1.19 Å) but in the range of 1.5-1.3 Å its length rises abruptly up to 2.14 Å (i.e. by 84%) and ΔH is reduced from 46.2 to 17.1 kcal. The presence of such a sharp transition in the bond length and the significant positive difference in ΔH in this case testify the great probability of this reaction pathway for the reaction between ozone and water. According to this route the reaction products should be H_2O_2 and O_2. In fact some amounts of hydroperoxide have been experimentally observed in water ozonolysis.

5.2.2. Methanol

The ozone reaction with methanol may proceed through the formation of the following transition states:

$HOCH_2$-H--O_3 (1) CH_3O-H --O_3 (2) $CH_3(H)$=O--O_3 (3)

through attack on the C-H bond -TS1, through attack on the O-H bond - TS2 and attack on the oxygen atom - TS3. The first one presumes the formation of $HOCH_2$·and HO_3· (HO· + O_2). This mechanism supposes also the formation of $HOCH_2OOOH$, $HOCH_2OH$ or $HOCH_2OO$· in the second step. In case of TS2 first will be formed CH_3O·, HO_3·(HO· + O_2) and then CH_3OOH. For TS3, $HOCH_3OOOH$ will be obtained. The quantum chemical calculations on the three routes are presented in Fig.5.3.

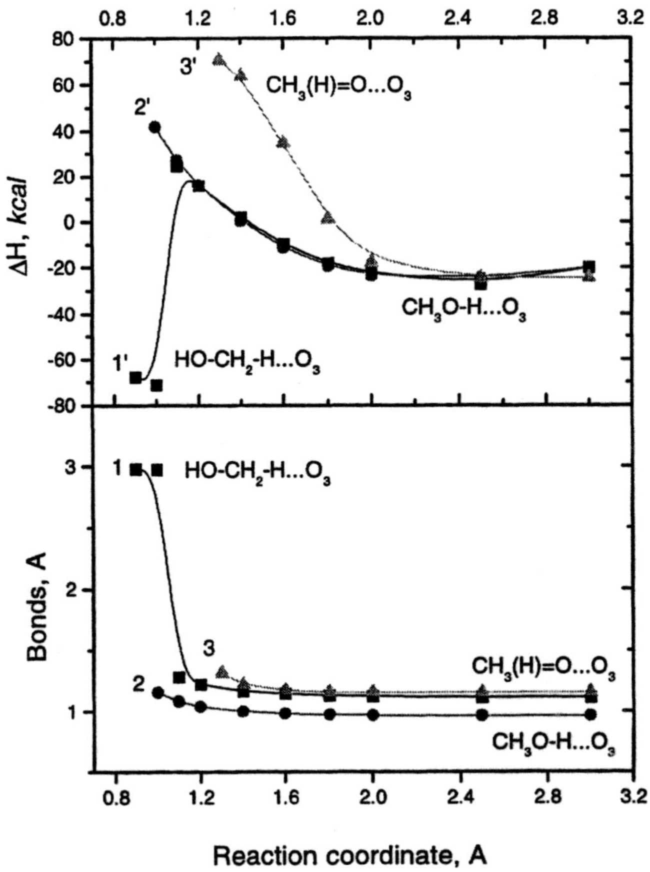

Fig.5.3. Ozone interaction with methanol.- below - dependence of the C-H bonds lenght at TS1 - 1, O-H at TS2 and O-O_2 at TS3 on the reaction coordinate; above - dependence of the corresponding heats of formation 1', 2' and 3' on the reaction coordinate

The analysis of the results show that TS1 is the preferred route for this reaction. At RC variation from 2.5 to 1.1Å the C-H bond length changes smoothly from 1.12 to 1.226 Å (i.e. by 9%). When the RC reaches the value of 1.0 Å the C-H bond length rises sharply to 2.973 Å and ΔH drops abruptly thus suggesting a C-H bond break. However, at this moment no change in the O-O_2 bond length is observed on the TS1 diagram. Thus two particles, i.e. $HOCH_2\cdot$and $HO_3\cdot$ are obtained in this step. The variations which are observed at TS2 and TS3 are fluent without any indications of qualitative changes in the system. This suggests that these two routes are quite inefficient. The probability for TS1 has been already experimentally proved in Chapter 2. Similar results were obtained on calculations of ozone reaction with ethyleneglycol.

The procedure for TS1 search is given in Table 5.3.

Table 5.3

Reaction of methanol with ozone - localization of TS1

GRADIENT NORM MINIMIZED USING NS01A
SCF FIELD WAS ACHIVED

HEAT OF FORMATION = 17.940267 KCAL/MOL
ELECTRONIC ENERGY = -4287.852106 EV
CORE-CORE REPULSION = 2831.052188 EV
GRADIENT NORM = 0.087353

RMS FORCE	=	0.019062
DIPOLE	=	3.63216 DEBAY
NO. OF FILLED LEVELS	=	16
IONIZATION POTENTIAL	=	9.681064 EV
MOLECULAR SYMMETRY:		C1
MOLECULAR WEIGHT	=	80.040
SCF CALCULATIONS	=	157
COMPUTATION TIME	=	1434.06 SECONDS

FINAL GEOMETRY OPTIMIZATION
AM1 PRECISE NS01A T=60M PULAY ITRY=1200
METHANOL + OZONE
T.S. LOCALIZATION

										CHARGE
H	0.000000	0	0.000000	0	0.000000	0	0	0	0	0.2684
O	0.982484	1	0.000000	0	0.000000	0	1	0	0	-0.2345
C	1.344494	1	110.628859	1	0.000000	0	2	1	0	-0.0344
H	1.103293	1	119.800700	1	7.734395	1	3	2	1	0.1408
H	1.107941	1	112.091391	1	156.622107	1	3	2	1	0.1446
H	1.477202	1	100.900700	1	-101.173304	1	3	2	1	0.1470
XX	1.000000	0	90.000000	0	0.000000	0	6	3	2	
O	1.137345	1	63.905024	1	-146.142569	1	6	7	3	-0.2217
O	1.241895	1	110.317690	1	38.642660	1	8	6	7	0.2668
O	1.183390	1	113.205484	1	56.471877	1	9	8	7	-0.4670
0	0.000000	0	0.000000	0	0.000000	0	0	0	0	

The transition state shows that the vibration with pseudo frequency is along the C-H bond. It is raised by 37% that corresponds to the $HOCH_3\cdot$ and $HO_3\cdot$ formation and α-H-atom cleavage.

5.2.3. Ethyleneglycol

The introduction of second OH-group does not change the preferred reaction route - C-H bond break (Fig.5.4).

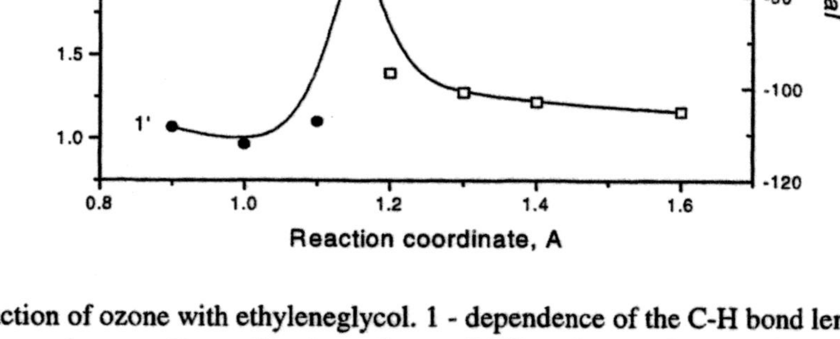

Fig.5.4. Reaction of ozone with ethyleneglycol. 1 - dependence of the C-H bond length (TS1) on the reaction coordinate; 1' - dependence of ΔH on the reaction coordinate

5.2.4. Formaldehyde and Acetone

The procedure for searching the most favorable TS for ozone reaction with formaldehyde and acetone is depicted in Tables 5.4 and 5.5.

Table 5.4

Reaction of formaldehyde with ozone - localization of TS

GRADIENT NORM MINIMIZED USING NS01A
SCF FIELD WAS ACHIVED

HEAT OF FORMATION	=	46.477233 KCAL/MOL
ELECTRONIC ENERGY	=	-3849.714576 EV
CORE-CORE REPULSION	=	2421.463447 EV
GRADIENT NORM	=	0.028602
RMS FORCE	=	0.007385
DIPOLE	=	2.75136 DEBAY
NO. OF FILLED LEVELS	=	15
IONIZATION POTENTIAL	=	9.318082 EV
MOLECULAR SYMMETRY:		C1
MOLECULAR WEIGHT	=	78.024
SCF CALCULATIONS	=	43
COMPUTATION TIME	=	277.15 SECONDS

FINAL GEOMETRY OPTIMIZATION
AM1 PRECISE NS01A T=60M PULAY ITRY=1200
FORMALDEHYDE + OZONE
T.S. LOCALIZATION

					CHARGE
O	0.000000 0	0.000000 0	0.000000 0	0 0 0	-0.1762
C	1.200896 1	0.000000 0	0.000000 0	1 0 0	0.1458
H	1.095919 1	135.459847 1	0.000000 0	2 1 0	0.1698
H	1.412302 1	116.529164 1	167.377984 1	2 1 3	0.1580
O	1.222269 1	138.224492 1	93.827421 1	4 2 1	-0.2356
O	1.235351 1	104.033312 1	-28.230674 1	5 4 2	0.2906
O	1.194135 1	112.289186 1	51.238553 1	6 5 4	-0.3524
0	0.000000 0	0.000000 0	0.000000 0	0 0 0	

It is seen that the length of the C-H bond in the transition state is increased by 29% resulting in its rupture and formation of OHC· and HO_3· - radicals.

Table 5.5

Reaction of acetone with ozone - localization of TS

GRADIENT NORM MINIMIZED USING NS01A
SCF FIELD WAS ACHIVED

HEAT OF FORMATION	=	40.786156 KCAL/MOL
ELECTRONIC ENERGY	=	-6149.223501 EV
CORE-CORE REPULSION	=	4409.650496 EV
GRADIENT NORM	=	0.031238
RMS FORCE	=	0.005446
DIPOLE	=	2.71796 DEBAY
NO. OF FILLED LEVELS	=	21
IONIZATION POTENTIAL	=	9.286127 EV
MOLECULAR SYMMETRY:		C1
MOLECULAR WEIGHT	=	106.078
SCF CALCULATIONS	=	142
COMPUTATION TIME	=	2611.29 SECONDS

FINAL GEOMETRY OPTIMIZATION
AM1 PRECISE NS01A T=60M PULAY ITRY=1200
ACETONE + OZONE
TS LOCALIZATION

										CHARGE
O	0.000000	0	0.000000	0	0.000000	0	0	0	0	-0.3010
C	1.239801	1	0.000000	0	0.000000	0	1	0	0	0.2262
C	1.495467	1	122.540300	0	0.000000	0	2	1	0	-0.2620
C	1.466488	1	121.581600	1	178.781883	1	2	1	3	-0.2715
H	1.118573	1	109.193989	1	116.963962	1	3	2	1	0.1163
H	1.116646	1	109.978078	1	-123.626602	1	3	2	1	0.0976
H	1.116895	1	110.546095	1	-3.045697	1	3	2	1	0.1235
H	1.099567	1	117.606477	1	-165.264690	1	4	2	1	0.1349
H	1.108390	1	115.688326	1	52.082983	1	4	2	1	0.1373
H	1.541413	1	113.400998	1	-39.639822	1	4	2	1	0.2274
O	1.130087	1	141.632512	1	-69.485801	1	10	4	2	-0.2428
O	1.254625	1	105.153145	1	26.878349	1	11	10	4	0.2996
O	1.181268	1	112.192736	1	-51.302198	1	12	11	10	-0.2731
0	0.000000	0	0.000000	0	0.000000	0	0	0	0	

In the transition state the length of the α- C-H bond is extended by 40% which leads to its cleavage producing CH_3COOCH_2·- and HO_3·- radicals.

5.2.5. Dimethylether

The ozone interaction with dimethylether may take place both through TS1 and TS3. Fig.5.5 demonstrates the results from the calculations with TS1 - the ozone attack on C-H bond. The calculations on TS3 are not given because they are similar to those carried out on methanol with TS3. It should be noted that in this case the rise of the energy is smooth and the change of the $O-O_2$ bond length in ozone molecule is negligible. The latter is even smaller than the equilibrium values for the O=O bond.

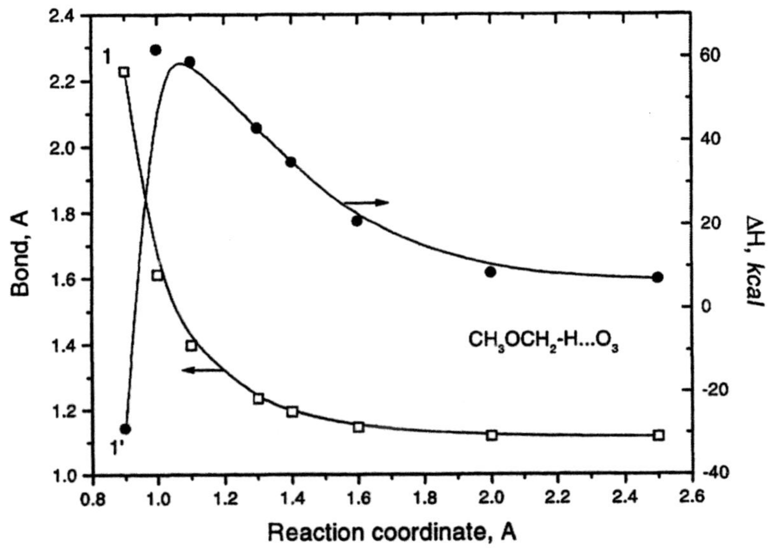

Fig.5.5. Ozone reaction with dimethylether: 1 - dependence of the C-H bond length (TS1) on the reaction coordinate; 1' - dependence of ΔH on the reaction coordinate

The experimental data from the study on the mechanism of this reaction shown in Chapter 2 reveals that it takes place via α-hydrogen atom abstraction. The calculations illustrated by Fig.5.5. with TS1 correspond completely to the concepts developed in Chapter 2.

The procedure for searching the TS of the reaction dimethylether + ozone gives the following results (Table 5.6).

Table 5.6

Reaction of dimethylether with ozone - localization of TS

```
GRADIENT NORM MINIMIZED USING NS01A
SCF FIELD WAS ACHIVED

HEAT OF FORMATION        =     28.955315 KCAL/MOL
ELECTRONIC ENERGY        =   -5470.166525 EV
CORE-CORE REPULSION      =    3858.306637 EV
GRADIENT NORM  =     0.039366
RMS FORCE      =     0.007187
DIPOLE    =    3.91911 DEBAY
NO. OF FILLED LEVELS     =     19
IONIZATION POTENTIAL     =    9.228045 EV
MOLECULAR SYMMETRY:   C1
MOLECULAR WEIGHT         =     94.067
SCF CALCULATIONS         =     116
COMPUTATION TIME         =     1720.04 SECUNDS

FINAL GEOMETRY OPTIMIZATION
AM1 PRECISE NS01A T=60M PULAY ITRY=1200
DIMETHYLETHER + OZONE
```

TS LOCALIZATION										CHARGE
C	0.000000	0	0.000000	0	0.000000	0	0	0	0	-0.1013
O	1.433392	1	0.000000	0	0.000000	0	1	0	0	-0.1533
C	1.354323	1	115.333687	1	0.000000	0	2	1	0	-0.1314
H	1.115735	1	110.385725	1	57.868975	1	1	2	3	0.0795
H	1.118623	1	109.048981	1	-64.464029	1	1	2	3	0.0949
H	1.118623	1	103.796319	1	177.135163	1	1	2	3	0.1235
H	1.101094	1	119.564651	1	-19.535388	1	3	2	1	0.1209
H	1.102459	1	111.353778	1	-168.722445	1	3	2	1	0.1735
H	1.550694	1	98.692414	1	79.895322	1	3	2	1	0.1816
O	1.103082	1	134.122537	1	22.683035	1	9	3	2	-0.2696
O	1.261974	1	109.117689	1	16.943166	1	10	9	3	0.2668
O	1.183390	1	113.205484	1	56.471877	1	11	10	9	-0.3853
0	0.000000	0	0.000000	0	0.000000	0	0	0	0	

In the transition state the length of the α- C-H bond is stretched by 41% which leads to its cleavage to CH_3COOCH_2·- and HO_3·- radicals.

5.3. Sulphur Containing Compounds

The study of the reactions of sulphur containing compounds with different reagents, including ozone, is of great importance both from practical point of view for environment protection, atmospheric processes, treatment of drinking and waste water, aging and stabilization of elastomers and from theoretical interest for the reaction ability theory, namely for ozone chemistry. Applying the approach for calculating the interaction between oxygen containing compounds and ozone we will follow this reaction with sulphur containing compounds.

5.3.1. Hydrogen Sulphide

The ozone reaction with hydrogen sulphide may take place via TS2 and TS3.

HS-H...O_3	H_2S...O-O_2
2	3

In Fig.5.6. are presented the results form the calculations on this reaction.

In the transition state the length of the α- C-H bond is increased by 40% which leads to its cleavage producing CH_3COOCH_2·- and HO_3·- radicals.

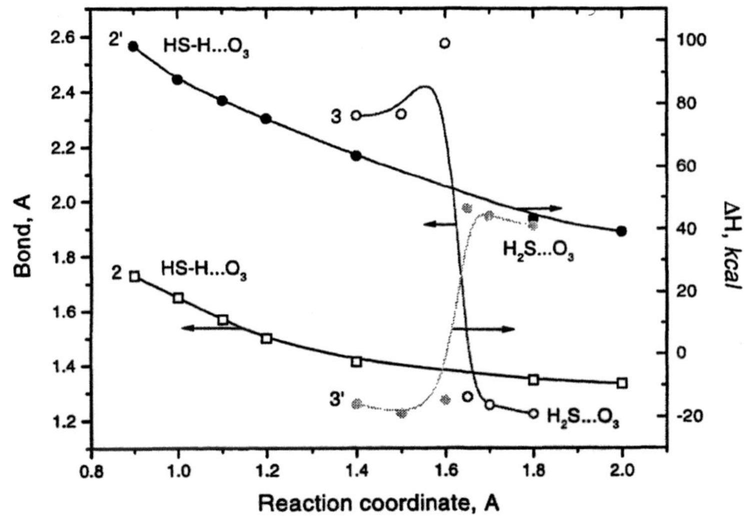

Fig.5.6. Ozone reaction with hydrogen sulfide: 2 - dependence of the S-H bond length (TS2) on the reaction coordinate; 3 dependence of the O-O_2 bond length on the reaction coordinate at TS3; 2' - dependence of ΔH (TS2) on the reaction coordinate and 3' - dependence of ΔH (TS3) on the reaction coordinate

The course of curves 2 and 2' stand for the change in the S-H bond length and the value of the energy of TS2 formation evidences the inefficiency of this reaction route. There is a constant rise of the formation energy at insignificant change in the bond length. Thus at change of the reaction coordinate from 2.0 to 0.9Å the S-H bond length is increased by 30% and the heat of formation rises by 150%. The calculations with TS3 (curves 3 and 3') give a quite different picture. The RC variation from 1.65 to 1.6 Å results in qualitative sharp change in the O-O_2 bond length - from 1.285 to 2.574 Å and the heat of formation drops from 46.4 to -15 kcal. Thus it can be concluded that this pathway is actually the preferred reaction route. The compound containing oxygen is most probably an unstable intermediate which is rapidly decomposed to water and elemental sulfur as pointed below:

$$H_2S + O_3 \rightleftharpoons [H_2S...O\text{-}O_2]^{\#} \longrightarrow H_2O + S + O_2$$

$$[H_2S...O\text{-}O_2]^{\#} \downarrow H_2O + SO_2$$

This mechanism is in agreement with the easy oxidability of hydrogen sulfide to sulfur or sulfur dioxide (see Chapter 6).

5.3.2. Methylsulphide

Upon ozonolysis of methylsulphide the reaction may proceed via the formation of three different transition states:

$HSCH_2\text{-}H...O_3$	$CH_3S\text{-}H...O_3$	$CH_3(H)S...O\text{-}O_2$
1	2	3

Figure 5.7 exemplifies the results from the calculations of the ozone reaction with methylsulphide considering the three probable transition states.

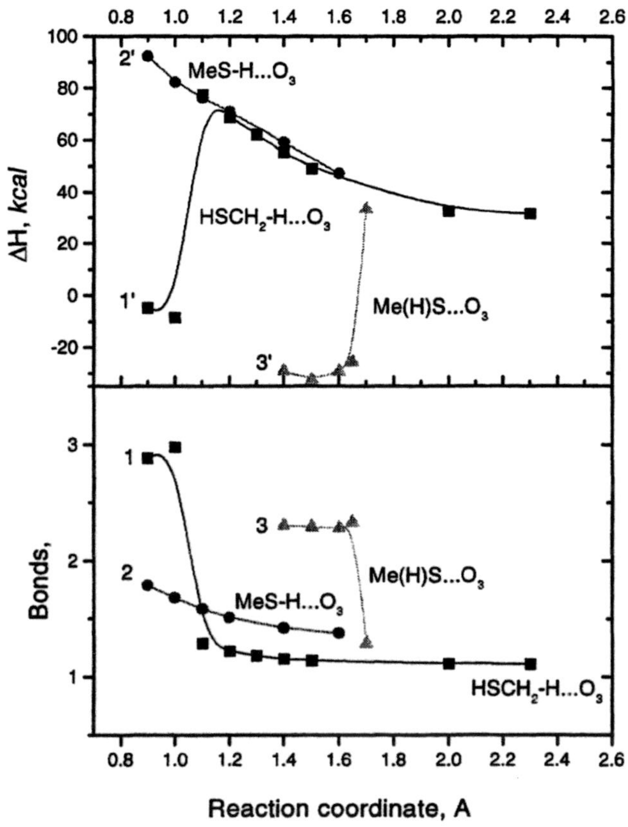

Fig.5.7. Ozone reaction with methylsulphide: below - dependence of the C-H bond length (TS1) - 1, S-H (TS2) - 2 and O-O_2 (TS3) - 3 on the reaction coordinate; above- dependence of the corresponding heats of formation 1', 2' and 3' on the reaction coordinate.

It is obvious that routes - 2 and 2' are inefficient. At RC variation from 1.6 to 0.9 Å the S-H bond is stretched by about 30% and ΔH is increased by 100%. Upon TS1 and TS3-based calculations a qualitative different pattern is observed. In the first case (TS1) at RC change from 1.1 to 1 Å the C-H bond approaches the value of 2.982 Å and the heat of energy is reduced from 77.4 to -8.3 kcal. As to the latter case (TS3) the RC variation from 1.7 to 1.65 Å leads to increase in the O-O_2 length with 1.289 Å and the heat of formation is sharply reduced from 33.3 to -25.6 kcal. Thus the both transition states appear to be quite probable as reaction routes. However taking account of the RC and ΔH magnitudes at which the qualitative changes occur in the systems one can assume that the route with TS3 is more favorable. Here the interaction between the two molecules (ozone and methylsulphude) starts at a greater distance (1.7 Å) and the value of ΔH at the beginning is lower though the gain from the transition - 58.9 kcal is smaller than that at TS1 equal to 85.7 kcal. The preference for TS3 is additionally supported by the results from thiosemicarbazides ozonolysis whereby the changes in the system were followed by SO_2 evolution.

5.3.3. Dimethylsulphide

The reaction in this case can take place through two transition states - TS1 and TS3.

$$\underset{1}{CH_3SCH_2\text{-}H...O_3} \qquad \underset{3}{(CH_3)_2S...O\text{-}O_2}$$

The calculations on the two reaction routes are presented in Fig.5.8.

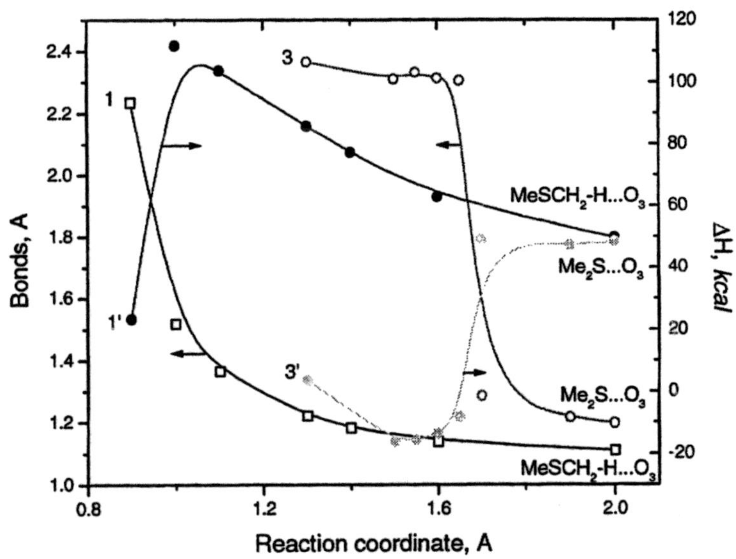

Fig.5.8. Ozone reaction with dimethylsulphide: 1 - dependence of C-H bond length (TS1) on the reaction coordinate; 3 dependence of $O\text{-}O_2$ bond length on the reaction coordinate at TS3; 1' - dependence of ΔH (TS1) on the reaction coordinate and 3' - dependence of ΔH (TS3) on the reaction coordinate

Based on these results one can assume the proceeding of the reaction via the both transition states. Actually at TS1 the RC change from 1 to 0.9Å leads to C-H bond break with ΔH drop from 111.6 to 23.3 kcal and at TS3 the RC change from 1.7 to 1.65 Å results in $O\text{-}O_2$ cleavage and ΔH drop from 48.8 to -8.4 kcal. If we apply the same approach as upon the methylsulphide ozonolysis then we should give preference to the route with TS3 since the ozone interaction with the S atom begins at 1.7 Å and ΔH=48.8 kcal. In Table 5.6 are outlined the results of the search procedure for TS3.

Table 5.6

Reaction of dimethylsulphide with ozone - localization of TS

```
GRADIENT NORM MINIMIZED USING NS01A
SCF FIELD WAS ACHIVED

HEAT OF FORMATION      =      25.184250 KCAL/MOL
ELECTRONIC ENERGY      =    -4933.327276 EV
CORE-CORE REPULSION    =     3548.879170 EV
GRADIENT NORM =     0.299787
RMS FORCE       =   0.054787
DIPOLE    =    1.39911 DEBAY
NO. OF FILLED LEVELS   =     19
IONIZATION POTENTIAL   =    8.077920 EV
MOLECULAR SYMMETRY:  C1
MOLECULAR WEIGHT       =     110.128
SCF CALCULATIONS       =     60
COMPUTATION TIME       =    28943.49 SECUNDS

FINAL GEOMETRY OPTIMIZATION
AM1 PRECISE NS01A T=60M PULAY ITRY=1200
DIMETHYLSULPHIDE + OZONE, S - ATTACK
TS LOCALIZATION                                              CHARGE
 C 0.000000 0       0.000000 0    0.000000 0      0 0 0  -0.3423
 S 1.747691 1       0.000000 0    0.000000 0      1 0 0   0.1713
 C 1.746493 1 103.745290 1        0.000000 0      2 1 0  -0.3428
 H 1.114349 1 109.532289 1      108.120998 1      1 2 3   0.1078
 H 1.117283 1 108.566800 1     -133.275373 1      1 2 3   0.1047
 H 1.113394 1 113.329057 1      -13.388482 1      1 2 3   0.0946
 H 1.121340 1 106.783666 1      151.793107 1      3 2 1   0.0913
 H 1.112606 1 110.777722 1      -89.026801 1      3 2 1   0.1004
 H 1.113669 1 113.544434 1       33.875997 1      3 2 1   0.1173
 O 1.813371 1      99.140829 1   91.044800 1      2 1 6  -0.2567
 O 1.238066 1 122.537804 1      179.670447 1     10 2 1   0.4984
 O 1.189194 1 115.514231 1       83.962040 1     11 10 2 -0.3437
 0 0.000000 0       0.000000 0    0.000000 0      0 0 0
```

5.4. Nitrogen Containing Compounds

5.4.1. Ammonia

The ozone reaction with ammonia can occur via two reaction pathways, i.e. TS2 and TS3.

$$\underset{2}{H_2N\text{-}H...O_3} \qquad \underset{3}{H_3N...O_3}$$

The calculations on the two reaction routes are presented in Fig.5.9.

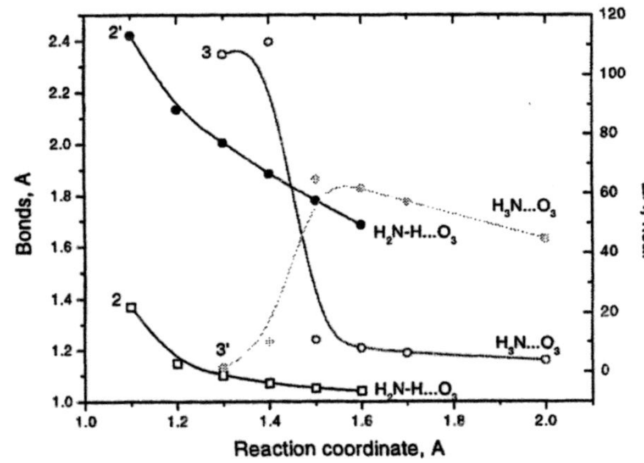

Fig.5.9. Ozone reaction with ammonia: 1 - dependence of N-H bond length (TS2) on the reaction coordinate; 3 - dependence of O-O_2 bond length on the reaction coordinate at TS3; 2' - dependence of ΔH (TS2) on the reaction coordinate and 3' - dependence of ΔH (TS3) on the reaction coordinate

For TS2 route the RC variation from 1.6 to 1.0 Å leads to increase of the N-H bond length by about 31% and ΔH - with 130% while for TS3 route the RC change from 1.5 to 1.4 Å causes the break of O-O_2 bond and ΔH is reduced from 64.7 to 10.1 kcal. Thus the preferred route appears to be via TS3 which is accompanied by nitrogen oxide formation:

$$H_3N + O_3 \rightleftharpoons [H_3N...O_3]^{\#} \longrightarrow H_3N\rightarrow O + O_2$$

5.4.2. Methylamine

Figure 5.10 illustrates the results from the calculations on ozone reaction with methylamine.

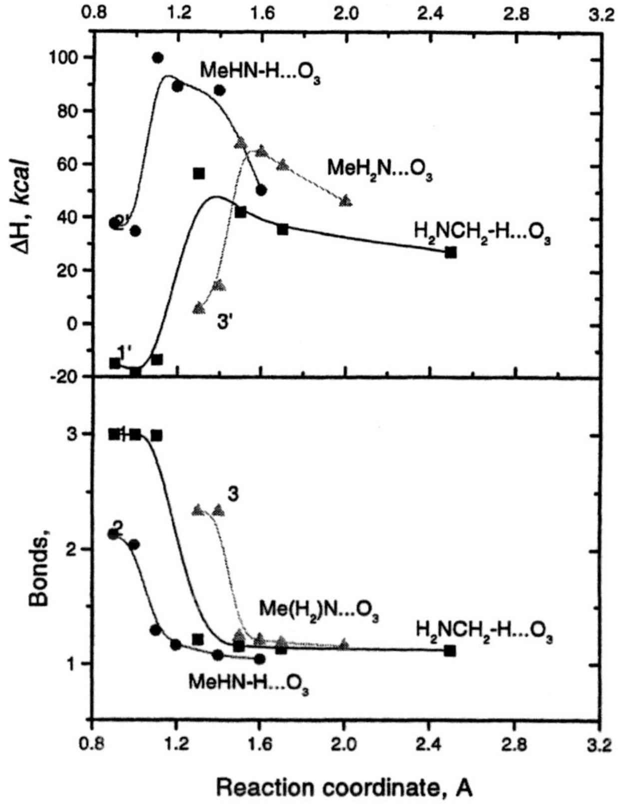

Fig.5.10. Ozone reaction with methylamine: below - dependence of the length of C-H bond (TS1) - 1, N-H (TS2) - 2 and O-O_2 (TS3) - 3 on the reaction coordinate; above- dependence of the corresponding heats of formation 1', 2' and 3' on the reaction coordinate.

Analysing the results obtained it may be concluded that the three transition states lead to qualitative changes in the system but when ozone attacks the nitrogen atom the interaction between the two molecules begins at the biggest reaction coordinate.

The TS1 geometry, after applying the procedure for TS localization, is shown in Table 5.7. However it does not allow geometry optimization for the other two transition states.

Table 5.7

Reaction of methylamine with ozone - localization of TS

GRADIENT NORM MINIMIZED USING NS01A
SCF FIELD WAS ACHIVED

HEAT OF FORMATION =	90.540787 KCAL/MOL
ELECTRONIC ENERGY =	-3759.668722 EV
CORE-CORE REPULSION =	2403.735661 EV
GRADIENT NORM =	0.064749
RMS FORCE =	0.013217
DIPOLE =	1.92064 DEBAY
NO. OF FILLED LEVELS =	16
IONIZATION POTENTIAL=	9.137121 EV

MOLECULAR SYMMETRY:		C1
MOLECULAR WEIGHT	=	79.055
SCF CALCULATIONS	=	206
COMPUTATION TIME	=	2391.24 SECONDS

FINAL GEOMETRY OPTIMIZATION
AM1 PRECISE NS01A T=60M PULAY ITRY=1200
METHYLAMINE + OZONE, C-H - ATTACK

TS LOCALIZATION											CHARGE
H	0.000000	0	0.000000	0	0.000000	0	0	0	0		0.1364
N	1.000903	1	0.000000	0	0.000000	0	1	0	0		-0.3447
C	1.431355	1	111.011420	1	0.000000	0	2	1	0		-0.1134
H	1.001973	1	108.692875	1	-121.676995	1	2	1	3		0.1499
H	1.122637	1	109.544687	1	64.286981	1	3	2	1		0.0806
H	1.121132	1	109.393412	1	-176.353838	1	3	2	1		0.0896
H	1.136511	1	114.333359	1	-54.714275	1	3	2	1		0.0517
O	1.924459	1	105.179765	1	-72.092692	1	7	3	2		-0.3928
O	1.683558	1	169.260487	1	170.308011	1	8	7	3		0.2709
O	1.080573	1	108.404019	1	80.752208	1	9	8	7		0.0717
0	0.000000	0	0.000000	0	0.000000	0	0	0	0		

The data of Table 5.7 indicates that even in the case with TS1 the optimization of its geometry does not allow precise conclusions for the reaction route. Most probably, the calculation, method with AM1 Hamiltonian and the parameterization procedure are nor quite suitable for this reaction though the use of MINGO/3, MNDO and PM3 do not provide also satisfactorily results.

5.4.3. Dimethylamine

The calculation results are presented in Fig.5.11.

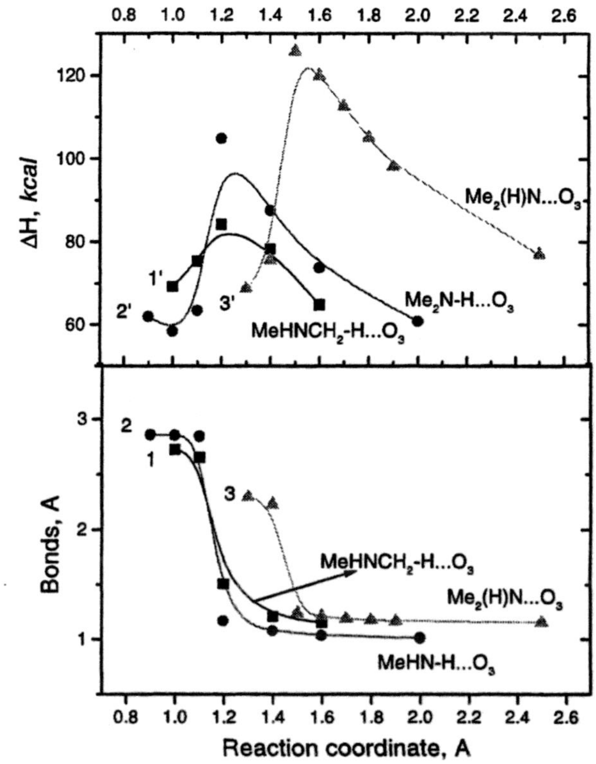

Fig. 5.11. Ozone reaction with dimethylamine: below - dependence of the length of C-H bond (TS1) - 1, N-H (TS2) - 2 and O-O_2 (TS3) - 3 on the reaction coordinate; above- dependence of the corresponding heats of formation 1', 2' and 3' on the reaction coordinate.

It should be noted that in this case, similar to methylamine ozonolysis, three transition states are possible. The analysis of the results indicate that the three transition states lead to qualitative changes in the system but among them the most probable are those related to the ozone attack on the nitrogen atom and the C-H bond. For the former it is due to the highest RC value and for the latter to the lowest heat of formation.

Table 5.8

Reaction of dimethylamine with ozone - localization of TS

```
GRADIENT NORM MINIMISED USING NS01A
SCF FIELD WAS ACHIVED

HEAT OF FORMATION       =       61.781140 KCAL/MOL
ELECTRONIC ENERGY       =     -5217.677611 EV
CORE-CORE REPULSION     =      3704.959820 EV
GRADIENT NORM =       0.037013
RMS FORCE         =   0.006443
DIPOLE     =     5.27252 DEBAY
NO. OF FILLED LEVELS    =        19
IONIZATION POTENTIAL    =     8.149698 EV
MOLECULAR SYMMETRY:   C1
MOLECULAR WEIGHT        =       93.082
SCF CALCULATIONS        =       37
```

COMPUTATION TIME = 462.23 SECUNDS

FINAL GEOMETRY OPTIMIZATION
AM1 PRECISE NS01A T=60M PULAY ITRY=1200
DIMETHYLAMINE + OZONE, C-H - ATTACK
TS LOCALIZATION CHARGE

H	0.000000	0	0.000000	0	0.000000	0	0 0 0	0.2218
N	0.999125	0	0.000000	0	0.000000	0	1 0 0	-0.2429
C	1.428077	1	115.318063	0	0.000000	0	2 1 0	-0.1341
C	1.361194	1	116.972078	1	-153.160480	0	2 1 3	-0.1043
H	1.121377	1	109.122271	1	44.679142	1	3 2 1	0.0979
H	1.120953	1	110.381140	1	165.086646	1	3 2 1	0.0948
H	1.140249	1	110.486763	1	-74.783096	1	3 2 1	0.1381
H	1.110186	1	116.406114	1	-179.454319	1	4 2 1	0.1365
H	1.111330	1	115.502029	1	-37.635213	1	4 2 1	0.1295
H	1.437328	1	109.086276	1	70.120314	1	4 2 1	0.1227
O	1.149069	1	154.869200	1	-21.025365	1	10 4 2	-0.2502
O	1.236603	1	112.786814	1	82.411015	1	11 104	0.2400
O	1.193909	1	115.359758	1	93.890320	1	12 11 4	-0.4560
0	0.000000	0	0.000000	0	0.000000	0	0 0 0	

The TS1 optimization procedure points to C-H bond elongation by 30% which obviously results in its break producing $HO_2\cdot$-radical. However the proceeding of the reaction via ozone attack on the nitrogen atom with biradical or bipolar transition state could not be excluded.

5.4.4. Trimethylamine

The calculations on the ozone reaction with trimethylamine are given in Fig.5.12.

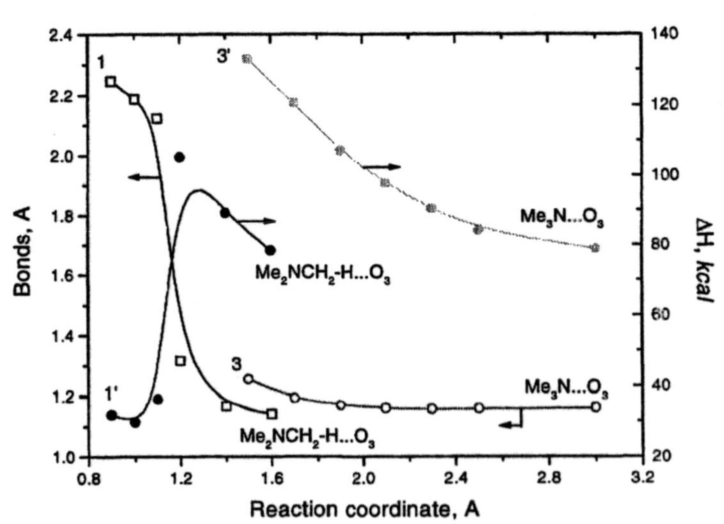

Fig.5.12. Ozone reaction with trimethylamine: 1 - dependence of C-H bond length (TS1) on the reaction coordinate; 3 - dependence of $O\text{-}O_2$ bond length on the reaction coordinate at TS3; 1' - dependence of ΔH (TS1) on the reaction coordinate and 3' - dependence of ΔH (TS3) on the reaction coordinate

As seen only TS1 results in qualitative changes with decrease of the reaction coordinate. TS3 is characterized by a constant rise of the energy and no change in the bond lengths of the complexes is observed. Principally this result is confusing as ozone acts as electrophilic agent in its reactions, and in this case the three electrodonor methyl groups should favor the attack on the nitrogen atom which is however not observed. It should be noted that the occurrence of steric hindrances at this interaction could be regarded as a reasonable explanation of this fact.

5.5. PHOSPHOROUS CONTAINING COMPOUNDS

5.5.1. Phosphine

The results from the quantum chemical calculation on ozone interaction with phosphine are shown in Fig.5.13.

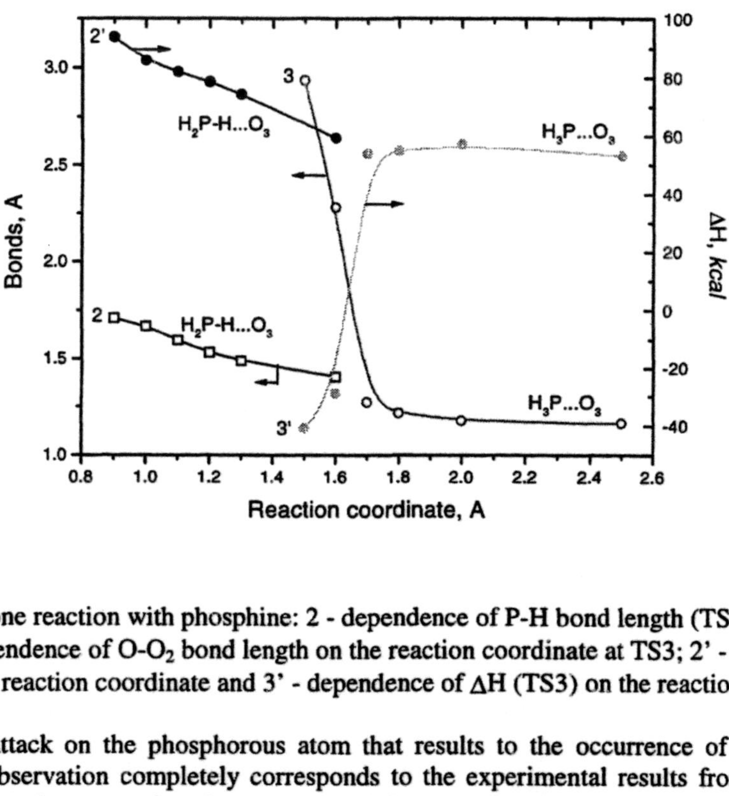

Fig.5.13. Ozone reaction with phosphine: 2 - dependence of P-H bond length (TS2) on the reaction coordinate; 3 - dependence of O-O_2 bond length on the reaction coordinate at TS3; 2' - dependence of ΔH (TS2) on the reaction coordinate and 3' - dependence of ΔH (TS3) on the reaction coordinate

It is only the attack on the phosphorous atom that results to the occurrence of an saddle point at RC=1.72 Å. This observation completely corresponds to the experimental results from the study on the interaction mechanism between phosphorous - containing compounds and ozone. In all cases the basic product is phosphine oxide.

5.5.2. Methylphosphine

The results from the calculations are given in Fig.5.14.

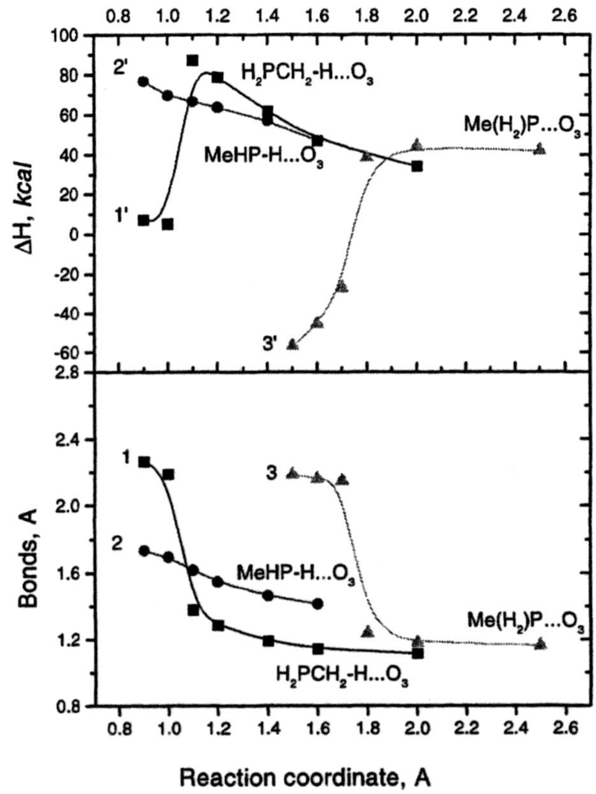

Fig. 5.14. Interaction between ozone and methylphosphine: below - dependence of the length of C-H bond (TS1) - 1, P-H (TS2) - 2 and O-O_2 (TS3) - 3 on the reaction coordinate; above- dependence of the corresponding heats of formation 1', 2' and 3' on the reaction coordinate.

It should be emphasized that the route via TS3 again causes the appearance of an inflection point on the potential surface but at RC=1.82 Å, i.e. the methyl group favors the interaction between ozone and P to be carried out at a longer distance.

The procedure on TS localization is demonstrated by Table 5.8.

Table 5.8

Reaction of methylphosphine with ozone - localization of TS

GEOMETRY OPTIMISED USING EIGENVECTOR FOLLOWING (EF)
SCF FIELD WAS ACHIEVED

HEAT OF FORMATION = 42.443592 KCAL
ELECTRONIC ENERGY = -3686.869478 EV
CORE-CORE REPULSION = 2408.444761 EV
DIPOLE = 3.49638 DEBYE
NO. OF FILLED LEVELS = 16
IONIZATION POTENTIAL = 9.587821 EV
MOLECULAR SYMMETRY: C1
MOLECULAR WEIGHT = 96.022
SCF CALCULATIONS = 202
COMPUTATION TIME = 128.078 SECONDS

FINAL GEOMETRY OBTAINED
AM1 PULAY T=500M TS
METHYLPHOSPHINE-OZONE TS
GEOMETRY OPTIMIZATION CHARGE

P	0.0000000 0	0.000000 0	0.000000 0	0 0 0	.1509
C	1.7256227 1	0.000000 0	0.000000 0	1 0 0	-.3316
H	1.1086171 1	112.321176 1	0.000000 0	2 1 0	.0713
H	1.1097753 1	108.414256 1	-121.846850 1	2 1 3	.0906
H	1.1114342 1	106.630222 1	120.000361 1	2 1 3	.0939
H	1.3641962 1	99.851673 1	61.538462 1	1 2 3	-.0398
H	1.3645814 1	100.087818 1	-36.138920 1	1 2 3	-.0282
XX	1.00000 0	90.000000 0	180.000000 0	1 2 3	
O	2.5445821 1	79.329204 1	-96.026745 1	1 8 2	-.2737
O	1.1626121 1	173.730888 1	-95.253083 1	9 1 8	.5424
O	1.1596079 1	121.040862 1	-36.666604 1	10 9 8	-.2757

0

It is seen that the results from this procedure does not show the distinct cleavage of ozone O-O bond and the P-O bond formation though this fact has been experimentally confirmed.

5.5.3. Dimethylphosphine

Figure 5.15 examplifies the results from the calculations on ozone interaction with dimethylphosphine.

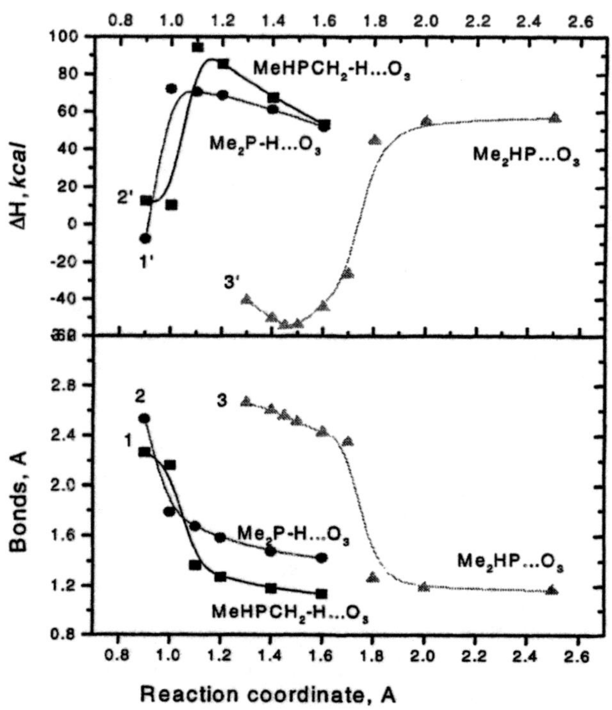

Fig. 5.11. Ozone reaction with dimethylphosphine: below - dependence of the length of C-H bond (TS1) - 1, P-H (TS2) - 2 and O-O$_2$ (TS3) - 3 on the reaction coordinate; above- dependence of the corresponding heats of formation 1', 2' and 3' on the reaction coordinate.

It is clearly seen that the addition of a new methyl group does not result in any substantial change in the reaction route but RC=1.85Å.

5.5.3. Trimethylphosphine

The results from the calculations on the ozone interaction with trimethylphosphine are given in Fig.5.16.

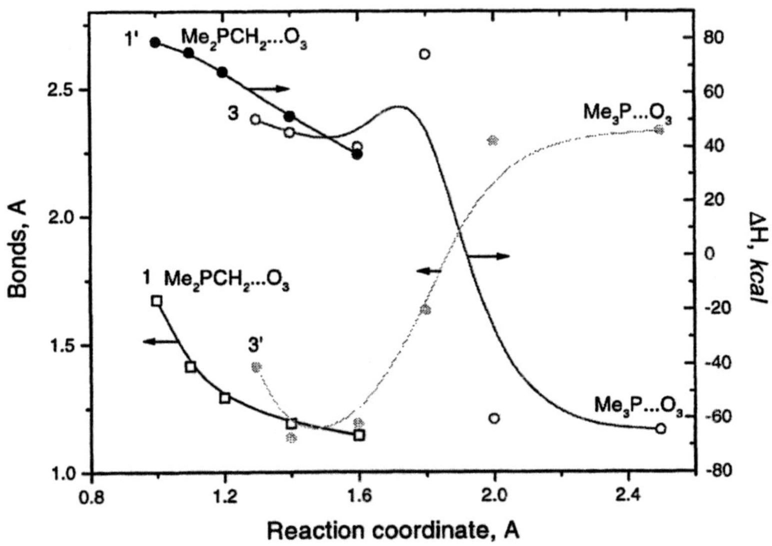

Fig.5.16. Ozone interaction with trimethylphosphine: 1 - dependence of C-H bond length (TS1) on the reaction coordinate; 3 - dependence of O-O_2 bond length on the reaction coordinate at TS3; 1' - dependence of ΔH (TS1) on the reaction coordinate and 3' - dependence of ΔH (TS3) on the reaction coordinate

The data in Fig.5.16 indicates that the further introduction of a methyl group does not affect the reaction route but the value of RC is 2.10 Å. Summarizing the results obtained for ozone reaction with phosphine and its methyl derivatives one can conclude that the P-atom is the reaction center in these compounds, the ozone attack is electrophilic and promoted by the presence of electrondonor groups.

REFERENCES

1. Banichevich, A, SD Peyerimhoff. Theoretical Study of the Ground and Excited States of Ozone in Its Symmetric Nuclear Arrangement. *Chemical Physics*, 174: 1, 93-109, (1993).
2. Koch, W, G Frenking, G Steffen, D Reinen, M Jansen, W Assenmacher. The Low Lying Electronic States of O_3. *J.Chem.Phys.*, 99: 2, 1271-1277, (1993).
3. Boursier, C, C Boulet, F Menardbourcin, JM Flaud, C Camypeyret. Semi-Classical Calculation of Rotational Relaxation Rates in Ozone. *Europhys.Let.*, 24: 4, 259-262 (1993).

4. Gonzalezluque, R, M Merchan, P Borowski, BO Roos. On the Theoretical Determination of the Electron Affinity of Ozone. *Theor.Chim. Acta,* 86: 6, 467-476 (1993).
5. Leininger, ML, HF Schaefer. Molecular geometry and vibrational frequencies of ozone from compact variational wave functions explicitly including triple and quadruple substitutions. *J.Chem.Phys.,* 107: 21, 9059-9062 (1997).
6. Olzmann, M, E Kraka, D Cremer, R Gutbrod, S Andersson. Energetics, kinetics, and product distributions of the reactions of ozone with ethene and 2,3-dimethyl-2- butene. *J.Phys.Chem. A,* 101: 49, 9421-9429 (1997).
7. Gutbrod, R, E Kraka, RN Schindler, D Cremer Kinetic and theoretical investigation of the gas-phase ozonolysis of isoprene: Carbonyl oxides as an important. for OH radicals in the atmosphere. *J.Am.Chem.Soc.,* 119: 31, 7330-7342 (1997).
8. Mariey, L., Lamotte, J., Hoggan, P., Lavalley, J.C., Boulanine, K., A. Tsyganenko. On the basicity of ozone, *Chem Lett,* 8, 835-836 (1997).
9. Lu, Z.M., Kellman, M.E., Phase space structure of triatomic molecules, *J Chem Phys,* 107, 1, 1-15 (1997).
10. Anderson, SM, P Hupalo, K Mauersberger Rotational Structure in the Near-Infrared Absorption Spectrum of Ozone. *J.Chem.Phys.,* 99: 1, 737-739 (1993).
11. Cacace, F, M Speranza. Protonated ozone: Experimental detection of O_3H^+ and evaluation of the proton affinity of ozone. *Science,* 265: 5169, 208-209 (1994).
12. Lee, C, HR Kim. A classical trajectory study of $O+O_2$ collision. *Chemical Physics Letters,* 233: 5-6, 658-664 (1995).
13. Joens JA. Potential Energy Surfaces of Ozone in Its Ground State and in the Lowest-Lying 8 Excited States - Comment. *Chemical Physics Letters,* 227: 6, 688-689 (1994).
14. Banichevich, A, SD Peyerimhoff. Theoretical Study of the Ground and Excited States of Ozone in Its Symmetric Nuclear Arrangement. *Chemical Physics,* 174: 1, 93-109 (1993).
15. Koch, W, G Frenking, G Steffen, D Reinen, M Jansen, W Assenmacher. The Low Lying Electronic States of O_3-. *Journal of Chemical Physics,* 99: 2, 1271-1277 (1993).
16. Boursier, C, C Boulet, F Menardbourcin, JM Flaud, C Camypeyret. Semi-Classical Calculation of Rotational Relaxation Rates in Ozone. *Europhysics Letters,* 24: 4, 259-262 (1993).
17. Rîuse, R.A. *J.Am.Chem.Soc.,* 95, 34600 (1973); 96, 5095 (1974).
18. Hiberty, P.C. *J.Am.Chem.Soc.,* 98, 6088 (1976).
19. Rouse, R.D. *Int.Quantum Chem.Symp.,* 7, 289 (1973).
20. Goddard III, W.A., T.H. Dunning, Jr., W.J. Hunt, P.J. Hay. *Acc.Chem.Res.,* 6, 368 (1973).
21. Harding, L.B., W.A. Goddard III, *J.Am.Chem.Soc.,* 97, 6293, 6300 (1975); 98, 6093 (1976).
22. Bodrowicz, F.M., W.A. Goddard III.-In:*Modern Theoretical Chemistry; Methods of Electronic Structure Theory,* 3. H.F. Schaefer (ed.), N.Y., Plenum Press, 1977, 79.
23. Harding, L.B., W.A. Goddard III. *J.Am.Chem.Soc.,* 100, 7180 (1978).
24. Cremer, D. Theoretical Determination of Molecular Structure and Conformation. 7. Stereoselectivity of the Ozonolysis Reaction. *J. Am. Chem. Soc.,* 103, 13, 3619-3626 (1981).
25. Bailey, P.S., T.M. Ferrell. *J.Am.Chem.Soc.,* 100, 899 (1978).
26. "Computational Strategies for the Optimization of Equilibrium Geometry and Transition-State Structures at the Semiempirical Level", Peter L. Cummings, Jill E. Gready, *J. Comp. Chem.,* 10, 939 (1989).
27. "An Efficient Procedure for Calculating the Molecular Gradient, using SCF-CI Semiempirical Wavefunctions with a Limited Number of Configurations", M. J. S. Dewar, D. A. Liotard, *J. Mol. Struct. (Theochem),* 206, 123-133 (1990).
28. J. Baker, *J. Comp. Chem.,* 7, 385 (1986).
29. "Atomic Charges Derived from Semiempirical Methods", B. H. Besler, K. M. Merz, Jr., P. A. Kollman, *J. Comp. Chem.,* 11 431 (1990).
30. "Ground States of Molecules. 38. The MNDO Method. Approximations and Parameters.", M.J.S. Dewar, W.Thiel, *J. Am. Chem. Soc.,* 99, 4899, (1977).
31. H, C, N, O - M.J.S. Dewar, W. Thiel, *J. Am. Chem. Soc.,* 99, 4907, (1977).
32. Li - Parameters taken from the MNDOC program, written by Walter Thiel, Quant. *Chem. Prog. Exch. No. 438;* 2, 63, (1982)
33. Be - M.J.S. Dewar, H.S. Rzepa, *J. Am. Chem. Soc,* 100, 777, (1978)

34. B - M.J.S. Dewar, M.L. McKee, *J. Am. Chem. Soc.*, 99, 5231, (1977).
35. F - M.J.S. Dewar, H.S. Rzepa, *J. Am. Chem. Soc.*, 100, 58, (1978).
36. Al - L.P. Davis, R.M. Guidry, J.R. Williams, M.J.S. Dewar, H.S. Rzepa *J. Comp. Chem.*, 2, 433, (1981).
37. Si, P, S, Cl - M.J.S. Dewar, M.L. McKee, H.S. Rzepa, *J. Am. Chem. Soc.*, 100, 3607 (1978).
38. M.J.S. Dewar, J. Friedheim, G. Grady, E.F. Healy, J.J.P. Stewart, *Organometallics*, 5, 375 (1986).
39. S - M.J.S. Dewar, C. H. Reynolds, *J. Comp. Chem.*, 7, 140 (1986).
40. Cl - M.J.S. Dewar, H.S. Rzepa, *J. Comp. Chem.*, 4, 158, (1983)
41. Zn - M.J.S. Dewar, K. M. Merz, *Organometallics*, 5, 1494 (1986).
42. Ge - M.J.S. Dewar, G.L. Grady, E.F. Healy, *Organometallics*, 6, 186 (1987).
43. Br - M.J.S. Dewar, E.F. Healy, *J. Comp. Chem.*, 4, 542, (1983)
44. I - M.J.S. Dewar, E.F. Healy, J.J.P. Stewart, *J. Comp. Chem.*, 5, 358, (1984)
45. Sn - M.J.S. Dewar, G.L. Grady, J.J.P. Stewart, *J. Am. Chem. Soc.*, 106, 6771 (1984).
46. Hg - M.J.S. Dewar, G.L. Grady, K. Merz, J.J.P. Stewart, *Organometallics*, 4, 1964, (1985).
47. Pb - M.J.S. Dewar, M. Holloway, G.L. Grady, J.J.P. Stewart, *Organometallics*, 4, 1973, (1985).
48. Part XXVI, Bingham, R.C., Dewar, M.J.S., Lo, D.H, *J. Am. Chem. Soc.*, 97, (1975).
49. "AM1: A New General Purpose Quantum Mechanical Molecular Model", M.J.S. Dewar, E.G. Zoebisch, E.F. Healy, J.J.P. Stewart, *J. Am. Chem. Soc.*, 107, 3902 (1985).
50. B - M.J.S. Dewar, C Jie, E. G. Zoebisch, *Organometallics*, 7, 513 (1988).
51. F, Cl, Br, I - M.J.S. Dewar, E.G. Zoebisch, *Theochem*, 180, 1 (1988).
52. Al - M.J.S. Dewar, A.J. Holder, *Organometallics*, 9, 508 (1990).
53. Si - M.J.S. Dewar, C. Jie, *Organometallics*, 6, 1486 (1987).
54. P - M.J.S. Dewar, C.Jie, *Theochem*, 187, 1 (1989)
55. Zn - M.J.S. Dewar, K.M. Merz, Jr., *Organometallics*, 7, 522 (1988).
56. Ga, Hg - M.J.S. Dewar, C. Jie, *Organometallics*, 8, 1544 (1989).
57. "Optimization of Parameters for Semi-Empirical Methods I-Method", J.J.P. Stewart, *J. Comp. Chem.* 10, 221 (1989).
58. "The Dynamic 'Level Shift' Method for Improving the Convergence of the SCF Procedure", A. V. Mitin, *J. Comp. Chem.* 9, 107 (1988).
59. "Ground States of Conjugated Molecules. IX. Hydrocarbon Radicals and Radical Ions", M.J.S. Dewar, J.A. Hashmall, C.G. Venier, *J.Am.Chem.Soc.*, 90, 1953 (1968).
60. "Triplet States of Aromatic Hydrocarbons", M.J.S. Dewar, N. Trinajstic, *Chem. Comm.*, 646, (1970).
61. "Semiempirical SCF-MO Treatment of Excited States of Aromatic Compounds" M.J.S. Dewar, N. Trinajstic, *J. Chem. Soc., (A)*, 1220, (1971).
62. "Convergence Acceleration of Iterative Sequences. The Case of SCF Iteration", Pulay, P., *Chem. Phys. Lett.*, 73, 393, (1980).
63. "Fast Semiempirical Calculations", Stewart. J.J.P., Csaszar, P., Pulay, P., *J. Comp. Chem.*, 3, 227, (1982).
64. "A New Rapid Method for Orbital Localization." P.G. Perkins and J.J.P. Stewart, *J.C.S. Faraday (II)* 77, 000, (1981).
65. Beppu, Y., *Computers and Chemistry*, Vol.6 (1982).
66. "Molecular Orbital Theory for the Excited States of Transition Metal Complexes", D.R. Armstrong, R. Fortune, P.G. Perkins, and J.J.P. Stewart, *J. Chem. Soc., Faraday II*, 68 1839-1846 (1972)
67. Broyden, C. G., *Journal of the Institute for Mathematics and Applications*, 6, 222-231 (1970).
68. Fletcher, R., *Computer Journal*, 13, 317 (1970).
69. Goldfarb, D. *Mathematics of Computation*, 24, 23 (1970).
70. Shanno, D. F. *Mathematics of Computation*, 24, 647 (1970).
71. Shanno, D. F., *J. of Optimization Theory and Applications* 46, 87 (1985).
72. "Calculation of Nonlinear Optical Properties of Molecules", H. A. Kurtz, J. J. P. Stewart, K. M. Dieter, *J. Comp. Chem.*, 11, 82 (1990).
73. "Semiempirical Calculation of the Hyperpolarizability of Polyenes", H. A. Kurtz, *I. J. Quant. Chem. Symp.*, 24, xxx (1990).

74. "Ground States of Molecules. 44 MINDO/3 Calculations of Absolute Heat Capacities and Entropies of Molecules without Internal Rotations." Dewar, M.J.S., Ford, G.P., *J. Am. Chem. Soc.*, 99, 7822 (1977).
75. Komornicki, A., McIver, J. W., *Chem. Phys. Lett.*, 10, 303, (1971).
76. Komornicki, A., McIver, J. W., *J. Am. Chem. Soc.*, 94, 2625 (1971)
77. "Valency and Molecular Structure", Gopinathan, M. S., Siddarth, P., Ravimohan, C., *Theor. Chim. Acta,* 70, 303 (1986).
78. "Bond Indices and Valency", Armstrong, D.R., Perkins, P.G., Stewart, J.J.P., *J. Chem. Soc., Dalton*, 838 (1973).
79. Gopinathan, M. S., and Jug, K., *Theor. Chim. Acta*, 63, 497 (1983).
80. "Location of Transition States in Reaction Mechanisms", M.J.S. Dewar, E.F. Healy, J.J.P. Stewart, *J. Chem. Soc., Faraday Trans. 2*, 3, 227, (1984)
81. "Molecular Quadrupole Moments", A.D. Buckingham, *Quarterly Reviews*, 182 (1958 or 1959)
82. "MNDO Cluster Model Calculations on Organic Polymers", J.J.P. Stewart, *New Polymeric Materials*, 1, 53-61 (1987).
83. "Calculation of Elastic Moduli using Semiempirical Methods", H. E. Klei, J.J.P. Stewart, *Int. J. Quant. Chem.* 20, 529-540 (1986).

Chapter 6

PROPERTIES, PREPARATION, ANALYSIS AND DECOMPOSITION OF OZONE

6.1. PROPERTIES

Ozone is an allotropic modification of oxygen. The ozone molecule in the ground state has an obtuse angle of $116^0 45' \pm 5'$ and an oxygen-oxygen bond length of 0.1278±0.0003 nm. In the gas phase the ozone molecule is a singlet biradical while in liquid phase it acts in most cases as a bipolar ion. In strong alkaline medium it is converted in to an ion-radical O_3^- which is spectrally identified through the registration of the intensive band at 435.8 nm with $\varepsilon = 1770\ M^{-1}.s^{-1}$ and by the ESR data with g_1, g_2, g_3 and g_{av} equal to 2.0036, 2.0411, 2.0179 and 2.0108, respectively with half-time conversion at -50°C - 12±1 min. The ion-radical reacts with H_2O producing 2HO· and O_2. The $HO^\bullet$ - radical thus formed participates further in the reaction process [1].

The ozone molecule is stable in acid medium and in order to achieve maximum rate to its oxidation it is necessary to use catalysts - transition metals ions (most often with one electron and sufficiently high redox potential as Ce^{+3}/Ce^{+4}, Co^{+2}/Co^{+3}, etc. [2].

It is assumed that the ozone stability in neutral and acid medium is related to the formation of HO_3^+- the hydrate shell of which retards the reaction [3]:

$$O_3 + H_2O \rightarrow 3/2O_2 + H_2O$$

In water ozone molecule is decomposed via a radical mechanism in water [4]:

Scheme 6.1

$$O_3 + H_2O \rightarrow 2HO^\bullet + O_2 \text{ (initiation)}$$
$$O_3 + HO^\bullet \rightarrow HO_2^\bullet + O_2$$
$$O_3 + HO_2^\bullet \rightarrow HO^\bullet + 2O_2 \text{ (propagation)}$$
$$2HO^\bullet \rightarrow H_2O_2 \text{ (termination)}$$
$$O_3 + H_2O_2 \rightarrow HO^\bullet + HO_2^\bullet + O_2$$
$$HO^\bullet + HO_2 \rightarrow H_2O + O_2 \text{ (termination)}$$

In alkaline medium the ozone decomposition is accelerated by hydroxyl anions [5]:

Scheme 6.2

$$O_3 + HO^- \rightarrow O_3^- + HO^\bullet$$
$$HO^\bullet + HO^- \rightarrow H_2O + O^-$$
$$O_3 + O^- \rightarrow O_2^- + O_2$$
$$O_2^- + O^- + H_2O \rightarrow O_2 + 2HO^-$$
$$O_3^- \rightarrow O^- + O_2$$
$$2O^- + H_2O \rightarrow O + 2HO^-$$
$$O_2^- + H_2O \rightarrow HO_2^- + HO^-$$
$$HO_2^- \rightarrow HO^- + 0.5O_2$$

The ozone decomposition in water is a complex multistep process. The rate of its decomposition depends strongly on pH of the medium with a twofold increase at change from 9.6 to 1.2 pH.

The ozone decomposition in water is also accelerated by the presence of active carbon, SiO_2, transition metal ions, some oxides (P_2O_3) and peroxides (BaO_2) [6-8]. Fluoro- and chlorocontaining polymers, many metals, etc. however, do not affect the ozone stability [9, 10].

Ozone is readily soluble by a good solubility in water and organic solvents - a wide range of chlorocontaining hydrocarbons, freons, acetonitrile, acetic acid, etc. Its solubility obeys the Henry' s law and the Henry' s constant for H_2O at 25°C is $9.46.10^4$ at/(mol/ml) [11, 12].

Some data on ozone solubility in water at various temperatures are given in Table 6.1.

Table 6.1

Ozone solubility in water (S) in dependence on temperature [13]

t, °C	0	10	20	30	40	50	60
S, g/l	1.09	0.78	0.57	0.40	0.27	0.19	0.14

Ozone is a highly toxic substance. The prolonged contact with atmosphere containing 4.10^{-4}% ozone results in headache, eyes inflammation and respiratory tract inflammation. The medical dose is 0.1 mg/m^3 [14].

In most of its reaction with inorganic compounds ozone reacts with participation of one oxygen atom, and the other two are released as O_2. Usually the elements are oxidized to their highest oxidation state. For instance, Mn is oxidized to MnO^{4-}, the halogen metals to oxides (ClO_2, Br_2O_5), ammonia - to NH_4NO_3, the lower nitrogen oxides are converted into N_2O_5 [15-17].

Ozone induces the following transformations of hydrogen sulfide 18]:

$$H_2S + O_3 \rightarrow H_2O + SO_2$$
$$3H_2S + 4O_3 \rightarrow 3H_2SO_4$$

The simple and complex cyanides react readily with ozone giving ammonium carbonate and carbamide according to Scheme 6.3 [19]:

Scheme 6.3

$$CN^- + O_3 \rightarrow OCN^- + O_2$$
$$OCN^- + 2H^+ + H_2O \rightarrow CO_2 + NH_4^+$$
$$NH_4^+ + OCN^- \rightarrow NH_2C(O)NH_2$$

The oxidative action of ozone on organic compounds in aqueous medium can take place through three routes [20]:

- direct oxidation with partcipation of one oxygen atom;
- addition of a whole ozone molecule;
- catalytic effect increasing the oxidative role of oxygen.

It has been found that the ozone interaction with aqueous alcohols solutions results in carboxylic acids formation as end products [21].

The oxidation of acetone by ozone in water-alkaline medium (pH= 10±12) leads to formaldehyde and acetate ion formation [22]. The di- and tri-ethylamine ozonolysis in aqueous medium gives formic acid and nitrate [23]:

$$(CH_3)_2NH + 8/3\ O_3 \rightarrow 2HCOOH + HNO_3 + H_2O$$

In nonaqueous medium first is obtained $R_3N^+OOO^-$, whose further conversion depends on the amine structure and medium nature [24].

Sulphurcontaining compounds (R_2CS) are oxidized by ozone to SO_2 and the corresponding ketone [25].

The ozonation of aromatic compounds takes place via benzene ring cleavage. Harries [26] has established that benzene is oxidized through glyoxal to oxalic acid. The presence of hydroxyl group in the aromatic ring accelerates strongly the rate of interaction with ozone. The initial phenol is rapidly converted into aldehydes and carbonic acids. The products of the deep phenol oxidation with ozone are practically nontoxic and nonaccumulative compounds. In phenols with unsaturated side chain the oxidation proceeds via the side chain [27].

Polynuclear hydrocarbons are destroyed much more readily as compared with benzene derivatives. The rate constants of the reaction grow in the sequence: benzene > naphthaline > phenanthrene > pyrene > anthracene [28].

Similar dependence is observed at the transition from phenol ($k=2.3.10^2\ M^{-1}.s^{-1}$) to α-naphthol ($k=6.7.10\ M^{-1}.s^{-1}$) [29].

The ozone action on naphthaline results in phthalic acid formation, and on pyrene - in phenanthrene and aldocabonic acid formation. Dioxyneindol and indoxyl are identified among the products of ozonized aqueous solution of indol [30].

Ozone possesses strong bactericide properties. Due to its high oxidizing ability it destroys the bacteria oxidation-reduction system and protoplasma while chlorine affects only the ferments of the microbe sell. This is the reason for the higher antivirus action of ozone [31]. The disinfecting action of ozone is manifested 10-15 times faster than with chlorine. There is a critical doze of ozone (0.4-0.5 mg/l) whose exceeding ensures maximum bactericide action. Upon maintaining of this dose for 4 min the degree of virus inactivation including stimulators of poliomyelitis, stomach infections and enterivirus reaches 99.9% [32].

Since ozone is almost insoluble in water the disinfection process depends on the concentration of ozone-air mixture, the way of mixing with water, the contact time, etc. In dependence on the water quality ozone is used through two different schemes: 1) for polluted water, particularly in spring and autumn, the water is doubly treated at the purification station inlet through the so called process of pre-ozonation and at the station outlet - post-ozonation and 2) for pure water, usually in summer and winter, lake and highmountain sources is applied only post-ozonation or they are not ozonized at all [33].

Since the bactericide effect of ozone is preserved for a short time the water-supply system during disinfection should be protected from secondary pollution before reaching the consumer. Ozone in a concentration of 0.5-1.0 mg/l destroys algae and unicellular organisms. The mollusk larvae die at doses of 3 mg/l. The destroy of cyclops, oligochaeta and daffnis occurs at 2 mg/l dose [34-39].

6.2. PREPARATION

Ozone for industrial purposes is prepared by thermal, photochemical, chemical, electrochemical, in electric discharge and under the effect of particles flux.

Ozone synthesis takes place via the reactions presented in Scheme 6.4.

Scheme 6.4

$$O_2 + (e^-, h\nu, T) \rightarrow 2O\ (O_2^*)$$

0

$$O_2 + O + M \leftrightarrow O_3 + M$$

1

$$O_2^* + O_2 \rightarrow O_3 + O.$$

2

where: M is any third particle

At low temperatures the gas is mainly composed of oxygen molecules and at higher temperatures of atomic ones. It should be noted that the ozone partial pressure is insignificant at all temperature ranges under normal pressure. The maximum equilibrium pressure which is observed at temperatures up to 350K amounts only to 9.10-7 bar. The values of the equilibrium constant of reaction (1) at various temperatures are summarized in Table 6.2 [40].

Table 6.2

Equilibrium constants (K_e) of reaction 1 on temperature

T, K	1500	2000	3000	4000	5000	6000
K_e, M	$1.662.10^{-11}$	$4.413.10^{-7}$	$1.264.10^{-2}$	2.104	48.37	382.9

At higher temperatures when the content of atomic oxygen is high the equilibrium of reaction (1) is shifted to the left and the ozone content is low. At lower temperatures the equilibrium is moved to the right but because of the lower atomic oxygen concentration the ozone content is negligible. In order to increase the ozone yield one should bear in mind the following two conditions: 1) low temperature and 2) formation of superequilibrium concentrations of atomic oxygen. The achivement of superequilibrium concentrations of atomic oxygen at low temperatures can be realized by using nonthermal processes such as oxygen dissociation by particles flux, electrons, hv, electrochemical and chemical affects. Ozone will be always generated at low temperatures if a process of oxygen dissociation occurs. Ozone in higher concentrations can be generated using thermal methods which guarantee the preservation ('temper") of the superequilibrium concentrations of atomic oxygen at low temperatures.

6.2.1. Thermal Methods

Superequilibrium concentrations of atomic oxygen under thermal conditions may be obtained in two cases: 1) jump cooling and 2) establishment of a sharp temperature limit through which the atomic oxygen diffuses from the hot to the cold zone. It has been reported that ozone in considerable concentrations (100-1000 orders higher than the equilibrium) can be generated by passing oxygen through a heated up to 1500-1200°C source and its concentration rises with temperature [41, 42].

Ozone can be also obtained by the fast cooling of a heated in plasma oxygen with liquid oxygen or cooling of an argon plasma with oxygen [43-45].

6.2.2. Photochemical Methods

The photochemical generation of ozone is carried out at the radiation of a gaseous 1-69 bar oxygen or liquid oxygen with UV-light at λ<210 nm of a quantum yield 2 [46, 47]. The latter does not depend on the wavelength up to λ=200 nm, pressure, the mixing with nitrogen and temperature [48]. The mechanism of photochemical ozone generation is not clear enough but it is assumed that absorbing a quantum light oxygen is first dissociated to atoms and after that ozone is formed:

Scheme 6.5

$$O_2 + h\nu \rightarrow O + O$$
$$O_2 + O + M \rightarrow O_3 + M$$

This mechanism explains the quantum yield of 2 and most likely is valid at λ<175 nm when oxygen dissociated to one oxygen atom in ground state and one in excited (^{3}P) and (^{1}D). At λ>175 nm the energy is insufficient for the proceeding of such a decomposition. At the same time oxygen dissociation to two oxygen atoms in ground state requires energy corresponding to λ=243 nm, but ozone is not produced at $\lambda \geq$210 nm.

It is assumed [50] that ozone generation in the range of 175<λ<210 nm can be related to the formation of excited oxygen molecules:

$$O_2 + h\nu \rightarrow (O_2^*) \quad (0')$$

$$O_2^* + O_2 \rightarrow O_3 + O$$

(2)

$$O_2 + O + M \leftrightarrow O_3 + M$$

(1)

Ozone is also formed from oxygen in the presence of mercury vapors by light radiating with $\lambda = 253.7$ nm (mercury resonance band). It is supposed that mercury exerts sensibilizing properties [51]. The photochemical generation of ozone plays a very important role in the atmospheric processes but it can cot compete with the industrial ones mainly due to the higher energy costs (32 kWh/kg ozone) for obtaining a short-wave high energy radiation.

6.2.3. Electric-Discharge Methods

The industrial preparation of ozone is carried out mainly in electric-discharge, by passing gaseous oxygen through high voltage (8-10kV) electrodes. At the collisions of electrons with oxygen molecule part of the kinetic energy is converted into energy of bond dissociation and exciting. In the former cases one oxygen molecule gives rise to two oxygen atoms and in the latter two oxygen molecules produce one oxygen atom and one molecule of ozone.

Scheme 6.6

$$O_2 + e \rightarrow 2O\ (O_2^*)$$

(0)

$$O_2 + O + M \leftrightarrow O_3 + M$$

(1)

$$O_2^* + O_2 \rightarrow O_3 + O$$

(2)

The schematic presentation of the basic types of discharge is depicted in Fig.6.1.

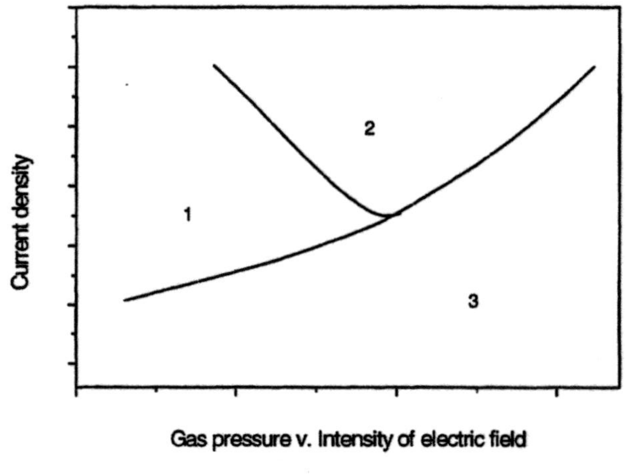

Fig.6.1. Zones of existence of the basic types of discharges. 1 - smolder discharge; 2 - arc discharge ; 3 - silent discharge. On the abscise axis is the ratio of the gas pressure to the electric field intensity and on the ordinate axis is the current density.

Varburg [62-64] assumes the ratio between the intensity of the discharge current and its voltage as the main characteristic of the discharge. The effect of the electric discharge on chemical compounds depends on the discharge nature, which is defined by the electrode potential difference, the pressure in the discharge zone and the current density.

Smolder discharge is observed at low gas pressures, up to several tenths bars, though at electrodes cooling it can be also realized under atmosphere pressure. In silent discharge at 25 mm Hg 0.5% wt ozone is produced. In the electrodes environment ozone is not formed and its concentration drops with increase in pressure, voltage [65] and the addition of other gases. The maximum yield of ozone at normal temperatures is 23 g ozone/kWh. When the pressure is 3.5 mm Hg and the electrodes are cooled by liquid air the ozone amount can reach 150 g/kWh.

Silent discharge is usually observed under atmospheric pressures at relatively high potential differences. It represents an independent discharge caused by the gas conductivity on account of its residual ionization. Varburg [59] refers the discharge from edge and discharge in ozonator as a silent type discharge. The discharge in ozonator takes place between dielectric electrodes at pressures close to atmospheric one.

It should be pointed out that silent discharge, due to the high values of gas pressure and low temperature in-between the electrode space, is very suitable for carrying out chemical reactions as the products are with higher molecular weight that the initial compounds. One of these reactions is in fact the ozone synthesis.

The ozone synthesis is practically conducted in discharge tubes - ozonators. One of the most applied ozonators are Siemens ones (Fig.6.2).

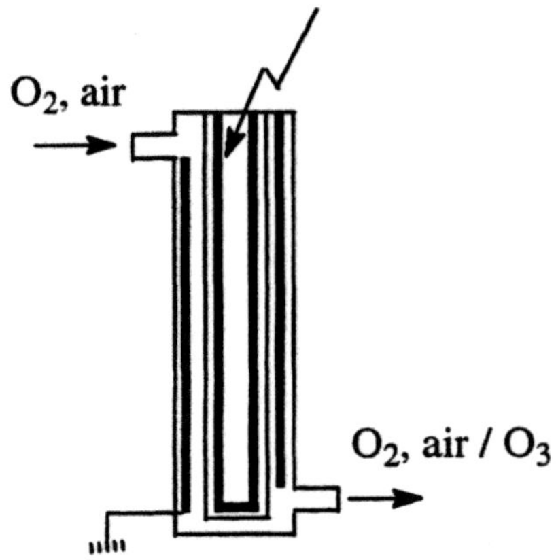

Fig.6.2. Principal scheme of Siemens ozonator

This device represents two welded one into another coaxial tubes with impregnated internal surface of the inner tube and external surface of the outer tube by electrical conductive coatings on the base of aluminum, silver, copper, etc. At the supply of high voltage (up to 20 kV) and dry and pure oxygen ozone is produced at the opposite ozonator end. The ozone concentration depends on the current parameters - voltage, frequency and power, on the ozonator properties - thickness, length and type of the glass tubes, electrodes distance and temperature.

We will further discuss some of its characteristics and the mechanism of ozone generation but we should mention that the theoretical yield of ozone in it can reach up to 450 g/kWh. The edge discharge and the process of ozone formation from it are described in Ref. 59. Provided the data of this study, i.e. ozone yield of 50 g/kWh, one may conclude that this process is energetically less favorable than the ozonator one.

Crown discharge - particular form of silent discharge arising at higher voltage and electrodes curvature. It is characterized by the appearance of a shining gas layer around the electrodes. Beam ionization of the gas is accomplished in the crown and the maximum yield of ozone from air is found to be 23 g/kWh [72].

Spike discharge - it appears at still higher values of the voltage than at the crown one and is characterized by a periodic repetition of formation and fading of spike formation. The reason for this effect is the drop in the resistance between the electrodes during the spike formation time which results in voltage drop and the spike goes down. Then the resistance begins to rise and simultaneously the voltage increases and the spike is lighted again. Up to the mid- last century this method was the main method for ozone synthesis. It is extremely ineffective and gives up to 50 times smaller ozone amounts as compared to the other discharge methods [75, 76].

Arc discharge - with increase in the current density and voltage the temperature of the electrodes increases and the smoldering discharge gradually passes into an arc whereby an intensive evaporation of the electrode metal occurs. This type of discharge is referred to the processes of isothermal plasma. The reactions taking place thereby have thermal character. Thus in the zone of the arc discharge insignificant amounts of thermodynamically equilibrium ozone concentrations may be formed. At definite conditions, however, up to 12 g/kWh ozone can be obtained [74].

6.2.3.1. Theory of the Methods

On the basis of the foregoing survey it can be concluded that the ozone preparation in a silent discharge (ozonator) is practically the only convenient and economical industrial method for ozone synthesis. Depending on the ozone content in the ozonized air at the ozonator outlet (for a given ozonator type) the ozone yield varies from 30 to 50 g/kWh that corresponds to an energy of 60-36 eV per one ozone molecule. [94]. Since the formation of one ozone molecular requires 34.5 kcal, i.e. 1.5 eV, the theoretical yield of ozone would amount to 1200 g/kWh. However for industrial ozonators the yield of ozone is not more than 30-60 g/kWh which is 20-40 times smaller that the theoretical one. The main reason for this great difference between the theoretical and practical yield of ozone is the ozone reaction with the electron (or with M- any third particle):

$$O_3 + e^-(M) \rightarrow O_2 + O + e^-(M),$$

leading to its decomposition. The mechanism of ozone formation includes the following two processes:

$$O_2 + O + M \rightarrow O_3 + M \qquad (1)$$
$$O_2 + O_2^* \rightarrow O_3 + O \qquad (2)$$

Reaction 2 is endothermic by 4 eV (93.6 kcal) and it can take place only if the energy for excitation of O_2 is higher than 4 eV. The lowest excitation state of oxygen is the metastable $^3\Sigma_u^+$ with 4 eV energy of excitation, but since this is a forbidden transition the probability for ozone generation through it is very negligible. This is also the lowest possible value of the energy of the bombarding electron which from a theoretical standpoint could make possible the ozone generation since the dissociation energy of O_3 is 5 eV and the ionization energy is 12.1 eV.

Upon applying the principle of Frank-Condon [95] it has been established that during the electron transition in the ozone molecule from ground ($^3\Sigma_g^-$) to metastable ($^3\Sigma_u^-$) state the lowest value of the transition energy is about 7 eV which is sufficient for ozone dissociation (5.1 eV) (Fig. 6.3).

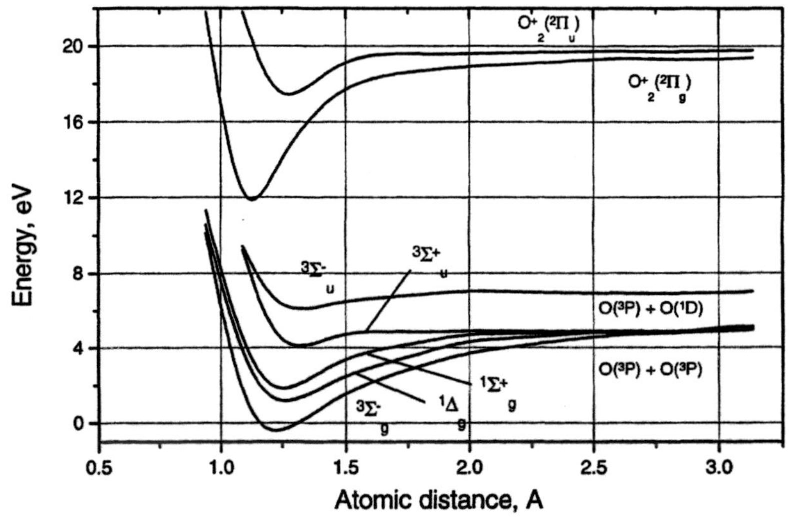

Fig.6.3. Energetic levels of O_2 and O_2^+ and the energies of the most probable transitions from ground state of the ozone molecule after Frank-Condon principle.

This fact is a serious argument in favor of the conclusion that ozone activation occurs through dissociation of its molecule. Moreover, the transition $^3\Sigma_g^- \rightarrow {}^3\Sigma_u^-$ is allowed in contrast to the $^3\Sigma_g^- \rightarrow {}^3\Sigma_u^+$ transition. Since the average value of energy for the ($^3\Sigma_g^- \rightarrow {}^3\Sigma_u^-$) transition is 8 eV the effective energy of electrons should be equal to this value. This means that the theoretical yield of ozone should be smaller that the thermodynamical one 8:5.1 = 1.57 and as the dissociation of one ozone molecule gives two oxygen atoms the decrease should be two times higher, i.e.3. In this case the theoretical yield must be 1200:3 = 400 g/kWh. The real yields, in particular those obtained in air atmosphere, are by about 10 times lower that this magnitude. For that reason the scheme for ozone formation is supplemented by the following reaction:

$$O_3 + M \rightarrow O_2 + O + M \qquad (1)$$

where: M is any third particle including the ozonator wall.

It has been practically found that ozone yield of reactions carried out under low pressures and temperatures can approach value of 150 g/kWh. The formal scheme of the reaction of ozone electrosynthesis from pure oxygen can be presented as follows:

$$O_2 + e^- \rightarrow 2O\ (O_2^*) + e^- \qquad (0)$$
$$O_2 + O \rightarrow O_3^*$$
$$O_2^* + O_2 \rightarrow O_3 + O \qquad (2)$$
$$O_3^* + M \rightarrow O_3 + M$$
$$O_3 + O \rightarrow 2O_2$$
$$O_3 + e^- \rightarrow O_2 + O + e^-$$
$$O_3 \rightarrow I\text{-}O_2$$
$$O \rightarrow I\text{-}O_2$$

where: $I\text{-}O_2$ is the portion of oxygen absorbed on the electrode which in fact does not go in the gas phase.

The introduction of heterogeneous steps (the last two reactions) into the reaction scheme reflects more reliably the lower efficiency of the electron energy via the reactions with the ozonator solid surfaces.

The kinetic scheme for ozone generation from air includes some new reactions:

$$O_2 + e^- \rightarrow 2O\ (O_2^*) + e^- \quad (0)$$
$$O_2 + O \rightarrow O_3^*$$
$$O_2^* + O_2 \rightarrow O_3 + O \quad (2)$$
$$O_3^* + M \rightarrow O_3 + M$$
$$O_3 + O \rightarrow 2O_2$$
$$O_3 + e^- \rightarrow O_2 + O + e^-$$
$$N_2 + e^- \rightarrow N_2^* + e^-$$
$$N_2^* + N_2 \rightarrow 2N_2$$
$$N_2^* + O_2 \rightarrow N_2 + 2O$$

This mechanism differs from the preceding one by introduction of reactions with nitrogen participation. The first one describes the electron excitation of nitrogen during its collision with electron and the other two accounts for the dissipation of the energy of excitation.

6.2.3.2. Ozonators

Depending on the electrodes type the ozonators are divided into two groups: ozonators with coaxial and plain electrodes. The ozonators of the first type, usually consist of 200-1000 calibrated glass pipes of 3-5 mm wall thickness, diameter - 20-100 mm and 1200-1500 mm length. The glass is of boron type with minimum alkaline ions content. Ozone generators with metal enameled pipes are also patented [96-98]. Ozonators with plain electrodes are constructed similarly to filter presses from frames which provide cooling and discharging spaces. Glass, ceramics, enamels, etc. are used as dielectrics. The enamel and ceramic dielectrics are usually thinner but due to the high dielectric constants allow the usage of higher voltages as compared to some low quality glasses. The electrodes are supplied with alternating voltage (6-15 kV) with 50-10000 Hz frequency. The ozone concentration in the gas phase reaches a maximum up to 5-7 % vol. France, Germany, Russia, USA, Japan and China are the main manufacturers of high-duty ozone generator systems. We have invented an ozonators with plain electrodes [99]. One kg ozone is obtained from 16-18 kWh oxygen and 32-36 kWh air. In some cases the electric energy consumption is reduced to 8-9 kWh per 1 kg ozone. If the initial gas is cooled up to -183°C then the consumption of electric energy is the lowest (3.3 kWh/kg O_3.).

The modern devices have an efficiency of about 22.6 % and so far some improvement of this index is not expected. The contemporary ozonators have a capacity of 200-50 kg O_3/h.

The theory of ozone formation pointed above states that a higher efficiency could be reached if oxygen or air are preliminary purified from the gaseous impurities - H_2O, H_2, hydrocarbons, etc. This can be done by drying of oxygen or air at their cooling below the dew-point (-45°C) followed by adsorption. Some of the accompanying gases like N_2, Ar, He, CO_2 can increase the ozone yields at other equal conditions as they act as effective quenchers of the O_3* excited molecules. The presence of nitrogen in the discharge space however, produces always definite amounts of nitrogen oxides.

In order to prevent the decomposition processes of ozone in the ozonator the time of the gas stay in the ozonator is decreased by increasing the inlet rate of the gas mixture. The ozone yield is raised but its concentration in the gas flow is decreased. In many cases, however, the goal is to obtain more concentrated ozone mixtures at lower rates of the gas mixtures thus reducing the apparatus dimensions and decreasing the swept away of the reagents and solvents. Usually the rate of the gas flow inlet is optimized on the basis of economical considerations.

We will focus our attention on the new ozonator with plane electrodes designed by us. (Fig.6.4).

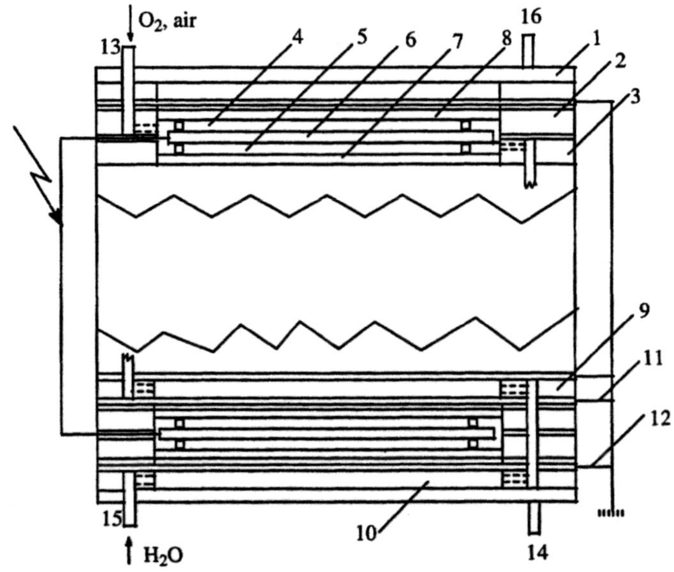

Fig.6.4. Column ozonator with flat-parallel electrodes

The ozonator consists of pressing plates 1 which play the role of grounded electrodes in the first and last ozonizing element in one ozonizing column, pressing mechanism (not shown in the figure) and self-tightening frames and electrodes placed in the following consequence: frame 3, creating together with frame 3 (an mirror image of frame 2) the discharging chambers 4 and 5, high voltage electrode 5, surrounded form the both sides by dielectric 7 and 8, frame 9 which comprises the cooling chamber 10, whereby the cooling fluid is circulating,and the grounded electrodes 11 and 12 are placed between frames 3 and 9 and 9 and 2. Thus one ozonizing element contains grounded electrode 12, frame 2, dielectric 7, high voltage electrode 6, dielectric 8, frame 3, grounded 11. Frames 2, 3 and 9 in each next element are turned to 180°C. The cooling chambers 10, constructed by frame 9, may be placed not only after each ozonizing element but it can also serve several ozonizing elements thus making the apparatus more compact. Each ozonizing column is built from several ozonizing elements but not less than two. The air or oxygen inlet and outlet is carried out through orifices 13 and 14, respectively. The cooling fluid is fed through orifice 15 and is removed away by orifice 16. The pressing plates are of steel, frames 2, 3 and 9 are made from placticized polyvinylchloride or silicon rubber or some other self-packing material of high resistance to the action with high voltage and ozone, and the electrodes are on the base of aluminum or copper.

The operation pattern of the device is as follows: the air or oxygen to be ozonized are fed through orifice 13 to the gas discharging chamber 4 and 5 whereby under the action of the barrier discharge (as a result of the high voltage between 1, 11 and 6 electrodes) ozone is produced. Further the air or oxygen pass into the second ozonizing element and are additionally enriched in ozone and so on until the ozone concentration reaches the equilibrium for the definite temperature which is ensured by the cooling fluid coming though orifice 15 and leaving by orifice 16.

We have also made some modifications of the ozonators: i.e. portable ozonator P-1 for medical and human purposes and P2-15 with 15 g ozone/h capacity for the purposes of organic synthesis, ecology, pharmaceutics, agriculture, food-stuff industry, water treatment, etc. In 1991 our devices were graduated with golden medals at the World exposition EXPO'92 - Plovdiv.

The P-1 ozonator appearance is depicted in Fig.6.5.

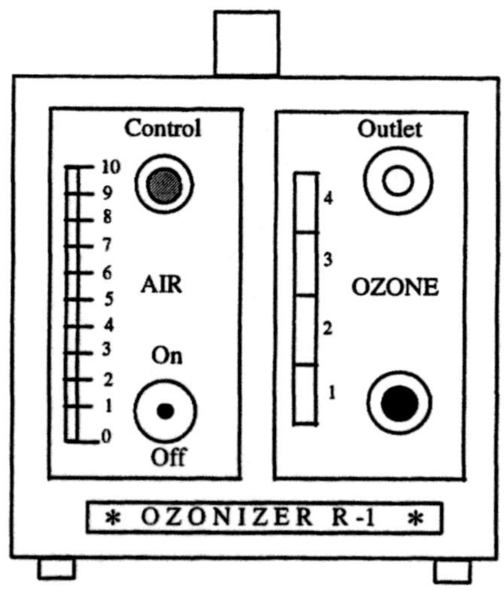

Fig.6.5. View of P-1 portable ozonator for medical purposes

It is designed on the basis of a glass discharge device shown in Fig.6.6.

Fig.6.6. Glass discharge device

The glass discharge device is undetachable and is prepared from low alkaline glass. It consists of inner cylindrical pipe with 2 mm wall thickness, 22 mm inner diameter, 150 mm total length, and 120 mm length of the cylindrical part on which surface is placed the grounded electrode 6 with assembly hoop 5 and current-carrying rim 7, air or oxygen inlet 2 and outlet 3 with inner diameter of 6 mm, hermetized inlet of the high voltage electrode 11, glass ampoule 4 100 mm long diameter 15 m, wall thickness of 2 mm which surface is covered by the high voltage electrode 8 consisting of 20 copper tinned conductors with 1.5 mm cross-section. The latter are disposed at equal distances between each other and parallel to the longitudinal axis welded to rings 9 and 10, through ring 9 and high voltage electrode being connected with outlet 11. The ampoule is placed in tube 1.

The gas scheme of the device is demonstrated in Fig.6.7.

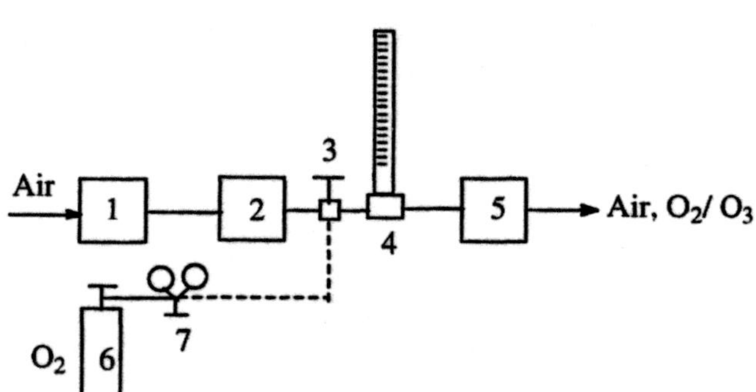

Fig.6.7. Scheme of the gas lines of P-1: 1. Filter, 2. Compressor, 3. Crane. 4. Rotameter, 5. Ozonator, 5. Oxygen bottle and 7. Manometer.

The air is fed by micro compressor 2 through filter 1, a junction crane and rotameter 4 into the ozonator 5 or in case we aim at higher ozone concentration at the ozonator outlet an oxygen from the oxygen bottle 6 through manometer 7, crane 3 and rotameter 4 it is fed into ozonator 5.

The electric scheme of P-1 is presented in Fig.6.8.

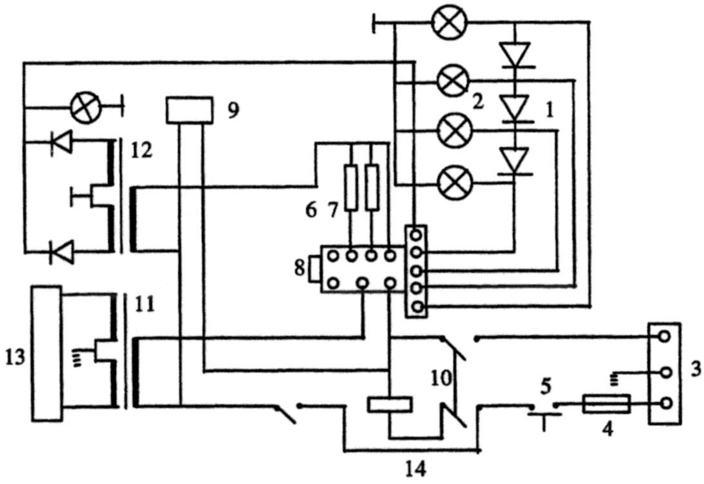

Fig.6.8. Electric scheme of P-1 ozonator: 1 - diodes, 2 - lamps, 3 - conector, 4 - safety unit, 5 - micro switch, 6 - resistor, 8 -packet switch; 9 - aquarium pump, 10 - switch, 11 - E 3-8, 12 - transformer, 13 - ozonator, 14 - relay.

Ozonator P-2-15 (Fig.6.9) is constructed from a battery of six consecutively connected glass coaxial ozonators of 1200 mm length, 300 mm inner diameter, 2.5 mm wall thickness and electrode distance of 5 mm. It is cooled by running water and electric supply of 220V/20kV/50Hz. The modules for air preparation and the reactors for mixing of ozone with gas and liquid reagents are developed separately.

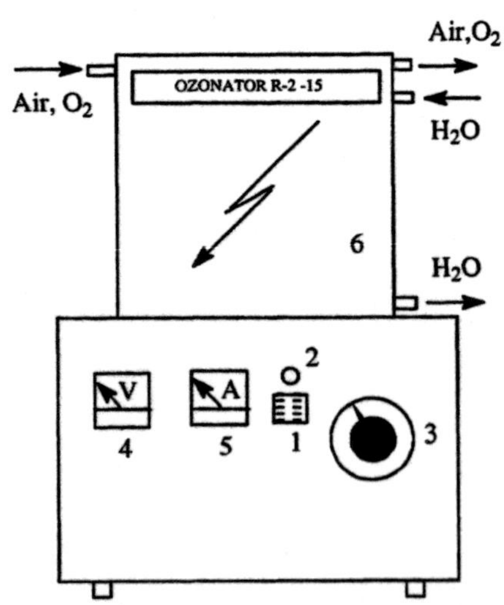

Fig.6.9. View of ozonator for semiindustrial purposes P-2-15.

6.2.4. Particles Flow

Here we will review briefly the methods of ozone generation with participation of monochromatic flux of slow electrons, cathode rays or radiation-chemical action on oxygen.

Slow electrons. Anry [79] has shown that the formation of ozone starts at energy of 8.9 eV (one of oxygen exciting potentials) after which the rate of the process sharply rises at 17.9 eV (the double excitation potential) and 21 eV (critical potential of oxygen dissociation to an atom and positive ion). At 14 eV (ionization potential of oxygen) no changes in the process of ozone formation have been registered. In other studies [80] it is stated that the low limit of ozone formation is 6.4 eV. These results suggest that the excited oxygen molecules play more important role in ozone synthesis than the ionization.

Cathode rays - upon the action of high energy electrons the ozone yield amounts to 3-5 kg/kWh [82].

6.2.5. Radiation-Chemical Methods

The ozone generation in gaseous and liquid oxygen during radiation exposure is investigated in details by Pshezchezkii [83]. The energetic yield of ozone from gaseous oxygen is 1.5-1.7 molecules of O_3 per 100 eV [84] corresponding to 27-31 g O_3/kWh.

6.2.6. Electrochemical Methods

The expectations that the processes of anode oxidation with oxygen participation would lead to high energetic ozone yields fail completely so far. The most efficient electrochemical methods for ozone generation are the electrolysis of solutions of hydrochloric and sulfuric acid solutions [85, 87, 90]. In the first case ozone is yielded in concentration of 58 % wt and energetic yield of 24 g/kWh. The efficiency of the electrochemical methods depends on a number of factors such as temperature, electrolyte concentration, current density, electrode materials and type. One of the ways to improve the efficiency of these methods is through applying high current densities by using flat platinum electrodes [88]. Under laboratory conditions ozone is generated in a special device [89].

6.2.7. Chemical Methods

Ozone is generated in some chemical reactions accompanied by liberation of atomic oxygen. Ozone is formed during oxidation of phosphorus, decomposition of oxygen containing salts [92] and the interaction of fluor with salts solutions [93].

6.3. ANALYSIS

There exist three methods for analysis of ozone concentration ranging from 0.1 ppm to 100 %:

- **Physical** - based on measuring the physical characteristics of ozone as absorption in the UV-VIS and IR range of the spectrum, increase of the pressure during its thermal decomposition, etc.
- **Physico-chemical** - based on monitoring the physical effects arising from the ozone interaction with various reagents, like chemiluminescence, heat, electro-driving force, etc.
- **Chemical** - based on determination of the amount of the reagent reacted with ozone at a known reaction stoichiometry.

Ozone concentrations varying from several ppb to 100% can be iodometrically determined. The determination is carried out when a definite volume (V_1) of gas mixture containing ozone passes through a buffered at pH = 6.8 (0.029 M KH_2PO_4 + 0.0425 M K_2HPO_4) saturated aqueous solution of KI:

$$V_1 = v.t \qquad \text{(Eq.6.1)}$$

where: v - the rate of the gas flow; t - time of gas passing.

The following reaction takes place in the solution:

$$2I^- + O_3 + 2H^+ = I_2 + O_2 + H_2O$$

The reaction is accompanied by a quantitative liberation of iodine and the solution gets yellow. The released iodine is titrated by $Na_2S_2O_3$ with known normality (N_2):

$$I_2 + 2Na_2S_2O_3 = 2NaI + Na_2S_4O_6$$

The iodine content and respectively the ozone concentration is determined from the volume of the consumed thiosulphate (V_2) and (N_2).

$$N_1 = V_2.N_2/V_1 \quad \text{(Eq.6.2)}$$

As one molecule ozone gives one molecule of iodine which is titrated by 2 thiosulphate molecules with normality of Mw/2, the ozone concentration will be equal to N_1.

The iodine quantity can be also determined colorimetrically by following the absorption in the visible spectrum at 470 nm, which method was developed by us and discussed in Chapter I.

Quantitative estimation of ozone can be also performed by using its reaction with olefins, for instance with trans-stilbene. Ozone reacts with stilbene in stoichiometric amounts and it can be determined either from the kinetics of ozone uptake or gaschromatographycally by following the kinetics of stilbene consumption.

Determination of small ozone concentrations under working and dwelling conditions is carried out by monitoring the change of the color of two-valent iron [100], fluorescein, o-toluidine, imdigocarmine [101-103], N-phenyl-2-naphthylamine [104] sulphophthalein [105], diacetyldihydroxytoluidine [106], N,N'-dimethyl-p-phenylenediamine [107] and $MnCl_2$ [108] or the quenching of the bright blue luminescence of luminol [109].

The quantitative luminescent method involves the paper impregnated with alkaline solution of luminol [110] or of luminol and hematin [111] is placed in front of a photomultiplier. Upon admission of ozone it quenches the luminescence and the photomultiplier records this change. The methods sensitivity is 0.005 μg/l.

Usually one of the chemical methods, mostly iodometric one, is applied for calibration of the spectral methods of analysis: spectrophotometric (245-300 nm), infra-red (1100 cm-1), chemiluminiscent, calorimetric, colorimetric, coulometric, etc.

Here one should outline the coulometric methods of analysis as promising, highly sensitive and easy for automatization [112]. Fig.6.10. demonstrates the scheme of such a device [113].

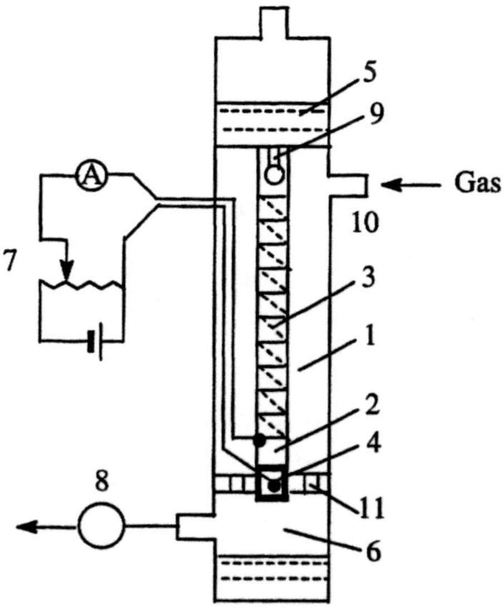

Fig.6.10. Scheme of a device for coulometric analysis of ozone
1 - ring-shaped space; 2 - electrodes base; 3 - cathode; 4 - anode; 5 - electrolytes volume; 6 - vessel for the used electrolyte; 7 - battery; 8 - pump; 9 - electrolytes inlet; 10 -gas inlet; 11 - gas and electrolyte outlet.

It contains a vessel for the electrolyte 5, base of electrodes 2, on which the cathode 3 in a form of multiwinding spiral is placed and anode 4 as one winding. The electrolyte solution is supplied through inlet 10 in the narrow ring-shaped space 1 whereby the gas interacts with the solution on the electrode base and together with the solution leaves via outlet 11. By means of battery 7 a voltage of 0.25 V is applied to the electrodes and the arising current is measured by amperemeter (if necessary it is preliminary accelerated). The ozone content is determined by using its reaction with KI as shown above. The iodine is reduced at the cathode and at the anode the iodide anion results in iodine formation. The electrolyte contains 3 M NaBr, 0.001 M NaI, 0.1 M Na_2HPO_4 and 0.1 M NaH_2PO_4.

As ozone and oxygen possess different half-wave potentials their position changes depending on the phone. In Fig.6.11 is shown a polarogram which is obtained at passing of ozone-oxygen mixture through a device with rotating platinum electrode. The limit current is proportional to the ozone concentration in the whole range of 0.05-26 % concentrations.

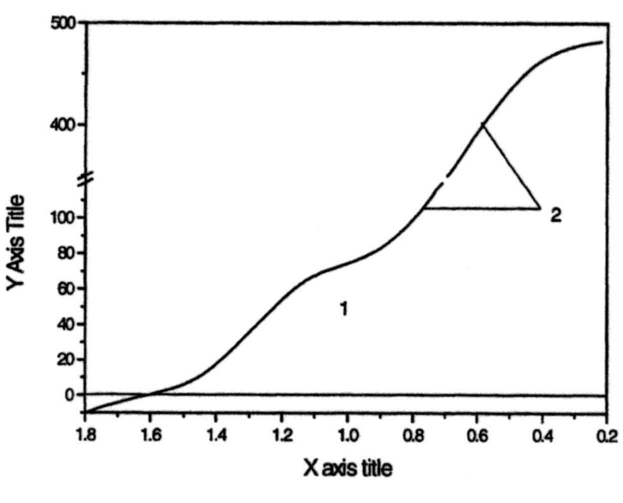

Fig.6.11. Polarogram O_3/O_2 on the phone of 1 M H_2SO_4,obtained from rotating platinum electrode

Chemiluminescence method is based on the interaction of ozone with NO, C_2H_4, rhodamine, etc. [114] generating electronically excited states. As a result of their relaxation to the ground state light emission is observed. The intensity of this light is proportional to the ozone concentration. The interaction of ozone with NO and ethylene leads to rise of the analysis sensitivity by ensuring a constant intensity of light emission irrespective of the analysis duration.

The calorimetric analysis is a method for measuring the amount of heat liberated during ozone decomposition in the presence of a catalyst. The range of concentrations which can be measured varies from 1-7 % to 1-0.1 ppm. At high ozone concentrations (1-7 %) the gas flow containing ozone is passed through over a catalyst Co/SiO_2 with a thermocouple. The decomposition is accompanied by an intensive heating and its measurement is registrated and monitored. In the cases when the ozone concentrations are very low (1-0.1 ppm) a more special apparatus is applied. Two thermistors with 0.01 MΩ resistance connected in a bridge scheme are disposed in two cylindrical channels of a massive aluminum block. One of the thermistor is covered by a hopcalite and the heat liberated by ozone decomposition on the hopcalite thermistor is measured.

The integral amount of ozone in the atmosphere is measured by applying a spectrophotometric method by measuring the intensity of the sunbeams at 310-330 nm which is the ozone absorption band and also at 356 nm where its absorption is negligible. Ozonator M-83 (Russia) operates at 341 and 369 nm.

The foregoing methods make possible the measurement of ozone concentrations in the range from 0.1 ppb to 7 % vol.

6.4. Decomposition

The following three basic routes of ozone decomposition are assumed:

1. Decomposition in the stratosphere
2. Decomposition during its purposeful application and during its spontaneous generation
3. Decomposition in chemical reactions

6.4.1. Stratospheric Ozone

The ozone in atmosphere was discovered in 1873 and its regular measurement dated as far as that year. In 1881 the Irish chemist Hartley by the absorption of UV-sunbeams discovered the presence of ozone in the upper atmosphere. The basic amount of atmospheric ozone is found in the stratosphere at a height of 20-25 km. At this height the relative quantity of ozone is 1 ppm and the total amount is 3.3 milliards tons. This amount corresponds to a 3 mm thick gas layer under normal pressure and temperature. Ozone is distributed irregularly around the Earth - at the tropics its average amount is smaller, at the poles the average amount is greater and in addition at all latitudes a season alteration of the ozone content is observed. In Fig.6.12 is shown the structures of the average atmosphere and ozone distribution by height in dependence on the changes of temperature with height.

The importance of stratospheric ozone is determined by its optical properties - its ability to absorb the UV-sun radiation with wavelength of 220-300 nm. The absorption spectrum of ozone in the UV and visible region of the spectrum is presented in Fig. 6.13.

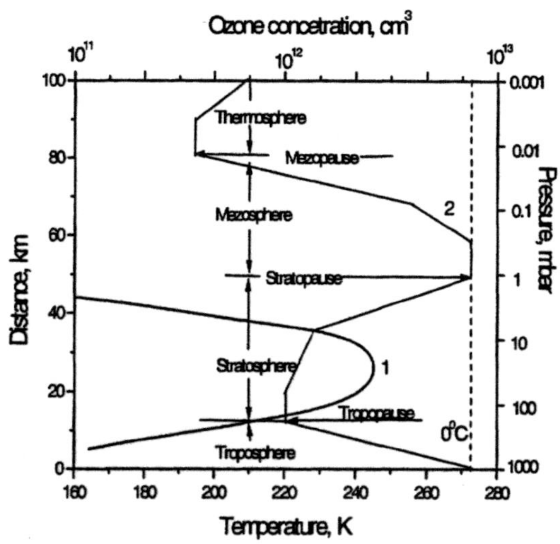

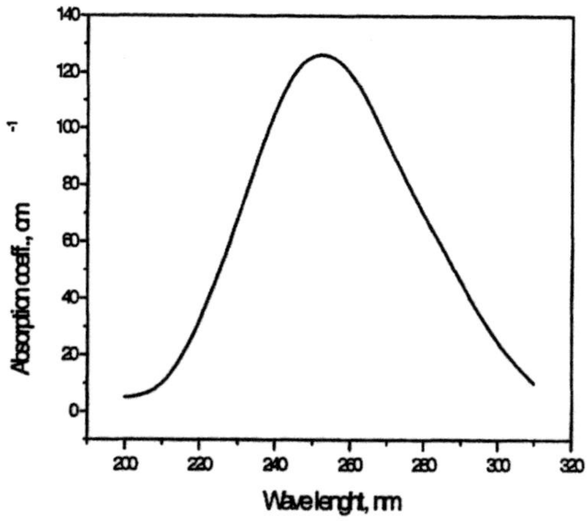

Fig.6.12 Vertical structure of the Earth atmosphere. 1 - ozone distribution; 2 - gas pressure and temperature distribution.

Fig.6.13. UV - absorption spectrum of ozone with maximum at 254 nm.

The basic absorption band named after Hartley is within the range of 200-300 nm. The maximum absorption is observed at a wave length of 253.65 nm which means that the ozone layer with 3 mm thickness at normal pressure reduces the sun radiation by 1040 fold. Ozone absorbs also weakly in the visible and IR range. The most interesting absorption band is at 9.57 μm wavelength. This is in fact the thermal radiation of the Earth and is used for measuring the total amount of ozone by satellites [115-117].

The strong absorbance of ozone in the UV-range protects the Earth's surface from the sun biologically active radiation in the range of 250-320 nm which destroys the most important biological elements - proteins and nucleic acids. Moreover the UV-absorption by ozone is accompanied by warming of the ozone containing layers in the stratosphere which to a great extent determines its thermal regime and thus the dynamic effects that occur in it.

The ozone layer has a substantial effect on the atmospheric properties and ensures conditions for the existing of Earth's flora and fauna.

The ozone layer is an important factor for the global ecology since it ensures a harmless (safe) for the living organisms level of UV-radiation close to the Earth' surface, participates in maintaining the already established climate on the Earth, control the content of anthropogenic and natural impurities in the atmosphere. It is considered that as result of the decreased ozone content in the atmosphere (this is potentially possible also under the effect of anthropogenic factors) the number of cases of skin cancer diseases is increased, in particular with people living at small longitudes and other not fully studied biological consequences for the Earth' s flora and fauna [118].

The stratospheric ozone is obtained from its basic component - molecular oxygen under the action of UV-radiation as a result of photophysical reactions. Actually all studies in the field of stratospheric photochemistry are related to stratospheric ozone [110-123]. The bases of the photochemical theory for stratospheric ozone generation were founded by the English chemist Chepmann [124, 125]. According to this theory ozone generation begins with photodissociation of molecular oxygen:

$$O_2 + h\nu \rightarrow O + O \qquad (1)$$

The energy of dissociation of molecular oxygen is 5.115 eV which corresponds to the energy of a light quantum with 242.4 nm wavelength. Upon participation of a third substance such as the accompanying oxygen and nitrogen molecules in the atmosphere, the following two competing reaction take place:

$$O_2 + O + M \rightarrow O_3 + M \quad (2)$$
$$O + O + M \rightarrow O_2 + M \quad (3)$$

The prevalence of one of the two reactions is defined by the ratio between the concentrations of the atomic and molecular oxygen. The recombination reaction (3) is characteristic for heights over 60 km whereby a considerable part of oxygen is dissociated. For the smaller heights reaction (2) resulting in ozone formation is more acceptable. In fact this reaction is considered as the basic source of ozone generation in the stratosphere. The decomposition of ozone after Chepmann proceeds via the following reactions:

$$O_3 + O \rightarrow O_2 + O_2 \quad (4)$$
$$O_3 + h\nu \rightarrow O_2 + O\,(^3P) \quad (5)$$
$$O_3 + h\nu \rightarrow O_2 + O\,(^1D) \quad (6)$$

giving again oxygen atoms in ground or excited state. This cycle of decomposition is called cycle of "odd oxygen".

Photodissociation of ozone (5, 6) takes place under the action of radiation with wavelength less than 1134 nm. The formation of ozone by reaction (2) occurs mainly at heights of 30-70 km since the lower "hard" UV-radiation which is necessary for the molecular oxygen dissociation, practically does not penetrate, while the ozone decomposition after reactions (5, 6) can be observed at all heights nearly to the Earth's surface.

The development of the experimental techniques, specification of the kinetic parameters and the detailed study of the ozone balance show that the photochemical generation of ozone in the stratosphere exceeds substantially its destruction via reaction (4). It has been found that 20 % of the stratospheric ozone is decomposed by this reaction [126].

The attempts to estimate the vertical distribution and total content of ozone by reactions (1) - (6) has always given enhanced results which suggests [127] that these reactions are far from all processes that take place with ozone participation. In the atmosphere containing oxygen and hydrogen one should bare in mind that at least 35 various chemical reactions may occur [128] but only 9 of them are of importance for the processes of ozone generation and decomposition.

6.4.1.1. Climate-forming Factor

The influence of the ozone layer on the climate is related to light absorption which occurs not only in the UV but also in the IR-region of the spectrum. Absorbing the IR radiation from the Earth's surface ozone increases the hothouse effect [129, 130]. The problem about its connection with the climate is very complex due to the diversity of physical and chemical factors which affect the content of atmospheric ozone [131]. The effect of ozone on the Earth's temperature and climate depends on the height at which the ozone content is altered as a result of the absorption of short- and long-wave radiation from the Sun and the Earth. In Fig. 6.14 based on the data of NASA [132], is shown the probable variation of Earth's temperature if the ozone content at the different heights increases in such a way that it would reach 3 % through the whole column.

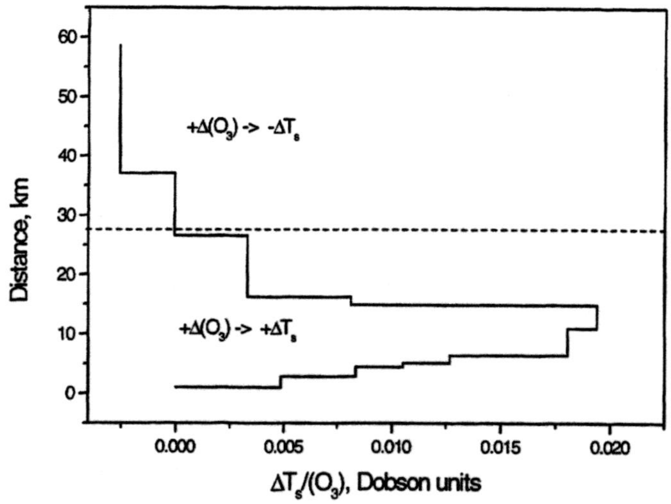

Fig. 6.14. "Radiation" variation of the Earth's surface temperature in dependence on the change of the total ozone content in pole by 3% at the expense of local alteration of ozone on different elevations

It is seen that at 30 km height the growth of the ozone content would lead to Earth' s surface warming and at height over 30 km - to its cooling. The maximum effect is observed at change of the ozone content in the region of the tropopause. The effect at these heights is explained by the absorption and re-emission of the Earth' s radiation with 9-10 μkm wavelength. A maximum hothouse effect will be reached at maximum temperature difference between the absorbed and emitted radiation and when the temperature in the tropopause approaches its lowest values. "Cooling", mostly because of the increased content of ozone at heights over 30 km, is associated with absorption of the Sun radiation.

In Fig.6.15 is presented the probable change of temperature on Earth,s surface (according to the data of the World Meteorology Organization) at 2040 year, as a result of the variation in atmospheric microcomponents composition, including ozone, at growing industrial and agricultural production.

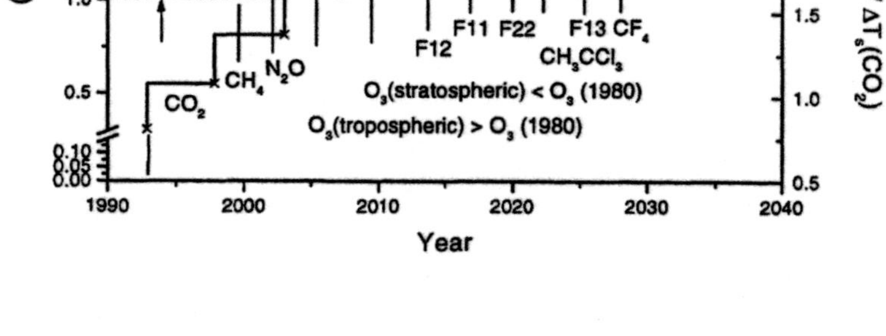

Fig.6.15. Probable change of temperature on Earth's surface at 2040 resulting from the rise of various atmospheric microcomponents

Fig.6.15 shows that the maximum possible climatic effect of ozone probably will be stronger than that of carbon dioxide.

6.4.1.2. Interaction with Atmospheric Components

The short-wave, chemically active Sun radiation with wavelength < 300 nm does not penetrate into the troposphere and thus from photochemical point of view it can be assumed that eternal night reigned there. Ozone absorbs the light with longer wavelength and makes possible the occurrence of various chemical processes in the short-wave UV region [118]:

Scheme 6.7

		Live time, years
	H_2 ($H_2O + H_2$	2
	CH_4 ($CH_3 + H_2O_2$	5 - 10
hv(<310) + O_3	RH (H_2O + Products	
↓		
	CH_3Cl ($H_2O + CH_2Cl$	1,5
$O(^1D) + H_2, CH_4, H_2O$	CH_3Br ($H_2O + CH_2Br$	1,7
↓		
	Freones (Products	
$OH^{\cdot}$ ⟶	CO ($H + CO_2$	0,3
	H_2S (H_2O + SH	0,001
	SO_2 (Products	0,001
	HCHO (CHO + H_2O	0,001

This mechanism demonstrates the effect of anthropogenic and natural atmospheric components on ozone in the troposphere. Ozone photolysis results in formation of metastable excited oxygen radicals in ^1D state with energy of 1.97 eV and radioactive lifetime of 110 s (i.e. much greater than the collision times in troposphere). The reaction of radicals $O(^1D)$ with water vapors, hydrogen and methane leads to hydroxyl radicals generation which in its turn interacts with a number of anthropogenic and natural impurities.

This mechanism is an important step of the overall mechanism for removal of troposphere contaminants and it is its existence that guarantees the protection of the ozone layer from some freons.

Similar processes take also place in the stratosphere but here namely the short-wave radiation affects the freons. Moreover, from ecological standpoint these processes are not so important for the processes occurring in the troposphere.

It is to be noted that among all potential harmful anthropogenic factors the freons are the most hazardous.

In Fig.6.16 is given the change of the ozone content under the effect of freons and other anthropogenic factors. It is seen that the ozone content is reduced substantially with the increasing production of freons (mostly that of freon 11 $CFCl_3$ and freon 12 CF_2Cl_2).

The dependence in Fig.6.16 is also confirmed by the data obtained from the American satellite Nimbus 7. It has been monitored that in the period 1978-1985 the ozone content at the various longitudes was reduced by 2.6 - 10 % with increase of the chlorine content. The discovery of the so called Antarctica ozone hole in 1985 which is characterized by a reduced ozone content with about several percents in the spring is another argument for the harmful effect of freons on the ozone layer.

A number of international agreements concerning the protection of the ozone layer have been accepted - the Vienna convention for protection of the ozone layer (23.05.1985), the Montreal report of the convention (1987). The main purpose in this respect is the future rejection of the ozone hazard freons production and their replacement by more safe compounds or freons which possess the same physicochemical and biological properties but do not destroy the ozone layer.

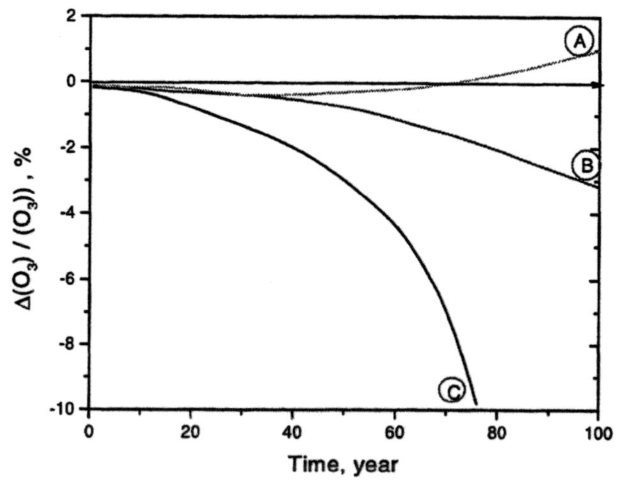

Fig.6.16. Change of the total ozone content in dependence on the grow of anthropogenic contaminants in the atmosphere: A - increase of the CH_4 content by 1 % per year, of N_2O - by 0.25 % per year, of CO_2 - by 0.5 % per year (compared with the 1980 level) without increasing the freons content; B - A + rise of the freons amount by 1.5% per year (as compared with 1980 level): C - A + increase of freons content by 3% per year (compared with 1980 level).

The question why some freons are dangerous and other safe for the ozone layer is a matter of great interest and provokes intensive research in this respect.

The ozone destruction in the atmosphere is associated with catalytic cycles of the following type:

Scheme 6.8

$$\begin{array}{l} X + O_3 \rightarrow XO + O_2 \\ XO + O \rightarrow X + O_2 \\ \hline O_3 + O \rightarrow O_2 + O_2 \end{array} \qquad \begin{array}{l} X + O_3 \rightarrow XO + O_2 \\ XO + O_3 \rightarrow X + 2O_2 \\ \hline 2O_3 \rightarrow 3O_2 \end{array}$$

where: X is OH, NO or Cl which are formed during freons photodissociation in the atmosphere [133].

The cycles of ozone decomposition with participation of OH, NO and Cl is named hydrogen, nitrogen and chlorine, respectively. Here should be mentioned the bromine atoms which appear to be particular efficient in the process of ozone decomposition. However their amount in the atmosphere is much smaller than that of the chlorine atoms. Many papers are devoted to the study of the stratospheric cycles of ozone destruction but the most fundament contributions are provided in [126, 134]. The efficiency of the action of molecules - catalysts in the cycles of ozone destruction is determined by their concentration in the atmosphere which depends on the rates of regeneration and going out of the cycle [135].The ratio of the decomposition rate of the odd oxygen to the rate of the catalyst going out of the cycle determines the chain length and corresponds to the number of O and O_3 molecules decomposed by one catalyst molecule. The number of decomposition steps in one catalytic center can reach the values of 106. Thus, upon the chlorine cycle one Cl atom present in the stratosphere can decompose 100 000 molecules ozone. This is actually the base of the potential danger of freons for the ozone layer.

According to the modern concepts [136] the hydrogen cycle plays the main role during ozone degradation in the upper and lower stratosphere while the nitrogen and, to a smaller extent, the hydrogen and chlorine cycles are more efficient in the middle stratosphere.

The following cycles are characteristic for the upper and middle stratosphere:

Scheme 6.9

$$\begin{array}{l} OH + O \rightarrow H + O_2 \\ H + O_2 + M \rightarrow HO_2 + M \\ HO_2 + O \rightarrow OH + O_2 \\ \hline O + O_3 \rightarrow 2O_2 \end{array}$$

$$\begin{array}{l} OH + O \rightarrow H + O_2 \\ H + O_3 \rightarrow OH + O_2 \\ \hline 2O \rightarrow O_2 \end{array} \qquad (1)$$

In the areas of maximum concentration of the ozone layer its decomposition takes place via the nitrogen cycle:

Scheme 6.10

$$\begin{array}{l} NO + O_3 \rightarrow NO_2 + O_2 \\ NO_2 + O \rightarrow NO + O_2 \\ \hline O + O_3 \rightarrow 2O_2 \end{array} \qquad (2)$$

At a definite ratio between the phone atmospheric concentrations of the components from the so-called families of "odd" nitrogen and hydrogen radicals zero cycles can take place:

Scheme 6.11

$$\begin{array}{l} OH + O_3 \rightarrow HO_2 + O_2 \\ HO_2 + NO \rightarrow NO_2 + OH \\ NO_2 + h\nu \rightarrow NO + O \\ O_2 + O + M \rightarrow O_3 + M \end{array}$$

$$\begin{array}{l} OH + O_3 \rightarrow HO_2 + O_2 \\ HO_2 + NO + M \rightarrow HNO_3 + M \\ HNO_3 + h\nu \rightarrow NO_2 + OH \\ NO_2 + h\nu \rightarrow NO + O \\ O_2 + O + M \rightarrow O_3 + M \end{array}$$

The cycles of the zero type play the role of a peculiar buffer since they reduce the efficiency of the ozone decomposition through bounding the active nitrogen and hydrogen components.

In the lower stratosphere the oxidation of CO catalyzed by nitrogen oxides is an important photochemical source for ozone generation:

Scheme 6.12

$$\begin{array}{l} CO + OH \rightarrow CO_2 + H \\ H + O_2 + M \rightarrow HO_2 + M \\ NO + HO_2 \rightarrow OH + NO_2 \\ NO_2 + h\nu \rightarrow NO + O \\ O + O_2 + M \rightarrow O_3 + M \\ \hline CO + 2O_2 \rightarrow O_3 + CO_2 \end{array} \qquad (3)$$

The appearance of CO in the lower layers of the stratosphere is related to the atmospheric masses transport from the troposphere whereby its formation is a result of methane oxidation (the oxidation of one molecule of CH_4 yields 5 O_3 molecules). The efficiency of this ozone source depends on the phone concentrations of $\cdot OH$ and $HO_2\cdot$ [136].

The relative part of the various cycles of ozone decomposition and formation varies markedly depending on the conditions (height, time, day, etc.). Thus at the nitrogen cycle in dependence on the ratio between the reaction rates can take place the reaction of ozone decomposition (2) or generation (3). For that reason the nitrogen oxides in the middle stratosphere catalyze the ozone decomposition and in the lower stratosphere these very oxides being catalysts of CO or hydrocarbons oxidation contribute to its regeneration.

The ozone decomposition through the chlorine cycle:

Scheme 6.13

$$Cl + O_3 \rightarrow ClO + O_2$$
$$O + ClO \rightarrow Cl + O_2$$
$$\overline{O + O_3 \rightarrow 2O_2}$$

is the most substantial at great heights (30-50 km). If the level of the emitted in the atmosphere pollutants is preserved, the maximum decomposition of ozone occurring through this cycle is estimated to 70% [123]. An example of such a decomposition of ozone are the discovered over the Antarctic" ozone holes" [137]. For the first time a decline of total ozone concentration over the Antarctica was registered in 1979. Then it was only 10 % and in 1986 it was already more than 50 %. This decrease in ozone concentration is mainly attributed to the proceeding of the chlorine cycle [138-140].

The chain termination in the cycles can be achieved either as a result of the interaction between the catalysts themselves:

Scheme 6.14

$$OH + HO_2 \rightarrow H_2O + O_2$$
$$OH + NO_2 + M \rightarrow HNO_2 + M$$
$$ClO + NO_2 + M \rightarrow ClONO_2 + M,$$

where: M is an atmospheric molecule,

or during the reaction of the catalyst species with any third particle in the gas phase, aerosols or water droplets. In many cases the termination of the chain is due rather to turbulent transfer resulting in the removal of the molecule-catalyst from the stratosphere than to chemical reactions.

The decrease of the ozone concentration in the stratosphere causes changes in the concentrations of other gaseous components, interacting with ozone as a result of the shift of the photochemical equilibrium of the whole system.

It can be concluded that the problem of the catalytic ozone decomposition in the atmosphere, from theoretical point of view has received a satisfactorily explanation. The basic catalytic systems and the natural and anthropogenic factors that promote them are quite well established and studied. The observation that 70% of the amount of decomposed ozone is due to anthropogenic action (manufacture of freons) makes the problem of ozone layer protection in the atmosphere a global and demands the joint efforts of all nations in this respect.

6.4.2. Catalysts for Decomposition

Upon studying the oxidation process it has been established that small amounts of ozone to oxygen accelerate the proceeding of many oxidation reactions and promote their proceeding under more favorable conditions. Ozone finds wide application in such important industrial processes like: purification of drinking water, bleaching of textiles, oxidation of sulfurous gas, complete oxidation of the exhaust gases from the production of nitric acid, at the production of many organic compounds, etc. The industrial application of ozone is related to the development of effective methods for the electrosynthesis of ozone [147].

It should be outlined that although ozone possesses a number of advantages as an oxidizing agent they also induce great difficulties during experimental performance and kinetic studies of chemical reactions with ozone participation The ozone molecule is very unstable and sensitive to various contaminants as well as to the construction materials of experimental and industrial devices. All this predetermines the great interest of the researches for investigating the reactions of homogeneous and heterogeneous ozone decomposition and the oxidation processes in its presence.

6.4.2.1. Decomposition of Gaseous Ozone

The reaction of gaseous ozone decomposition can be described by the following stoichiometric equation:

$$O_3 = 1.5\ O_2 + 35\ \text{kcal}$$

This equation supposes one sufficiently simple mechanism of the process. The presence of only one chemical element in the system, i.e. oxygen, limits the number of the active species which might participate in the separate elementary steps of the reaction. The unstability of the ozone molecule, most probably, accounts for the assumption of a simple monomolecular mechanism. However, the numerous experimental data fail to confirm this assumption [142, 143].

Some authors [124, 125] consider the process of gaseous ozone decomposition as a bimolecular reaction that takes place at temperature of 100°C. Analyzing the literature experimental results on ozone decomposition reaction Jahn, Benson and Axworthy [144, 145] reveal that the main reason which impedes the study on the kinetics of the gaseous ozone decomposition is the occurrence of two concurrent pathways - first order heterogeneous reaction on the reactor' s walls and a bimolecular reaction in the bulk. At lower temperatures the heterogeneous reaction is the prevalent while at temperatures in the range of 50-100°C the homogeneous process is the basic route.

The complexity of the process, the lack of fit between experimental data and theoretical calculations have caused many authors [146, 147] to assume a radical chain mechanism for reliable describing of the reaction kinetics.

Benson and Axworthy summarized the experimental observations on the thermal decomposition of ozone and propose three types of mechanism to account for the kinetics of the process.

Atomic mechanism:

$$O_3 + M \rightarrow O_2 + O + M \qquad (k_1, k_2)$$
$$O + O_3 \rightarrow 2O_2 \qquad (k_3)$$

Provided that the oxygen atoms concentration is stationary it can be described by the following kinetic equation:

$$-d[O_3]/dt = 2k_1k_3[O_3]^2[M] / k_2[O_2][M] + k_3[O_3] \qquad \text{(Eq.6.1)}$$

where: $[M] = [O_3] + 0.44\ [O_3] + 0.41\ [N_2] + 1.06\ [CO_2] + 0.34\ [He]$, and the coefficients stand for the relative efficiency of the various gases per the ozone molecule efficiency

The calculation of the equilibrium constant of the reaction allows determination of the rate constants of the elementary steps when $[M] = [O_3]$:

$$k_1 = 4{,}61.10^{12} \exp(-24000/RT)\ M^{-1}.s^{-1}$$
$$k_2 = 6{,}00.10^{17} \exp(600/RT)\ M^{-1}.s^{-1}$$
$$k_3 = 2{,}96.10^{10} \exp(6000/RT)\ M^{-1}.s^{-1}$$

However, the negative value of the activation energy of the backward reaction, evidences that the atomic mechanism does not describe completely the real performance of the reaction.

Mixed mechanism taking account of the energetic chains is more probable:

$$O_3 + M \rightarrow O_2 + O + M \qquad (k_1,k_2)$$
$$O + O_3 \rightarrow 2O_2^* \qquad (k_3)$$
$$O_2^* + O_3 \rightarrow O_2 + O_2 + O \qquad (k_4)$$
$$O_2^* + M \rightarrow O_2 + M \qquad (k_5)$$

This mechanism can be demonstrated by the following kinetic expression:

$$- d[O_3]/d = 2k_1k_3[O_3]^2[M] / k_2[O_2][M] + k_3[O_3](1 - 2/k_5[M] / k_4[O_3])) \qquad (Eq.6.3)$$

where: M has the same meaning as in the atomic mechanism

Atomic-monomolecular mechanism

The atomic mechanism is further supplemented by the help of atomic-molecular mechanism and accepting that the reaction is monomolecular it may be presented in the following manner:

$$O_3 + M \rightarrow O_2 + O + M$$

Benson and Axworthy [145] propose the following step for this mechanism:

$$O_3 + M \rightarrow (O_3)_i^* + M \qquad (a_i,b_i)$$
$$(O_3)_i^* \rightarrow O_2 + O \qquad (c_i,d_i)$$
$$O + O_3 \rightarrow 2O_2 \qquad (k_3)$$

where: $(O_3)_i^*$ denote the excited ozone molecules (their energy is equal or greater that the dissociation energy).

Thus the kinetic equation becomes:

$$-d[O_3]/dt=2k_3[O_3]^2[M].(\Sigma a_ic_i/b_i[M] + c_i)/ (\Sigma b_i[O_2][M](\Sigma b_id_i/\Sigma d_i/b_i[M]+c_i) + k_3[O_2])) \qquad (Eq.6.4)$$

At sufficiently low pressures it may be assumed that $b_i[M] < c_i$ and then the equation is transformed into:

$$-d[O_3]/dt = 2k_3[O_3]^2[M] (\Sigma a_i)/[O_2][M] (\Sigma b_id_i/c_i) + k_3[O_2])) \qquad (Eq.6.5)$$

The comparison of the kinetic equations for the atomic and atomic-molecular mechanism gives the following ratios:

$$k_1 = \Sigma a_i;\ k_2 = \Sigma b_id_i / c_i \qquad (Eq.6.6).$$

Kondrat'ev [150] reported similar results upon studying the thermal ozone decomposition in the temperature range of 136-226°C and relates the experimental data to the atomic mechanism of ozone decomposition. The measured rate constant of this process amounts to $(2.9\pm0.9)10^2\exp(-3700\pm300/RT)$ cm^3 M^{-1} s^{-1}.

6.4.2.2. *Kinetic Stability*

The homogeneous reaction of pure gaseous ozone decomposition is characterized by a definite rate at sufficiently high temperatures. The decomposition of ozone or its mixtures with oxygen at room temperature is associated with heterogeneous reaction that takes place at the reactor walls. Harper and Gordon [151] have observed that the concentration of 95% O_3 under atmospheric pressure and room temperature drops to 80 % for 24h.

A measurable homogeneous decomposition of ozone at ambient temperature could be observed only under the effect of mercury vapors, nitrogen oxides or traces of organic compounds.

The study of the kinetics of ozone thermal decomposition is also complicated by the fact that over a definite critical temperature the stationary kinetic decomposition is transformed into explosion further transferred in detonation. In many papers treating the thermal ozone decomposition it is stated that the stationary homogeneous decomposition of gaseous ozone may be followed only to temperature of 100°C. After Thorp' s data [152] ozone detonates at 105°C. The value of the critical temperature of ozone detonation is closely related to the concentration of the initial ozone. Depending on it four types of ozone decomposition are distinguished [152]: 1 - "normal" - slow decomposition resulting from the exothermity of the process itself at concentrations varying from 0 to 20 % wt [152-155[; 2 - at high concentrations the half- selfdecomposition of ozone is induced in some points of the reactor and gradually is distributed in the whole volume: 3 - at concentration of 48 % wt explosion occurs and 4 - at concentrations > 54 % wt detonation takes place.

The explosion limits of gaseous ozone-oxygen mixtures are demonstrated in Fig.6.17. The relationship is linear but it should be noted that the usage of metal reactors results in 25% enhancement of the mixture stability. The increase in pressure also results in rise of stability [156].

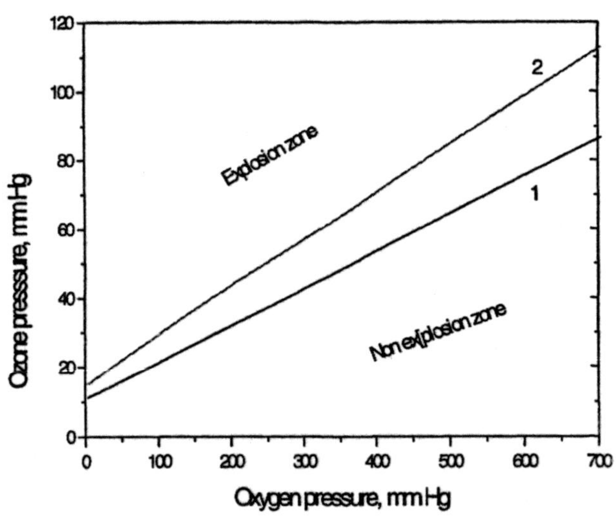

Fig.6.17. Limits of explosion of ozone-oxygen mixtures in 1 - metal vessels and 2 - glass vessels.

Other authors have obtained similar results [157, 158]. There is no doubt that the process of gaseous ozone explosion follows the mechanism of thermal explosion.

The kinetic stability of liquid ozone is of great importance because of its wide application as oxidizer in jet engines.

Upon boiling point of the liquid ozone 100 % ozone can be stored in reactors of Mo glass for 2.5 days and for that time its concentration drops to 97 % [156].

Kobozev, Yastrebov and Pitzhelauri [155, 159] have confirmed that the explosion sensitivity of concentrated liquid ozone and its oxygen solutions decreases abruptly with rise of the oxygen concentration with respect to the thermal impulses. Ozone-oxygen solution with ozone concentration C does not explode if the energy of the thermal impulse is less than that calculated by the equation: E = 1.86exp(-c/6.9).

The transition to explosion decomposition of the gaseous ozone-oxygen mixture in liquid ozone-oxygen solution has been studied by Strakhov et al. [160] under various pressures - 150, 300 and 500 mm Hg. It has been found that with rise of the pressure the stability towards explosion from the gas to the liquid phase is moved to lower concentrations of ozone in the gas phase. Under pressures higher than 500 mm Hg the systems with ozone concentrations < 50 % vol in the gas phase and < 25% vol in the liquid phase appear to be explosion-proof.

6.4.2.3. Catalytic Decomposition

The first works on catalytic decomposition of gaseous ozone [161] show that the ozone decomposition is accelerated by the presence of Pt, Pd, Ru, Cu, W, etc. The catalytic action of metals in ozone decomposition reactions has been studied by Kaschtanov et al. [142] who outlined that Ag exerts a higher catalytic activity in this reaction as compared with that of Cu, Pd, Sn. Schwab [163] investigating the catalytic properties of oxides and hydroxides of metals belonging to the I-IV group in this reaction has concluded that the catalytic properties of the oxides rise with increase of the positive valence on the metal. The authors present the following sequence of the elements and their oxides depending on their catalytic action:

$$Cu < Cu_2O < CuO;\ Ag < Ag_2O < AgO;\ Ni < Ni_2O_3;\ Fe < Fe_2O_3;\ Au < Au_2O_3;\ Pt < \text{colloidal Pt}.$$

The relationship obtained can be explained by the higher activity of the element ions than the elements themselves and that the ion charge plays the main role in the catalysis.

In a series of works Mehandjiev and Naydenov have studied the ozone decomposition over α-Fe_2O_3 in the temperature range of 23-65°C and come to the conclusion that the water vapors in small concentrations inhibit slightly the process and the nitrogen oxides formed during the electrochemical synthesis of ozone from air act as catalyst poison. The scheme of the decomposition process of ozone and the catalyst deactivation is:

$$O_3 + Z \rightarrow ZO + O_2$$
$$O_3 + ZO \rightarrow Z + 2O_2$$
$$O_3 + NO_x \rightarrow ZNO_x$$

At ozone decomposition and CO ozonolysis over CeO_2 it is found that ozone is adsorbed on the surface interacting with the surface Ce^{+3}-ions and forming $[Ce^{+4}O_3]$. The latter are transformed into $[Ce^{+4}O] + O_2$. Upon blowing of CO through the solution CO_2 is obtained via the reaction:

$$CO + [Ce^{+4}O^{\cdot}] \rightarrow CO_2 + Ce^{+3}$$

It is proved that the surface form of ozone $[Ce^{+4}O_2]$ is not active in the reaction of CO oxidation.

Generation of active species of the type [ZO] on the surface has been assumed when MnO_2 is applied as a catalyst. It is known that it is an efficient catalyst for CO and benzene oxidation.

The complete oxidation of benzene in the presence of ozone and catalysts takes place at temperatures lower by about 160-300°C as compared with the oxidation of oxygen which points ozonolysis as a promising method for cleaning of exhaust gases.

6.4.2.3.1. Metal Catalysts

Emel'yanova et al. [167, 168] have investigated the catalytic decomposition of gaseous ozone at temperatures varying from -80°C to +80°C and ozone concentration of 8.8% vol. They have used the elements of the Pt group: Pt, Pd, colloidal Ir and Ru, NiO and Ni_2O_3 as catalysts. The last two catalysts were used to make comparison with the results of Schwab. The experiments were carried out in a flow reactor at a flow rate of 5l/h. The typical kinetic curve for the decomposition of gaseous ozone in the presence of Rh-colloid and Ir-colloid at temperatures of +3°C and +20°C are shown in Fig.6.18.

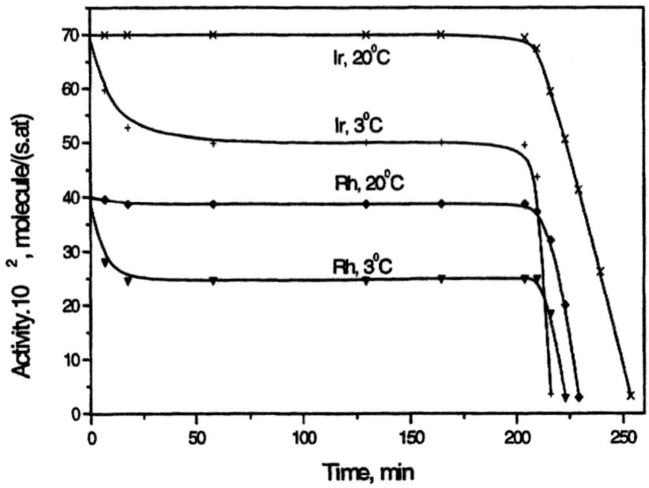

Fig.6.18. Activity variation of Ir- and Rh-colloid during the catalytic decomposition of gaseous ozone.

All curves are characterized by the presence of a region of constant activity which at 20°C is established in the very first minutes of the reaction while at 3°C it takes place in the course from 10 to 40 min. After the first up to the fourth hour of work (depending on the catalyst nature) the process is abruptly retarded and the catalytic activity drops to zero.

The activation energy of the catalytic decomposition process of gaseous ozone at temperatures varying from -80°C to +20°C has been estimated to 1000 cal/mol.

The reaction of ozone catalytic decomposition is highly exothermic. This fact affects strongly the proceeding of the process as a whole and in particular leads to intensive development of the homogeneous processes. In order to elaborate the role of the homogeneous factor some experiments in the presence of silica gel supported metal colloids have been carried out [156]. The results obtained show that the addition of silica gel does not only reduce the catalyst activity but in some cases it is enhanced by 2-3 times. The pure silica gel practically does not decompose ozone at temperatures from -80°C to +20°C. At temperature of +80°C 5% of ozone is decomposed at a flow rate of 5 l/h. The amount of the silica gel additive exceeds the amount of the initial catalysts by 5 and more fold and irrespective a retardation of the process is not observed. This result confirms the heterogeneous character of the process though the presence of regions of stationary catalytic activity could not be explained without assuming a mixed character of the process.

The catalytic activity of nickel oxides, studies in the temperature interval from +20°C to +80°C is characterized by the same relationships as with metals of the Pt group. The authors confirm the observed by Schwab high catalytic activity of nickel oxides at temperatures higher than the room one. In Table 6.3 are listed the results for the catalytic efficiency of the catalysts investigated.

Table 6.3

Catalytic activity expressed by the number of decomposed ozone molecules per catalyst unit mass per unit time at various temperatures

Catalyst	Pt - colloid	Pd - colloid	Rh - colloid	Ir - colloid	Ni_2O_3	NiO
-80°C	$3,07.10^{19}$	$7,90.10^{19}$	$1,58.10^{19}$	$0,58.10^{19}$	not active	$8,00.10^{19}$
20°C	$6,64.10^{19}$	$15,7.10^{19}$	$4,29.10^{19}$	$3,16.10^{19}$	$3,80.10^{19}$	$3,14.10^{19}$

As seen the catalytic activities of nickel oxides are of the same order as those of the Pt-colloid ($3.8.10^{19}$ molecules O_3/sg are decomposed on NiO_2, and $6.64.10^{19}$ molecules O_3/sg on Pt-colloid). Upon low test temperatures the metals belonging to the Pt group change slightly their activity with a decrease in the temperature while the nickel oxides lose it completely. At -80°C N_2O_3 is absolutely inert and even the addition of greater catalyst amounts could not cause ozone decomposition. The activity of NiO drops slowly with temperature drop and at - 80°C about 5 % of ozone is decomposed. The authors assume that the reducing of the oxide catalysts activity with the decrease in the temperature is related to the fact that at low temperatures the active centers do not possess enough energy to accomplish the reaction. The conclusion that can be drawn form these investigations is that the metal catalysts preserve their high catalytic activity in a wide temperature range while the oxide catalysts are effective only at sufficiently high temperatures (higher than 20°C).

Sudak and Vol'fson synthesized a silver-manganes catalyst for ozone decomposition which can be applied for removing ozone from various gas mixtures at low temperatures (the decomposition of ozone begins at temperatures much lower than room temperature). The industrial reactors working with this catalyst operate for a long time at maintaining the catalyst activity.

The search for other highly efficient metal catalysts has resulted in development of special technologies and supports which could ensure the proceeding of the ozone decomposition reaction in the kinetic regime and the obtaining of a maximum numbers of active catalytic centers. Here should be outlined some investigations carried out by Japan researchers.

For instance the authors of Ref. 170 reported the preparation of a catalyst containing 2 % wt Pt on a fibers-base support. It is prepared from 5% SiO_2 and 95% γ-Al_2O_3 of 3 μk thickness and specific area of 120 m^2/g impregnated by chloroplatinic acid at 80°C and subsequently dried at 120°C and 400°C in air atmosphere. At a working temperature of 20°C, flow rate of 20 $000h^{-1}$ and ozone concentration in the purified gas of 10 ppm, these catalysts show 95 % activity while in the control test with granule-supported catalyst the activity is only 54 %.

According to Ref. 171 activated Al_2O_3 is impregnated with Pd, Rh, Ce nitrates, dried and heated up to 600°C, then is dissolved in diluted HNO3 and a ceramic material of " honeycomb" is treated. After drying at 130°C a catalyst containing 73% wt Al_2O_3, 7% wt CeO, 0.9% wt Pd and 0.1% wt Rh is obtained The test reaction of ozone decomposition has the following parameters: O_3 - 200 ppm; volume rate - 57000 h^{-1}, operating temperatures from 20 to 100°C. The catalytic activity of this catalyst is by 20% superior to that without CeO.

Another catalyst containing TiO_2, SiO_2 and Pt [172] is proposed and it decomposes 94, 97 and 99% of air ozone at temperatures 20, 50 and 100°C, respectively.

Terui and co. [174] have deposited metals (Fe, Mn, Co, Ni, Zn, Ag) or their oxides from 0 to 60% wt or Pt, Pd, Rh from 0 to 10% wt or TiO_2 - SiO_2; TiO_2 - ZrO_2; TiO_2 - SiO_2 - ZrO_2 on polyurethane colloid with 400 m^2/g specific surface. The catalysts are placed in pipe reactors and their activity with respect to 0.2 ppm O_3 is 99%.

The same authors have modified MnO_2 by one or more elements such as Fe, Co, Ni, Cu, Ag, Au, Pt Rh. This catalyst is applied for decomposition of ozone in waste gases.

Microporous polytetrafluroethylene film with 80% porosity and 0.2 μm average pores diameter is first treated with alcohol solution of circonium butylate, then with water, further dried and activated at 260°C. The catalyst thus prepared exerts high catalytic activity for ozone decomposition [175].

6.4.2.3.2. Oxide Catalysts

Because of the high-expensive metals of the platinum group much more research is being focused on the development,study and application of oxides and deposited oxide systems as catalysts for decomposition of ozone.

In these respect the oxides of Al, Si, Ti, Zr, Mn, Fe, Co, Cu, Ag are the most wide by applied and studied as catalysts of ozone decomposition.

Martinov et al. [176] have investigated the influence of oxide type catalysts (CuO, CoO, NiO) on the processes of ozone decomposition. The catalysts were prepared from aqueous solutions of the corresponding nitrates which are deposited on the walls of cylindrical reactor (l=25 cm, D=1.6 cm). The reactor is heated at 370°C for 5 h. In order to get reproducible results prior the measurements the catalysts are subject to ozone treatment (concentration - 0.5 % vol., volume rate 100 l/h for 6h).

Solving the diffusion-kinetic equation the values of the ozone decomposition coefficients, as well as the diffusion and kinetic constants of the decomposition reaction were determined.

The results form these calculations for the separate samples are given in Table 6.4 where: W - the volume rate, γ - coefficient of ozone decomposition, k_{exp} and k_{kin} are the experimental and theoretical rate constants, respectively.

Table 6.4

Kinetic parameters of ozone decomposition reaction on oxides

Catalyst	W, l/h	k_{exp}, s^{-1}	$\gamma.10^5$	k_{kin}, s^{-1}
CuO	20	1,0	4,8	2,8
CoO	100	0,9	4,1	2,4
CoO	200	1,3	6,3	3,6
NiO	20	0,9	4,4	2,6
NiO	100	1,0	4,8	2,8
NiO	200	1,7	8,1	4,4

It is seen that the values standing for the catalyst activity at W = 20 and 100 l/h are almost the same while at W - 200 l/h the value of the coefficient of ozone decomposition is by about 1.5 times greater that that in the presence of NiO and CuO. This difference could be explained by the proceeding of a catalytic reaction in the miscible zone. The values of the rate constants for ozone decomposition in the kinetic regime are nearly 3 fold higher.

Vol'fsson and co. studying the process of ozone decomposition onto promoted and unpromoted by Ag β-MnO_2 have established that the decomposition rates for the two samples in the absence of moisture are quite similar. However, the presence of moisture in the reaction mixture leads to a sharp decrease in the catalytic activity of the specimen containing MnO_2 which is probably related to the blocking of the active catalytic centers by the water molecules. The promoted samples preserve their catalytic activity for a longer time. This observation is explained by the fact that the addition of Ag increases the number of the active centers capable of chemisorption of the oxygen in its atomic form and correspondingly increases the electrondonor properties of the catalyst surface thus preventing the adsorption of water molecules on the surface. According to the proposed mechanism the rate limiting stage in the heterogeneous-catalytic decomposition of ozone is the cleavage of the bond between the chemisorbed oxygen and the catalyst surface [177].

Tarunin et al. [178] investigate the catalytic decomposition of ozone in the presence of γ-Al_2O_3, hopcalite (MnO_2, CuO, bentonite), silver-manganese oxide catalyst (2.2% AgO, 97.8% MnO_2), alumo-palladium catalyst (2% Pd/98% γ-Al_2O_3) and alumo-platinum catalyst (0.64% Pt/99.36% γ-Al_2O_3). The catalytic activity was evaluated at temperature of 60°C and 2.7.10-5mol/l initial ozone concentration. The rate constants and the activation energy were also determined and are listed in Table 6.5.

Table 6.5

Activation energy and rate constants of the decomposition reaction of ozone over some heterogeneous catalysts.

Catalyst	E, kJ/mol	Activity, molecules O_3/(g.s)	k, s^{-1}
γ-Al_2O_3	45,2	$5,75.10^{16}$	13,3
Hopcalite	36,8	$1,2.10^{18}$	67,0
AgO/MnO_2	25,1	$2,47.10^{18}$	140,0
Pd/γ-Al_2O_3	42,7	$7,2.10^{17}$	21,8
Pt/γ-Al_2O_3	-	$1,95.10^{17}$	-

It has been found that the values of the activation energy for the different catalysts do not depend on their dispersity but it affects markedly their catalytic activity. Among the catalysts studied the sample containing 2.2% Ag + 97.8% MnO_2 is the most efficient, the activities of the alumo-palladium and alumo-platinium are similar and the addition of small amounts of Pd (2%) and Pt(0.64%) to γ-Al_2O_3 increases its activity substantially, The authors provide also data that the specific surface of the specimen does not depend on the dispersity thus concluding that the heterogeneous reaction of the catalytic decomposition of ozone takes place mainly on the "outer" surface of the catalysts granules, i.e. in the outer kinetic region.

Table 6.6

The effect of the catalyst surface and mass on the kinetic parameters of the ozone decomposition reaction over hopcalite catalysts.

Indicator	Series 1, Sample 1	Series 1, Sample 2	Series 1, Sample 3	Series 2, -
S, cm^2	0,2	1,1	2,9	0,15
m, g	0,0083	0,1218	0,5648	-
$k.10^2$, s^{-1}	0,97	4,61	8,98	0,2
$k.10^2/S$	4,1	4,0	3,1	1,3
$k.10^2/m$	1,17	0,34	0,16	-
$\gamma.10^4$	7,0	7,0	5,3	2,3

Note: γ - a coefficient of ozone decomposition which is determined by the ratio between the number of decomposed ozone molecules at contact with the surface per the total number of collisions with the surface {$\gamma = 4Vk/Sc$: V - reactor volume; k - experimentally measured first order rate constant; c - thermal rate of ozone molecules motion = 360 m/s; S - specific catalysts surface}.

The first series of samples are prepared in the form of pills or granules, air preheated at 200°C for 3h. The second series are finely powder hopcalite which is deposited as thin film on glass pellets (globules) through a silicon acid gel and dried in air at temperature of 200°C.

Other authors [180] compare the activity of an industrial hopcalite catalysts with that of hopcalite fibers prepared in the following way: industrial hopcalite catalyst with composition of MnO_2 - 60%, CuO - 25%, bentonite clay - 15%, specific surface of 60 m^2/g and pores size of 10 nm is ground, pressed and treated for a couple of hours at 500°C and at 900°C. Then it is melted by a laser beam and fibers of 20-30 µmk diameter and surface of 24 m^2/g are stretched.

The catalytic activity of the specimen is evaluated by the values of the rate constant related to a unit mass ($s^{-1}.g^{-1}$) and the coefficient of ozone decomposition γ. The data are presented in Table 6.7.

Table 6.7

Activity of the samples during ozone decomposition per unit mass

Catalyst	k, 1/(s.g)	(
Hopcalite	8,4	$1,8.10^{-2}$
Hopcalite fibers	0,03	$3,1.10^{-8}$
Fibers treated with acid	0,013	$1,6.10^{-7}$

Taking into account the heterogeneity of the surface and ignoring side reactions the ozone decomposition can be presented as:

$$O_3^V + S \rightarrow O_3' \rightarrow O_3'' \rightarrow \text{Degradation} \qquad (k_1, k_2, k_3)$$

where: O_3^V- the ozone concentration in the volume; S - the outer surface on which the ozone molecules are adsorbed, O_3^- and O_3^+ - the ozone amount on the inner and outer surface.

It is assumed that ozone is adsorbed directly on the outer surface and after that diffusing in the pores bulk it decomposes. For the cases when the amount of ozone molecules on the outer and inner surface changes very slowly the following expression holds:

$$- d[O_3]/dt = k_3[O_3]'' = k_1[S].[O_3]^V \qquad \text{(Eq.6.7)}$$

i.e. the rate of ozone decomposition is proportional to the outer surface.

Popovich [181] continues the investigations on the catalytic decomposition of ozone comparing the catalytic activity of various supported and nonsupported catalysts and inert materials. Alumosilicate in the form of balls with dimensions 2-3 mm, composition - 88% SiO_2, 12% Al_2O_3, specific surface 300 m^2 and pore size of 4-5 nm. has been applied as a support. The heterogeneous catalysts (5Ni - 2.5 Ag; 5Ni - 2.5Ag - 2.5 Co and only Co) are deposited by using aqueous solutions of nitric acid salts of the corresponding metals. The samples are heated in air atmosphere at 200°C for 3h [192]. Hopcalite has the following composition: MnO - 60%, CuO - 25% and bentonite clay - 15%. The relationships between the changes in the catalyst activity and the outlet ozone concentrations with time are demonstrated in Figs 6.19 and 6.20.

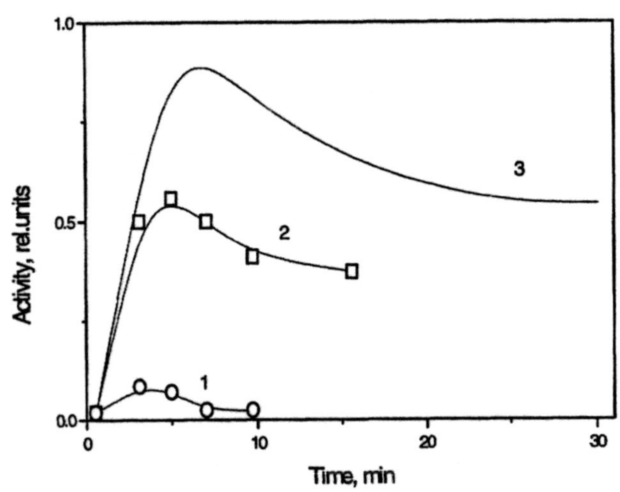

Fig.6.19 Dependence of catalysts activity on time: 1 - alumosilicate; 2 - 5Ni - 2.5 Ag; 3 - hopcalite.

Fig.6.19 presents a typical dependence of the ozone decomposition coefficient on time. It is seen that the catalyst activity drops after reaching the maximum rate of ozone decomposition and attains a nearly stationary value.

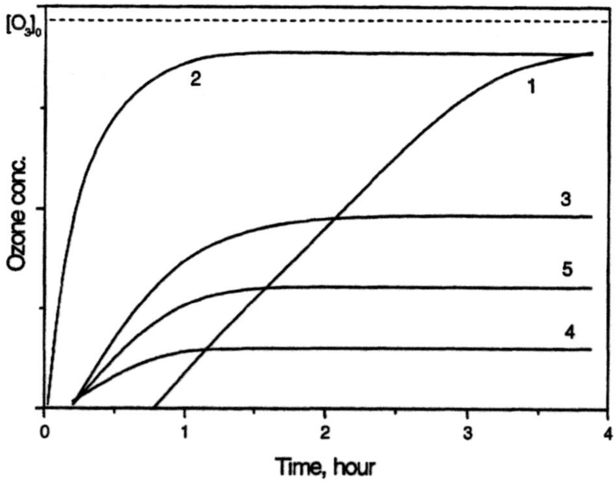

Fig. 6.20. Dependence of outlet ozone concentration on time.
1 - glass reactor, S = 26.4 cm^2, inlet ozone concentration = $1.5.10^{12}$ molecule/cm^3, gas flow rate (ω) = 12 l/h; 2, 3, 4 -alumosilicate 1.67 g, catalyst Ni-Ag-Co 0.76 g, hopcalite 0.68 g, respectively, inlet ozone concentration = $1.25.10^{12}$ molecule/cm^3, ω= 88 l/h; 5 - hopcalite plate 3.8 cm^2, inlet concentration $3.75.10^{14}$ mol/cm^3, ω = 4.2 l/h. Scale is defined by the initial ozone concentration - dash line.

As seen from Fig.6.20 the reaching of stationary regime depends on one hand on the flow rate, ozone concentration and the catalyst mass and on the other, to a much greater extent, on the change of the surface state.

The data of the studies carried out by different authors on the catalytic and inert systems are outlined in Table 6.8.

Table 6.8

Coefficients of ozone decomposition on the surface of various catalytic and inert systems [180-185]

System	γ	System	γ
Hopcalite	$5,4.10^{-4}$	Sulfur	$3,0.\ 10^{-8}$
5Ni-2,5Ag/Alumosilicate	$1,4.10^{-4}$	Alumosilicate	$1,1.\ 10^{-8}$
CoO	$1.\ 10^{-4}$	Glass (non passivated)	$< 10^{-8}$
Co/Alumosilicate	$1.\ 10^{-5}$	Quartz (non passivated)	$1.\ 10^{-8}$
5Ni-2,5Ag-2,5Co/ Alumosilicate	$1,2.\ 10^{-3}$	Quartz (passivated)	10^{-11}
Sea salt	$1,2.\ 10^{-9}$	Ammonium sulfide	$6.\ 10^{-8}$
Vulcanic ashes, Allaid (1972)	$3,1.\ 10^{-8}$	Ammonium persulfate	$3,4.\ 10^{-8}$

The values of the coefficients for ozone deactivation are sufficiently high for the catalysts but one should not ignore those of the inert supports. Considering the high adsorbing capacity of the inert materials it may be expected that falling in the atmosphere in the presence of metal particles on their surface they can be converted in effective decomposers of atmospheric ozone. In this sense it is important to know that these materials which at high ozone concentrations lose rapidly their activity may be quite efficient, in decomposing ozone in the atmosphere. The high values of the coefficient of ozone decomposition obtained for inert materials should be considered always when the aerosols influence on the ozone layer is evaluated. It can not be excluded that aerosols similar to the studied catalysts, i.e. metals or metal oxides supported on relatively inert metal support, may be generated in the atmosphere from the evaporation of metal meteorites in the atmosphere.

Aktyasheva and Eme;'yanova compare the activity of Al_2O_3 (1 - 8.10^{16} mol O_3/sg) with that reported by other authors for the activities of Pt- and Pd-colloid and Ni_2O_3 and reveal that Al_2O_3 is actually a relatively weak catalyst for ozone decomposition.

The decomposition of gaseous ozone on ferric-oxide system - Fe_2O_3/SiO_2 has been the object of investigations in Ref.185. The samples were prepared by impregnation of SiO_2 with water solutions of $Fe(NO_3)_2$ or of $FeCl_3$, drying and air heating at 500°C for 5h. The Fe_2O_3-based catalysts are superior to those prepared from $FeCl_3$. It has been established that the kinetics of the ozone decomposition reaction is described by a first order equation and the activation energy of the process has the magnitude of 12.6 kcal/mol. The high value of the latter evidences occurrence of the reaction in the kinetic region.

A series of papers deals with elaboration of the mechanism of ozone decomposition on NiO and in a model reaction of cumene oxidation in the presence of NiO. The NiO samples were prepared via heating of nickel hydroxide in air at temperature of 500°C.

The dependence of the ozone decomposition on the catalyst amount and the inlet and outlet ozone concentration is shown in Fig.6.21.

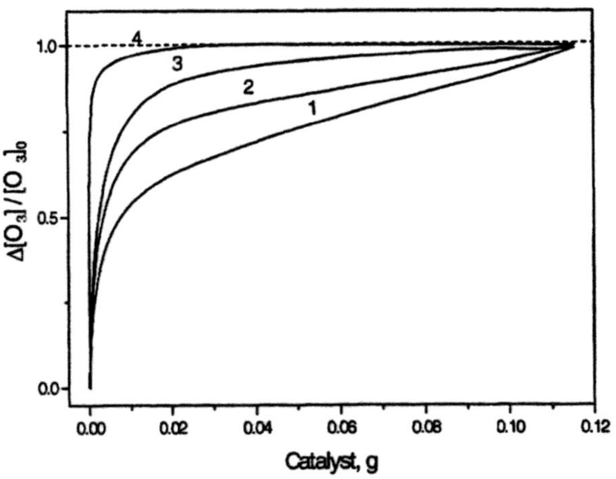

Fig.6.21. Relative amount of decomposed ozone in dependence on the catalyst quantity at inlet and outlet ozone concentrations of: 1 - 3.45.10^{-4} M; 2 - 4.40.10^{-4} M; 3 - 2.60.10^{-4} M; 4 - 0.80.10^{-4} M.

The data in Fig.6.21 reveal that the lower the initial ozone concentration, the higher is the catalyst activity. On the bases of the data on ozone geometry [188] it has been proposed that ozone molecules are decomposed on the surface in the following manner:

One ozone molecule occupies a surface of 3.14.10^{-20} m^2. According to the assumed pattern of ozone molecules disposition on the surface they should interact in-between but the rate of ozone decomposition depends on the concentration (reaction of first order), and thus the limiting step of the decomposition reaction will be either the ozone adsorption on the surface or the electron transfer from the catalyst to ozone. The surface Ni^{2+} ions will be oxidized to Ni^{3+} ions generating ion radical $O_3^{\cdot}$. This fact is confirmed by the ESR data and the UV spectra of NiO (the absorption band at 320 nm disappears).

Upon oxidation of cumene a linear dependence of the initial rates of dimethylphenylcarbinol (DMPC), cumenehydroxide (CHP) formation on the amount of NiO is registered. This fact confirms the occurrence of a reaction between NiO and O_3 resulting in $O_3^{\cdot}$ formation as shown in Scheme 6.15.

Scheme 6.15

$$Ni^{+2} + O_3 \rightarrow [\, Ni^{+3} + {}^{\cdot}O_{3\,ads} \,] \rightarrow {}^{\cdot}O_3 + Ni^{+3}$$

$$\downarrow RH$$

$${}^{\cdot}O_3 + RH \rightarrow R^{\cdot} + {}^{\cdot}OH + O_2$$

$$Ni^{+3} + RH \rightarrow R^{\cdot} + Ni^{+2} + H^{+}$$

The catalytic effect of NiO is relatively weak since the total rate of products formation is only from 5 to 17% greater than the rate of ozone consumption, i.e. the concentration of ozone anion radicals is negligible small.

It has been found that only 5% of the reaction takes place through a mechanism of electron transfer, and the rest is via molecular pathway:

$$NiO + 2O_3 \rightarrow [2O_3...NiO]_{ads} \rightarrow NiO + 3O_2 + Q$$

Studying the reaction of ozone decomposition on γ-Al_2O_3 supported Ni, Co, Cu and Ag it has been established that the most efficient catalyst is that containing 6% Ni [187]. In Table 6.9 are outlined the results of the catalytic investigations on the catalysts mentioned above.

Table 6.9

Ozone decomposition level (DL) on the studied catalysts

Catalyst	6% Ag, 400°C	6% Ni, 400°C	6% Cu, 400°C	6% Co, 400°C	6% Ni, 300°C	3% Ni, 400°C
DL	500	530	300	200	200	400

The test with blank sample containing γ-Al_2O_3 shows that ozone decomposes also on the support itself. The reaction mechanism has been elaborated applying four series of catalysts - two with 3% wt Ni prepared at 300 and 400°C, respectively, and two with 6% wt Ni prepared at the same temperatures. In the presence of the first two catalysts 50% of the γ-Al_2O_3 surface has been occupied by nickel oxides, and the ratio Ni_2O_3:NiO is respectively 4:1 and 1.24:1 at 300 and 400°C. For the second series the total surface of γ-Al_2O_3 is covered by nickel oxides and the ratio Ni_2O_3:NiO is 1.24:1 at 400°C and as seen from the data in Table 6.9 it is the most active one. In the calcinated sample the ratio Ni_2O_3 : NiO is 1:9 but its catalytic activity is lower that of the noncalcinated, most probably, because of the formation of inactive mixed compounds between Al and Ni. Moreover the pure (unsupported) NiO exerts also lower activity as compared with the latter which can be related to its smaller (about 6 fold) surface than that of the supported catalyst. In our previous study on cumene oxidation by ozone the formation of Ni^{3+} under ozone action and reduction of Ni^{3+} to Ni^{2+} under hydrocarbon action has been observed. Thus the catalytic cycle is closed. The data from the X-ray analysis shows the presence of areas of Ni^{2+} transition to Ni^{3+} upon decomposition of ozone on the catalysts above. However this transition can occur only if 530 ozone molecules are available which clearly evidences the occurrence of catalytic cycles in ozone decomposition. The suggested mechanism is given in Scheme 6.16.

Scheme 6.16

$$NiO + O_3 \rightarrow ONi^{3+}(O_3)$$
$$ONi^{3+}(O_3) + NiO \rightarrow O{=}Ni{-}O{-}Ni{=}O + O_2$$

$$2NiO + O_3 \rightarrow O{=}Ni{-}O{-}Ni{=}O + O_2 \text{ fast}$$
$$O{=}Ni{-}O{-}Ni{=}O + O_3 \rightarrow O{=}Ni(O_3){-}O{-}Ni{=}O$$
$$O{=}Ni(O_3){-}O{-}Ni{=}O \rightarrow 2NiO + 2O_2 \text{ slow}$$

The survey on the available patent information reveals that after the period of application of Pt metals as catalysts for ozone decomposition the researchers focus all their attention to using oxide catalysts of transition metals. More that 90% of the patent literature for the period after 1986 up to now is devoted to the development of oxide catalysts for ozone decomposition. Most of the patents are by Japan inventors which is closely related to the fact that Japan is the basic manufacturer of ozonegenerating units, their application in cleaning of drinking, waste and used water and gases, microelectronics, Xerox machines and laser printers production. In the proposed catalytic systems some high efficient groups could be outlined. These are the catalysts synthesized on the base of - Fe, Co, Ni, Cu, Ag, Mn [189-194] or their compositions supported or simultaneously acting with Al_2O_3, SiO_2, TiO_2, ZrO, Cr_2O_3, MgO, MoO_2, V_2O_3 [194-197]. In most of the cases the preparation of the support is of particular interest. It is produced in the form of granules, sticks, cylinders, fibers, textiles, etc., but mostly as infinite hexagons in the form of honeycomb [192-202]. Besides ceramics, special attention is paid to the possibilities for application of organic materials for support preparation on the base of specially treated perfluoroethylene for reaching a maximum surface or spongy polyurethanes, etc.

Fujiata and co., [203] have prepared a relatively cheap catalysts on the base of Fe_2O_3-hydrate. Starting from $FeSO_4$ and $NaCO_3$ a finely dispersed Fe_2O_3 - hydrate of a particle size of 0.25 μm is obtained. This catalyst decomposes 10 ppm ozone at 120°C almost completely (99%). The treatment of ZrO by surface OH with $CoSO_4$ gives more effective catalyst [204] and those obtained from the NiO deposition on ceramics support - honeycomb promoted by alkali metals [205, 206] exhibit very good catalytic properties at lower temperatures. Some examples of improved conventional catalysts on the base of Ag or AgO and Cu or CuO are reported in Refs. 207 and 208.

It should be noted that the catalysts on the base of MnO_2 deposited on various supports - ceramics or organic [209-212] or of its mixtures with TiO_2 [213, 214], Al_2O_3 and SiO_2 [215, 216], NiO [217, 218] or simply specially treated MnO_2 are actually the most widely described and discussed in literature.

Another large group of catalysts are those in which MnO_2 is combined with one or more active components such as Cu [220-222], Fe [223] or Ag [224-226].

The mechanism of ozone decomposition on these catalysts is accomplished via two routes:

1. Molecular - when the ozone surface compound on the metal oxide is decomposed without formation of active species and with heat release distributed in the support' s crystal lattice. This mechanism is dominant because of the low energy of the O-O_2 bond and the high value of the heat of ozone thermal decomposition equal to 35 kcal.

2. Mechanism with electron transfer from the low valence state of the oxide metal to ozone.

6.4.2.3.3. Carbon Catalysts

Ozone interacts with various carbon materials such as activated carbon, soot, graphite, carbon fibers, textiles, etc. The character of these interactions depends both on the nature of the carbon surface and the temperature [227]. Two types of interactions have been investigated: complete oxidation with gaseous carbon oxides formation and partial oxidation generating oxygen containing functional groups on the surface [227, 228]. With temperature the ratio between the two processes is moved to the direction of the complete oxidation. The latter is accompanied by an oxidative destruction of the surface with formation of gaseous carbon oxides. In addition a process of catalytic decomposition of ozone tales place on the carbon surface.

In a series of papers [229-231] the physico-chemical properties and the kinetics of CO and CO_2 release after ozone treatment of activated carbon have been studied.

The authors of [232] have followed the kinetics of low temperature decomposition of ozone on various trades activated carbon samples possessing different structural-adsorptive characteristics. The data in Table 6.10 demonstrate the influence of the nature of activated carbon (ACb) on the decomposition level of ozone at 25°C.

Table 6.10

The effect of the activated carbon nature on the decomposition rate (W) of ozone at $[O_3]_o = 2.10^{-7}$ M at 25°C.

Activated carbon	$[O_3]_f.10^8$, M	η, %	$W.10^{-15}$, molecules O_3/(g.s)
SKN-K	3,0	86	0,8
ART	0,8	96	1,4
KAD - ioden	1,2	94	1,3
AG - 3	0,3	99	1,3

where: $[O_3]_f$ is final ozone concentration; η - the degree of ozone decomposition.

It is seen that all types of CB result in high degree of ozone decomposition gradually decreasing to a constant value. For the CB types with greater mezopores size the rate and the level of ozone decomposition rise.

The interaction of ozone with activated carbon texture on the base of polyacrylonitrile with highly developed surface (1900 m^2/g) has been studied at temperatures of 80, 20 and -80°C. The level of the ozone decomposition at these temperatures was found to be 5.0, 3.0 and (1.0-2.0).10^{15} molec.O_3/(g.s), respectively. It has been shown that the carbon texture, similar to other carbon materials with highly developed surface, possess high adsorptive ability towards ozone but its desorption is almost impossible. It has been established that the role of the catalytic decomposition is dominant upon the proceeding of the three processes. In the temperature interval from -80 to + 80°C 80-100% of the inlet ozone is decomposed.

The catalytic properties of activated carbon texture with respect to ozone decomposition are similar to those of Al_2O_3. Thus at 80°C, 1.10^{18} mol.O_3/(g.s) is decomposed on Al_2O_3 and on carbon texture it is 0.5.10^{18} mol.O_3/(g.s). However it is substantially less active that Pt, Pd and NiO containing catalysts which leads to 10^{18}-10^{20} mol.O_3/(g.s) decomposition [124, 125]. These comparisons, however do not reflect the real catalytic activity of the separate catalytic systems since the specific surfaces of the samples are not considered. For instance, the specific surface of aluminum oxide reaches up to 200 m^2/g, that of metals to 50 m^2/g while that of the carbon texture reaches 1900 m^2/g.

Studying the process of ozone catalytic decomposition on fibrous carbon materials with various structural characteristics it has been found that the rate and decomposition conversion do not depend on the linear rate of ozone supply, but increase with rise of the initials ozone concentration and decrease the ratio between the volume of the materials micro- and mezopores [234].

The results from the ozone decomposition on laminar graphite compounds (LGC) by various metal chlorides are presented in Table 6.11 [235]. It is seen that the investigated materials in dependence on their activity show the following order:

LGC-$CoCl_2$ = LGC-$PdCl_2$ = LGC-$MoCl_2$ = LGC-$FeCl_3$ > LGC-$ZnCl_2$ > LGC-$CuCl_2$ = LGC-$FeCl_2$ - graphite

Upon ozone decomposition on metal chlorides or their mixtures with graphite in all cases the activity of LGC with Fe, Co, Ni, Zn, Cu, Pd, Mo chlorides is much higher that those of the salts or their mixtures with graphite.

Recently some novel types of catalysts containing carbon black, carbon fibers [236], carbon-base texture or systems including activated carbon (5-90%) and mixtures of different oxides - Ti-Si mixture, Ti-Si, Ti-Zr-Si [237] or carbon black as rods with hardness B and surface of 150-500 m^2/g are proposed:

Table 6.11

Comparative characteristics of the activity of LGC-MCL_2 and MCl_x in mixture with graphite during ozone catalytic decomposition at 80°C

M	x	t_{st}	η_{O3}, %	r_H	$A_{st}.10^{-12}$	t_{st}	η_{O3}, %	$A_{st}.10^{-12}$
	2	LGC					mix.	
Pd			98	4.3	5.0		desact.	
	2	60						
Co			99	3.1	5.0	20	99	5.5
	2	20					non act.	
Cu			14	2.3	0.7			
	2	15						
Fe	3	45	58	6.3	2.9	60	22	1.1
Mo	5	50	58	3.3	2.9	30	14	0.7
Zn	2	30	18	3.8	0.9	30	14	0.7
Ni	2	20	23	4.0	1.2	30	14	0.7
Fe	2	15	14	2.3	0.7		non act.	

where: t_{st} - the time for reaching stationary regime [min]; η - ozone decomposition degree in the stationary region; r_i - the initial reaction rate [mol.O_3/min]; A_{st} - catalytic activity [mol.O_3/(g.s)].

The preparation of catalysts of "honeycomb" type based on activated carbon, MnO_2, TiO_2, Ag_2O, Fe_2O_3, CuO is described in [239, 240]. These catalysts have shown high catalytic activity up to 72-90%. The granulated catalysts containing 30-70% wt activated carbon, 15-50% wt SiO_2, 1-10% wt Al_2O_3, 2-30% wt MgO filled and with 'honeycomb" structure decomposes 20 ppm O_3 at a rate of 1m/s with 76% efficiency [241].

The catalytic decomposition of liquid ozone appears to be very specific. In 1954 it was Shumacher who made the first attempt to carry out the decomposition of liquid ozone [153]. The ozone-oxygen mixture containing from 7 to 30% vol O_3 is blown over various metal plates (Al, Fe, Cu, Zn). The results have quantitative character and reveal a weak decomposition of ozone over Cu while at the contact with Zn the liquid ozone explodes.

Emel'yanova and Strahov [156] have studied systematically ozone decomposition reactions at -183 and - 195.6°C applying Pt-, Pd-, Ir-, Rh- and Ag-as colloid gels and Ni, Fe, Cu as metals and their oxides. The colloidal metal catalysts are prepared by the reduction of the corresponding metal salts [242, 243] and the oxide catalysts by the thermal decomposition of the corresponding metal salts and hydroperoxides. The experiments were conducted with 100% or diluted by nitrogen or oxygen, ozone [167]. The results from the studies on the decomposition of 100% liquid ozone at -196.5°C are listed in Table 6.12.

Table 6.12

Catalysts activity during decomposition of 100% liquid ozone

Catalyst	Pt - colloid, -195,8^{0}C	Pt - colloid, -183^{0}C	Pd - colloid	Ag - colloid
Activity x 10^{-16}, molecules O_3/s^{-1}	3,12	7,46	4,53	5,00

Note: the catalysts under study, Ir-colloid, Rh-colloid, Fe, FeO, Cu, CiO and NiO are inactive in the liquid ozone decomposition under the experimental conditions.

The data in Table 6.12 shows that Pt and Pd-colloids are efficient catalysts for the decomposition of liquid ozone. The colloid silver also exerts some catalytic activity, particularly, during the first hours of ozone exposure. The authors explain this phenomenon by the formation of oxide layer on the metal surface at its contact with ozone. This layer prevents the further access of ozone.

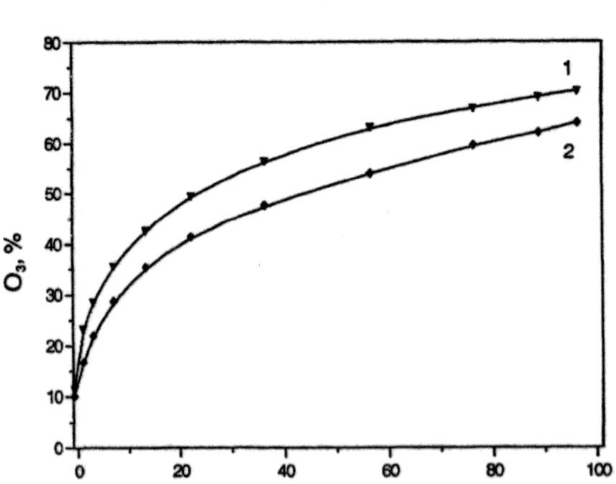

Fig.6.22. Kinetic curves of liquid 100% ozone decomposition at 195.6^{0}C over the colloid Pd (1) and Pt (2).

The kinetic curves of ozone decomposition in Fig.6.22 show that the process of ozone decomposition proceeds with a high rate only during the first hours (up to 10 h) after which the rate decreases. The activation energy $E_a = 960$ cal/mol, the reaction order n = 3 and $k_{Pt} = 0.576.10^{-48}$ mol.s^2/h and $k_{Pd} = 0.78.10^{-48}$ mol.s^2/h have been estimated. The authors related the high order of the reaction to the deactivation of the active metal surface in the course of its oxidation. The massspectral and X-ray data of the treated samples confirms also this assumption [168]. The accumulation of both molecular and atomic oxygen on the catalyst surface has been also established. While the molecular oxygen is mainly physically adsorbed, the atomic one may be absorbed on the surface as unstable oxygencontaining compounds. It has been found that the more developed is the surface, the greater is the catalyst activity [167]. The good agreement between the calculated values for the reaction order and the rate constants of diluted (from 0.5 to 7% vol) and concentrated liquid ozone suggest that the ozone decomposition does not depend on its concentration and is a purely surface process [243].

The conclusions put forward about the mechanism of catalytic action during liquid ozone decomposition relate the decomposition with the formation of active intermediate compound of MeO_3 type. Unlike the physicochemical decomposition of ozone which is characterized by the formation of short energetic chains the energy necessary for the catalytic decomposition is provided by the adsorption activation of the molecule. Under the action of this energy the intermediate compound MeO_3 can decompose to MeO_2 and MeO or react bimolecularly with the neighbouring molecule MeO_3 giving MeO_2 and O_2. The MeO formation deactivates irreversibly the catalyst [244].

6.4.2.3.4. Other Chemical Reagents

The high energy molecule of ozone and its high affinity for electron characterize it as the second powerful oxidizer after fluor. Most of the elements of the Mendeleev Table can be oxidized to their highest oxidation state under the action of ozone. Ozone can participate in these reactions either via one of its atoms or with the all three atoms. The oxidation processes with ozone participation take place intensively independently on the aggregate state of the reacting compounds, for example:

$$3SO_2 + O_3 \rightarrow 3SO_3$$
$$NO + O_3 \rightarrow NO_2 + O_2$$
$$2NO_2 + O_3 \rightarrow N_2O_5.+ O_2$$

lead is oxidized to PbO_2, silver to AgO and Ag_2O_3; in aqueous solutions Mn^{2+} is oxidized to MnO^{4-}; Co^{+2} to Co^{+3} ; As to As^{+5}, etc.

The interaction of ozone with the gas phase yields such compounds as:

$$I(IO_3)^{3-}; Br_2O_5; ClO_2; Cl_2O_7, \text{etc.}$$

One should outline the products of ozone reaction with hydroperoxides of alkali metals - ozonides of alkali metals. Thus at low temperatures the following reactions occur:

$$3MeOH + 2O_3 \rightarrow 2MeO_3 + MeOH.H_2O + 1/2O_2,$$

where: M - may be Na, K, Rb, Cs.

Ozonides decompose spontaneously with the formation of higher metal oxides and evolve oxygen according to the reaction:

$$2MeO_3 \rightarrow 2MeO_2 + O_2$$

The reactions of ozone with organic compounds are of great practical importance. According to the modern concepts for the mechanism of these reactions]245-247] ozone reacts with unsaturated organic compounds via its addition to the cleaved double bonds giving ozonides [248-250]. The various methods of ozonides treatment such as hydrolysis, reduction, oxidation, etc. results in their decomposition yielding aldehydes, ketones or acids. The latter can further participate in reactions of polymerization, addition or conversion into esters [251].

The effect of ozone on polymers is a well-known and investigated phenomenon. Thus the polyethylene oxidation by ozone gives aldehydes and ketones. Ozone is one of the basic degradating agents of rubber products [254].

Ozone is widely applied for elaboration of the structure of complex organic compounds [245].

Ozone finds also application in medicine [246].

The catalytic decomposition of ozone in the atmosphere has a satisfactorily theoretical explanation. The basic catalytic cycles, the natural and anthropogenic factors which promote it are defined. The fact that 70% of the amount of the decomposed ozone is due to anthropogenic activity (freons production) makes the problem of the ozone layer protection in the atmosphere global from ecological point of view.

The kinetics of ozone decomposition in the gas phase takes place through a atomic-molecular mechanism.

The values of the critical temperatures for ozone explosion are closely connected with its concentration: from 0 to 20% - normal, as a result of the process exothermity; from 20 to 48% - self independently; from 48 to 54% - explosively and over 54% - detonating.

The explosion sensitivity of concentrated liquid ozone and its solutions in oxygen drops sharply with rise of oxygen concentration with respect to the thermal impulses described by the equation: $E = \exp(-C/6.9)$.

Among the metals catalysts used in the catalytic decomposition of gaseous ozone those of Pt, Ru, Rh, Pd, Cu, Ag-supported or as colloids are the most efficient.

The catalytic properties of oxide catalysts are improved with the increase of the positive valence of the metal and their sequence with respect to their catalytic activity is:

$$Cu < Cu_2O < CuO;\ Ag < Ag_2O < AgO;\ Ni < Ni_2O_3;\ Fe < Fe_2O_3;\ Au < Au_2O_3;\ Mn < MnO_2.$$

The mechanism of ozone decomposition on oxide catalysts includes two reaction pathways:

- molecular - when the ozone surface compound on the metal oxide is decomposed without formation of active species and with heat liberation being distributed in the crystal lattice of the support.
- Mechanism with electron transfer from the low valence state of the oxide metal to ozone

The catalytic decomposition of ozone on fibrous carbon materials with different structural characteristics does not depend on the ozone inlet rate but increases with rise of the initial ozone concentration and decrease of the ratio between the volume of the materials micro- and mezopores.

The mechanism of the catalytic decomposition of liquid ozone is associated with the formation of active intermediate compound of the MeO_3 type which can further decompose to MeO_2 and MeO or react bimolecularly with a neighbouring MeO_3 molecule giving MeO_2 and O_2. The formation of MeO deactivates irreversibly the catalysts. The energy necessary for the catalytic decomposition is provided by the adsorption activation of the molecule.

Ozone is widely used for producing valuable organic compounds, both independently, and in the presence of homogeneous and heterogeneous catalysts.

REFERENCES

1. Nebel, C. Ozone. Kirk Othmer Encyclopedia of Chemical Technology (third edition), 16, 650-713 (1981).
2. Horwath, M, L Vitzky, G Huttner. Ozone. Budapest Academy Kiadoi, 1985.
3. Gorbenko-Germanov, DS. Doctor of Scienses Theses. Moscow, 1974.
4. Hoigne, J, et al. Water Res., 17, 173, 185 (1983); 19, 993 (1985).
5. Tkachenko, SN, VI Demidyuk, LA Kireeva, MP Popovich, EZ Golosman. Catalytic Properties of Oxide-Alumocalcium Systems. 1. Ozone Decomposition. Zhurnal Fizicheskoi Khimii, 67: 5, 1076-1078, (1993).

6. Martynov, VI, VI Demidyuk, SN Tkachenko, MP Popovich. Kinetics of Ozone Decomposition on Oxide Catalysts. Zhurnal Fizicheskoi Khimii, 68: 11, 1972-1976 (1994).
7. Dhandapani, B, ST Oyama. Gas phase ozone decomposition catalysts. Appl Catal B-Environ, 11, 2, 129-166 (1997).
8. Dhandapani, B, ST Oyama. Kinetics and mechanism of ozone decomposition on a manganese oxide catalyst. Chemistry Letters, : 6, 413-414 (1995).
9. Gross, Z., Simkhovich, L. Ozone as primary oxidant in iron(III). porphyrin catalyzed hydroxylation of hydrocarbons. J Mol Catal A-Chem, 117, 1-3, 243-248 (1997).
10. Basiuk, VA. Organic reactions on silica surface: Synthetic applications. *Uspekhi Khimii,* 64: 11, 1073-1090 (1995).
11. Schutt, WS, ME Sigman, YZ Li. Fluorimetric investigation of reactions between ozone and excited state aromatic compounds on silica gel. *Analytica Chimica Acta,* 319: 3, 369-377 (1996).
12. Bulanin, KM, JC Lavalley, AA Tsyganenko.Infrared study of ozone adsorption on CaO. *J Phys Chem B,* 101, 15, 2917-2922 (1997).
13. Ufimkin, DP, VV Lunin, LV Sabitova, LN Burenkova, VA Voblikova, SN Tkachenko. The effect of ozone on heterogeneous-catalytic oxidation of carbon monoxide in the presence of cement-containing catalyst. *Zhurnal Fizicheskoi Khimii,* 69: 11, 1964-1967 (1995).
14. Kazakov, DV, NN Kabalnova, VV Shereshovets. Decomposition of triphenyl phosphite ozonide catalyzed by pyridine. *Reaction Kinetics and Catalysis, Letters,* 54: 2, 439-444 (1995).
15. Kutsuna, S, M Kasuda, T Ibusuki. Transformation and Decomposition of 1,1,1- Trichloroethane on Titanium Dioxide in the Dark and Under Photoillumination. *Atmospheric Environment,* 28: 9, 1627-1631 (1994).
16. Naydenov, A, R Stoyanova, D Mehandjiev. Ozone decomposition and CO oxidation on CeO_2. *Journal of Molecular Catalysis A - Chemical,* 98: 1, 9-14 (1995).
17. Rakitskaya, TL, EK Vasileva, AY Bandurko, VY Paina. Kinetics of Ozone Decomposition on Activated Carbons. *Kinetics and Catalysis,* 35: 1, 90-92 (1994).
18. Rakitskaya, TL, AY Bandurko, AA Ennan, VV Litvinskaya. Kinetics of the Low-Temperature Decomposition of Ozone by Carbon Fiber Materials. *Kinetics and Catalysis,* 35: 5, 705-707 (1994).
19. Chtourou, H, B Riedl, BV Kokta, A Adnot, S Kaliaguine. Synthetic Pulp Fiber Ozonation - An ESCA and FTIR Study. *Journal of Applied Polymer Science,* 49: 2, 361-373 (1993).
20. Peng, TS, Y Wang, AM Arif, JA Gladysz. Synthesis, Structure, and Reactivity of eta 2-1,3- Diene and Enyne Complexes of the Chiral Rhenium Lewis Acid [(eta 5-C_5H_5)Re(NO)(PPh_3).]+ - Ozonolysis Within a Metal Coordination Sphere. *Organometallics* 12: 11, 4535-4544 (1993).
21. Reguera, E, J Fernandezbertran, J Duque. On the Interactions of Ozone with Manganous Hexacyanoferrates. *Polyhedron,* 13: 3, 479-484 (1994).
22. Reguera, E, J Fernandezbertran. Study of the Interaction of Ozone with Solid Ferrocyanides. *European Journal of Solid State and Inorganic Chemistry,* 31: 11, 1021-1027 (1994).
23. Galstyan, GA, TM Galstyan, LI Mikulenko. The Kinetics and Mechanism of the Catalytic Reaction of Ozone with Methylbenzenes in Acetic Acid. *Kinetics and Catalysis,* 35: 2, 231-236 (1994).
24. King ,DE. Oxidation of gold by ultraviolet light and ozone at 25 degrees C. *Journal of Vacuum Science, & Technology A - Vacuum Surfaces and Films,,* 13: 3 Part 2, 1247-1253 (1995).
25. Gesser, HD, G Zhu, NR Hunter. Conversion of ethane to methanol and ethanol by ozone sensitized partial oxidation at near atmospheric pressure. *Catalysis Today,* 24: 3, 321-325 (1995).
26. Belskikh, LI, LE Gorlenko, GI Emelyanova, VV Lunin. Effect of thermo-oxidative treatment on the composition, structure and catalytic activity of monolith porous catalyst. *Zhurnal Fizicheskoi Khimii,,* 69: 5, 807-810 (1995).
27. Takata, S, T Tomikawa, S Ishii, N Yasuda, N Yago. Effect of the pore structure of honeycomb-type catalysts on the ozone decomposition performances. *Nippon Kagaku Kaishi,* : 7, 557-560 (1995).
28. Shimizu, T, H Okushi. The properties of a metal oxide semiconductor junction prepared using a high-purity ozone surface treatment. *Applied Physics Letters,* 67: 10, 1411-1413 (1995).
29. Gekhman, AE, NI Moiseeva, II Moiseev. Decomposition of hydrogen peroxide catalyzed by vanadium(v). compounds: The pathways for the formation of ozone. *Russian Chemical Bulletin,* 44: 4, 584-598 (1995).

30. Vilter, H. Vanadium-dependent haloperoxidases. *Metal Ions in Biological Systems,* 31, 325-362 (1995).
31. Mccoustra, MRS, AB Horn. Towards a laboratory strategy for the study of heterogeneous catalysis in stratospheric ozone depletion. *Chemical Society Reviews,* 23: 3, 195-204 (1994).
32. Moiseeva, NI, AE Gekhman, II Moiseev. Metal complex catalyzed oxidations with hydroperoxides: Inner-sphere electron transfer. *J Mol Catal A-Chem,* 117, 1-3, 39-55 (1997).
33. Starek, J, A Zukal, J Rathousky. Comparison of the Adsorption of Humic Acids from Aqueous Solutions on Active Carbon and Activated Charcoal Cloths. *Carbon,* 32: 2, 207-211 (1994).
34. Fernandezbertran, J, E Reguera. Mechanochemical reactions in alkali halide pressed disks. *Solid State Ionics,* 93, 1-2, 139-146 (1996).
35. Itoh, N, V Filmanovich, M Koshiba, K Kuribayashi. Behavior of 316 stainless steel and Al_2O_3-SiO_2 ceramics in ozone environment. *Nippon Kagaku Kaishi,* 4, 283-289 (1997).
36. Hon, YS, SW Lin, YJ Chen. A Convenient and Efficient Workup of Ozonolysis Reactions Using Triethylamine. *Synthetic Communications,* 23: 11, 1543-1553, (1993).
37. Hon, YS, L Lu. The Reactions of the Ozonides with Secondary Amines - An Efficient and Novel Way to Prepare Tertiary Amine from Mono-Substituted and 1,1-di-substituted Alkenes via Corresponding Ozonides. *Tetrahedron, Lett.,* 34: 33, 5309-5312 (1993).
38. Hon, YS, SW Lin, L Lu, YJ Chen. The mechanistic study and synthetic applications of the base treatment in the ozonolytic reactions. *Tetrahedron,* 51: 17, 5019-5034 (1995).
39. Hon, YS, Yan. Chemoselective Reactions of a Compound Containing Both Aldehyde and Ozonide Groups. *Tetrahedron Lett.,* 35: 11, 1743-1746 (1994).
40. Hon, YS, JL Yan. The Ozonlytic Cleavage of Cycloalkenes in the Presence of Methyl Pyruvate to Yield the Terminally Differentiated Compounds. *Tetrahedron, Lett.,* 34: 41, 6591-6594 (1993).
41. Hon, YS, YL Yan. Syntheses of bifunctional compounds from cycloalkenes via ozonide intermediates, *Tetrahedron,* 53, 14, 5217-5232 (1997).
42. Hon, YS, L Lu. An extremely efficient way to prepare conjugated carbonyl compounds from terminal alkenes via the reactions of ozonides, triethylamine and stable phosphorus ylides. *Tetrahedron,* 51: 29, 7937-7942 (1995).
43. Wu, HJ, JH Chern. Synthesis of 4-oxo- and 4-anti-formyl-8,10,12,13- tetraoxapentacyclo [5.5.1.0(2,6).0(3,11).0(5, 9).]tridecanes. *Tetrahedron,* 53: 52, 17653-17668 (1997).
44. Wu, HY, JH Chern, CY Wu. Ozonolysis of 2-endo-7-anti-diacylnorbornenes. A new entry for the synthesis of 2,4,6,13-tetraoxapentacyclo[5.5.1.0(3,11).0(5,9).0(8, 12).]tridecanes. *Tetrahedron,* 53: 7, 2401-2414 (1997).
45. Li, SG, P Deslongchamps. Experimental Evidence for a Synperiplanar Stereoelectronic Effect in the Ozonolysis of a Tricyclic Acetal. *Tetrahedron Letters,* 34: 48, 7759-7762 (1993).
46. Joshi, BV, CB Reese. Ozonolysis of 6-N-benzoyl-9-(5-deoxy-2,3-O- isopropylidene-beta-D-rrythro- pent-4- enofuranosyl) adenine and Related Compounds. *Tetrahedron, Lett.,* 34: 36, 5807-5810 (1993).
47. Deninno, MP, KE Mccarthy. The C-14 radiolabelled synthesis of the cholesterol absorption inhibitor CP-148,623. A novel method for the incorporation of a C-14 label in enones. *Tetrahedron,* 53: 32, 11007-11020 (1997).
48. Pothion, C, M Paris, A Heitz, L Rocheblave, F Rouch, JA Fehrentz, J Martinez. Use of ozonolysis in the synthesis of C-terminal peptide aldehydes on solid support, *Tetrahedron Lett,* 38, 44, 7749-7752 (1997).
49. Zvilichovsky, G, V Gurvich. Products of ozonolysis of L-1,4-cyclohexadienylalanine. Intramolecular cyclization and cyclization with hydroxylamine. The synthesis of two isomers of L-isoxazolylalanine, *Tetrahedron,* 53, 12, 4457-4468 (1997).
50. Claudia, C, E Mincione, R Saladino, R Nicoletti. Oxidation of Substituted 2-Thiouracils and Pyrimidine- 2-Thione with Ozone and 3,3-Dimethyl-1,2-Dioxirane. *Tetrahedron,* 50: 10, 3259-3272 (1994).
51. Tochtermann, W, S Bruhn, C Wolff. A Facile Approach to 4,9-Epoxy-Octahydroazulene Lactones. *Tetrahedron Lett.,* 35: 8, 1165-1168 (1994).
52. Barrero, AF, E Alvarezmanzaneda, J Altarejos, S Salido, JM Ramos. Synthesis of biologically active drimanes from, (-)- sclareol. *Tetrahedron Lett.,* 35: 18, 2945-2948 (1994).

53. Palomo, C, JM Aizpurua, F Cabre, JM Garcia, JM Odriozola. Synthesis of beta-Alkylserine-N-Carboxyanhydrides Through beta-Lactams via Cycloaddition Reaction of Alkoxyketenes to Chiral alpha-Alkoxyaldehyde-Derived Imines. *Tetrahedron Lett.*, 35: 17, 2721-2724 (1994).
54. Ghosh, A, MJ Miller. Synthesis of novel hydantoin analogs of 5'-nor carbocyclic nucleosides: Versatility of a chiral acylnitroso cycloaddition. *Tetrahedron Letters*, 36: 36, 6399-6402 (1995).
55. Barton, DHR, DK Taylor, C Tse. An improved synthesis of (-).-dodecahydro-3a,6,6,9a-tetramethylnaphtho[2,1-b]furan via ozonolysis of (-).- sclareol. *Tetrahedron Letters*, 35: 51, 9505-9508 (1994).
56. Saladino, R, C Crestini, F Occhionero, R Nicoletti. Ozonation of thionucleosides. A new chemical transformation of 4-thiouracil and 6-thioguanine nucleosides to cytosine and adenosine counterparts. *Tetrahedron*, 51: 12, 3607-3616 (1995).
57. Zvilichovsky, G, V Gurvich. Ozonolysis of 2-(1,4-cyclohexadienyl).glycine. *Tetrahedron*, 51: 18, 5479-5490 (1995).
58. Spurgeon, D. End agreed for ozone-destroying pesticide. *Nature*, 389, 6649, 319 (1997).
59. Saturn's satellites Rhea and Dione, *Nature*, 388, 6637, 45-47 (1997).
60. Simonich, SL, RA Hites. Importance of vegetation in removing polycyclic aromatic hydrocarbons from the atmosphere. *Nature*, 370: 6484, 49-51 (1994).
61. MKW Ko, ND Sze, MJ Prather. Better Protection of the Ozone Layer. *Nature*, 367: 6463, 505-508 (1994).
62. Brune, W. Ozone depletion - There's safety in numbers. *Nature*, 379: 6565, 486-487 (1996).
63. Schwarzbach, SE, Ozone depletion - CFC alternatives under a cloud. *Nature*, 376: 6538, 297-298 (1995).
64. Jones, AE, JD Shanklin. Continued decline of total ozone over Halley, Antarctica, since 1985. *Nature*, 376: 6539, 409-411 (1995).
65. Butler, JH. Ozone depletion - Methyl bromide under scrutiny. *Nature*, 376: 6540, 469-470 (1995).
66. Reams, GA, PC Vandeusen, AA Lucier. Ambient ozone and loblolly pines. *Nature*, 378: 6556, 449-450 (1995).
67. Mclaughlin, SB, DJ Downing. Ambient ozone and loblolly pines - Reply. *Nature*, 378: 6556, 450-451 (1995).
68. Neeb, P, O Horie, GK Moortgat. The *Nature*, of the transitory product in the gas-phase ozonolysis of ethene. *Chemical Physics Letters*, 246: 1-2, 150-156 (1995).
69. Manney, GL, L Froidevaux, JW Waters, RW Zurek, WG Read, LS Elson, JB Kumer, JL Mergenthaler, AE Roche, A Oneill, RS Harwood, I Mackenzie, R Swinbank. Chemical depletion of ozone in the Arctic: Lower stratosphere during winter 1992-93. *Nature*, 370: 6489, 429-434 (1994).
70. Feder, T. Germany eyes new ozone controls. *Nature*, 373: 6514, 466 (1995).
71. Odinokov, VN, GY Ishmuratov, OV Sokolskaya, RI Galeeva, RR Muslukhov, GA Tolstikov. Ozonolysis of Alkenes and Study of Reactions of Polyfunctional Compounds. 52. Selective Ozonolysis of, (E).-2,7-Octadienals and, (E).-3,7-Dimethyl-2,6- Octadienals in the Synthesis of Insect Enzymes. *Zhurnal Organicheskoi Khimii*, 29: 1, 24-30, (1993).
72. Odinokov, VN, OS Kukovinets, NI Sakharova, GA Tolstikov. Ozonolysis of Alkenes and Study of Reactions of Polyfunctional Compounds. 53. Synthesis of Esters of 2,3-Dihydro Derivatives of Triprenic, Diprenic and Monoprenic Acids. *Zhurnal Organicheskoi Khimii*, 29: 1, 31-38, (1993).
73. Odinokov, VN, OS Kukovinets, NI Sakharova, EY Tsyglintseva, GA Tolstikov. Ozonolysis of Alkenes and Study of Reactions of Polyfunctional Compounds. 55. Synthesis of Acyclic Precursors of Ferrulactone-1 from 1,5,9-Trimethyl-1E, 5E,9E-Cyclododecatriene. *Zhurnal Organicheskoi Khimii*, 29: 2, 346-349 (1993).
74. Odinokov, VN, OS Kukovinets, VG Kasradze, AV Dolidze, VR Akhmetova, EP Serebryakov, GA Tolstikov. Ozonolysis of Alkenes and Study of Reactions of Polyfunctional Compounds. 56. Selective Ozonolysis of, (S).-(+).-Dihydromyrcene and Synthesis of Enzymes of Callosobruchus-Chinesis and Danaus-Chysippus. *Zhurnal Organicheskoi Khimii*, 29: 1, 39-43, (1993).
75. Odinokov, VN, OS Kukovinets, RA Zainullin, VG Kasradze, GA Tolstikov. Ozonolysis of alkenes and the study of reactions of polyfunctional compounds. 57. Synthesis of echinolone based on the product of 1-methyl-1-cyclohexene ozonolysis. *Zhurnal Organicheskoi Khimii*, 31: 1, 103-105 (1995).

76. Odinokov, VN, VR Akhmetova, , RG Savchenko, MV Bazunova, AA Fatykhov, AY Zapevalov. Ozonolysis of perfluoroalkenes and perfluorocycloalkenes, *Russian Chemical Bulletin,* 46, 6, 1190-1191 (1997).
77. Odinokov, VN, VR Akhmetova, RG Savchenko, AA Fatykhov, AY Zapevalov. Synthesis of perfluoroheptanal by ozonolysis of perfluoro-1-octene. *Russian Chemical Bulletin,* 43: 6, 1084-1085 (1994).
78. Ishmuratov, GY, RY Kharisov, VN Odinokov, GA Tolstikov. The ozonolysis of unsaturated compounds in the synthesis of insect pheromones and juvenoids. *Uspekhi Khimii,* 64: 6, 580-608 (1995).
79. Odinokov, VN, VR Akhmetova, RG Savchenko, AA Fatykhov, AY Zapevalov. Synthesis of perfluoroheptanal by ozonolysis of perfluoro-1-octene. *Russian Chemical Bulletin,* 43: 6, 1084-1085 (1994).
80. Odinokov, VN, VR Akhmetova, RG Savchenko, MV Bazunova, AA Fatykhov, AY Zapevalov. Ozonolysis of perfluoroalkenes and perfluorocycloalkenes, *Russian Chemical Bulletin,* 46, 6, 1190-1191 (1997).
81. Gross, A, GD Billing. Rate Constants for Ozone Formation and for Isotopic Exchange Reactions. *Chemical Physics,* 173: 3 393-406, (1993).
82. Hegeler, F, H Akiyama. Spatial and temporal distributions of ozone after a wire-to-plate streamer discharge. *IEEE Transactions on Plasma Science,* 25: 5, 1158-1165 (1997).
83. Kitayama, J, M Kuzumoto. Theoretical and experimental study on ozone generation characteristics of an oxygen-fed ozone generator in silent discharge. *J Phys-D-Appl Phys,* 30, 17, 2453-2461 (1997).
84. Hegeler, F, H Akiyama. Ozone generation by positive and negative wire-to-plate streamer discharges. *Jpn J Appl Phys Pt 1,* 36, 8, 5335-5339 (1997).
85. Popescu, S, D Vaju. Improvement of the corona discharge in dry air in order to obtain ozone with maximum efficiency. *J Anal Atom Spectrom,* 12, 9, 1091-1093 (1997).
86. PC Foller, GH Kelsall. Ozone Generation via the Electrolysis of Fluoboric Acid Using Glassy Carbon Anodes and Air Depolarized Cathodes. *Journal of Applied Electrochemistry,* 23: 10, 996-1010 (1993).
87. Voblikova, VA, YV Filippov, VB Gavrilyuk. The Effect of the Electrode Cooling on the Ozone Synthesis in Oxonizers of Different Systems, Dielectric-Dielectric and Dielectric-Metal. *Vestnik Moskovskogo Universiteta Seriya 2 Khimiya,* 34: 4, 367-370 (1993).
88. Voblikova, VA, YV Filippov, EI Sobitov, VE Pustovalov. Distribution of Temperature Along the Gas Flow in High Frequency Ozonizer. *Vestnik Moskovskogo Universiteta Seriya 2 Khimiya,* 34: 3, 255-259 (1993).
89. Tatapudi, P, JM Fenton. Synthesis of Ozone in a Proton Exchange Membrane Electrochemical Reactor. *Journal of the Electrochemical Society,* 140: 12, 3527-3530 (1993).
90. Vigdorovich, VN, YA Ispravnikov, EA Nizhadegavgani. Ozonogenerator as Computer Integrated Pneumoelectrochemical Monoblock. *Doklady Akademii Nauk,* 335: 1, 47-51 (1994).
91. Bystrova, TV, AY Stolyarevskii, YL Chizhov, GP Chugunova. Plasmadynamic method for ozone preparation. *Khimicheskaya Fizika,* 14: 4, 119-125 (1995).
92. Yagi, S, M Kuzumoto. Silent discharges in ozonisers and CO2 lasers. *Australian Journal of Physics,* 48: 3, 411-418 (1995).
93. Katoh, M, Y Nishiki, S Nakamatsu. Polymer Electrolyte-Type Electrochemical Ozone Generator with an Oxygen Cathode. *Journal of Applied Electrochemistry,* 24: 6, 489-494 (1994).
94. Gibalov, VI. Ozone Synthesis in Barrier Discharge. *Zhurnal Fizicheskoi Khimii,* 68: 6, 1136-1141 (1994).
95. Kostyuchenko, SV, NN Kudryavtsev, SM Starikovskaya, AV Tretyakov, IV Filyugin. Ozone Generation in High-Voltage Nanosecond Discharge Plasma in Oxygen at Low Pressures. *Khimicheskaya Fizika,* 13: 10, 71-87 (1994).
96. Eliasson, B, W Egli, U Kogelschatz. Modelling of Dielectric Barrier Discharge Chemistry. *Pure and Applied Chemistry,* 66: 8, U1766-U1778 (1994).
97. Evtushenko, AM, IP Chikhacheva, GV Timofeeva, SD Stavrova, VP Zubov. Graft Polymerization of Acrylic Acid into Ozonated Siloxane Matrixes. *Vysokomolekulyarnye Soedineniya Seriya A & Seriya B,* 35, B312-B315, (1993).

98. Boochathum, P, M Shimizu, K Mita, Y Tanaka. Structure of Solution-Grown trans-1,4-Polyisoprene Crystals. 1. Determination of Stem Length and Fold Surface Structure by Ozonolysis gpc Measurement. *Polymer,* 34: 12, 2564-2568, (1993).
99. Chtourou, H, B Riedl, BV Kokta, A Adnot, S Kaliaguine. Synthetic Pulp Fiber Ozonation - An ESCA and FTIR Study. *Journal of Applied Polymer Science,* 49: 2, 361-373, (1993).
100. Tanaka, Y, S Kawahara, T Ikeda, H Tamai. Analysis of the Sequence Distribution of 1,2 Units in Polybutadiene by an Ozonolysis-GPC Method. *Macromolecules,* 26: 19, 5253-5255 (1993).
101. Ebdon, JR, SA Rimmer. A one-pot method for the preparation of latices of telechelic oligomers, by ozonolysis of latices of polymers containing main-chain unsaturation, *J Polym Sci A-Polym Chem,* 35, 15, 3255-3262 (1997).
102. Couve, J, R Dobrevaschue, J Sledz, F Schue, P Nicol. Characterization of star-branched styrene-butadiene copolymers by ozonolysis. *Eur Polym J,* 33, 2, 137-141 (1997).
103. Razumovskii, SD, VV Podmasterev. Conformationally-nonequilibrium Macromolecules, and their reaction ability: Kinetics of ozone addition to deformed and undeformed polydienes, *Vysokomol. Soedineniya ser.A,* 38, 12, 2049-2052 (1996).
104. Gatenholm, P, T Ashida, AS Hoffman. Hybrid biomaterials prepared by ozone-induced polymerization.1. Ozonation of microporous polypropylene. *J Polym Sci A-Polym Chem,* 35, 8, 1461-1467 (1997).
105. Kim, WS, TS Huh. Ozonolysis of 2-methylthio-3-methyl-2-butene on polyethylene: A new access to tetramethyl-1,2,4-trioxolane. *Bull Kor Chem Soc,* 18, 5, 543-544 (1997).
106. Reyx, D, I Campistron. Controlled degradation in tailor-made Macromolecules, elaboration. Controlled chain-cleavages of polydienes by oxidation and by metathesis. *Angew Makromol Chem,* 247, 197-211 (1997).
107. Rimmer, S, JR Ebdon, MJ Shepherd. Polymer-supported reagents as aids in controlled degradation of polymers containing ozonides in the main chain to give telechelic oligomers. *Reactive & Functional Polymers,* 26: 1-3, 145-155 (1995).
108. Kang, ET, KG Neoh, X Zhang, KL Tan, DJ Liaw. Surface modification of electroactive polymer films by ozone treatment. *Surface and Interface Analysis,* 24: 1, 51-58 (1996).
109. Fargere, T, M Abdennadher, M Delmas, B Boutevin. Synthesis of Graft Polymers from an Ozonised Ethylene- Vinyl Acetate Copolymer.2. Determination of the Efficiency and of the Decomposition Rate of Macroinitiators. *Macromolecular Chemistry and Physics,* 195: 5, 1673-1685 (1994).
110. Varghese, S, B Kuriakose, S Thomas. Short Sisal Fibre Reinforced Natural Rubber Composites - High-Energy Radiation, Thermal and Ozone Degradation. *Polymer Degradation and Stability,* 44: 1, 55-61 (1994).
111. Giurginca, M, T Zaharescu, A Meghea. Degradation of ethylene-propylene elastomers in the presence of ozone. *Polymer Degradation and Stability,* 50: 1, 45-48 (1995).
112. Hong, SG, CM Liao. The surface oxidation of a thermoplastic olefin elastomer under ozone exposure: ATR analysis. *Polymer Degradation and Stability,* 49: 3, 437-447 (1995).
113. Boochathum, P, M Shimizu, K Mita, Y Tanaka. Structure of Solution-Grown trans-1,4-Polyisoprene Crystals.1. Determination of Stem Length and Fold Surface Structure by Ozonolysis gpc Measurement. *Polymer,* 34: 12, 2564-2568 (1993).
114. Chtourou, H, B Riedl, BV Kokta. Surface Modification of Polyethylene Pulp Fiber by Ozone Treatment - An Analytical and Thermal Characterization. *Polymer Degradation and Stability,* 43: 1, 149-156 (1994).
115. Lovill J, T Sullivan, J Knox, J Kornev. *Proc. Symp. on Atmos. Ozone*, Berlin, 2, 67 (1977).
116. Prabhakara, C. *Mon. Weath. Rev.*, 97, 307 (1969).
117. Prabhakara, C, E Rodgers, B Conrath. *J. Geophys. Res.*, 81, 6391 (1976).
118. Tal'roze, V, I Larin. *Uspekhi Khimii,* 60, 506, (1991).
119. WMO Report (1981). The Stratosphere 1981 Theory and Measurements. WMO global Ozone Research and Monitoring Project Report 11.
120. Perov, S, A Hrigian. *Gidrometeoizdat*, 287 (1980).
121. Johnston, H, J Podolski. *Rev. Geoph. Space Phys.*, 16, 491 (1978).
122. Washington, D. *National Acad. Press.*, 337 (1982).

123. WMO (1986). Atmospheric Ozone 1985, WMO Global Ozone Research and Monitoring Project, Report 16.
124. Chapman, S. *Phil. Mag.* Ser. 7, 10, 369 (1930).
125. Chapman, S. *Met. Roy. Soc.*, 3, 103 (1930).
126. Johnston, H. *Rev. Geoph. Space Phys.*, 13, 637 (1975).
127. Hunt, B. *J. Atm. Sci.*, 23, 88 (1966).
128. Nicolet, M. *Rev. Geophys. Space Phys.*, 13, 593 (1975).
129. NASA Reference Publication, 1162, N.Y. Acad. Press (1986).
130. Gerchenson, Yu, S Zvenigorodskii, V Rozenstein. *Uspekhi Khimii*, 59, 1601 (1990).
131. Johnston, H. *Photochemistry in the Stratosphere*, UCLA, Berkley, 20 (1975).
132. Tal'roze, V, S Zvenigorodskii, A Poroikova. *Izv. Akad. Nauk SSSR*, 14, 355 (1978).
133. Crutzen, P, R Smalcel. *Planet Space Sci.*, 31, 1009 (1983).
134. Farmen, J, B Gardiner, J Shanklin. *Nature*, 315, 207 (1985).
135. Solomon, S, R Garcia, F Rowland, P Wueblles. *Ibid*, 321, 755 (1986).
136. McElroy, M, R Salaurth, S Wofsy, J Lodan. *Ibid*, 321, 759 (1986).
137. Molina, T, M Molina. *J. Phys. Chem.*, 91, 433 (1986).
138. Vupputuri, R. *Atm. Environ.*, 22, 2809 (1988).
139. Kondrat'ev, K. *Meteorologia i Klimatolgia*, 19, 212 (1989).
140. Kondrat'ev, K. *Uspekhi khimii*, 59, 1587, (1990).
141. Filippov, Yu, Yu Emil'ianova, I Semiokhin. *Sovremenie problemi fisicheskoi khimii*, 2 14 (1968).
142. Raidler, K. *Chemical Kinetics*, 1 (1950).
143. Harteck, P, S Dondes. *J. Chem. Phys.*, 31, 2241 (1953).
144. Jahn, S. *Z. Anorgan. and Allgem. Chem.*, 48, 260 (1906).
145. Benson, S, A Axworthy. *Z. Chem. Phys.*, 26, 1718 (1957).
146. Glissman, A, H Schumaher. *J. Phys. Chem.*, B 21, 323 (1933).
147. Lewis, B. *J. Chem.*, 37, 533 (1933).
148. Axworthy, A, S Benson. *Advances Chem.*, 21, 388 (1959).
149. Benson, S, A Axworthy. *Advances Chem.*, 21, 398 (1959).
150. Kondrat'ev, VN. *Kinetics of Gase Phase Reactions.* Nauka, Moscow, 1958.
151. Harper, S, W Gordon. *Advances Chem.*, 21 28 (1959).
152. Thorp, C. Bibliography of Ozone Technology, 2, 30 (1955).
153. Schumacher, H. *An. Ascon. guini, Argentina*, 41, 198 (1953).
154. Yastrebov, V, N Kobozev. *Zh. Fiz. Khim.*, 33, 1701 (1959).
155. Yastrebov, V, E Pitskhelauri, N Kobozev. *Zh. Fiz. Khim.*, 33,1209 (1959).
156. Kamenetskaya, S, S Pshezhetskii. *Zh. Fiz. Khim.*, 32, 1122 (1958).
157. Pshezhetskii, S, N Morozov,S Kamenetskaya,V Siryatskaya, E Gribova. *Zh. Fiz. Khim.*, 33, 2306 (1959).
158. Yastrebov, V, N Kobozev. *Zh. Fiz. Khim.*, 33, 649 (1959).
159. Strakhov, B, V Lebedev, N Kobozev. *Vest. Moskovskogo Universiteta seria khim.*, 4, 31 (1961).
160. Emel'yanova, G, B Strakhov. *Sovremenie Problemi Fiz. Khim.*, 2,149 (1968).
161. Monhot, W, WKampschulte. *Ber.*, 40, 2891 (1907).
162. Kashtanov, L, N Ivanova, B Rizhov. *Zh. Prikladnoi Khim.*, 9, 2176 (1936).
163. Schwab, G, C Hartman. *Z. Phys. Chem.*, 6, 72 (1964).
164. Emel'yanova, G, V Lebedev, N Kobozev. *Zh. Fiz. Khim.*, 38, 170 (1964).
165. Emel'yanova, G, V Lebedev, N Kobozev. *Zh. Fiz. Khim.*, 39, 540 (1965).
166. Sudak, A, V Vol'fson. *Catalytic Ozone Purification of Air.* Naukova Dumka, Kiev, 87 1983.
167. Hata, K, M Horiuchi, T Takasaki. Jpn. Pat., *CA*, 108, 61754u (1988).
168. Tchihara, S. Jpn. Pat., *CA*, 108, 192035h (1988).
169. Kobayashi, M, M Mitsui, K Kiichiro., Jpn. Pat., *CA*, 109, 175615a (1988).
170. Terui, S, K Sadao, N Sano, T Nichikawa. Jpn. Pat., *CA*, 112, 20404p (1990).
171. Terui, S, K Sadao, N Sano, T Nichikawa. Jpn. Pat., *CA*, 114, 108179b (1991).
172. Oohachi, K, N Fukutake, T Sunao. Jpn. Pat., *CA*, 119, 119194g (1993).
173. Martinov, I, V Demiduk, S Tkashenko, M Popovish. *Zh. Fiz. Khim.*, 68, 1972 (1994).
174. Vol'fson, V, A. Sudak, V Vlasenko, G Kalmikova. *Kinet. Katal.*, 23, 84 (1982).

175. Tarunin, B, M Perepletshikov, M Klimova, Yu Aleksandrov, Yu Prokop'ev. *Kinet. Katal.*, 22, 431 (1981).
176. Kupenko, O, M Popovish, G Egorova, Yu Filipov. *Vestn. Moskovskogo Universiteta, ser. khim.*, 27, 162 (1986).
177. Tkachenko, S, V Demiduk, V Kolesnik, V Frolov, N Smirnova, M Popovish. *Vestn. Moskovskogo Uneversiteta, ser. khim.*,31, 614 (1990).
178. Popovich, M, *Vestn. Moskovskogo Uneversiteta, ser. khim.*, 29, 29 (1988).
179. Popovich, M, N Smirnova, L Sabitova, Yu Filipov. *Zh. Fiz. khim.*, 59, 740 (1985).
180. Sabitova, L M Popovich, N Smirnova, T Nugaev, Yu Filipov. *Zh. Fiz. khim.*, 59, 1636 (1985).
181. Pitskhelauri, E, M Popovich, L Sabitova, N Smirnova, T Nugaev, Yu Filipov.*Vestn. Moskovskogo Uneversiteta, ser. khim.*, 23, 174 (1982).
182. Popovich, M, N Smirnova, L Sabitova, Yu Filipov. *Vestn. Moskovskogo Uneversiteta, ser. khim.*, 26, 167 (1985).
183. Egorova, G, M Popovich, Yu Zhitnev, V Zhuravlev, S Tkachenko, Yu Filipov. *Zh. Fiz. Khim.*, 56, 2532 (1982).
184. Egorova, G, N Mordkovich, M Popovich, S Tkachenko, Yu Filipov. *Zh. Fiz. Khim.*, 58, 1756 (1984).
185. Olshyna, K, R Cadle, R De Pena. *J. Geophys. Res.*, 84, 1771 (1979).
186. Zvenigorodskii, S, V Sirota, A Skoblikova, S Khvorostovskii. *Zh. Prikladnoi Khim.*, 58, 929 (1985).
187. Popovich, M, N Smirnova, L Sabitova, Yu Filipov. *Vestn. Moskovskogo Uneversiteta, ser. khim.*, 28, 548 (1987).
188. Aktyasheva, L, G Emel'yanova. *Zh. Fiz. Khim.*, 47, 2113 (1973).
189. Mali, M, J Ross, D Brikmaun. *Phys. Rev. B. Condans. Matter.*, 36, 3888 (1987).
190. Rubashov, A, B Strakhov. *Zh. Fiz. Khim.*, 47, 2115 (1973).
191. Rakovsky, S, D Cherneva, D Shopov. *Proc. 4th Symp. Heterogenous Catal.*, Varna, 2 (1979).
192. Rakovsky, S, D Cherneva, A Spozhakina, K Novakov. *Proc. 8th Symp. Heterogenous Catal.*, Varna, 1 (1996).
193. Razumovskii, S, S Rakovskii, D Shopov. *Ozone and Its Reactions with Organic Compounds* (in Russian), Publ. House of Bulgarian Academy of Sciences, Sofia, 1983.
194. Nakatsuji, T, N Tadao, K Yoshimoto. Jpn. Pat., *CA*, 122, 141314 (1995).
195. Sasaoka, E, K Kaneshima, N Massashi. Jpn. Pat., *CA*, 122, 321417 (1995).
196. Arai, N, A Kiyoshi. Jpn. Pat., *CA*, 124, 155518 (1996).
197. Naydenov, A, R Stoyanova, D Mehandjiev. *J. Mol. Catal. A. Chem.*, 98, 9 (1995).
198. Arai, N, A Kiyochi. Jpn. Pat., *CA*, 123, 348322 (1995).
199. Nakatsuji, T, N Tadao. Jpn. Pat., *CA*, 123, 348324(1995).
200. Nakatsuji, T, N Tadao, Egawa, Takeshi. Jpn. Pat., *CA*, 123, 321090 1995).
201. Atobe, Hitoshi, Kaneko, Toraichi, Yono, Jpn. Pat., *CA*, 123, 121871 (1995).
202. Dhaudapani, Balamurgan, Ogama, Shigeo. *Chem. Lett.*, 6, 413 (1995).
203. Kamisaka, Kenichi, Shimizugawa. Jpn. Pat., *CA*, 124, 14566 (1996).
204. Obata, Takeji, Sato, Haruo. Jpn. Pat., *CA*, 124, 65232 (1996).
205. Kazuhiko, Ishiguro, Tatsuya. Eur. Pat, Apl. EP 695, 578 (1996).
206. Sakurai, Toshihiko, Atsushi. Jpn. Pat., *CA*, 124, 240821 (1996).
207. Mizunon, Yamaoka, Yoshio Jpn. Pat., *CA*, 125, 40686 (1996).
208. Fujiata, Hiroshi, Yamamoto, Shigehise. Jpn. Pat., *CA*, 115, 165637u (1991).
209. Yokoyama, Shigeo, Sawazarri, Tetsuo. Jpn. Pat., *CA*, 114, 231413w (1991).
210. Terada, Isso, Sugito, Naoto, Kobayashi. Jpn. Pat., *CA*, 115, 238823p (1991).
211. Shimada, Yasuo. Jpn. Pat., *CA*, 121, 163041 (1994).
212. Yamamura, Seiichiro, Ikebata, Masaaki. Jpn. Pat., *CA*, 114, 48705h (1991).
213. Sazonov, V, Yu Ostrovskii, G Zabortsev, V Popovskii. *React. Kinet. Catal. Lett.*,38, 369 (1989).
214. Tatsushima, Masaru, Sakura, Makoto. U.S. US 4, 871, 709 (1990).
215. Kuwabara, Haruo, Fujita, Hiroshi. Jpn. Pat., *CA*, 114, 234340e (1991),
216. Massafumi, Nakatsuji, Tadao, Nagano. Jpn. Pat., *CA*, 115, 286310 (1991).
217. Yoshimoto, Massafumi, Nakatsuji, Tadao. Jpn. Pat., *CA*, 115, 286315w (1991).
218. Nakatsushi, Tadao, Yoshida, Kimihiko. Jpn. Pat., *CA*, 120, 330204 (1994).

219. Yoshimoto, Massafumi, Tadao, Nagano. Jpn. Pat., *CA*, 121, 163037 (1994).
220. Sasaoka, Eiji, Kamashima, Massashi. Jpn. Pat., *CA*, 122, 321417 (1995).
221. Nakatsuji, Tadao, Nagano. Jpn. Pat., *CA*, 123, 348324 (1995).
222. Dhaudapani, Balamurgan, Ogama, Shigeo. *Chem. Lett.* 6, 413 (1995).
223. Kamisaka, Kenichi, Shimizugawa. Jpn. Pat., *CA*, 124, 14566 (1996).
224. Mizuno, Nakatsushi, Yamaoka, Yoshio. Jpn. Pat., *CA*, 125, 40686 (1996).
225. Suzuki, Mitsuaki, Hoshi, Yasuhisa. Jpn. Pat., *CA*, 115, 238814m (1991).
226. Tamyama, Yoshuki, Koizumi, Yukio. Jpn. Pat., *CA*, 110, 29549m (1989).
227. Yoshimoto, Kimihiko, Massafumi. Eup. Appl. EP 388, 094 (1991).
228. Trui, Sadao, Yokota, Yoshiyuki. Jpn. Pat., *CA*, 114, 286316k (1991).
229. Sakakibara, Yoshinobu. Jpn. Pat., *CA*, 120, 14073j (1994).
230. Arai, Kiyoshi, Yokota, Hisaaki. Jpn. Pat., *CA*, 108, 227540y (1988).
231. Falke, Holger, Strauss, Guehther. Eur. Pat. Appl. EP 403, 984 (1991).
232. Atale, Hitoshi, Kaneko, Taraichi, Yano. Jpn. Pat., *CA*, 123, 121871 (1995).
233. Mori, Katsushiko, Hasimoto, Akira. Jpn. Pat., *CA*, 118, 153488v (1993).
234. Kobayashi, Motonobu, Mitsui, Kiichiro. Jpn. Pat., *CA*, 110, 120511d (1989).
235. Rakitskaya, T, E Vasil'eva, A Bandurko, V Paina. *Kinet. Katal.*,35, 103 (1994).
236. Aktyacheva L, G Emel'yanova. *Vestn. Moskovskogo Univ. Ser. khim.*, 31, 21 (1990).
237. Aleinikov, V, B Donster, T Burashkina. *Khim. Techn.*, 1, 12 (1989).
238. Magne, P, N Dupont-Pavlovsky. *Carbon*, 26, 249 (1988).
239. Aktyacheva L, G Emel'yanova. *Vestn. Moskovskogo Univ. Ser. khim.*, 24, 462 (1983).
240. Aktyacheva L, L Gorlenko, T Lazarova.*Vestn. Moskovskogo Univ. Ser. khim.*, 28, 122 (1987).
241. Aktyacheva L, G Emel'yanova. *Zh. Fiz. Khim.*, 63, 2606 (1989).
242. Rakitskaya, T, A Bandurko, A Ennan, A Litvinskaya. *Kinet. Katal.*, 35, 763 (1994).
243. Osintseva, S, L Gorlenko, G Emel'yanova, L Kvacheva, Yu Novikov, M Vol'pin. *Zh. Fiz. Khim.*, 63, 3228 (1989).
244. Inone, Akira, Kobayashi, Motonobu, Nitsui. Jpn. Pat., *CA*, 108, 43231s (1988).
245. Obata, Takeji, Sato, Haruo. Jpn. Pat., *CA*, 124, 65232(1996).
246. Yoshimoto, M, T Nakatsuji, Nagano. Eur. Pat. Appl. EP 501,003 (1993).
247. Tanaka, Eiji, Tsushima, Tetsuya. Jpn. Pat., *CA*, 115, 95381e(1991).
248. Arai, Kiyoshi, Gekkan. Jpn. Pat., *CA*, 118, 239681g (1993).
249. Arai, Kiyoshi, Gekkan. Jpn. Pat., *CA*, 108, 100073t (1988).
250. Emel'yanova, G, V Lebedev, N Kobozev. *Vestn. Moskovskogo Univ., Ser. Khim.*, 6, 7 (1960).
251. Emel'yanova, G, V Lebedev, N Kobozev. *Vestn. Moskovskogo Univ., Ser. khim.*, 6, 31 (1961).
252. Pankratov, A, L Bobrisheva. *Zh. Fiz. Khim.*, 35, 969 (1961).
253. Schumaher, H. *Z. Phys. Chem.*, B13, 353 (1957).
254. Bailey, P. *Ozonation in Organic Chemistry*, 39-II, Acad. Press., N.Y. (1983).
255. Cornelis, A. *In Preparative Chemistry Using Supported Reagents*, Acad. Press., San Diego (1987).
256. Mazur, Y. *Pure Appl. Chem.*, 41, 145 (1975).
257. Keinan. E, Y Mazur. *Synthesis*, 523 (1976).
258. Neeb, P, O Horie, G Moortgat. *Tetrahedron Lett.*, 37, 9297 (1996).
259. Horie, O, G Moortgat. *Atmospheric Environment*, 25A, 1881 (1991).
260. Horie, O, G Moortgat. *Geophys. Res. Lett.*, 21, 1523 (1994).
261. Kaden, D, J Warren, L Ryan, G Boorman, P Mellick. *Inhalation Toxicology*, 8, 213 (1996).
262. Fotty, R, S Rakovsky, M Anachkov, I Marchevsky. *Oxidat. Comm.*, 2 (1997).
263. Basuk, V. *Uspekhi khim.*, 64, 1073 (1995).
264. Kefely, A, S Rakovsky, D Shopov, S Razumovskii, K Rakovsky, G Zaikov. *J. Polym. Sci. Polym. Chem. Ed.*, 19, 2175 (1981).
265. Blinov,N, A Popov, S Rakovskii, A Stoyanov, D Shopov. *Visokomol. Soedinen. Ser. A*, 31, 2212 (1989).
266. Rakovsky, S. *Polymeric Materials Encyclopedia*, Ed. J. C. Salamone, CRS Press., Inc. (1996).

Chapter 7

OZONE APPLICATION

The ozone application is based on its powerful oxidative action. The ozonolysis of chemical and biochemical compounds, as a rule, takes place with high rates at low temperatures and activation energies [1]. This provides a basis for the development of novel and improved technologies, which find wide application in ecology [2], chemical, pharmaceutical and perfume industries, cosmetics [3, 4], cellulose, paper and sugar industries, flotation, microelectronics, veterinary and human medicine, agriculture, foodstuff industry and many others [5-8]. At present it is very difficult to imagine a number of high technologies without using ozone. Here should be mentioned the technologies for purification of waste gases, water and soils, manufacture of organic, polymer and inorganic materials, disinfecting and cleaning of drinking and process water, disinfecting of plant and animal products, sterilization of medical rooms and instruments, resolving cosmetics problems, sterilization and therapy in veterinary and human medicine, surface cleaning and functionalization in the manufacture of polymer and microelectronics articles, flotation, deodorization and decolourization of gases, liquids and solid substances, oxidation in the intermediate stages of various technologies, etc.

7.1. ECOLOGY

7.1.1. Waste Gases

More than 100 contaminants of the atmospheric air have been identified. Among them SO_2, CO_2, nitrogen oxides, various hydrocarbons and dust constitute 85 %.

The main sources of harmful substances emissions, i.e. dust, SO_2 and CO_2 in the air, are the processing plants for coke, briquettes, coals, the thermal power stations, air, water, and road transport [9].The exhaust gases contain also, CO, organic and inorganic compounds, etc.[10]. Dust, sulfur dioxide, carbon dioxide, nitrogen oxides and organic compounds, etc. are the main pollutants released in the environment from metallurgy, the manufacture of fertilizers and petrochemistry. The most typical air pollutants emitted by some chemical productions are listed in Table 7.1.

Table 7.1.

Main air contaminants emitted by some chemical productions

No.	Production	Pollutants	No.	Production	Pollutants
1.	Nitric acid	NO, NO_x, NH_3	12.	Ammonium nitrate	CO NH_3, HNO_2, NH_4NO_3 - dust
2.	Sulfuric acid: a) Nitroso b) contact	NO, NO_x, SO_x, H_2SO_4, Fe_2O_3, dust	13	Superphosphate	H_2SO_4, HF, dust
3.	Hydrochloric acid	HCl, Cl_2	14	Ammonia	NH_3, CO
4.	Ocsalic acid	NO, NO_x, $C_2H_2O_2$, dust	15	Calcium chloride	HCl, H_2SO_4, dust
5.	Sulfamidic acid	NH_3, H_2SO_4, $NH(SO_3NH_4)_2$,	16	Chlor	HCl, Cl_2, Hg

6.	Phosphor	P_2O_3, H_3PO_4, HF, $Ca_5F_4(PO_4)_2$, dust	17	Caprolactam	NO, NO_2, SO_2, H_2S, CO
7.	Phosphoric acid	P_2O_3, H_3PO_4, HF, $Ca_5F_4(PO_4)_2$, dust	18	PVC	Hg, $HgCl_2$, NH_3
8.	Acetic acid	CH_3CHO, CH_3OH	19.	Artificial fibers	H_2S, CS_2
9.	Nitrogen fertilizers	NO_2, NO, NH_3, HF, H_2SO_4, HNO_3	20.	Mineral pigments	Fe_2O_3, $FeSO_4$
10.	Carbamide	NH_3, CO, $(NH_2)_2CO$, dust	21.	Electrolysis of NaCl	Cl_2, NaOH

The technologies with ozone participation are very promising for SO_2 utilization as sulfate, CO as carbonate, nitrogen oxides as nitrates.

Scrubbers used previously for removing only acids, for example HCl and HF, etc., now can be used for SO_2 separation via the addition of an oxidizing agent into the water intended for gas treatment. Ozone as compared with the conventional oxidizers such as hydroperoxide, chlorine, sodium hypochloride, perchlorate , shows appreciably higher oxidizing efficiency [11].

The rate constant of ozone reaction with NO is about 1010 $cm^3/(mol.s)$, and with NO_2 it is ~ 107 $cm^3/(mol.s)$ whereby the oxidation of NO by ozone in the liquid phase is characterized by a higher absorption rate and rise in the concentration of the obtained HNO_3. Moreover the oxidation of NO can be accomplished in the exhaust gases or in the course of absorption [12].

The CO oxidation by ozone is carried out in the presence of Fe, Ni, Co and Mn oxides. In most cases ozonation is more appropriate than the conventional methods and sometimes it appears the only possible way for its preparation [13].

The purification of exhaust gases from burners working on liquid and gas fuel is accomplished by using oxidation catalysts of Perrovski type oxides - ABO_3 combined with ozonation. Ozone is injected prior the waste gases flow. A may be La, Pr or other alkali earth element; B may be Co, Mn or other transition metal. A may be also partially substituted by Sr, Ca or any other alkali earth element. A catalyst deposited on ceramic support of honeycomb type may contain $SrCo_{0.3}Mn_{0.7}O_3$. It is oxidized 24% CO in the absence of ozone and 80% in ozone atmosphere [14].

The removal of nitrogen oxides (NO_x) from waste gases is carried out through ozone oxidation in charged vertical column. The lower part of the column is loaded by 5-15% $KMnO_4$, and ozone is blown through the lower and upper part of the column and the waste gases are fed to the center of the column charge The inlet concentration of 1000 ppm NO_x in the waste gas is reduced to 50 ppm in the outlet gas flow. Another method for cleaning of exhaust gases from NO_x involves the application of plasma generator and ozonator [15] which practically leads to the complete removal of nitrogen oxides.

Sulfur containing compounds are removed from gas mixtures by silicon oil scrubbing and ozone oxidation of absorbed pollutants. A model system providing that the gas mixture contains methandiol and deimethylsulphide is fed at a rate of 9-120 m^3/h into a silicon oil charged scrubber followed by ozone oxidation of the absorbate. The oxidates are extracted with water and the silicon oil is regenerated. Thus the purified gas mixtures practically do not contain any sulfur [16].

A special apparatus is designed for decomposing acyclic halogenated hydrocarbons in gases. It includes a chamber for mixing of the waste gases with ozone coupled with UV-radiation, ozonator, and inlet and outlet units. This method is very appropriate for application in semiconducting industry whereby acyclic and halogenated hydrocarbons are used as cleaning agents [17].

The mechanisms of ozonolysis of volatile organic compounds such as alkenes and dienes are discussed and the products output is determined by matrix isolation FTIR spectroscopy [18].

The synthesis of a material from zeolite via pulverization, granulation and drying at 500-700°C , cooling to 100-200°C , electromagnetic radiation or ozone treatment is described in [19]. The material thus obtained is suitable for air deodorization, drying and sterilization.

The purification of gases containing condensable organic pollutants can be carried out by gas treatment with finely dispersed carbon, TiO_2, Al_2O_3, Fe_2O_3, SiO_2 and H_2O_2 and ozone as oxidizers [20].

A wet-scrubbing process for removing total reduced sulfur compounds such as H_2S and mercaptans, as well as the accompanying paramagnetic particles from industrial waste gases is proposed [21]. For this purpose a water-absorbing clay containing MnO_2 is used. The collected clay is regenerated by ozone oxidation.

The removal of mercury from waste gases is carried out by catalytic ozonation [22]. The used gases mixed with ozone are blown over a zeolite supported Ni/NiO catalyst. This procedure results in 87% conversion of mercury into mercuric oxide which is isolated by filtration. Pt and CuO_x/HgO system is proposed as another catalyst for this process [22].

7.1.2. Waste Water

In contrast to the cleaning of natural water by ozonation which is experimentally confirmed as the most appropriate, the purification of wastewater by ozone is still an area of intensive future research. This could be explained by the great diversity of pollutants in the used water and the necessity of specific approach for each definite case.

The recycling of water from cyanide wastewater (copper cyanide contaminant) is accomplished by ozone oxidation combined with UV radiation and ion exchange method [23]. Further, the oxidate is passed through two consecutive columns charged with cationite and anionite. The ozonolysis priority over the conventional chlorinating method is demonstrated by the fact that ozonation fails to yield chlorides whose removal requires additional treatment.

The ozonolysis of CN-ions leading to CNO^- -ion formation combined with UV-radiation proceeds at 3 fold higher rate as compared with that without radiation employment. Upon exchange in NaOCN, the Na^+ is substituted by H^+ followed by its decomposition to CO_2 and NH^{4+} via hydrolysis in acid medium. The ammonium ion is absorbed by the cation-exchange resin. The solution electroconductivity on the anion-exchange resin becomes lower that 10 muS/cm and cyanides, cyanates or copper ions have not been monitored. On the basis of these experimental results, a recycling method and apparatus for detoxification of solutions containing cyanides is proposed. The recycled water from the cyanide wastewater may be re-used as deionization water in gold - plating. The process ion-exchange resins are regenerated after conventional methods. Actually, this method is not accompanied by the formation of any solid pollutants [23].

The wastewater from the electrostatic and galvanic coatings containing CN^- -ions and heavy metals is treated consecutively by ozone and CO_2. The cyanides and carbonates being accumulated in the precipitate are already biodegradable [24].

Wastewater containing Cr(III) is oxidized by ozone combined with UV-radiation [25]. The radiation reduces the oxidation time about three-fold. The oxidate is consecutively passed through columns charged with cationite and anionite resulting in the obtaining of wastewater with electroconductivity below 20 mu.S/cm. This deionized water can be successfully used for washing of coated articles. The Cr(IV) concentrated solution from the anionite regeneration can be further treated by appropriate ion-exchange method yielding a high purity Cr(IV) solution [25].

Purification of wastewater containing $ClCH_2COOH$ and PhOH is carried out by ozone treatment and UV-radiation in the presence of immobilized photocatalyst. Thus the rate of decomposition is higher than the rate obtained only in the presence of ozone and UV-radiation [26].

Used water containing small amounts of propyleneglycol nitrate or nitrotoluenes is a subject to combined oxidation with ozone and H_2O_2, under pressure and heating to the supercritical point. Thus a nominal conversion of the contaminants higher than 96.7% is achived [27].

In laboratory experiments for reduction in residual COD in biologically treated paper mill effluents it is subject to ozonation or combined ozonation and UV radiation at various temperatures and pressures [28]. At a ratio of the absorbed ozone to COD less that 2.5:1 g/g, the elimination level with respect to COD and DOC is up to 82% and 64%, respectively. The ozone consumption is essentially higher in the case of the UV combined ozonation at pH>9 and elevated temperature to 40°C.

The decolourization and destruction of wastewater containing surface active substances is performed by ozone treatment at ozone concentration of about 80 mg/l. It has been found that for wastewater which has undergone partial biological treatment by fluent filters or anaerobe stages up to COD concentration > 500 mg/l and ratio 30D5/COD>0.2 ozonolysis with ozone concentrations up to 1.8 g/l does not change

CBR, while after the complete biological treatment at COD<500 mg/l ozonation results in rise of the ratio 30D5/COD from <0.05 to 0.37. This evidences the increase of CBR [28].

The purification of wastewater from paper manufacture characterized by enhanced biodegradability, as well as the removal of COD and halogenated compounds is carried out by a method based on ozonation and biofiltration [29].

Wastewater from the paper pulp production are clean up by treatment with ozone and activated acid tar. The high values of pH favor the lignin decomposition and carboxylic acids formation since in this case the $^{\bullet}OH$ - radicals and not ozone are the oxidizing agent. The application of activated acid tar has a positive effect on the dynamics of the microbiological growth, substrate consumption and CBR of organic acids. For example, maleic and oxalic acids are decomposed completely at ozonation of wastewater in the presence of activated acid tar. For immobilization of biological culture the use of polyurethane foams appear to be very suitable [30].

The used water in collectors of a plant for paper manufacture are ozonized in a foamy barbotating contactor. The results indicate that ozone is very efficient in the oxidation of coloring and halogenated compounds. Its activity is proportional to the amount of absorbed ozone regardless of the variations in the gas flow rate, the inlet ozone concentration and contactor design. The amount of consumed ozone depends on the operating conditions and wastewater characteristics. The absorption rate rises in the presence of stages including ozone decomposition, particularly at higher rates [31].

In biological granular activated carbon (GAC) columns the effect of pretreated ozonation on the biodegradability of atrazine is investigated. The metabolism of isopropyl-^{14}C-atrazine gives higher amount of $^{14}CO_2$ than the ring UL-^{14}C - atrazine which shows higher rate of dealkylation that the process of ring cleavage. The pretreatment with ozone increases the mineralization of the ring UL-^{14}C - atrazine and consequently rises the GAC-columns capacity. 62% of the inlet atrazine is transformed into $^{14}CO_2$ in columns charged by ozonized atrazine and water. However, in columns with nonozonized atrazine and water only 50% of the inlet atrazine is converted into $^{14}CO_2$ and in columns supplied only with nonozonized atrazine only 38% of it are converted to $^{14}CO_2$ [32].

The wastewater from petrochemical industry contains substantial amounts of phenol. The phenols mixtures appear to be more toxic that phenol itself and possess a synergetic effect. Their decomposition to nontoxic products is important and modern problem. The oxidation of phenols such as *p*-cresol, pyrocatechin, rezorzine and hydroquinone by O_3 in aqueous medium appears to be a promising method for their degradation. At concentrations of 100 mg/l and pH = 11.5 their complete oxidation is performed for 30 min.

The phenol content in used water is reduced from 145-706 mg/l to 2.5 mg/l at ozone consumption of 1.1 - 2.6 g/l. In real conditions the ozone consumption varies from 5 to 10 g/l per g phenol.

The results from the ozonolysis of wastewater from coke processing are shown in Fig.7.2.

Table 7.2

Results from the ozonation of wastewater from coke processing. $[O_3]$ = 5 mg/l, flow rate 6 l/min and treatment time - 4 h

Pollutant, mg/l	Before ozonation	After ozonation
Monophenols	710	0.8
Polyphenols	380	198
Cyanides	3	29
Thiocyanides	384	0
Thiosulfates	538	0
Sulfides	43	2
H_2O_2	0	12
Bases	114	27

The treatment of concentrated aqueous solutions of phenol (1.0g/l) with ozone causes the appearance of a yellowish coloring after absorption of 1 mol O_3 per 1 mol phenol which gradually fades away. In this case the higher phenols undergo slight oxidation and H_2O_2 is identified in the water after the treatment.

The tests demonstrate that upon varying content of phenols in the used water the ozonation is one of the most promising method for their removal. The ozonation was found to be applicable for the decomposition of rhodanide in neutral and weak acid medium in the temperature range of 9-25°C. The ozone consumption in this case is 2 mg/mg. The ozone uptake for the oxidation of the cyanide ion (CN^-) is 1.8 mg/mg.

The complex Zn cyanides are oxidized similarly to the simple soluble cyanides. Regardless of the higher resistance of Cu cyanide complexes, the ozone treatment reduces the CN^- content substantially (traces). Copper carbonate is precipitated as a reaction product.

The ozonation can be successfully employed as a method for cleaning the wastewater from the manufacture of ammonium nitrate containing CN^-, S^{-2}, CSN^- ions. The purification level is 83, 98 and 95% respectively at O_3 consumption of 0.7 g/l/g water for 10 min contact.

The ozonation of wastewater from sulfate-cellulose production results in 60% purified used water which is returned in the process cycle.

The cleaning of wastewater from the petrochemical plants by ozone is efficient under wide range of pH (5.8 - 8.70 and temperatures (5 - 50°C). Thus the level of petrochemicals in the wastewater at pH 5.8, ozone consumption of 0.52 mg/mg and 10 min contact is reduced from 19-33 mg/l to 2 mg/l. In Table 7.3 is shown the efficiency of various treatment procedures for deodorization of petrochemicals contaminated water.

Table 7.3

Result from deodorization of petrochemicals contaiminated water (grades/dilution)

Petroleum products	Before treatment	Coagulation and filtration	Charcoal and filtration.	Chlorination	Ozonoation	Coagulation, filtration and ozonation
Petrol	5/80	5/60	5/50	5/80	2÷3/10	1/0
Kerosin	5/100	5/80	5/70	5/100	4÷5/25	1/0
Petroleum	5/100	5/60	5/60	5/100	5/20	2/5
Oil	5/10	1/0	1/0	5/10	1÷2/2	1÷0/0

Ozonation is very economical and highly efficient method for decomposing cancerogenic substances such as 3,4-benzopyrene, etc., particularly after the biological purification of wastewater.

The application of ozone is also advisable for mercury oxidation in chlorine producing, for decontamination of the used water from the poisonous tetraethyllead, for destroying pesticides traces, etc.

The highest efficiency of cleaning of petrochemicals contaminated water is achieved at the following sequence: primary cleaning (mechanical), secondary (biological) and complete purification (ozonation).

Ozone is also successfully applied for the complete purification of wastewater form dyes manufacture. Their color decreases about 10 000 times at exposure to 0.22 g/l ozone and the biological activity of the reaction products is reduced almost to zero. The semi-industrial experiments in CP "Kistenetz" were carried out on the dyes manufactured in the plant:

1. Direct Congo red - $C_{32}H_{26}O_6N_6S_2Na_2$, Mw=698

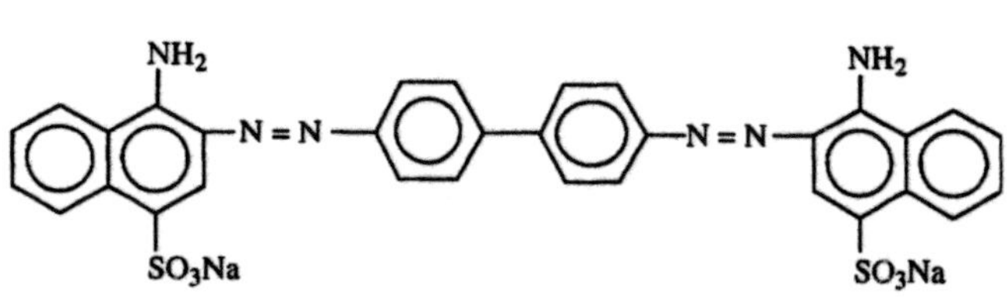

2. Direct blue KM - $C_{34}H_{30}O_5N_5S_1Na_1$, Mw=643

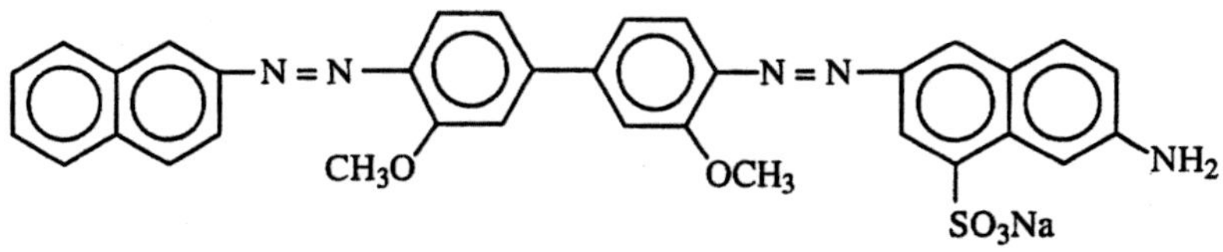

3. Acid blue ZK - $C_{26}H_{20}O_{10}N_3S_3Na_3$, Mw=699

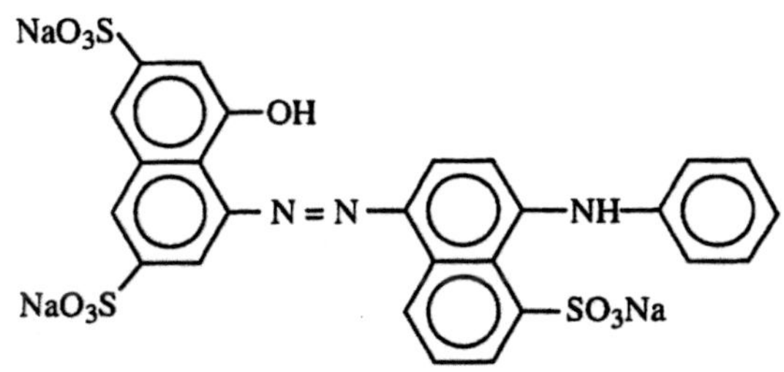

4. Acid chromium green - $C_{16}H_{11}O_{10}N_4S_2Na_2$, Mw=529

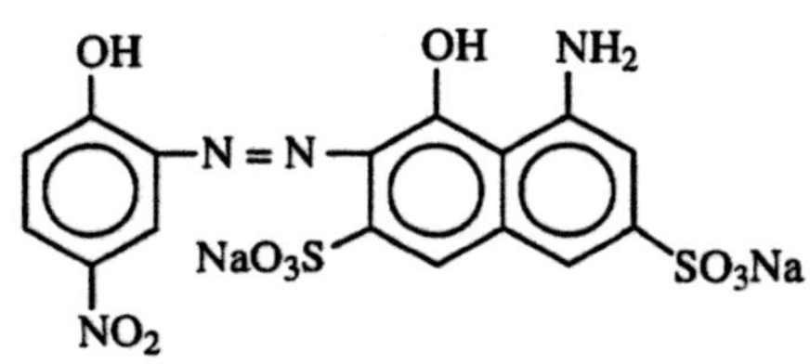

5. Acid black ATT - $C_{22}H_{16}O_9N_6S_2Na_2$, Mw=618

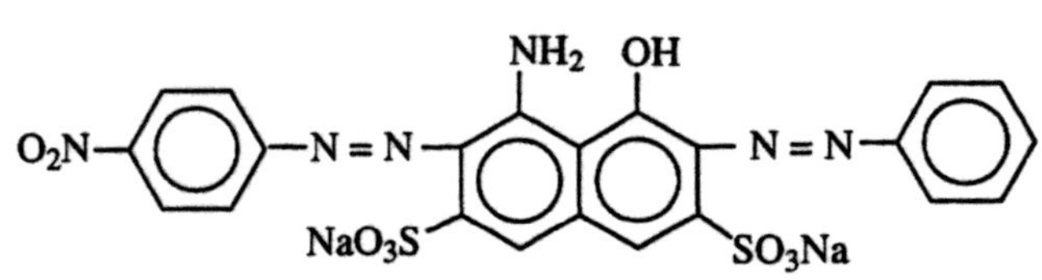

6. Acid sulfonic blue - $C_{32}H_{25}O_6N_5S_2Na_2$, Mw=685

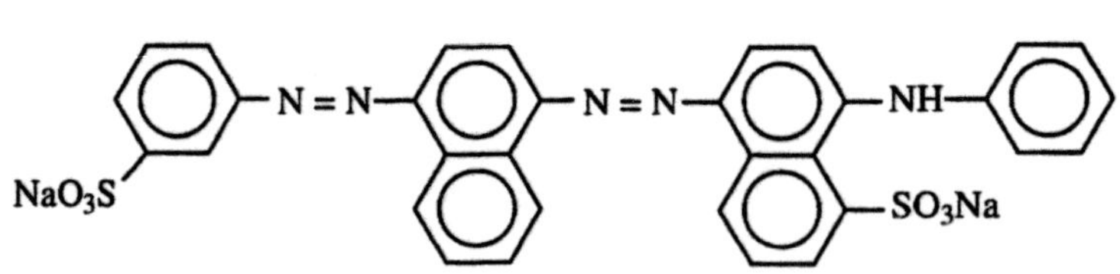

7. Acid chromium black C - $C_{22}H_{19}O_5N_6S_1Na_1$, Mw=502 è $C_{16}H_{13}O_7N_4S_1Na_1$, Mw=428

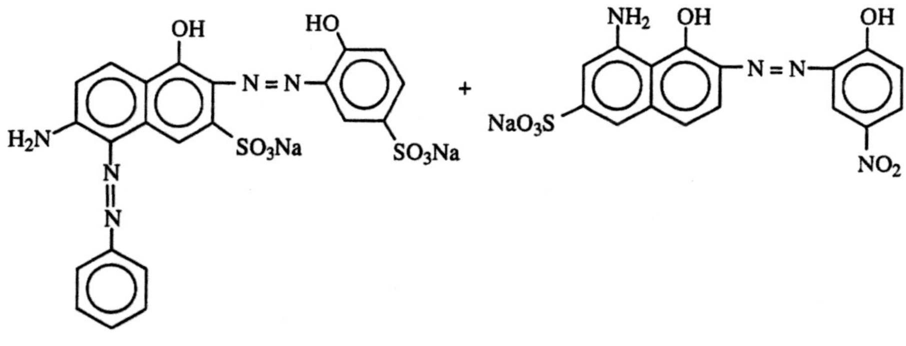

8. Direct black Z - $C_{34}H_{26}O_7N_8S_2Na_2$, Mw=768

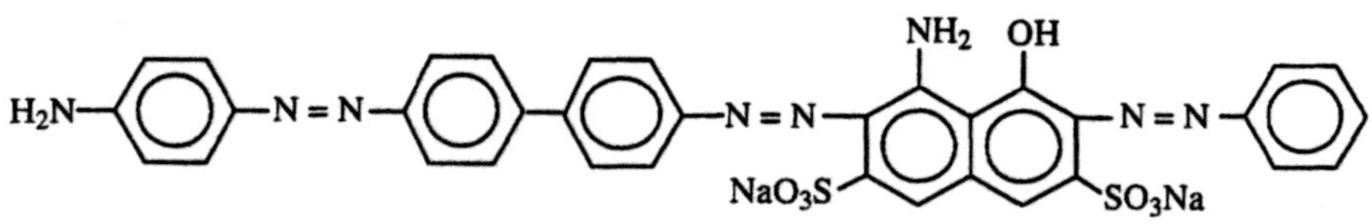

9. Acid chromium yelow R - $C_{17}H_{14}O_3N_3S_1Na_1$, Mw=367

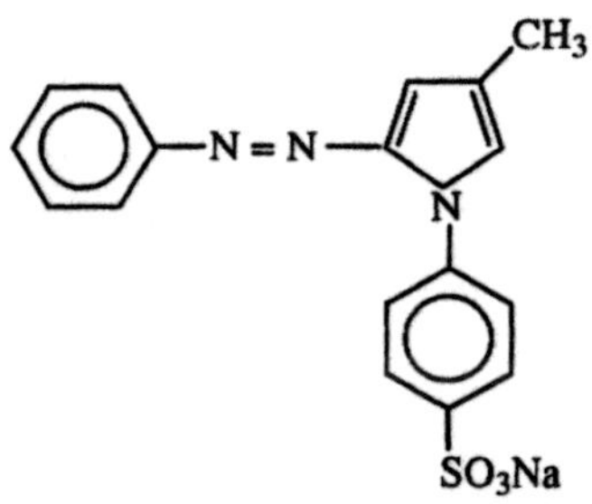

In the process of developing a technology for water purification from various organic contaminants, we have proposed a method and apparatus for quantitative determination of chemical compounds, separated by means of liquid chromatography. Special attention is paid to those compounds susceptible to ozone attack such as olefin hydrocarbons and their derivatives, phenols, amines, thiocarbamates, inorganic compounds, etc. The characterization of the separated compounds is carried out by spectral, mass spectral, refractometer, electrochemical analyses, but the detectors in most of these equipments are rather expensive and complex units without sufficient identification capabilities, particularly for the compounds pointed above. In this sense the apparatus designed by us allows to overcome the disadvantages of the known methods and equipment. The scheme of the ozone detector applicable for identification and liquid chromatography separation of chemical compounds is presented in Fig.7.1.

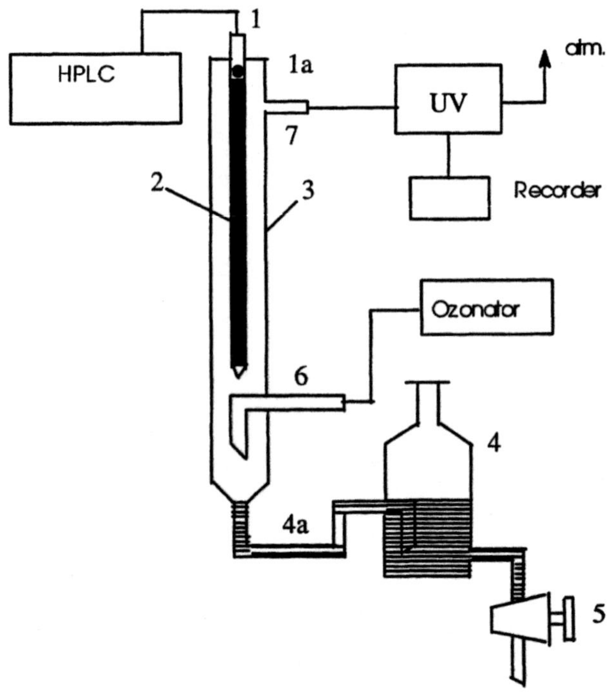

Fig.7.1. Set-up of ozone detector for identification of separated by liquid chromatography chemical compounds. 1 - eluent access; 1a - entry for pouring out the elate as thin film on 2 - solid glass stick; 3 - outer glass jacket; 4 - vessel for eluent accumulation; 4a - hydraulic détente; 5 - tap for pouring out the eluate; 6 - ozone inlet; 7 - ozone outlet.

The ocean power stations operate on the basis of the temperature difference between the upper and lower water layers. Taking advantage of the one-step solution of ozone in depth, the COD is decreased which prevents the befouling in the piping systems found in the upper layers. The ozone-enriched water from the lower layers is supplied to a cyclic filter and is directly injected into upper seawater for control of the sea microorganisms growth [33].

The removal of benzofurenes from ash, clays, soils, water and oils by combustion, ozonolysis of supercritical water, cracking-processes of petroleum, as well as thermal, photochemical and biological decomposition, has been the subject of many investigations [34]. The direct blowing of ozone through water pipers results in slimes removal from the inner pipers walls [35].

The ozone blowing at a rate of 2.87 mg/min through aqueous solutions of 15 pesticides with 10 μg/l concentration leads to their transformation and their concentrations are reduced by about 20% for 60 min. The identified intermediates from the rearrangement of EPN, phenitrothione, malathion, diazinon, izoxathion and chloropiriphos are the corresponding oxones, while for phenthione they are the corresponding sulfoxide, sulfone and sulfooxide oxone analogues, and for disulfothione the intermediate is its sulfo analogue. 2-Amino-4-ethylamino-6-chloro-1,3,5-thriazine is identified as symazin and its 4-isopropylammonium derivative is atrazine intermediate. The isopropylthiolane results in obtaining of its 1-oxo-derivative. Upon replacement of ozone by air changes are actually not registered [36].

At the purification of process water in a soil decontamination plant containing pesticides the removal of organophosphorous compounds such as Thiometon and Disulphoton is of particular importance. In this connection four methods have been applied: 1) ozonolysis at pH = 2.5; 2) ozonolysis at pH = 8; 3) UV/H_2O_2 treatment and 4) oxidation by Fenton reagent [37]. The laboratory experiments were carried out with pure compounds soluted in buffered deoionized water and in process water which is extracted by solid-liquid extraction of the contaminated soil. The use of ozone in an acid medium turns to be ineffective since the Thimeton reaction practically stops after the formation of PO-derivatives. In all cases when the oxidation is carried out by HO - radicals, a sufficient removal of the pesticides and their metabolites is achieved. Oxadixyl, a cyclic nitrogen containing compound which is present in high concentrations in the process water is the most stable one, thus being the main soil contaminant. The investigated experimental conditions and the results obtained show that these methods are quite acceptable for universal application [37].

The water-soluble agrochemicals ASULAM and MECOPROP are decomposed for 30 min at ozone blowing (flow rate - 5.1 mg/min and concentration of 10 ppm)) through their aqueous solution (200 ml). The decrease in the flow rate of ozone retards the decomposition process. Similar treatments for 5h results in ACEPHATE and DICAMBA decomposition to 20%, the rate of decomposition being promoted with H_2O_2 addition [38].

The primary ozonation products of organophosphorous pesticides in water such as diazinone, phenthione (MPP) and ediphenphos (EDDP) [39], are identified by means of GC-MS analysis. The massspectra of the ozonation products of 17 organophosphorous pesticides point oxones as the primary reaction products. This fact is also confirmed by the SO_4^{2-} generation resulting from the ozonation of their thiophosphoryl bonds. Oxones are relatively stable towards ozone attack but they are further hydrolyzed to trialkylphosphate and other hydrolysis products. However, with MPP the thiomethyl radicals are first oxidized to thiophosphoryl bonds giving MPP-sulfoxide,. MPP-sulfone, MPP-sulfoxide-oxone are also obtained. Two main oxidation products have been identified at the oxidation of bis-dithio-type ethiones. Phosphate type EDDP is stable towards ozonolysis but its oxidation products are identified after hydrolysis [39].

Pesticides and their degradation products which are not mineralized to CO_2, NH_3, H_2O and inorganic salts can damage and contaminate the water piping. The treatment by photolysis or ozonation may substantially increase the mineralization rate. The photodestructive products of S-triazoles, chloroacetanylide and parakaute are declorinated and/or oxidized. The ozonation of these herbicides results in formation of products whose side alkyl chains are oxidized or removed, and the aromatic ring is oxidized or soluted but dechlorination does not occur. In some cases the medium may cause microbiological effect [40].

A deodorization method for treatment of ill- smelling air containing ammonia, hydrogen sulfide or amines from refrigerator chambers, lavatories, cattle-sheds, etc. involves the ozone contact with air in the presence of porous catalyst of 120 m^2/g specific surface. The used gas passes through a layer of carbon black for adsorption of the residual ozone and the nondegradated smelling substances. The active component of the catalyst may be: transition metal or its oxides supported on porous honeycomb type supports [41].

The contaminated smelling gases in drain and fecal water are dried to 60-30 % humidity, followed by ozone treatment on an oxidizing catalyst. For example, contaminated gas containing 30 ppm methylsulfide is dried to 30% humidity and then is treated with ozone in the presence of honeycomb type catalyst containing 83:12.5 = TiO_2:SiO_2:MnO_2 with a volume rate of 50 000 h^{-1}. In this case the deodorization efficiency is 99 % while at 100 % humidity it drops to 80% [42].

7.1.3. Natural (Drinking) Water

The basic sources of drinking water usually contain various organic admixtures in mg/l: carbon - 30, nitrogen - 0.8, fatty acids - 30, highmolecular organic acids - 0.2. naphthenolic acids - 1.5, phenols - 1.2, luminescent substances: neutral resins - 17%, oils, humus - 55%, acid resins - 19%, hydrocarbons, naphthenolic acids - 4-7%.

The coloring of natural water varying from pale yellow to brown is due to the presence of humus substances, Their decolorization is carried out by adsorption on coagulated $Al(OH)_3$ or $Fe(OH)_3$. The increasing demands for fresh water requires the use of more colored water which however is pre-chlorinated and then subject to adsorption up to their complete decolorization.

The ozone treatment removes the bad taste and odor of water, the resistant phytoplankton being oxidized to 20%. The ozone consumption for decolorization of natural water varies in broad limits - from 1 to 18 mg/l and depends on the humus composition. The process of humus compounds oxidation by ozone is presented in Scheme 7.1.

Scheme 7.1

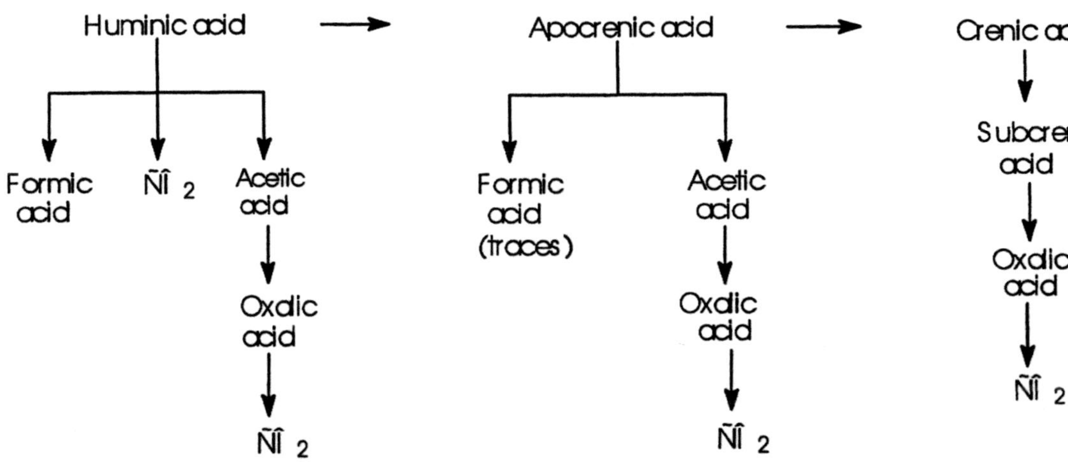

The various biological processes lead to formation of substances causing the bad taste and smell of water - mercaptanes, sulfides, alcohols, carbonyl compounds, acids, terpenes, amines, etc.

The deodorization effect of ozone is preserved in wide range of pH values, temperatures and ion composition. Due to its high oxidation potential, ozone can also degrade the resistant toxic substances. It is known that during the blooming in the water reservoirs are accumulated biologically active substances among which algotoxine is particularly harmful for the human health. Its complete degradation is achieved by using ozone concentration of 14 mg/l and 50 min contact time.

At ozone concentration of 8 mg/l in water, only 6% of *Scendezmus* algae remain alive after 10-15 min. However is to be noted that the ozone action on the various algae is quite different and specific. Thus *Astetionella* type algae are very ozone resistant.

The oxidative, disinfective, decolorizative and deodorizative properties of ozone have found wide application for preparation of drinking water. Many world corporations such as the French Taileygas, Degramont, Ozonia, the German Fisher, AEG, Siemens, the Swiss Braun & Bovery, the American General Electrics, General Motors etc., are among the biggest producers of drinking water. Twenty of the biggest world ozonator stations for drinking water treatment are listed below (Table 7.5).

Table 7.5

Some of the world biggest ozonation stations for drinking water treatment

City	State	Water, m^3/day	Ozone, kg/h
Moscow	Russia	1 200 000	200
Montreal	Canada	1 200 000	150
Ashford Common	GB	-	146
Coppermills	GB	-	144
Neuilly-sur-Marne	France	600 000	120
Choisy-le-Roi	France	900 000	120
Masan	Korea	-	112.5
Kiev	Ukrain	400 000	80
Helsinki	Finland	495 000	63
Manchester	GB	480 000	50
Orly	France	320 000	50
Singapore	Singapore	230 000	48
Minsk	Belaruss	200 000	47
Mery-sur-Oise	France	300 000	45

Lodz	Poland	190 000 290 000	40
Chiba	Japan	270 000	38
Nizhni Novgord	Russia	350 000	30
Bodenzee	Germany	480 000	30
Nant	France	380 000	30
Amsterdam	Holland	125 000	25
Brussels	Belgium	250 000	24
Quebec	Canada	220 000	19
Wrotzlaw	Poland	180 000	17

Ozonation stations for drinking water preparation work in Portugal, Greece, Bulgaria, USA, etc. all over the world.

Recently water treatment stations in Pancharevo and Bistritsa in Sofia, Bulgaria, of capacity 33 and 66 kg O_3/h, respectively are built.

The technology of drinking water ozonolysis leading to its deodorization, disinfecting and decolorization includes gas units for dry air producing, ozonator, barbators or other mixing units, apparatus for HO•-radicals generation, promoting the ozone formation and filter-catalyst for decomposition of the residual ozone. The preferable materials for barbators are zeolites or ceramics on the base of aluminum and/or silicon with pores diameter of 5-100Å, and the catalysts should contain one or more of the following components: Pt, MnO_2, CuO, Ni_2O_3, Zn, YiO_2 and SiO_2 [43].

For desinfection of drinking water, the ozone-containing air is mixed with water in the water pipe, afterwhich the bubbled water is supplied into a special designed chamber filled by porous ceramics charge for intensification of ozone dissolution into water. Some problems concerning the optimization of the UV stimulated oxidation of organic contaminants in soil water with ozone and hydroperoxide is discussed [44].

The disinfection of drinking water by ozonation can be also carried out at the outlet of various water supplying units before and after consummation, such as taps, fountains, water heaters, dental units [45], etc. In this connection are used small dimensional and low capacity personal ozonators coupled or installed additionally to this equipment.

The attempts for treatment of drinking water with ionization radiation appear to be ineffective mainly because of nitrates and H_2O_2 formation. These problems, however, do not occur when the radiation is performed in the presence of sufficient ozone amount [46]. The addition of ozone before and in the course of radiation converts the radiation into a purely oxidation process, a modification of the so called - Advanced Oxidation Process (AOP). The combined ozone-ionization radiation method is unique since two processes for HO-radicals generation are simultaneously induced. This method yields higher concentrations of HO-radicals as compared with other AOP. This in its turn results in precipitation of smaller amounts at equal ozone concentrations or reduces the ozone consumption at one and the same amount of precipitates. This makes the process particularly suitable for remediation of weakly contaminated soil water. The ozone concentrations at this process are higher than those used for drinking water disinfection. Thus, the drinking water subject to this treatment is disinfected more rapidly. The purification of wastewater using this method is more efficient than the UV-radiation treatment only. This combination reduces COD without causing the rise of BOD. In this sense this method appears to be more attractive than the conventional two-stage processes of ozonation and biological treatment. The injection of gaseous ozone into the radiation chamber improves the water flow turbulence and substantially increases the efficiency of contaminants degradation.

The European organization "European regulation" has allowed the use of ozone for iron and manganese removal from mineral water, this initiating studies on secondary products formation, amount and composition , particularly during ozonation of bromine water. This water, usually contain BrO_3^- ions and various halogenated organic compounds (HOC). Some projects regulate a standard for PDK??? of HOC of 25 μg/l BrO_3^--ions, while after other it is from 10 to 0.5 μg/l. The present standard for LD of HOC is 100 μg/l [47]. Upon ozonation of bicarbonate mineral water - one sodium-rich water, and the other, calcium rich, it has been established that: 1) the process depends on the initial Fe and Mn concentrations and the Na and Ca concentrations; 2) HOC and bromate ions are secondary products and 3) the specific components characterizing a definite mineral water affect the ozonolysis. This requires definition of the conditions for Fe and Mn removal for each concrete type of mineral water. The removal of iron and manganese is efficient when the amount of ozone is three times greater than the stoichiometric one (0.43 mg O_3/mg Fe and 0.87 mg O_3/mg mn).

The disinfection of swimming pool water is carried out by controlled addition of compounds- bromine and iodine donors (NaBr or NaI) and water oxidation by hypochloride, ozone or potassium peroxymonosulfate [49]. We have designed a compact module system (stationary and mobile) which includes module for air preparation, ozonator, absorption column and unit for the complete decomposition of the residual ozone. Its capacity is 12-15 g/h and it is very suitable for sanitation of drinking water and used water from swimming pool (3.5 l/s capacity).

7.1.4. Soils

The decontamination of dredging wastes containing spilled petroleum is carried out by using a system for the direct injection of ozone (100-1000 ppm) containing air into pipes at the bottom; means for pumping out the spilled petroleum to a tank for its storage; units for its mixing with ozone for decomposition of organic matter in the tank; units for returning of the processed petroleum for the complete degradation of the residual organic matter by means of aerobic bacteria. The efficiency of organic matter decomposition is by about 20% higher than that by using the conventional methods including only treatment by ozone [49].

Another technology for treatment of contaminated soils [50] involves the ozonolysis of the water slurry of the contaminated soil layer, in the presence of H_2O_2 or in its absence and consecutive biological decomposition. The priority of the proposed technology over the conventional ones is that the removal of organic contaminants is by *on-site* treatment of the soils.

The *on-site* recyclation of petroleum contaminated soils by ozonation requires the good knowledge of the geological setting and the recycling system. Usually the recycling stages include: 1) system for air evacuation from the soil; 2) means for injection of ozone combined with air evacuation from the soil; 3) development of means for ozone generation; 4) screening of soils contaminants; 5) use of systems for safety control; 6) organization of the cleaning units - pumping and injecting drills; 7) decision for air injection into the soil; 8) methodology for measurement and regulation of ozone concentration; 9) hydraulic purification of the highly contaminated soil areas, etc. [51].

The treatment by ozone can be successfully applied for decomposing mineral oils, polycyclic aromatic compounds (PACs), phenols and some pesticides to biodegradable, nontoxic compounds. It should be noted that although ozone decomposes a great number of microorganisms, the soil microflora can be easily remediated after this treatment [52]. The latter can be carried out on- and out of the contaminated area. A number of problems related to the ozonolysis of organic compounds, the PACs degradation products, the effect of ozone on the soil microflora, the laboratory and pilot results, etc. are discussed.

A method and apparatus for decontamination of wastes such as soil, porous granulated sludge, are suggested. The polluted materials are placed in a special vessel and ozone is bubbled through. The waste gases obtained from the ozone oxidation are further treated with ozone [53].

A method for ozone treatment both *in-situ* and *on-site* has appeared to be applicable for the degradation of organic substances adsorbed on the solid surface [54].

Wet-scrubber method (based on MnO_2) for removing total reduced S-containing compounds such as H_2S, mercaptanes and non-magnetic particles from industrial waste gases using ozone, is suggested [21].

7.2. Industry

7.2.1. Manufacture of Organic Compounds

We will focus our attention to some of the numerous ozone applications in organic polymer and inorganic productions, passivation, cleanup and preparation of surfaces for electronics, superconductors, etc.

The reaction of ozone with olefins is the main reaction for ozonides preparation (1,2,4-trioxalanes) which can be converted into a mixture of carbonyl compounds, the exact composition of which depends on the olefin structure. This reaction is used for synthetic purposes as with small exceptions the C=C bond is cleaved quantitatively under very mild conditions. Usually the ozonolysis is carried out at low temperatures (as a rule at -70 ÷ -30°C) as demonstrated in Chapters 2 and 3. Mostly, ozonolysis of olefins, except in solid phase, is carried out in various solvents such as paraffins, halogenated hydrocarbons, oxygen, sulfur and nitrogen containing hydrocarbons like ahcohols, glycols, aldehydes, ketones, acids, ethers, esters, amines, amides, nitriles, sulfoxides, mineral acids, bases and mixtures from them, etc. In most cases the reaction products are not separated from the reaction mixture except for the purposes of ozonide obtaining, and the oxidates are subject to hydrolysis, reduction or oxidation.

The hydrolysis is applied for preparation of aldehydes [55] and ketones [56]. The reductive decomposition is carried out in the presence of Zn/AcOH, SO_3^{-2}, HSO_3^{-1}, $LiAlH_4$, $SnCl_2$, ArP_3, Ph_3P, SO_2, $(CH_3)_2S$, H_2/catalyst. The reduction by metal hydrides leads to alcohol formation in the decomposition products mixture while the use of other reductors yields carbonyl compounds. Actually there are no clear requirements for selection of the most appropriate reductor, but it turned out that dimethylsulfoxide is probably [57] the most preferable reductor in methanol solution. The ozonolysis in the presence of tetracyanoethylene yields directly carbonyl moieties omitting the reduction step [58]. A novel modification of the conventional methods of ozonolysis is exemplified by the ozonolysis of silica gel adsorbed alkenes (acetylenes) [59] and selective ozonolysis of polyalkenes, controlled with the help of dyes introduced in the reaction mixture [60]. Some examples of ozonolysis using various reductors for ozonides decomposition are given in Table 7.4.

Table 7.4

Preparation of aldehydes, ketones and alcohols during alkenes ozonolyis with consecutive reduction

Nî.	Substrate	Conditions	Product	Yield, %
1.	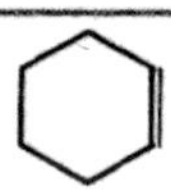	1. EtOAc, -20⁰- -30⁰C 2. $Pd/CaCO_3/H_2$	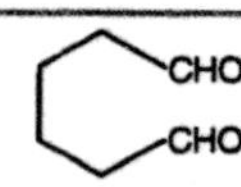	61
2.	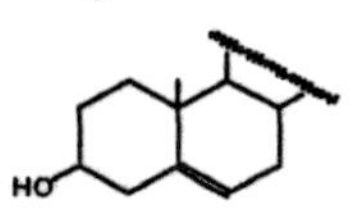	1. n-C_6H_{14} 2. Zn/HOAc	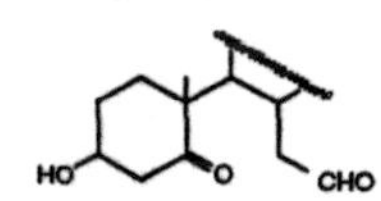	59
3.	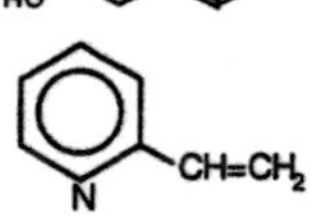	1. MeOH/-40⁰C 2. Na_2SO_3/H_2O	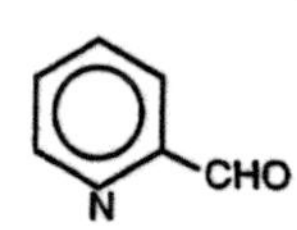	65
4.	$Me(CH_2)_5CH{=}CH_2$	1. MeOH/-30⁰÷ -60⁰C 2. Me_2S	$Me(CH_2)_5CHO$	75
5.	n-Bu-OOCCH=CHCOO-Bu-n	1. MeOH/-65⁰÷ -70⁰C 2. $(MeO)_3P$	n-Bu-OOCCHO	78
6.	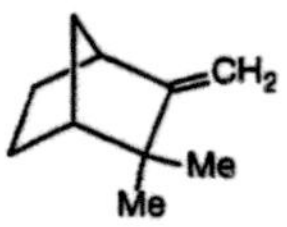	1. MeOH/-35⁰÷ -75⁰C 2. NaI/MeOH/HOAc	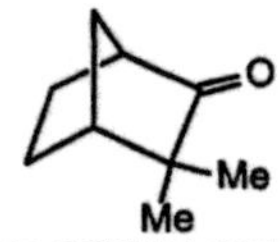	75
7.	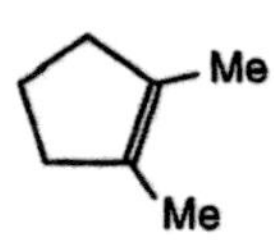	EtOAc/$(CN)_2C{=}C(CN)_2$, -70⁰C	$MeCO(CH_2)_3COMe$ + 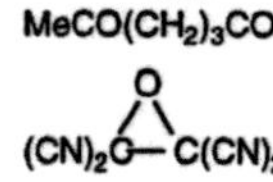	61
8.	$Me(CH_2)_5CH{=}CH_2$	1. n-Pentane/-38⁰ - -42⁰C 2. $LiAlH_4/Et_2O$	$Me(CH_2)_5CH_2OH$	93
9.	$H_2C{=}CH(CH_2)_8COOH$	1. MeOH/0⁰C 2. $NaBH_4$/NaOH/ EtOH/H_2O	$HOCH_2(CH_2)_8COOH$	91

Ozonolysis is a powerful chemical method for economical and ecologically pure preparation of various carbonyl compounds [69].

The ozonolysis of cyclic alkenes in protic solvents followed by reductive decomposition of the hydroperoxides formed is a classical method for dialdehydes synthesis [70]. This method is applied for the preparation of 3-ethoxycarbonylglutaric dialdehyde by ozonolysis of ethyl 3-cyclopentenecarboxylate. The dialdehyde is an intermediate in the synthesis of dolazetrone mesitilate which is an active medical substance in ANZEMET anti-emetik. Zn/AcOH, phosphines, amines and sulfides such as 3, 3'-thiodipropionic acid and its salts are the most suitable reductors for this system. Their efficiency is comparable to that of methyl sulphide but without its shortcomings. The polymer immobilized 3,3'-thiodipropionic acid is also very active in this reduction reaction [70].

The efficient conversion of oximes into the corresponding carbonyl compounds can be also carried out through ozone oxidation [71].

The application of ozone is a new and convenient way for the preparation of cyanoacetylaldehyde (3-oxopropylonitrile) (1) and its stable dimethyl acetal (3,3-dimethoxypropylonitrile) (2) used as valuable intermediates for organic synthesis [72]. For this purpose the ozonolysis of (E)-1,4-dicyano-2-butene or 3-butylonitrile is carried out. Then the oxidates are treated by DMS yielding 1. Further, compound 1 can be used either directly in the next reactions or is transformed into 2. The 2 output amounts to from 67 to 71%. The acetal can be again hydrolyzed to 1 by treatment with ion-exchange resin Amberlyst-15 [72].

The easily available allylphenyl ethers are ozonized at -40°C, treated with DMS giving solutions of the corresponding phenoxyacetaldehydes which are purified by column chromatography. Their reaction with 1-methyl-1-phenylhydrazine yields the corresponding hydrazones [73].

Quinolinealdehyde derivatives can be synthesized by ozonolysis of the respective quinolineolefins at -60 - -72°C in methanol or ethanol solution with consecutive reduction with DMS at the same temperatures. Then the reaction mixture is heated to room temperature for 1 h. The yield of the products amounts to 29%. [74].

6-Chloro-2-hexanone is prepared from 1-methylcyclopentane via three-step scheme: a) ozonolysis of 1-methylcyclopentane in carboyilic acid solution to 1-methylcyclopentanol; b) conversion of 1-methylcyclopentanol into 1-methylcyclopentyl hypochloride using NaClO; c) cyclization of 1-methylcyclopentyl hypochloride in 6-chloro-2-hexanone [75].

Ozonation is a stage in the synthesis of optically pure (R)-(+)-4-methyl-2 cyclohexen-1-one from (R)-(+)-pulegone and hydroxy ketone from (-)-*cis*-pulegol [76].

Polycarbonyl compounds and aldehyde-acids are the ozonolysis products of (+)-4a-[1-(triethylsiloxy)-ethenyl]-2-carene [77].

Carbonyl oxides obtained from vinyl ethers ozonolysis undergo rapid [3+2] cycloaddition with imines giving the corresponding 1,2,4-dioxazolidines in yields of 14-97% [78].

Carbonyl oxides can be also used for the synthesis of 3-vinyl-1,2,4-trioxalanes (α-vinylozonides) by a [3+2] cycloaddition with α,β-unsaturated aldehydes. However the α,β-unsaturated ketones are practically inactive in this reaction. The reaction of 3-vinyl-1,2,4-trioxalanes with ozone leads to the formation of the corresponding dizonides [79].

The carbonyl oxides prepared from the ozonolysis of enol ethers (for example, 1-metoxy-4,8-dimethylnone-1,7 diene) undergo stereoselective intramolecular cycloaddition with inactivated alkenes yielding bicyclic dioxalanes. The latter are easily converted into β-hydroxycarbonyl species and 1,3 diols by catalytic hydrogenation thus providing a new approach to the synthesis of 1,3-oxygenated products [80].

The ozonolysis of vinyl ethers is discussed in detail in Ref. 81.

The synthesis of 1,2-dioxalanes is carried out by ozonolysis of 1,1-disubstituted nonactivated olefins. Thus the ozonolysis of cyclopropyl-1,1-disubsituted olefins does not produce carbonyl oxides, but formaldehyde oxide. The latter can react with the initial olefin leading to the formation of 2-dioxalanes with 10% yield. At the addition of "foreign" olefins or aldehydes other dioxalanes and normal ozonides can be obtained [82].

The cycloalkenes ozonolysis in the presence of methyl pyruvate results in tri-substituted ozonides formation. The latter contain three reaction centers (peroxides, proton on the ozonides cycle and methoxy-carbonyl group) accessible for various functionalization. The cyclohexene ozonolysis in the presence of methylpyrovate gives ozonide whose treatment with PPh_3 or Et_3N yields $CHO(CH_3)_4CHO$ and $CHO(CH_3)_4COOMe$ (after esterification), respectively. This method proved to be very convenient and practical way for the synthesis of linear compounds containing various terminal groups from symmetrical cycloalkenes, in two steps and with good yields [83].

The direct conversion of olefins into esters is accomplished during mono-, di- and thi-substituted olefins ozonolysis in 2.5 M methanol sodium hydroxide-dichloromethane solution . The methyl esters are obtained in high yields. Thus, 3-benzyloxy-1-nonene (5b) is transformed into 2-benzyloxyoctanoate (7b) in 78% yield [84].

The ozonolysis of acyclic alkenes including terpenes is reviewed in Ref. 85.

The role of ozonolysis as ecological process for the selective and specialized oxidation of petrochemical olefins and cyclic alkenes, for the manufacture of biologically active substances and normal organic compounds is thoroughly discussed in Ref.86.

In Ref. 87 are summarized the data on ozonolysis of acyclic and cyclic mono-, di- and trienes in the various stages of the synthesis of insect pheromones and juvenoids.

Here should be also mentioned the importance of fulerene (C-60) ozonolysis [70] resulting in the obtaining of mixture containing various oxidation products such as ketone, ester and epoxides species. The intermediate in this reaction can act as oxygen atom carrier thus yielding phenols [88].

N-acylated esters of (cyclohexa-1,4-dienyl)-L-alanine are ozonized aiming at the synthesis of novel unnatural amino acids. The combined reduction and ozonolysis followed by condensation with a suitable nucleophile results in transformation of the aromatic ring of L-Phe to isooxasolyl, N-phenylpyrazolyl and to bicyclic pyrazolo[1,5-a]pyrimidine groups. The preparation of heterocyclic alanine derivatives is reported [89].

Thymidine diphospho-6-deoxy-a-D-ribo-3-hexulose synthesized by D-glucose ozonolysis of methyl-glucophosphate. is used as a central intermediate in the biosynthesis of di- and tridioxy sugars [90].

The ozonolysis of vinyl halides followed by reductive regio- and stereo-controlled intramolecular cyclocondensation is a key stage in the synthesis of amino sugars [91].

The ozonolysis of pyrols, oxazoles, imidazoles, and isooxazoles demonstrates another application of this reaction to the organic synthesis. Pyrols are efficient protective groups of the amino functions in the synthesis of α-aminoalcohols, α-aminoketones, α-aminoaldehydes and some peptides. The oxazole ring is also known as a protective group in the peptide synthesis [92].

The 1-substituted imidazoles ozonolysis leads to the formation of the corresponding N-acylamides, which are important amine or acyl derivatives [94].

Some novel tetraacetal oxa-cages and complex tetraquinone oxa-cages are synthesized from alkylfuranes by three stage reaction. Oxo rings in the tetraacetals are obtained by the ozonolysis of *cis*-endo-1,4-diones (norbornene derivatives) in dichloromethane solution at -78°C and consecutive reduction with DMS. The tetraquinone oxa-cages are obtained in the cis-endo-1,4-diones ozonolysis in dichloromethane solution at -78°C and TEA treatment [94].

A method for cleavage and oxidation of C_{8-30} olefins to compounds containing terminal carboxylic acid groups is reported in Ref. 95. The process is promoted by oxidation catalysts, such as Cr, Mn, Fe, Co, etc.

The conversion of ethane to methanol and ethanol by ozone sensitized partial oxidation at near atmospheric pressure has been quantitatively studied. The effect of temperature, oxygen concentration in the inlet gas, contact time in the reactor and ozone concentration in oxygen has been evaluated. The selectivity in regard to ethanol, methanol, as well as the combined selectivity towards formaldehyde-acetaldehyde-methanol-ethanol is also discussed [96].

Upon studying the ozone-induced oxidative conversion of methane to methanol and ethane to ethanol it has been established that these reactions do not occur in the absence of ozone which clearly suggests that the partial oxidation is initiated by the oxygen atoms generated from the ozone decomposition [75, 97].

Ortho-selective nitration of acetanilides with nitrogen dioxide in the presence of ozone, at low temperatures results in ortho-nitro derivatives formation in high yields [98].

The ozone mediated reaction of nonactive arenes with nitrogen oxides in the presence of suitable catalysts is reported as a new method for the synthesis of the corresponding nitro derivatives with high yields [99].

The ozone induced reaction of polychloro benzenes and some related halogeno compounds with nitrogen dioxide is a novel non-acid methodology for the selective mono nitration of moderately deactivated aromatic systems [100]. In the presence of ozone and preferably with methanesulfonic acid as a catalyst, the polychloro benzenes undergo mononitration with nitrogen dioxide at low temperatures giving polychloronitrobenzenes in nearly quantitative yields [100].

The ozone mediated reaction of aromatic acetals and acylal with nitrogen dioxide is suggested as a novel methodology for the nuclear nitration of acid sensitive aromatic compounds under neutral conditions [101].

Mineral and acid-free nitro compounds are prepared from CH_2Cl_2 soluted pyridine or its derivatives treated with $NO_2/O_2/O_3$ - mixture for 8 h [102].

The nitration of aromatic compounds with nitrogen oxides in the presence of ozone is a catalytic process [103].

The stereoselective synthesis of vinyl ethers is accomplished by N - (arylidene (or alkylidene) amino) - 2-azetidinones reaction with ozone and $NaBH_4$ treatment resulting in di- and trisubstituted olefins derivatives [78].

The steroeselective synthesis of (2s, 3s) norstatine derivatives is carried out through aldehydes ozonolysis in the presence of lithium methoxyallene [82].

An interesting method of succinic acid preparation from butadiene rubber ozonolysis is suggested [104].

The preparation of α-phenyl ketone, ω-carboxylate-ended telehelic methyl methacrylate oligomers by the ozonolysis of regioregular methyl methacrylate-phenylacetylene copolymers is described in Ref. 105. The oligomers have a number molecular mass varying from 1600 to 4500 (with respect to the initial copolymers) and polydespersity less than 2.

Ozone is also very convenient agent in the manufacture of organic ceramics [106].

The ability of ozone to destroy the double C=C bonds in organic compounds is the reason for its wide application in the preparation of bifunctional compounds. This is the principle that lies in the organization of the manufacture of dodecanedicarboxylic acid (1) and azelaic acid (2). The initial substrate for the preparation of (1) is cyclododecane which is obtained from the trimerization of butadiene and partial hydrogenation and (2) is prepared on the base of oleic acid:

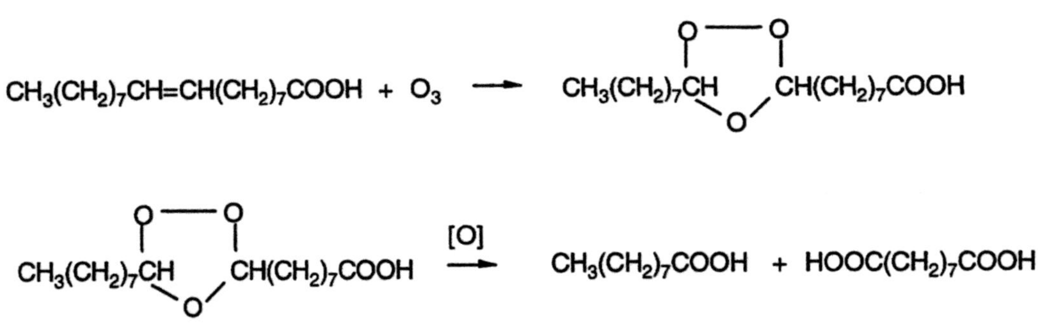

The main part of dicarboxylic acids is used in the manufacture of polyester fibers, and the azelaic acid esters (*n*-hexyl, cyclohexyl-, *iso*-octyl- and 2-ethylhexyl ester) are excellent plastisizers and synthetic lubricants.

The action of ozone is also used in pharmaceuticals in the preparation of valuable hormone products. Thus the C=C bond at C_{17} in the side chain of stigmasterol is destroyed by ozone yielding progesterone - an initial source for many hormones such as cortisone, male and female sex hormones, etc.

The method of selective decolorization of fabrics containing cellulose materials, such as cotton and oxidizing dyes [107], includes the following steps: application of oxidation blocking agent to the fibers; contact of the fibers with the oxidizing reagent in gas or evaporated state in the presence of moisture till the oxidation and decolorization of the dyer is carried out; interruption of this contact before the beginning of a substantial destruction of the fiber. The oxidizing agents are selected among ozone, chlorine and nitrogen oxides and sulfur dioxide flow. The application of ozone for decolorizing indigo-painted cotton jeans after this procedures results in jeans material which does not turn yellow for 6 months while the untreated goes yellow much more rapidly [107].

The changes in the composition and chemistry of UV/ozone modified wool fiber surfaces [108] is investigated by means of photoelectronic spectroscopy (XPS). The oxidation of the disulfide sulfur to sulfone groups ($^-SO_3H$) containing S^{6+} approaches almost 90% conversion. This is much more that the conversion levels by using oxygen plasma. Ozonolysis results in rise of C-O groups content, particularly of the carbonyl ones [108].

The process for producing cellulose fibers and moldings such as fibers, filaments, threads (yarn), films, membranes in the form of flat, pipe and empty fibrous membranes, etc. is carried out using ozone [109]. They are produced by extrusion of cellulose solutions in tertiary amine aminooxides and in some cases in water (particularly N-methylformaline N-oxide and water); regenerating bath and water for washing by treatment with hydrogen peroxide, peracetic acid, ozone or chlorine dioxide for the regeneration of tertiary amines oxides. The introduced compounds can be enzymatically or catalytically destroyed before solvent and water regeneration.

Pitch-based carbon or graphite fibers of high tensile strength are manufactured by primary treatment of the pitch-based fibers with high concentrated ozone for a short period; then the treatment by ozone-free gases follows; in the next step they are subject to carbonization of graphitization to give elliptical fibers of long, wool similar structure. Upon treatment in the absence of ozone the fibers stitched during graphitization [110].

The EPR analysis of thermal decomposition of peroxides in ozonized polypropylene fiber for grafting shows that the decomposition of the peroxide groups begins at about 70°C. The generation of several peroxides radicals is registered; the access of the fiber to the spin sample is enhanced through oxidation. Small amounts of RO_2 - radicals with lifetime of couple of weeks have been identified. The integral intensity of the EPR-signal rises with time and ozonation temperature [111].

Ozonolysis modifies the diffusion pattern of liquid monomers in polypropylene matrix as their addition occurs in the amorphous phase. The appearance of intra-morphological structure and peroxides localization upon ozonolysis of polypropylene (granules and fibers) is monitored using electron spin resonance and transmission electron microscopy [112].

Upon ozonolysis, UV-radiation and plasma treatment polymer peroxides are generated on the surface of films from polypropylene, polyurethanes and polyester fibers [113]. Their thermal and reductive-oxidative decomposition have been studied by means of functional analysis using peroxidase and iodide. However, their disposition in the polymer specimen is quite various thus impeding their analysis and depends on the treatment agent. For example, the treatment with plasma generates easily accessible peroxides in polyurethane films while the UV-radiation and ozonation leads to formation of peroxides uncapable of reacting with aqueous solution of peroxidase. The redox decomposition of the peroxide groups by ferro-ions at 25°C has shown that less than 50% of the peroxides may react with ferro-ions at rate constants similar to those of hydrogen peroxide in aqueous solution. The thermal degradation of peroxides does not follow first order kinetics, most probably because of the generation of various peroxide species, characterized by different rates of decomposition. The lowest rate constant of decomposition observed at 62°C is 3.10^{-3} /min which does not depend on the polymer nature and the method of peroxides generation [113].

Ozone-induced graft polymerization onto polymer surface is an important and convenient method for polymer modification [114].

Ozone-induced graft copolymerization of polyethylene glycol monomethyl ether methacrylate onto poly(etherurethane) improves the hydrophility and water absorption. The autoaccelerated effect in ozone-induced polymerization has been also discussed [115].

The modification of the surface properties of polypropylene and block copolymer is carried out by ozone treatment. Thus, the wettability of the polymers is improved - the contact angle of water becomes ≤ 67^0. The break of the polymer chain and carboxyl groups formation are accelerated by using high ozone concentrations. The cut-off fragments resulting from the ozone treatment are removed from the surface by ultrasound and organic solvents [116].

The low temperature nonelectrolytic nickel plating onto three types of polypropylene is carried out by substrate pretreatment with ozone. The latter modifies the polymer surface for galvanization while the combination of polar with anchor effect as a result of the ozone etching enhances the adhesive properties of the polymer surface. The washing of the material after ozonolysis is obligatory for ensuring a good adhesion of the material [117].

The surface properties of polypropylene (I), 100 μm films from (I), ethylene-propylene block polymer (II), or ethylene-propylene polymer prepared by the random method (III) can be improved by ozonation at concentrations of 1.38, 0.64 and 0.41 mol %, respectively. The adhesion of dyes, coloring agents, dye layers, on (I) is substantially improved and depends on the substrates in the following order: (II) > (III) > (I). It has been found that for each sample there are optimum conditions of ozonation. Reactive dyes such as epoxy and acrylo urethane resins impart a better adhesion force that the conventional nonreactive acrylic or vinylchloride resin [118].

Freshly extruded, 50 μm C_2H_4-ethyl acrylate-maleic anhydride copolymer film is treated by 500 ml/m^2 oxygen containing 10 g/m^2 ozone and is calendered with 200 μm monolayer C_2ClF_3 - polymer film at 15°C to produce moisture-proof packaging material with intra-layer adhesion of 800 g/15 mm and thermosealing force of 3.6 kg/15 mm against 250 and 1.3, respectively in the absence of ozone treatment [119].

Methylmethacrylate polymers with good thermal degradation resistance are prepared by ozonolysis [120] whereby the end unsaturated groups are converted into carbonyl or carboxyl ones. Ozone/air mixture is blown through a 10 g Acrypet VH solution in CH_2Cl_2 at -78°C followed by exposure only to air for 60 min and after solvent removal the solid residue is dried in vacuum at 297°C for 8 h. The polymer thus obtained is characterized by an initial temperature of thermal degradation of 315°C against 297°C for the polymer without ozone treatment [120].

The surfaces of propene polymer moldings are treated with ozone for improving their hydrophility. The ozone treatment of the surfaces for 8 h reduces the contact angle of water from 100 to 81^0 [121].

Ozonation of PVC latex is also carried out for removal of vinyl chloride residues. The aqueous dispersions of saturated polymers are treated with ozone and the vinyl monomers are removed [122].

Graft polymerization of vinyl monomers onto Nylon 6 fiber is carried out after ozone oxidation of the fibers or films from Nylon 6 with vinyl monomers such as acrylamide, methylmethacrylate and vinyl acetate. The molecular mass of Nylon 6 decreases slightly upon ozonation. For acrylamide graft polymerization system the preliminary treatment in air or vacuum by γ-rays radiation prior ozonation results in higher graft percent. For methyl methacrylate the apparent graft percent does not rise with the ozonation time. However the apparent graft percent for vinyl acetate is increased with the ozone time treatment [123].

The adhesion of PVC-, fluropolymer, or polyester-coated steel panels is improved by treatment with 5-50% solution of H_2O_2 or ozone [124].

Synthesis of water-soluble telehelic methyl-ketone-ended oligo- N,N-dimethylacrylamides by the ozonolysis of poly(N,N-dimethylacrylamide-stat-2,3-dimethylbutadiene)s is reported in Ref. 125.

The synthesis of telehelic methylmethacrylate and styrene oligomers with fruorophenyl ketone end groups is accomplished by the ozonolysis of copolymers containing 4-flurophenyl butadiene units [126].

Graft polymerization of acrylic acid is carried out into preliminary ozonized siloxane matrixes [127].

The manufacture of base discs for laser recording material is realized by covering of the plastic substrates surface with solid polymer layer possessing directing channels and/or signal holes. The surface of the plastic substrate is preliminary cleaned by UV/ozone exposure before the formation of the hardened polymer layer [128].

The controlled degradation of polymers containing ozonides in the main chain takes place in the presence of: 1) periodate suported on Amberlyst A26 (SPIR); 2) diphenylphosphine deposited on polystyrene (SPR); boron hydride supported on Amberlyst A26 (BER) [129]. Poly (butyl) methacrylate-copolymers are prepared by emulsion polymerization. These materials and the homopolymer poly (butyl) methacrylate are ozonized at various temperatures and treated by any of the reagents described above, thus giving telechelic oligomers in 99% yield. Molecular mass variations with ozonation temperature change has been observed. The end aldehyde groups are registered using ^{1}H and ^{13}C -NMR spectroscopy; the end hydroxyl groups are observed by means of FTIR and ^{13}C-NMR; the presence of hydroxyl groups is confirmed by tosylate formation; the carboxyl groups are identified by FTIR and ^{13}C-NMR and quantitatively determined by titration; in the SPR-generated oligomers the content of aldehyde groups constitutes about 80% of the end functional groups. The oligomers obtained in BER and SPIR contain 99% hydroxyl and carboxyl groups [129].

Polyethylene fibers are subject to ozone treatment for modifying their surface [130]. The analysis of the surface is carried out by means of X-ray photoelectron (ESCA) and IR (FTIR) spectroscopy. Carbon (C) and oxygen (O) were the main atoms monitored with ESCA (C-1s, O-1s areas) on the treated fibers. The analysis of C-1s peaks (C_1, C_2 and C_3) reveals that the oxidation level depends of the ozonation time. The components of 1s peak (O_o, O_1, O_2) are very useful for carrying out the surface analysis. They demonstrate the presence of carbonyl groups (1740-1700 cm^{-1}) even onto untreated fibers whose intensity rises with treatment time. Ozonolysis is directed from the surface to the fiber bulk. This is confirmed by the great enhancement of the carbonyl bond band after 3 h ozonation. The thermal analysis suggests structural and morphological changes of the fiber when ozonolysi time treatment exceed 2 h [130].

The processing of fiber-reinforced plastics is performed by blowing an ozone-oxygen mixture (flow rate of 0.4 l/min) for 5 h through a CH_2CL_2 solution containing glass reinforced fiber particles (1mm diameter) filled with $CaCO_3$ and unsaturated polyester. The fibers emerge on the surface till a fine powder is precipitated [131].

The manufacture of laminates through heat-sealing method involves the application of electric crown, UV-radiation or ozone exposure [132].

The manufacture of pour point depressants for oils , particularly useful for diesel oils, is carried out by oxidative degradation of waste polyethylene and/or polypropylene (I) with ozone at 30-150°C. 1000 g waste (I) is exposed to ozone action at 150°C for 5 h giving the pour point depressant. The addition of 1% depressant to diesel oil reduces the solidification of the oil from -20°C to -35°C and the temperature of filter plugging from -9 to -20°C [133].

7.2.2. Inorganic Productions

Stainless steel parts are treated by ozone for surface passivation. The parts are heated in oxidative or inert atmosphere at a temperature of condensation lower or equal to -10°C. The unreacted ozone is re-used [134].

The main passivating agent is oxygen in combination at least with ozone. The system is particularly appropriate for passivation of metal (for example, stainless steel, Ti, etc.) equipment, used in chemical plants and exposed to strong corrosion action at high temperatures and pressures [135].

For increasing the corrosion resistance of metals and alloys they were exposed from 1 s to 10 min in cold plasma under pressure 1-103 Pa and 100-5000 V in atmosphere containing O_2, O_3, N_2, H_2, air, CO_2, CO, N-oxides, H_2O (gas), combustible gas and/or inert gas. Thus, 17% C0-ferrite stainless steel is subject to plasma treatment for 4 min at 103 Pa, 100 mA and 250 V in nitrogen-oxygen mixture with 20% oxygen. The corrosion resistance was evaluated by treatment of the sample with a solution containing 17 ml 28% $FeCl_3$, 2.5 ml HCl, 188.5 ml H_2O and 5 g NaCl. The sample shows relatively good resistance as compared with the untreated one [136].

Electrochemical tests reveal the influence of dissolved ozone on the corrosion behavior of Cu-30 Ni and 304L stainless steel in 0.5N sodium chloride solutions [137]. These experiments include: measurements of the corrosion potential as a function of the time and ozone concentration; cyclic polarization experiments; isopotential measurements of the current density and study of the film components. The results of these experiments show that for Cu-30 Ni and 304L -stainless steel the corrosion potential is shifted to the more noble values (300 mV) at $[O_3] < 0.2$-0.3 mg/l. At higher concentrations it remains unchanged. The dissolved ozone reduces the corrosion level for Cu-30 Ni - alloys which is evaluated by the substantial decrease in the current density at constant applied potential. The improvement of the corrosion resistance should be related to the decrease of the thickness of the corrosion products film and to the higher oxygen fraction as compared with the chloride in the same film. For stainless steel the differences in the passivating films in ozonized and nonozonised solutions are negligible as it is shown by spectroscopy [137].

Laboratory experiments have been carried out to study the ozone application for acid oxidizing leaching of chalcopyrite with 0.5 M H_2SO_4. For evaluating the reaction mechanism the effect of particles size distribution , stirring and acid concentration, the dissolution reaction and reaction kinetics on the leaching have been investigated. The reaction rate is governed by ozone diffusion to the reaction mixture. Ozone is an efficient oxidizer and the process is most effective at 20°C [138].

The rate of acid leaching of chalcopyrite depends on the use of ozone as an oxidizing agent in sulfuric acid solutions. The leaching of chalcolyrite follows a parabolic law [139]. The use of ozone as an oxidizer provides conditions for the formation of elemental sulfur on the leached surfaces. The rate of leaching is reduced with temperature as the ozone solubility decreases with temperature. The results show that ozone is the best oxidizing agent for acid oxidizing leaching of chalcopyrite and may be applied in pilot plants for regular manufacture.

We have studied the possibility of using ozone for improvement Ag extraction form polymers deposits in a flotation plant in the town of Rudozem. It has been found that upon blowing of ozone (1% vol concentration, flow rate - 300 l/h) through the empty shaft of the flotation machine stirrer (5 m^3 volume) the degree of Ag extraction is increased by 1-2% [139].

The redox leaching of precious metals from manganese-containing ores carried out by other authors show also positive results [140].

Molybdenite flotation from copper/molybdenum concentrates by ozone conditioning results in relatively pure copper-free molybdenum. The process including multi-step ozone flotation proves to be a technical and economical profitable method [141].

The manufacture of potassium permanganate is accomplished by melting of Mn-containing compounds with KOH, dissolution of the melt and the solution oxidation. For reducing the energy consumption $Mn(NO_3)_2$ is used as a Mn-containing compound. It can be easily alloyed with KOH in a 1:5 - 1: 10 ratio at 250-300°C . Then the product is dissolved in 20-25% KOH solution, the solid residues is removed, soluted in 3-5% KOH and the solution is oxidized by ozone-air mixture [142].

The manufacture of silicon carbide ceramics is performed as melted organosilicon polymer is oxidized with 0.001% vol ozone. Polycarboxylstyrene is the prefered polymer. The ceramics obtained is characterized by high thermal resistance and acceptable physical properties [106].

Mixtures containing In- and Sn-compounds are molded and sintered in a furnace in air atmosphere containing ≥ 1000 ppm ozone. The ITO ceramics thus obtained are characterized by high density at lowtemperature sintering for a short time [143, 144].

A method for oxidizing carbonaceous material, and especially for bleaching gray kaolin for subsequent use as coating or filler for paper in the presence of ozone is reported in Ref. 145..

The manufacture of mercury(I)chloride includes the reaction of Hg with hydrochloric acid in the presence of water, subsequent removal and drying at 95-105°C. Ozone (0.1-0.1 g/g product) is bubbled through the reacting mass to increase the product yield and quality [146].

Arsenic acid is prepared from $(As_2Cl_6)_2$- -ions by ozonolysis [147].

The removal of color and organic matter in industrial phosphoric acid by ozone and the influence on activated carbon treatment is described in Ref.148. Industrial phosphonic acid containing 42-45% P_2O_5 and 220-300 mg/l organic matter (OM) is subject to combined treatment with ozone and activated carbon. The independent ozonation results in removal of the initial dark color of the acid and the organic matter. It is only through absorption on activated carbon that the level of OM could be reduced to 80% per 25 g/kg P_2O_5. The ozonation prior adsorption enhances the efficiency of the activated carbon effect and decreases its specific consumption [148].

The fabrication of high-Tc superconductors is carried out using ozone-assisted molecular beam epitaxy (MBE). It includes the simultaneous evaporation of the elementary components and application of ozone as a reactive oxygen source. The ozone priority over the other oxygen forms is that it is rather stable and could be produced and supplied to the substrate in a very pure state using simplified apparatus ensuring a well defined flow of oxidizing gas. Upon application of ozone the untempering through a thermal treatment for removing the inner stresses could be ignored. In order to prepare films with high temperatures of superconductivity the growth should be carried out at relatively low stresses in the system. In addition, the surface during film growth can be analyzed by reflectance high energy electron diffraction. The most recent improvements in ozone-assisted MBE for family of $YBa_2Cu_3O_7$- delta films are described. The results show that this technique is very appropriate for growth of high quality superconductive films and could be ideal for the manufacture of such structures [149].

The growth of superconductive oxides under vacuum conditions compatible with MBE requires the use of activated oxygen. The atomic oxygen or ozone appear to be such species [150]. The characteristics of a radio-frequency plasma source for molecular beam epitaxial growth of high-Tc superconductor films (200 Å) from $DyBa_2Co_3O_7$ on $SrTiO_3$ are described and discussed in Ref. 150.

An apparatus for the preparation of pure ozone vapor for use *in-situ* growth of superconducting oxide thin films is designed [151]. Pure condensed ozone is produced from distillation of diluted ozone-oxygen mixture at 77K. The condensed ozone is heated until the pressure of its vapors approaches the necessary for an adequate flow of ozone- gas to the chamber of thin films growth. The thin films from $YBa_2Cu_3O_7$ with zero resistance at temperatures of about 85K grow at ozone pressure in the chamber of 2.10^{-3} Torr even without subsequent untempering. It should be noted that in contrast to other highly reactive oxygen species ozone can be prepared and stored in very pure form which makes it very convenient for studies on the kinetics of growth and oxidation of films with a well defined gas [151].

A patent for preparation of oxides superconductors M-M^1-M^2-M^3 includes ozone oxidation where: M = elements of III B groups such as: Y, Sc, La, Yb, Er, Ho or Dy; M^1 = elements from II A group, such as: B, Sr, Ba or Ca; M^2 - Cu and one or more elements from I B group such as Ag or Au; M^3 = O and on or more elements belonging to VI A group like S or Se and/or elements from VII A group as F, Cl or Br. The superconductors thus prepared have high critical temperature and high critical current densities [152].

The manufacture of bismuth-, copper- alkaline earth oxide high-temperature superconductors is carried out through calcination and/or sintering in ozone-containing atmosphere with $Bi(OH)_2$, $Ca(NO_3)_2$ and $Sr(NO_3)_2$ and subsequent treatment by $CuCO_3(OH)_2$; the components are dried, calcinated in ozone containing air, molded and dried in air atmosphere. Their temperature of superconductivity is 107K and the critical current density is 405 A/cm^2 [153].

A lowtemperature method for preparation of superconductive ceramic oxides is described in Ref.154. It includes the treatment of the substrate heated surface by ozone to the complete evaporation of the other components thus forming a superconductive ceramic oxide.

The manufacture of rare earth barium copper oxide high-temperature superconductor ceramics is carried out by sintering at 930 - 1000°C in ozone atmosphere containing oxygen. Y_2O_3, $BaCO_3$ and CuO are mixed, calcinated, pulverized, pressed and sintered at 950°C in the presence of ozone-oxygen mixture after which they are gradually cooled to the critical superconductive temperature 94K [155].

A review devoted to the growth of co-evaporated superconducting yttrium barium copper oxide (Y Ba_2CuO_7) thin films oxidized by pure ozone is presented in [156].

The removal of organic pollutants from the surface of supports for microelectronics purposes is conducted by UV/ozone treatment [157].

Low temperature silicon surface cleaning is carried out by fluoric acid (HF) etching, washing , cleaning with deionized water, N--gas blowing, /UV-ozone treatment. This is a treatment preceding the process of silicon epitaxy [158 -160].

The method for total room temperature wet cleaning of silicon surface comprises the use of HF, H_2O_2 and ozonized water and is with 5% more economical than that the standard procedure of wet cleaning [161].

The removal of resist films supported on semiconductive substances is accomplished by using ozone [162].

The application of ozone for reducing the temperature and energy in the process of very large scale integration (VLSI) is described in [163, 164]. The main problems under discussion are as follows: cleaning of Si-support with ozone by two methods: dry process combined with UV and wet one - with ozonized water; stimulation of Si thermal oxidation with ozone; deposition of SiO_2 under atmospheric pressure and low temperature via chemicals evaporation using tetraethylorthosilicate and ozone; 4) deposition of Ta_2O_3 at atmospheric pressure and low temperature applying $TaCl_5$ and ozone; 5) tempering with ozone and UV-radiation of Ta_2O_5 films used for dielectric for memory units , this treatment reduces the current permeability into the film; 6) etching of photoresistant materials by $O_2/O_3/CF_4$ and the effect of ozonator charge and injection of CF_4 [163, 164].

The cleaning of synchrotonic radiation optics with photogenerated reactants has a number of priorities over the methods of discharge cleaning [165]. Upon discharge cleaning, the discharge particles should react with the surface contaminants until its sheilding by the rough discharge elements which may pollute or destroy the surface. Contrary, if the particles can be photon generated near the surface, the problem with the protection drops off and in some cases the cleaning can be more efficient. An estimation of the various methods for cleaning was made comparing the rates of polymethylmethacrylate films removal. A number of various light and geometry sources have been tested. The highest rate of cleaning was achieved upon using UV/O_3 cleaning method at atmospheric pressure. This method has been widely applied for cleaning semiconductive surfaces from hydrocarbon contaminants. It is noted that it is also effective in removing graphite-like pollutants from synchrotonic radiation optics. It proves to be more simple, economical and selective method as compared with other cleaning method. Although it requires the drilling of a hole in the vacuum chamber, the cleaning of the optics can be carried out without dissemble which saves much time. This method is successfully applied for cleaning grates and reflectors in several beams [165].

Native oxide growth and organic impurity removal on silicon surface is carried out by ozone-injected ultrapure water [166]. In order to manufacture high-efficient and reliable ULSI-units the further integration and minimization is in progress. The cleaning with $H_2SO_4/H_2O_2/H_2O$ is accompanied by serious problems: a great amount of chemical wastes is obtained which must be suitably treated. The cleaning technology with ultrapure water includes ozone injection in concentrations of 1-2 ppm. This method is very efficient in removing the organic impurities from the surface for a short time and at room temperature. The process wastes can be treated and recycled. The ozonized water components can be easily controlled [166].

The reaction mechanism of chemical vapor deposition using tetraethylorthosilicate and ozone at atmospheric pressure is reported and discussed in Ref.167.

Covering of semiconductor devices with silica films is carried out via CVD using $Si(OEt)_4$ and ozone. The deposition process is followed by heating in oxygen atmosphere with simultaneous UV radiation. The SiO_2-film thus formed can improve the water resistance of semiconductive devices [168].

The fabrication of nondoped silicate glass film with flat structure by O_3-TEOS deposition includes three stages: 1) substrate treatment; 2) formation of Al conducting layer on it using oxygen plasma through heating; 3) formation of SiO_2-film on the substrate by plasma using ozone and TEOS [169].

In Ref. 170 are reviewed the future trends for interlayer dielectric films production and their formation technologies in ULSI multilevel interconnections. The properties of the interlayer dielectric films and their preparation technologies should satisfy the following three requirements: 1) available for disposition (setting) on large surface; 2) low dielectric constant; 3) low deposition temperature. Two techniques have been developed for the selective deposition of SiO_2-films which is the best way for achieving a complete planirization of the interlayer surface of dielectric films: low temperature liquid phase deposition using a saturated aqueous solution of hydrofruorosilicon acid H_2SiF_6; 2) half-selective SiO_2-films deposition at 390°C with $Si(OC_2H_5)_3$ and ozone and preliminary treatment with tetraflurocarbonic (CF_4) plasma on TiW or TiN surfaces [170].

Lead zirconate and titanate thin films are successfully prepared by reactive evaporation. The elements Pb, Zr and Ti are evaporated in ozone-oxygen mixture. The films obtained have equal thickness and composition per a large surface (in the range of ±2% from the 4-inch support) [171].

Ceramic coatings on substrates are formed in the presence of ozone [172]; the substrates (silicon checks) are covered (dipped) in a solution of one or more (partially) hydrolyzed pre-ceramic silicon alholates with general forrmula $R_xSi(OR)_{4-x}$ ($R = C_{1-20}$ alkyl, aryl, alkenyl or alkinyl; x = 0-2). Further, the solvent is evaporated to form a coating which is subsequently heated in the presence of ozone up to 40-100°C thus converting into a ceramic coating. It in its turn may be covered also by a protective layer containing Si, or Si and C, or Si and N, or Si, C and N, or SO_2 and oxide. These coatings which are abrasive-, corrosive and thermo-resistant have also a small number of defects and suppress the diffusion of ionic contaminants such as Na and Cl ions and are particularly convenient for electronic units [172].

Method of forming zinc oxide light-shielding film for liquid crystals shields includes the injection of vapors of alkyl zinc compound with ozone or atomic oxygen in the activation chamber after which they pass through a chamber for ZnO film deposition at low temperature heating at about 200oC. The method is applicable for large scale production of these films [173].

An evidence for a new passivating indium rich phosphate prepared by UV/O_3 oxidation of indium phosphide, InP, is provided in Ref.174. The phosphate does not exist as crystal compound and its composition is $InP_{0.5}O_{2.75}$. The passivating ability of the latter with respect to InP surface is discussed.

For improving the light-absorption characteristics of oxide optical crystals they are heated in to ozone-oxygen atmosphere [175, 176]. Thus the light absorption from the optical crystal in a wave range different from that with which the crystal affects its own absorption is reduced to the most possible level. Devices using such oxide optical crystals with improved absorption characteristics work with high efficiency as optical amplifiers, optical isolators, optical recording medium, optical generators.

The manufacture of solar cell modules with transparent conducting film covered by amorphous Si layer and electrode on the backside linked to the transparent isolator layer is described in Ref.177. The preparation includes the application of a laser beam for electrode molding and exposure in oxidizing atmosphere containing 0.5-5% O_3 or • 10% O [177].

A method for strong oxidation in ozone atmosphere is proposed for surface activation of photoconductive PbS films [178].

Adhesion-producing materials for electroless plating and printing circuits contain particles of thermostable material slightly soluble in the oxidizing agent; the particles are dispersed in the thermostable resin which becomes almost insoluble in the oxidizing agent at hardening. These materials find application for printing circuits and are treated by an oxidizing agents (for example, chromic acid, chromate, permanganate or ozone) to create a concave material surface [179].

7.3 MEDICINE

Upon exposure of living organisms to ozone, depending on doses, the different vital systems can be suppressed or stimulated to a certain extent.

For the first time already in 1898 Binz [180] showed interest in the effect of ozone on the blood and plasma and found that ozone exhibits a somnolent effect. Brinkmson and Lambert [181] proved that ozone inactivates the cell enzymes and decreases the oxygen exchange between blood and tissues.

A thorough investigation of the effect of ozone on the human organism should allow to extend the range of its application in medicine. The operation mechanism of ozone at small (therapeutic) concentrations, up to 100 $\mu g/m^3$, in the human cell has been already studied. Ozone enhances the cell membrane resistance to damaging effects. Human blood, treated with ozone, is able to keep erythrocyte membrane intact for 5 to 15 days. A method for preparation of nonhemolyzed plasma has been developed which facilitates laboratory diagnostic investigations and preserves the blood without inducing hemolysis [182]. *In vivo* and *in vitro* experiments have shown that ozone stimulates erythrocytes gas-transportation function which provides a continuous optimum tissue oxygenation. Besides, ozone strengthens leukocytes antibacterial function, the high activity of the mieloperoxidase enzyme system being preserved for 15 days. Also, ozone could find application in disinfection of biological materials, blood in particular. Encouraging results of HIV inactivation by means of ozonolysis have been reported in Refs 183-189. Blood, plasma, infected with retrovirus, whose transcriptase (RT) activity was $1.4x10^5$ cmp/ml (enzyme for retrovirus replication) was separated into three fractions - A, B and C. Fraction A and fraction B (after cryoprecipation) were exposed to ozone, 12 $\mu g/m^3$, at 0°C for 30 min. Fraction C was not treated with ozone. Subsequently, after 13 days of incubation, the following changes in RT activity were observed: fraction A = 130 000; B = 0 and C = 175 000.

In acute massive blood losses the adaptive potentialities of the human organism are increased. Besides, the operating strength of the intracellular mechanism of hemoglobin antioxidative resistance is enhanced, as deduced from superoxide dismutase activation in erythrocytes. Phrophylaxis in hemorrhages by means of ozone inhalation is due to the activation of biological processes which provide erythrocytes gas-transportation function. The application of ozone in medicine opens prospects to invent means for correcting the respiratory systems by changing the molecular regulation mechanism of hemoglobin affinity to oxygen in various hypoxic conditions. Methods can be developed to preserve blood cell vitality and dynamic activity during conservation and transfusion. There is growing interest in looking for techniques for ozone use in cardiovascular diseases. Combined effects and hyperbaric oxygenation in conditions related to hypoxia have also been studied.

The ability of ozone to oxidize toxic products formed under various pathological conditions, has attracted clinicians attention. Ozone either directly oxidizes double bonds (in unsaturated compounds) or, by activating free radical processes, initiates a danger of enhancing the spontaneous process of lipid peroxide oxidation. Kontorshticova et al. [190] studied the lipid and protein spectra of blood plasma exposed to ozone. Blood, taken from a dog 15 min after suffering from hemorrhagic shock, was exposed for 5 min to oxygen-ozone mixture at a rate of 1.0 l/min and ozone concentration of 0.048 mg/l/ The total protein content and separate protein fractions were analyzed. The fatty acids content was determined by gaschromatography. The protein and lipid spectra reflect processes which take place in hypoxia and subsequent oxygenation in the organism of test animals. These are: a) elimination of lipid depots of nonesterified fatty acids; b) nonoxidized or partially oxidized products of lipid and protein nature; c) blood saturation with oxygen and activation of oxygenation processes. An increase (by 20 %) in α_2-globulin content zeruloplasmin and haptoglobin, which possess antioxidative properties, is observed in the protein spectrum. The content of fatty acids is decreased by about two fold. The total protein and β_2-globulin concentration also decrease. These changes are consistent with a direct effect of ozone on peptide bonds in proteins and on double bonds in unsaturated fatty acids.

We have studied the effect of ozone [191] on POL of rabbit brain and liver. In this connection the rabbits were exposed to ozone action in a special chamber for 2, 3, 10 and 15 h. The ozone concentration in the gas phase was varied in the range 50-250 ppm. We have found a substantial enhancement of the content of endogenous POL? fluorescence products (FP) and TBA-reacting products (TBA -rp) after 2 h exposure to 250 ppm ozone concentration. However, after 4 h exposure to the same ozone concentration the rabbits die. When the concentration of exposed ozone was 50 ppm the increase in FP content in the rabbits brain is increased by about 1.8 times after 10 h, and in the liver - by about fold 2 compared with that of the control test. The level of TBA-rp was 3.1 and 3 fold higher, respectively. It has been observed that the amount of FP in the brain between the 10th and 15th h does not change, but the FP content in the liver microsomes rises 2.6 times and that of TBA-rp - 3.6 fold. A series of test animals, which were preliminary injected with 56 mg/kg potassium salt of DL-α-tocopherol 10 h before the test, were also studied under the same conditions. The results from FP determination after 10 and 15 h of exposure reveal that their amount in the brain has increased by 1.09 and 1.3 times, respectively and in the liver microsomes - 1.5 and 1.9 fold. Simultaneously there is no marked difference in the DL-α-tocopherol content before and after exposure of the test animals as demonstrated by the HPLC data. These facts indicate that the brain is either more stable or more protected in regard to ozone action since DL-α-tocopherol protects first of all the biomembranes.

The effect of tocopherol and more than 20 its homologues and derivatives was investigated in model reaction of cumene oxidation and in enzymatic and enzyme-free POL in rabbits brain and liver microsomes [192-196]. It has been shown that the inhibitor efficiency depends rather on their diffusion coefficients through the membrane biolayer as well as their mobility than on the chemical nature of the substituents. A correlation between the rate of POL inhibition and the suppression of the luminol-dependent chemiluminescence in biomembranes has been established. However, no relationship was found between the action of 2,6-di-*tert*-butyl-phenol inhibitor and its derivatives and the chemiluminescence produced during the interaction between superoxide anion and luminol.

Fairly high doses of ozone can be employed to saturate blood with oxygen and deactivate nonoxidized and partially oxidized products. Therapeutic doses of ozone, 0.048 mg/l, are applied to correct pulmonary insufficiency in postreamination periods.

High concentrations of ozone, 0.2 mg/l, can be used to model pathological conditions which allow to study morphological changes in pulmonary tissue. The effect of ozone upon varying the mode of administration in organism has also been investigated [197].

An isotonic NaCl solution saturated with ozone was injected intracutaneously or a saturated aqueous solution was administered orally to rabbits. In animals, injected with ozonized water, the red blood cell level remained stable in the course of experiment (10 days). An increase in leucocyte level (2 fold) was observed at the expense of monocytes. Changes of the granulocytes were not pronounced. The functional activity of neutrophilic cells, evaluated by phagocytosis activity, increased by 1.5 fold. On intracutaneous administration of ozonized isotonic NaCl solution, the leucocyte level was diminished after the first hour, enhanced after the third hour and remained high by the end of experiment with respect to the initial level. By comparing the two methods, it was established that the toxic effect was more pronounced with the intracutaneous administration than the oral one. Thus, for example, the activity of the process of lipid peroxide oxidation, evaluated through malonic dialdehyde determination in erythrocytes, increased by 3-fold upon intracutaneous injection and did not change after oral administration. On exposure of blood from healthy donors aged 25-35 to 3 mg/l ozone the activity of the aspargataminotranspherase was increased whereas that of lactic dehydrogenase attained a maximum concentration and then fell down to the initial one.

Ozone therapy, being an effective means to treat a number of diseases, attracts the attention of practicing clinicians, dermatologists in particular [198]. Attempts have been made to treat various dermatoses; psoriasis, herpes, microbial eczemas, etc. [199]. The obtained results were positive. In this connection a joint project with the dermatology department at the Sofia Medical Academy is developed. Its object is to study the effect of ozone on dermatologically injured skin and development of ozone generating units.

The effect of ozone has focused scientists attention for more than two decades. Of particular interest is the oxidative effect of ozone on cells, especially tumor cells [200-202]. Cells from alveolar adenocarcinomas, breast adenocarcinomas and uterine carcinosarcoma and normal cells from lung diploid fibroplast as a reference were incubated in a special chamber. They were exposed to ozone (0.3 to 0.7 ppm) for 8 days. The number of cells was determined every 48 h. Ozone levels from 0.3 to 0.5 ppm inhibited the accumulation of tumor cells by 40-60% and did not affect the normal cells. The inhibition was 90% at a concentration of 0.8 ppm.

Many viruses are sensitive to ozone, for example *vesicular stomatitis*, *encefalomyocarditis*, *poliovirus type 2 and 3*, *coxsacknevirus*, *echovirus* and *adenovirus* [203-205]. The authors have studied the inactivation of rotaviruses HRV type 2 from humans and SA-11 from apes for various exposures of ozone

(0.05, 0.1, 0.2 mg/l) at 4°C. HRV are considerably more sensitive to ozone than SA-11. In this connection ozone might find therapeutic application in oncology.

7.3.1. Therapeutic Application

As a therapeutic means the so-called medical ozone is employed in concentrations from 0.05 to 5%. The types of application and the relevant concentration range of medical ozone/oxygen mixtures are specific for each pathological change (Table 7.6).

Table 7.6

Therapeutic doses

Types of applications and appropriate range of doses of ozone/oxygen mixtures for medical purposes

Colonic insufficiency and external exposure in closed system ↓

Intravenous
Intramuscular injection ↓
Hypodermic

Blood treatment↓
Stimulating treatment ↓

Production of ozonized water

Circulatory ↓

Cleaning of wounds and disinfection

0 10 20 30 40 50 60 70 80 90 100

Ozone concentration, μg/ml

As seen the ozone concentrations administered by injection reach 40 mg/l. Upon extracorporeal blood treatment, stimulation and circulatory treatment ozone is employed in concentrations up to 30 mg/l whereas for colonic insufflations, external gas application in closed system, cleansing of wounds and in ozonized water production it can be used up to 100 mg/l level.

Ozone therapy is successfully applied in a wide variety of diseases as shown in Table 7.7.

Table 7.7

Some examples of diseases and medicine fields of ozone application

Disease	Medical field
Anal fissure	-Dermatology / proctology
Acne	-Dermatology
Abcess	-Surgery / dermatology
Allergies	-General medicine, dermatology, -allergology
Virus infections	-AIDS, various special fields
Arterial circulatory disturbances	-Surgery, vascular surgery
Fistulare	-Proctology, dermatology, urology, gynecology

Mucous colitis	-Proctology, gastroenterology
Cystitis	-Urology
Cerebral sclerosis	-Gerontology, internal medicine and neurology
Decubitus (ulcers)	-Surgery, gerontology and dermatology
Furunculosis	-Dermatology
Gangrene	-Surgery
Ulcus cruris	-Dermatology
Hepatitis	-General medicine, internal medicine
Hypercholesterolaemia	-Internal medicine
Herpes (AIDS)	-Dermatology, internal medicine
Climacterium/menopause	-Gynecology
Cirrhosis of the liver	-Internal medicine
Raynaud's disease	-Surgery, vascular surgery
Constipation	-Internal medicine
Radiation scars (following irradiation/treatment)	-Radiology, dermatology
Osteomyelitis	-Surgery
Oncological additives	-Oncology and relevant fields
Parkinson's disease	-Neurology
Mycosis/fungus infections	-General medicine, dermatology
Polyarthritis	-Internal medicine, neurology
Spondylitis	-Orthopedics, surgery
Stomatitis	-Odontology, dentistry
Sudeck's disease	-Surgery, orthopedics
Thromboflebitis	-Internal medicine, dermatology
Wound healing disturbances	-Surgery

As a rule, the ozone-oxygen mixture is employed in cases of arterial circulatory disturbances. This method was first described by Dr. Lacoste in 1951 and since then has attracted the attention of many scientists. Rokitansky [207] was the first who made a statistical evaluation, according to which four out of ten patients with arterial-metabolic circulatory disturbances treated with ozone did not need operation. The results of other authors confirm his conclusions [208, 209].

Rectal application method (called ozone enema) was reported in 1935 and its successful application has been proved in cases of *mucous colitis* and *fistulae*. It was established that the oxygen content in the blood was increased [210, 211].

Ozone has shown an antalgic effect, this property being used through injections in case of pains of the spinal column [21]. Care should be taken of sensitive patients, since the initial pain on the injection is considerable. Usually 30-50 cm^3 ozone is injected intracutaneously in the lumber and sciatic region to relieve pain which is observed after 1-2 h and may continue for a couple of days.

Wolff [213[reported that as early as 1915 ozone was provided for local gas treatment in cases of *fistulae*, *dicubitus* (*ulcers*), *ulcus cruries*, *osteomyelitis* as well as badly healing wounds. For that purpose plastic bags filled with gas mixtures were employed. This therapy became particularly useful for inaccessible ulcers, fistulae and damage from X-radiation.

Payr [214] applied subcutaneous injection of ozone gas in cases of venous circulatory disturbances and varicosis. For the same cases, he also used intravenous injections. He also reported that Dr. Fisch had tested this method on himself without side effects.

The intraarterial application of ozone therapy is promising for orthopedic practice and for accident surgery. Intramuscular injection of oxygen-ozone mixture is mainly applied in some cases with obstinate carcinoma patients or as a supportive therapy together with the so-called 'minor" autohemotherapy. This is a successful therapy: painless, harmless and widely applied in inflammatory processes.

Dorstewitz [215] employed autohemotherapy to treat hepatitis. 50 - 100 ml blood from the patient are treated with ozone and then reinjected. 20 patients with herpes simplex and herpes zoster were treated in the same way for 5-10 days. A complete healing of skin lesions was observed. In one of the cases of extensive herpes in the trigeminal region with keratitis, healing was achieved within 10 days, although a persistent opacity of the keratic tissue remained. In six cases of herpes simplex, upon double treatment with ozone, a positive effect was observed after 28 h. It is desirable to start therapy at an early stage of illness.

Moreover, the so-called hematological ozone therapy (HOT) has found wide application [216]. The therapy has been employed since 1947. In a special hyperbaric chamber the air is enriched with ozone by means of UV-Cs lamp. It has been suggested that singlet oxygen is formed under these conditions. 60-70 ml blood from patients are placed in the HOT apparatus and enriched with oxygen-ozone. The blood thus treated is reinjected intravenously or intramuscularly into the patients. HOT therapy is applied in the following affections: chronic ulcers and gastritis, prophylaxis in recoveries from myocarditis; lipid metabolism disturbances hypercholesterinae; chronic liver trouble and recoveries from infectious hepatitis; prophylaxis in acute hepatitis, donor blood preliminary treated with ozone-mixture migraine, chronic nephritis with increased creatine and urea levels; acute artery retinae occlusions, chronic polyarthritis; scab treatment after burning; acute bleeding of the eye. Most likely, HOT therapy enhances prostoglandine synthesis. Following HOT therapy, the cholesterol level of 32 patients was examined (322 determination tests). It was found that the cholesterol level dropped at an average of 20% with respect to the initial values. It was also demonstrated that this therapy caused no toxic effect on the marrow by blocking the enzyme system.

Major autohemotherapy is applied if the artery in question is not accessible to palpation. However, in cases of milder circulatory disturbances, intramuscular and/or subcutaneous injections may be sufficient.

Rokitansky [217] widely recommended an external ozone gas therapy in the advanced stages of gangrene which prevents from further inflammation. He has indicated that as a result of intravascular therapy applied in gangrene of the toes of the foot, the average rate of amputations at the upper thigh is reduced from 15 to 8 % in stage III. In some cases during the post-operational period, an extensive formation of necrosis with purulent secretion might start to develop. However, by making use of intraarterial ozone injection and local ozone-oxygen gas treatment, it became possible to improve circulation and oxygenation of the tissue. Ozone therapy was also used to treat gas gangrene. Guinea-pigs were treated every day with 450 μg ozone. The death-rate decreased from 97 to 72 % [218]. Peripheral artery circulatory disturbances are treated with intraarterial ozone injections at a maximum concentration of 33 μg/l. In autohemotherapy the concentrations should not exceed 40 μg/l because of risks to develop hemolysis [219].

In cases where the microorganisms come into direct contact with ozone in a sufficient quantity, killing of the microbes most certainly take place via the oxidative decomposition of the capsid/protein shell of the virus, thus exposing their DNMA and RNA components to open attack; the presence of DNA and A catabolism products could be demonstrated empirically [220].

In the treatment of virally produced diseases such as herpes or hepatitis, however, the circumstances are quite different. In this case, a completely different method of application must be applied, which is safer for the patient, the choice here generally being in favor of "major autochaemotherapy" whereby the patients own blood receives an admixture of ozone upon withdrawal from the body and is immediately put back into system whereas during ozone therapy session lasting for 5 min a wound receives almost 800 mg of ozone, as acute case of virus infection needs only 10 mg of ozone in autohaemotherapy. It is for that reason that, as an active mechanism, a killing off of the virus does not come into question, such as is the case for wound treatment: instead there are two processes to be discussed here which probably take place simultaneously: 1) virus inactivation through ozone or ozone peroxides and 2) a peroxide intolerance of those body cells infected by viruses. Therefore, this means that in the ozone/oxygen therapy, not only the reaction of ozone itself, but also those of its reaction products must come under discussion. Ozone reacts with unsaturated fatty acids in the cell forming ozonides and peroxides. The intervention of ozone and/or its peroxides ought to take place at the acceptors of the free virion, whereby the virus-to-cell contact, and thus the entire reproductive cycle is interrupted. The free pair of electrons on the nitrogen of the N-acetylglucosamine probably represents the point of attack for the electrophilic ozone molecule or its peroxides, so that the virus is blocked for further reaction by oxidation, making it effectively inactive.

The infected cell, however, produces hydrogen peroxide as a defensive function. Apart form this, the electrophilic reaction of ozone can also take place with unsaturated fatty acids, as a membrane constituent of the virally infected body cell, and injected peroxides into the cell. The enhanced quantity of peroxide in the cell affect the cell processes through two ways: 1) the peroxides destroy the membranes of the human cell before reproduction cycle of the infected viruses has been completed or 2) produce a synergetic effect, destroying the microorganisms which have penetrated the cell. Intravascular ozone/oxygen injection using small quantities of ozone constitutes a direct intervention in the metabolic processes, particularly in that of erythrocytes, participating in the normal glucose and pentose phosphate cycle. The final product is 1,3-diphosphoglycerate which is of decisive importance for the deoxygenation of hemoglobin.

Under the effect of peroxides, obtained as a result of ozone reaction with unsaturated fatty acids in the erythrocytes membranes, the detoxifying mechanism of these substances is immediately put into action via the glutathione system. In order to maintain the glutathione redox system, the pentose/phosphate path must be accelerated in order to cover the ADPH debit as a reducer of GSSG. This means, simultaneously, an increased breaking down of sugar. The decisive product of the accelerated glucose metabolism is the presence of 2,3-diphosphoglycerate (2,3DPG) which represents a key compound in the curative effect of ozone:

$$HbO_2 + 2{,}3\ DPG \Leftrightarrow Hb^{\bullet}2{,}3\ DPG + O_2$$

Every increase in 2,3-DPG levels facilitates the release of oxygen through a shifting of the HbO/Hb balance in favor of deoxygenated hemoglobin. As hemoglobin is very stable, particularly in the case of diabetics, ozone therapy is particularly indicated for this decrease and brings about an improvement in the peripheral oxygen supply. It is possible, by measurement of the arterial and venous partial pressure of O_2, to observe *in vivo* the change in the oxygen situation during the course of treatment. However, not only the increase in arterial pO_2 but primarily the reduction in venous pO_2 is decisive for an increased supply of oxygen, which is equal to an increase in the arteria/venous pO_2 difference, even up to normal values of pO_2 = 60 mm Hg.

Immunoglobulins are proteins produced from lymphoid cell, which are unified according to similar general principles of their structural organization. The functional integrity of various classes of immunoglobulins is reduced to their participation in homeostasis, the most important factor of immunologic defense systems. Immunologic processes are associated with the presence of antibodies. Five classes of immunoglobulins are known: IgG, IgA, IgM, IgD and IgE. Changes of normal levels cause certain troubles, for example, increased levels of IgG lead to chronic liver deseases. The effect of ozone on the immune system was investigated by monitoring changes of immunoglobulins levels in the blood [221].

Nineteen patients with cervix carcinoma and 20 patients with ovarial carcinoma were entered into a study. They were injected intravenously with ozone-oxygen mixture (540 μg ozone) for 5 days. However, the therapeutic effect was not confirmed.

34 patients with chronic rheumatic troubles underwent autohemotherapy through intramuscular injections with 5 ml blood from patient enriched with 300-350 μg ozone.

1000 μg ozone were administered intravenously into 47 patients every week for 6 months (50-60 ml blood from patient enriched with ozone). Statistical data processing indicated insignificant changes of the IgA, IgG, IgC and IgM levels. Therefore, no immunodeficient effects were observed. Certain changes of IgG levels can be evaluated as immunostimulating.

The inhibiting effect of ozone on tumor tissue has been investigated by Sweet [222]. Zanker confirmed this effect, his attention being attracted to the effect of ozone on tumor metabolism.

Steiner et al. [223] studied the changes in the level of: DPG, DPN, F-6-PK, GAPDH, HBDH, ICDH, LDH and NAD in carcinomatous and healthy ovarial tissues (*in vivo*) under the influence of ozone. Cytostatics like Adriblastin and Xoloxan were also used for reference purposes. The lysolecithin levels decreased in the tumor tissue but did not change in the healthy one, which indicates that ozone has no influence on the tumor metabolism. However, the ozone therapy has a positive effect in employing ozone in tumor tissue.

Brinkman and Lamberts [224] were the first who reported on the effect of ozone on erythrocyte(RBC's) behavior. These authors exposed volunteer patients to the influence of ozone (1 ppm) for 10 min and then measured the oxygen evolved from hemoglobin. Upon prolonged ozone exposure oxygen is not bound with hemoglobin which causes blood-vessel occlusions in fingers and toes. Other authors [225] have observed changes in red blood cell biochemistry, the RBC's being taken from mice and apes. Brikley et al. [226] have investigated the morphological and biochemical changes in human red blood cells for ozone concentration of 0.5 ppm. This study indicated a 20% increase of erythrocyte shortness, a 20% drop in acetylcholine esterase activity, a 15% fall in glutathione levels (GSH) and a comparatively slight increase in the activity of lactic dehydrogenase (LDH). Some authors [227, 228, 230] have suggested that, due to inhaled ozone, the amount of oxygen free radicals is increased. As a result of this, hydrogen peroxide, which might be a notable oxidative metabolite, also increases its levels. The catalase activity in red blood cells is diminished by 50% and the intracellular hydrogen peroxide is formed upon ozone exposure 6-6 ppm.

Changes in erythrocyte membranes stability have been observed [229]. Ozone concentrations of 0.25 ppm from 1 to 6 h change the erythrocyte spherical shape which may change the normal capillary permeability. Schulz et al. [230] have found that exposures of 0.7 ppm ozone for 2 h increase the methemoglobin levels and decrease the total number of red blood cells. In some mice and Guinea-pigs the hemoglobin and hematocrit levels were increased at high concentrations of ozone as compensating response against the decreasing number of erythrocytes.

Many authors [231-233] have investigated the ability of vitamin E to protect erythrocytes and lung tissue from the influence of ozone because, being a biological antioxidant, ozone interferes with the metabolism and protects other vitamins, hormones and enzymes from oxidation. Chow and Kanes [231] found that under vitamin E deficiency in mice blood during a 7 day period with 0.8 ppm ozone, the activity of glutathione peroxidase, pyruvate kinase and LDH in the erythrocytes was enhanced while the total glutathione level was decreased. The latter maintains the normal functional state of erythrocytes, low redox potential of the cell constituents which preserves the thiol groups of a number of enzymatic proteins in reduced state. The principal function of glutathione is to eliminate hydrogen peroxide formed during direct oxidation of certain drugs. The toxic effect of hydrogen peroxide is due to its ability to oxidize hemoglobin to methemoglobin and to form hydroperoxides with unsaturated fatty acids which lead to changes in the lipids of the cell membranes.

The effect of ozone on erythrocyte membranes has been investigated with respect to the SH-groups in some enzymes. At an ozone concentration of 40 mmol/min over 5 ml membrane suspension, the activity of ATPase is totally lost. The latter can be regenerated by means of phosphatidilserine [234]. Peterson [235] observed changes in T-lymphocyte and B-lymphocyte levels for an ozone concentration of 784 $\mu g/m^3$ for 4 h.

Canada and Airriess [236] studied changes in the concentration of vitamin E, vitamin A and vitamin C (antioxidants) in the human serum after exposure to ozone concentration of 0.37 ppm for 2 h. These changes were insignificant. The alkali phosphotase in the serum was increased at ozone concentrations of 10 ppm for 1 h. This was explained by the fact that the blood tissues let the enzymes pass because of a visible danger to lungs at these concentrations. Other authors [57, 237, 58, 238] have followed the concentration levels of prostoglandines PGFa and PGEa during exposure of mice to ozone at concentrations of 4 ppm for 8 h. They observed an increase by 186 and 200% respectively, which was attributed to arachidonate oxidative induction by plasma membranes. Veninga and Wagenaar [239] established that creatine kinase (CPK) levels were increased after exposure to 0.02 ppm ozone for 2 h but, unexpectedly there were no changes at higher ozone concentrations. Probably, this effect is associated with some adaptation of the organism. There was response in serum lipids [240] at 1.1 ppm ozone for 24 h; the activity of lecithincholesterol transferase, free cholesterol, and lipoproteins was increased, however, there was a descending tendency in triglyceride levels. The bacterial effect of ozonized blood serum on gram negative bacteria (*Klebsiella*, *Pseudomonas* and *Salmonella*) has also been studied. This effect is dependent on the serum ozonation time, type of bacteria and bacteria membrane structure [241]. Prolonged exposures to ozone (0.4 to 1 ppm, 6 h daily) caused a drop in serum albumen levels and increased the α-globulin and β-globulin levels. The total protein level was not changed [242].

It has been established that ozone affects the central nervous system (CNS) [243]. Exposures to > 0.6 ppm ozone concentrations cause a headache and lethargic states. Changes in the visual sharpness (0.2 ppm) and eye muscle may occur [244]. A fall in the cerebral levels of catechol-*o*-methyl tranferase on exposure of dogs to 1 ppm of ozone (8, 12 or 18 h) was observed [245]. The reduction of enzyme activity was in a good correlation with a decrease in catechol amounts in the brain.

Takanashi and Miura have reported a considerable fall in the activity of metabolite enzymes in the liver: benzopyrene hydroxylase, 7-ethoxycumarine-*o*-diethylase and aniline hydroxylase. No effects were observed on *p*-nitroanisole-N-dimethylase on exposure to ozone of 0.8 ppm for 7 h. A 20% drop in cytochrome P-450, NADPH cytochrome, P-450 reductase, and cytochrome B5, included in mitochondria and/or ergastoplasma membranes, was established. Fluorescent pigments in the liver, being an end products in lipid oxidation, were increased upon exposure to 1.5 ppm ozone for 30-60 min. The effect of ozone concentration 1.0, 0.5 and $0.2.10^{-5}$ M has been investigated on mice for exposure time 2, 3 and 15 h [249, 250]. The quantity of cytochrome P-450 in liver microsomes at 1.10^{-5} M and 2 h exposure is 73.4 % with respect to the control level and B5 - 93.3 %, respectively. After 3 h under the same conditions these parameters are already 75 and 69.6 %, respectively and after 4 h the mice die. The cytochrome level at ozone concentration of $0.2.10^{-5}$ m and exposure time of 15 h is 91.3 and 93.2%, respectively. Simultaneously 10 h before the exposure of mice to ozone atmosphere the potassium salt of α-tocopherol is administered intraperitonitously - 56 mg/kg. This leads to increase in P-450 level to 158.9 % and that of B5 to 136.6%. In the control mice this leads to increase of their levels with 135 and 116%, respectively.. These data allow to assume that tocopherol forms adducts with cytochromes and ozone stimulates the adduct formation.

Zelac and Cromroy [251] have established that ozone is a better mutant than radiation. Humans exposed to 0.5 ppm ozone for 8 to 10 h manifested chromatid and chromosomal aberrations. No increase in chromosomal aberration was observed at 0.4 ppm ozone for 4 h.

In tracheotomy with dogs, with acquired interrupted respiration accompanied by hypertension and bradicardia during exposure to ozone (50 ppm, 4 min), the symptoms decreased due to stimulated *parasympathetic vagus nerves* [252, 253]. Exposure of rabbits to ozone (0.2 ppm, 5 h) gave rise to pathological changes without myocardial damage but cell membranes did not remain intact.

Ozone causes a multiplex effect on the endocrine system [254]. The authors report an inhibiting effect on ^{131}I(ozone 4 ppm, 5 h) which was observed even 12 days after exposure. Plasma thyrotropin was reduced by 50%. The thyroid gland increased its size. The authors suggest that decreased thyroid hormone levels appear to be prophylactic against ozone damage.

Gordon et al. [255] have reported that upon inhalation of ozone (0.8 ppm, 1 h) the activity of choline esterase was diminished not only in the blood, but also in the diaphragms of Guinea-pigs by 14% and lung tissue (16%). The investigations]256] of the effect of ozone on the immune system show that exposures of 0.6 ppm for 2 h did not affect the number of T-lymphocytes. However, an exaggerated spleen was observed during ozone exposure of 0.31 ppm for 103 h. Kidneys manifested dilated tubules.

7.3.2. Toxic Action

Mechanisms, according to which ozone damages biological tissue, are not clearly defined. However, several processes should be considered which lead to disturbances in normal cell functioning. Ozone oxidizes polyunsaturated fatty acids (PUFA) in the cell membranes according to Criege's mechanism [257]. Also, it contributes to the increase of malonic dialdehyde content and interacts with natural antioxidants (vitamin C, vitamin A and vitamin E). On the other hand, ozone reacts directly with aminoacids in tissues, proteins, and small peptides [240]. It inactivates the following enzymes: lysozymes, betaglucoronidase and acid phophatase, while urease is resistant to ozone.

Very likely, the cellular regulation of the electrolytes becomes worse, and in consequence of this, the enzymes become inactivated and the mitochondrial metabolism is changed. Normal cell functioning is disturbed which might lead to cell destruction [258].

Lung tissue, being vulnerable to ozone attack, has been the object of many investigations. Biochemical studies indicated changes of enzyme levels, the enzymes being an integral part of the intracellular protection from disturbances. These enzymes bind with free radicals, lipid peroxides, and hydrogen peroxide which are factors for changes of tissues. Chow and Hussain [259] have reported enhanced activity of glutathione peroxidase in mice inhaled with 0.8 ppm ozone for 3 days. Other natural antioxidants, such as superoxide dismutase, catalase, vitamin E, and vitamin C, destroy the free radicals, thus contributing to reduce the risk of permanent tissue damage due to ozone. For example, bronchia constriction, caused by ozone, fades away or disappears after treatment with vitamin C or with combination between vitamin E and vitamin C (synergistic effect). Complementary biochemical investigations of ozone toxicity involve measurements of the enhanced synthesis of collagen which induces fibrosal changes in lung tissue [260, 261].

Exposure of mice to small doses of ozone [262] gave rise to changes in two enzyme systems: 1) a decrease in the hepatic ascorbic acid (HAA) concentrations and 2) a change in the creatine phosphokinase (CPK). HAA was considerably changed for 30 and 120 min after cessation of the ozone exposure whereas CPK demonstrated changes only 15 min after that. Compared to reference animal, no significant changes were observed after 24 h for HAA and 30 min for CPK. Plasma histamine lactodehydrogenase was not affected with various doses of ozone. It is assumed that this is due to the ability of organisms to adapt themselves to the harmful agent, which, on the other hand, stimulates metabolite processes and enzyme activity. At higher doses of ozone the enzyme activity gradually falls down, probably due to some compensatory balanced conditions. At ozone concentrations of 0.8 ppm (8 h) the activity of the total lactate dehydrogenase (LDH) in the lungs was enhanced but not in the plasma and erythrocytes. The LDH-5 fractions from LDH isoenzymes considerably decreased in terms of their distribution in the lung and plasma, while fraction LDH-4 was increased. No change of isoenzyme activity in the lungs, plasma and erythrocytes was observed for ozone concentration of 0.5 ppm [263].

Saymor [264] has studied the behavior of mice upon prolonged exposure (120 days) to small doses of ozone. The DNA and RNA levels in the lung were decreased whereas the protein level was increased.

In vivo and *in vitro* high ozone concentrations are prerequisite for suppressed mitochondrial oxygen consumption and oxidative phosphorization and for increased membrane permeability. This can be associated with thiol groups oxidation and lipid oxidation [265].

Ozone exhibits a 100% bacteriological effect at concentrations of 0.023 mg/l without producing spores. Spores are destroyed at 0.61 mg/l [247]. At the same concentration, 87.5% of the hepatic virus B is inactivated. Large amounts of blood can be conserved without clotting by bubbling with oxygen-ozone mixture. Parallel to this, all the microbial materials in blood are oxidized, thus sterilizing it [266].

A detailed evaluation of the short-term influence of ozone on human health has been made in Refs 260-270. The transitory effect of ozone on the respiratory function was investigated by various respiratory measurements.

Miller et al. [271] have made comparative models of ozone operation regions in the respiratory tract, and Criebel and Smith suggested a new pharmacological model of ozone acute influence on human lung and have compared the effect of ozone on the lung (ozone 120 ppb, 2.5 h) in children and adults. They observed a considerable lowering of PFR and FEV values with adults compared to children. They succeeded in measuring the PFR baseline: 3.9 and 2.7% with adults and children, respectively. Adults made more symptomatic complaints than children. Stock and Kotchmar [272] reported that there is almost no risk of asthmatic attack in patients suffering from asthma at ozone concentrations over 120 ppb.

Many authors have investigated the influence of ozone on pulmonary lysosomes and mitochondria. Lysosomes contain quite many enzymes, mainly hydrolases, such as protease, lipases, nucleases, phosphatases, phosphodiesterases, etc. Dillar et al. [273] have studied the changes in lysosome levels upon exposure of mice to ozone (0.7 ppm). The activity of pulmonary lysosomal cathepsin A, cathepsin D, acidic phosphatase, β-N-acetylglucose aminidase, and benzyl arginine β-naphthylamide amidohydrolase was increased. This is due to inflammatory processes caused by ozone. The specific activity of protease and peptidase in the lung is enhanced when related to chronic obstructive genetic lack of α-1-antitrypsin factor.

Mitochondria are contained in every cell. A highly organized and specialized transportation chain, which transfers protons and electrons to molecular oxygen, is located in the internal mitochondrial membrane. It is composed by enzymes, redox systems, which organize proteins and lipids, and is called respiratory chain. The external membrane contains enzymes from the tricarboxylic acid cycle, fatty acid β-oxidation, etc. It has been established that at high ozone levels (2 ppm) the oxygen consumption of pulmonary mitochondria in rodents is decreased whereas this consumption is increased at 0.8 ppm ozone after inhalation for 20 days[274]. Mice, exposed to 0.8 ppm ozone, obtained an acute ozone intoxication due to blocking of the alveolar capillary function, followed by hemolysis and lethal hyloxia [275, 276]. The authors in Ref. 277 have studied in detail the cellular, biochemical, and functional effects of ozone on the lung.

Clearance of foreign materials (particles) from the lung and airways depends on the function of macrophages, ciliated cells and secretory cells, and on the physical and chemical properties of the alveolar cells. All these are affected by ozone exposure. Single acute exposure of animals and humans to ozone concentrations less than 0.6 ppm have been shown to accelerate clearance of particles from the tracheobronchial tree whereas acute exposures to ozone levels greater than 0.6 ppm caused a delay in particles clearance [79, 261]. Repeated exposures (2 h daily for 14 days) to 0.1 ppm ozone gave rise to acceleration in alveolar clearance of latex particles but had no effect on tracheobronchial clearance. These results, related to morphological studies, are consistent with certain adaptation [278-283].

The alveolar macrophage (AM) take part in defending the alveoli and airways against inhaled particles and pathogens. To carry out this important functions, AM maintain active mobility, phagocytic activity, membrane integrity and enzymatic capacity. Ozone exposures of AM *in vivo* has been shown to alter the number and morphology of AM, depending on dose and duration of exposure. A transient decrease in the absolute number of AM has been observed following exposures to high ozone concentrations [284, 285]. AM recovered from rats exposed for 16 h to 0.05, 0.10, 0.20 or 0.40 ppm ozone showed no differences in cell yield or cell viability. An increase in AM numbers has been also observed after *in vitro* ozone exposure [286]. Recruitment and persistence of AM result from chemotaxis and migration of cells into the lung alveoli and from local proliferation. A primary function of AM is the maintenance of sterility in the lung by phagocytosis of foreign particles. Changes in the phagocytic activity of AM has been observed following exposure to ozone. A marked depression of this activity was observed after exposure to 1.2 ppm ozone for 5 h [287] but enhanced phagocytic function has been observed after prolonged exposures for 20 days (3 times daily) to 0.8 ppm ozone. This difference in the phagocytic function of AM for various ozone concentrations and exposure conditions can be explained by changes in the cell populations. Ozone exposures also affect the secretory activity of AM and their ability to release factors which stimulate the migration of neutrophils and other cell types into the lung. A culture monolayer of rabbit AM was exposed for 2 and 6 h to 0.1, 0.3 or 1.2 ppm ozone. Biochemical functions of alveolar macrophage can also be affected. Ozone exposure caused a reduction in intracellular concentrations of acid phosphatase and β-glucoronidase, as well as in lyposomal enzymes [289, 290].

Exposure of the upper airways of mice to ozone at 0.5-0.8 ppm for 1-24 days has induced a two-fold increase in lgA^{+}B-lymphocytes. In the deep lung, exposure to 0.7 ppm ozone for 14 days causes an accumulation of T-lymphocytes [291, 292]. The number of leucocyte cells in controls, $3.8.10^5$, was increased to $7.5.10^5$ cells in exposed mice. Ozone exposure of lymph nodes of euthymic mice indicated that the histopathologic canges were more extensive in the athymic animals [293].

Changes in the immune function depend on the organ, dose and duration of ozone exposure. Concanavalin A (Con A) stimulates the cell-mediated arm of the immune system. Con A reactivity was unaffected at days 4 or 7 of 0.7 ppm ozone exposure (20h/day) but increased by 1.5 and 3 fold after 14 and 28 days, respectively. In contrast, splenic cells of mice manifested decreased reactivity to T-cell mitogens *in vitro* (phytohemagglutinin, PHA, 38% decrease, Con A, 75 5 decrease) but not to a B-cell mitogen (lipopolysaccharide, LPS0 or to alloantigens [294].

In rats, 7 days of exposure to 1 ppm ozone enhanced the reactivity of splenic cells to PHA, Con A and LPS [295]. Ozone inhalation gives also rise to nonspecific changes. Exposure to ozone at concentrations of 0.4 to 0.8 ppm can induce weight losses and/or thymic and splenic atrophy. A similar effect has been observed if steroids are injected into mice, i.e. ozone can produce steroid-mediated effects on lymphocytes.

The immune response of the human organism to a foreign agent involves the following cell quartet: macrophage, B lymphocytes, T helper, and T suppresser. Amoruso et al. [296] have determined the anion peroxide radical (O_2^-), produced by alveolar macrophage from mice exposed to ozone levels of 3.2 to 10.5 ppm. They found decreasing concentration levels. The authors proposed that this decrease in anion peroxide radical amount cannot be due to a direct effect of ozone on cell viability but rather to an infection caused by ozone, i.e. the immune system of experimentally tested animals enhances its activity. Alveolar macrophage from test animals, if stimulated by phorbal myristat acetate, also decreased the concentration of O_2^-. This explanation seems to a some extent confusing since the bactericide action of ozone which is very strong and in case of infection it would rather prevent than stimulate it.

Lung lesions observed after one week of exposure to 0.2, 0.5, 0.8 ppm ozone were more expressed than those observed after 3 months of continuous exposure. Some researchers [297, 298] suggested that the enzyme levels in the lung are increased which may account for the development of adaptation and tolerance to ozone.

Epithelial cels type I and squamous cells appear to be most sensitive to ozone [299]. An exposure to 0.5 ppm ozone increases the volume but reduces the area of epithelial cell types I [300]. These cells are regenerated to the normal condition after a week in fresh air and develop certain tolerance to ozone reexposure [3301, 302]. The authors have studied tracheal cells (volume, density) by electron microscopy. They have found that all changes due to chronic ozone exposure were reversible, the normal values being regained if test animals are exposed to fresh air.

Pulmonary fibroses, defined as abnormal accumulation of collagen, are developed after prolonged exposure to ozone levels greater than 0.5 ppm [303, 304]. Chronic ozone effects at concentrations close to that in atmospheric air are hardly noticeable and cannot be measured. Other scientists [305] have exposed mice to 1 ppm ozone for 40 days and found increased levels of total lung collagen but this effect rapidly faded away after 10 days of exposure to fresh air. Chronic exposure to 0.06 ppm ozone (13h/day, 5 days weekly, 18 months) caused changes in the lung function including a decrease in the general respiratory effect.

7.3.3. Ozone Solutions

Ozone disinfectant and sterilizing properties were used to freshen water in Germany at Wiesbaden yet in 1902. On this basis, in 1934 the dentists started utilizing ozonized water for inflammatory processes. This was reported at the Sixth World Congress on Ozone held in Washington [306]. The following properties of ozone are used in this case: a) disinfectant and sterilizing effect; b) haemostatic effect, especially in cases of running hemorrhages; c) accelerated wound healing, improved oxygen local supply and support to metabolic processes [307, 308]. Ozonized water, in the form of spray or stream, is used for the antrum in the following cases: gingivitis, paradontosis, white-gum, stomatitis. Preference is given to the spray mode, particularly in cases of caries for cleaning and disinfection, as well as for injuries and bleeding in the antrum. It is applied in cleaning nerve canals and pulpa tissue, in case of painful gingivitis and stomatitis. Ozonized water has also found wide application in orthopedy: artificial denture and crown cleaning prior to fixing, stopping of hemorrhages during fixing.

The ozone solutions in olive oils (ozone is present as ozonides) have found wide application in dermatology for treating: fistulae, decubitus and ulcer cruries. It is also successfully applied in mycosis and fungus finger and fingernail infections. Simply, patients has to put their fingers into a plastic sleeve covered by ozone-olive oil. Salzer et al. [309] have investigated the effect of ozonides olive oil on patients suffering from soorkolpitis. An especially positive effect was observed in its application in the vagina. The numbers of examined bacteria decreased very rapidly. These authors suggest that the positive effect of treatment is due to ozonides formed, as well as to tissue enrichment with oxygen. The same authors used autohemotherapy (100 ml blood enriched with oxygen-ozone; 1 ml contains 33 gammas ozone) in cases of chronic placenta insufficiency in pregnant women (20 patients, about thirtieth week). HPLC analysis indicated increased values. This therapy is also useful for weak bleedings during pregnancy.

Wermeister [310] described a technique, utilizing ozone gas of 80-100 μg/ml concentration, for wound cleansing and disinfection. Such an effect has not been observed during enzyme treatment. Often, the bacterially infected spot is washed out by ozone-oxygen mixture which give certain positive effect.

Ozonides that are prepared by means of the reaction between ozone and olefins do not cause inflammatory processes in skin and are not decomposed by catalase. Ozonides participate as an oxidizing agent without liberating peroxide groups which might oxidize cells. About 1000 subcutaneous injections of castor-oil ozonides have been made without tissue response [311]. A 4% solution of the ozonide of sorbic acid in ethyl diglycol (ozonide SV) is used in cosmetics to preserve face creams and as disinfectant. The lethal dose is 10 ml/kg. In wheat-germ oil, containing vitamin E, the ozonide keeps the vitamin E levels unchanged.

7.3.4. Sterilization

A number of methods have been developed, according to which ozone is utilized for sterilizing bottles, cans, containers and glass vessels that find wide application in the pharmaceutical industry and cosmetics [312-314].

Usually sterilization is performed at sufficiently high temperatures which may cause damage on same materials used in food processing industry, pharmaceutical industry and cosmetics. The effect of ozone on bacteria is well known, especially its toxicity and reactivity characteristics, that is why attention should be given to dosage and exposure time.

In [315} is discussed the application of ozone as sterilizing agent in the pharmaceutical industry. Physiological solution (0.9% NaCl), pH = 7, is treated with oxygen-ozone mixture containing 15% ozone for 15 min. The solution is allocated to glass or polyethylene banks which are sealed up. Analysis showed that after 2 h the microbe number was 0 and after 6 h the ozone was completely destroyed. Sterilization for 2 h with 0.5 ppm ozone caused a 100% lost of *Legionella pneumoniae* [278].

An apparatus has been designed to produce oxygen-ozone mixture for sterilization of milk, diary products, wines, butter and liquid medicines. The experimental set-up is described in detail in a patent [279]. The quality of wine, beer and liquor is improved after exposure to ozone. Ozone levels of 0.01 ppm and rates of 1l/min are employed in the pharmaceutical industry and food processing industry. Ozone has been employed fro cleaning dialyzers [316] and contact lens [317].

7.3.5. Ozonators

Portable ozonators have been designed for medicine purposes covering the whole range of terapheutical doses from 0 to 100 μg/l. These units are manufactured in "Shumen" plant Ltd., Shumen, Bulgaria as P-1 trademark.

REFERENCES

1. Razumovskii, S.D., S.K. Rakovski, D.M. Shopov, G.E. Zaikov. *Ozone and Its Reactions with Organic Compounds (in Russian)*. Publ. House of Bulgarian Academy of Sciences, Sofia, 1983.
2. Proskuryakov, VA, LI Shmidt. *Purification of the Sewage Waters from the Chemical Industry*. Khimya, Leningrad, 1977.
3. Bailey, P.S. "*Ozonattion in Organic Chemistry*", Academic Press, New York, v.I, 1978, v.II, 1982, *Adv. Chem. Ser. - A*, (Ed. HH Wasserman), 39 -I, 39 - II, 1978, 1982.
4. Kuznetzov, IE, TM Tritzkaya. *Protection of the Air Space Pollute by Dangerous Compounds*. Khimya, Moscow, 1979.
5. Itogi Nauki i Tekhniki - *Technology of Organic Compounds*, Moscow, VINITI, 7 (1983).
6. Itogi Nauki i Tekhniki -*Protection and Reproduction of Natural Resources*, Moscow, VINITI, 13 (1983).
7. Golubovskya, EK. *Biological Bases of Water Purification*. Moscow, Visshaya Shkola, 1978.
8. Bokris, OM (Ed.). *Environmental Chemistry*. Moscow, Khimya, 1982.
9. Pergut, EA, DO Gorelik. *Instrumental Methods to Control Air Pollution*. Leningrad, Khimiya, 1981.
10. Chanlett, ET. *Environmental Protection.*, McGraw-Hill, 1979.
11. Flaekt AB. HCl/SO_2 - scrubbing. *Res. Discl., 326, 453 (1991).*
12. W Ruck Ozonization of Biologically Purified Sewage Water. *Vom Wasser*, 80, 253-272 (1993).
13. Mccoustra, MRS, AB Horn. Towards a laboratory strategy for the study of heterogeneous catalysis in stratospheric ozone depletion. *Chemical Society Reviews*, 23: 3, 195-204 (1994).
14. Tabata, K., I.Matsumoto, Y.Fukuda. Oxidation catalysts for waste gases from burners *Jpn.Kokai Tokkyo Koho JP 01,270,928 [89,270,928] (Cl. B01D52), 30Oct 1989, Appl. 88/98,632, 21 Apt 1988.*

15. Pollo, I., J. Jarosszynska- Wolinska, J. Malicki, J. Ozonek, W. Wojcik. Removal of nitrogen oxides from waste gases. *Pol. PL 137,426 (Cl. B01D53/14), 31 Jul 1987, Appl. 245,514, 30 Dec 1983, 2pp.*

16. De Guardia A., A.Bouzaza, G.Martin, A. Laplanche. Aprocess of cleaning gases containing organic sulfur compounds (VOCs) by silicon oil scrubbing and ozone oxidation of absorbed pollutants.*Pollut.Atmos., 152, 82-92, (1996).*

17. Takeyama, Kenichi, K. Nitta. Apparatus for decomposing acyclic halogenated hydrocarbons in gases.*Jpn.Kokai Tokkyo Koho JP 01,236,925 [89,236,925] (Cl. B01D53/34), 21Sep 1989, Appl. 88/61,038, 15 Mar 1988, 3 pp.*

18. Moortgat, G.K., .Horie, B.C.Zahn. Ozonolysis of VOC: recent studies by matrix isolation FTIR-spectroscopy.*Pollut.Atmos., 29-44, (1991).*

19. Tomita, K. Material with adsorptive, oxidative and catalytic properties for air treatment. *Jpn.Kokai Tokkyo Koho JP 02,152,547 [90,152,547] (Cl. B01J20/18), 12 Jun 1990, Appl. 88/304,185, 02 Dec 1988, 4 pp.*

20. Vicard, J.F., G.Vicard. Method of cleaning gases containing condensable organic pollutants.*PCT Int.Appl. WO 92 19,364 (Cl. B01D53/34), 12 Nov 1992, Fr Appl. 91/5,853, 07 May 1991, 15pp.*

21. Iannicelli, J. Wet-scrubber process for removing total reduced sulfur compounds from industrial waste gases.*U.S. US 4, 923,688 (Cl. 423-224; B01D53/54), 08 May 1990, Appl. 799,494, 19 Nov 1985, 8pp.*

22. Buettner, F., P.Koch. Mercury removal from waste gases by catalytic ozonization. *Ger. offen DE 9,931,891 (Cl. B01D53/36), 04 Apr 1991, Appl. 25 Sep 1989, 3pp.*

23. Wada, H., T. Naoi, Y. Kuroda. Recycling of water from cyanide wastewater by ozone oxidation combined with UV radiation and ion exchange method.*Nippon Kagaku Kaishi, 9, 834-840, 1994*

24. Matt, K., H.G.Guttenberger. Process and apparatus for the detoxification of solutions containing cyanides and heavy metals. *Eur. Pat. Appl. EP 330,028 (Cl. C02F1/78), 30 Aug 1989, DE Appl. 3,805,906, 25 Feb 1988, 8pp.*

25. Waga, H., T.Naoi, Y.Kuroda. Recycling of water from chromium (VI) wastewater by ozone oxidation combined with UV radiation and ion exchange.*Nippon Kagaku Kaishi, 4, 306-313, (1995).*

26. Tanaka, K. Oxidation treatment promoted by photocatalyst and ozone.*PPM, 27(12), 10-15, (1996).*

27. Welch, J.F., J.D.Siehwarth. Method for processing of organic compounds. *U.S. US 4,861,497 (Cl. 210-759; C02F1/74), 30 Aug 1989, Appl. 169,851, 18 Mar 1988, 5pp.*

28. Oeller, HJ, I Daniel, G Weinberger. Reduction in residual COD in biologically treated paper mill effluents by means of combined ozone and ozone/UV reactor stages. *Water Sci.Technol., 35(2-3 Forest Industry Wastewaters V), 269-76, (1997), /Elsevier/.*

29. MoebiusC.H., M. Cordes-Tolle. Enhanced biodegradability by oxidative and radiative wastewater treatment. *Water Sci.Technol., 35(2-3 Forest Industry Wastewaters V), 245-250, (1997), /Elsevier/.*

30. Nakamura, Y., T.Sawada, F.Kobayashi, M.odliving. *Microbial treatment of kraft pulp wastewater pretreated with ozone.Water Sci.Technol., 35(2-3 Forest Industry Wastewaters V), 277-282, (1997), /Elsevier/.*

31. Zhou, H., D.W.Smith. Process parameter development for ozonation of kraft pulp mill effluents. *Water Sci.Technol., 35(2-3 Forest Industry Wastewaters V), 251-259, (1997), /Elsevier/.*

32. Huang, C.M., M.K.Banks. Effect of ozonation on the biodegradability of atrazine in granular activated carbon (GAC) columns.*J.Environ.Sci.Health, Part B, B31(6), 1253-1266, (1996).*

33. Sumimoto, H., T.Yamazaki. Injection of ozone into seawatwr piping for biofouling prevention in ocean thermal power plants. *Jpn.Kokai Tokkyo Koho JP 03,111,670 [91,111,670] (Cl. F03G7/05), 13 May 1991, Appl. 89/247,809, 22 Sep 1989, 3pp.*

34. Kawamoto, K. Techniques for decomposition of dioxins. *Kogaito Taisaku, 27(7), 617-27, (1991).*

35. Ukita, S. Removing slimes from water pipers by direct blowing of ozone. *Jpn.Kokai Tokkyo Koho JP 03 80,996 [91 80,996] (Cl. C02F1/50), 05 Apr 1991, Appl. 89/218,115, 23 Aug 1989, 4pp.*

36. Miriguchi, Y., H.Hayashi, T.Umetani. Laboratory experiment on decomposition of agrochemicals by ozonation.*Osaka-shi Suidokyoku Suishtsu Shikensho Chosa Kenkyu narabini Shiken Seiseki, 46, 10-16, (1994) /Pub. 1995/.*

37. Munz,C., R.Galli, R. Egli. Oxidative treatment of process water in a soil decontamination plant: I. Laboratory studies. *Chem.Oxid., 2, 247-63, (1992), /Pub. 1994/.*
38. Kojima, H., E.Katsura, H.Ogawa,.H.Kaneshima. Studies on prevention of envirovmental pollution by agricultural chemicals used in golf course I. Decomposition of water-soluble agricultural chemicals by ozone.*Hokkaidoritsu Eisei Kenkyushoho, 41, 71-73, (1991).*
39. Ohashi, N., Y.Tsuchiya, H.Sasano, A.Hamada. Ozonization products of organophosphorous pesticides in water.*Jpn.J.Toxicol.Environ.Health, 40(2), 185-92, (1994).*
40. hapeman-somish, c.j. Mineralization of pesticide degradation products. *ASC Symp.Ser., 459 (Pestic. Transform. Prod.: Fate Signif. Environ.), 133-47, (1991).*
41. Terui, s., K.Sano, K.Nishikawa, A.Inone. Deodorization method.*Jpn.Kokai Tokkyo Koho JP 02 139,017 [90,139,017] (Cl. BolD53/36), 29 May 1990, Appl. 88/293,929, 21 Nov 1988, 3pp.*
42. Terui, s., K.Sano, T.Kanazaki, K.Mitsui, A.Inoue. Deodorization of waste gases using ozone.*Jpn.Kokai Tokkyo Koho JP 01 56,124 [89 56,124] (Cl. B01D53/36), 03 Mar 1989, Appl. 87/209,305, 25 Aug 1987, 4pp.*
43. Ogawa, K., N.Seki. Apparatus for tap water treatment by ozone.*Jpn.Kokai Tokkyo Koho JP 04,243,597 [92,243,597] (Cl. C02F1/78), 31 Aug 1992, Appl. 91/7,486, 25 Jan 1991, 4pp.*
44. Hoshima, Y. Apparatus and method for dissolving ozone into water.*Jpn.Kokai Tokkyo Koho JP 02 40,289 [90 40,289] (Cl. C02F1/78), 09 Feb 1990, Appl. 88/190,318, 28 Jul 1988, 4pp.*
45. Filippi, A., F.Tilkes, E.G.Beck, H.Kirschner. Water disinfection in dental treatment units with ozone. *Dtsch.Zahnaerztl.Z., 46(7), 485-7, (1991).*
46. Gehringer, P. Radiation- induced oxidation for water remediation. *Oesterr. Forschungszent. Zeitbersdorf, [Ber.] OEFZS 1995, (OEFZS - 4738), 21pp.*
47. Charles, L., D.Pepin, Ph.Puig. The use of ozone for iron and manganese removal from mineral water: problems and control strategies. *J.Eur.Hydrol., 27(2), 175-191, (1996).*
48. Thornhill, R.W. Sanitation of swimming pool and other water. *Brit. UK Pat. Appl. GB 2,219,790 (Cl. C02F1/50), 20 Dec 1989, GB Appl. 88/14,222, 15 Jun 1988.*
49. Nitsuta, Y., Y.Nitta. Decomposition of organic matter in dredging wastes containing spilled petroleum.*Jpn.Kokai Tokkyo Koho JP 04,367,799 [92,367,799] (Cl. C02F11/00), 21 Dec 1992, Appl. 91/165,204, 11 jun 1991, 3pp.*
50. Gordon, G., G.E.Pacey. Slurry-ozone technology for on-site treatment of contaminated soils. *Chem.Oxid., 2, 230-46, (1992), /Pub. 1994/.*
51. Bruckner, F. *In situ* remediation of petroleum contaminated soils by ozone treatment. *ECOINFORMA 92, Int.Tag.Ausstellung Umweltinf.Umweltkommun., , 2, 211-22, (2nd 1992), /Pub. 1993/,*
52. Preusser, M., H.Ruholl, H.Schneidler, E.Wessling, C.Wortmann. Ozone treatment for the cleanup of soil contaminated with organic pollutants.*Wiss.Umwelt, 3, 135-40, (1990).*
53. Rimpler, M. Method and apparatus for decontamination of wastes such as soil or sludges by oxidation with ozone. *Ger. Offen. DE 19,512,448, (Cl. A62D3/00), 10 Oct 1996, Appl. 19/512,448, 03 Apr 1995, 9pp.*
54. Wessling, E. Experiences in development and application of the ozone method for the destruction of organic noxious substances at the solid phase- both in-situ and on-site.*Erzmetall, 44,*(4), *196-200, (1991).*
55. M Mori, H Yamakoshi, M Nojima, S Kusabayashi, KJ Mccullough, K Griesbaum, P Kriegerbeck, IC Jung. Ozonolysis of Cyclopentadiene Derivatives - Competitive Participation of [3+2] and [3+4] Cycloadditions of Carbonyl Oxide Moieties to alpha, beta-Unsaturated Carbonyl Groups. *Journal of the Chemical Society - Perkin Transactions* I : 12, 1335-1343, (1993).
56. KJ Mccullough, K Teshima, M Nojima. Unprecedented Formation of a Cyclic Tetramer from the Acidolysis of Indene Ozonide - Isolation and Characterisation of a Novel Dodecaoxacycloicosane Derivative. *J.Chem.Soc. - Chem.Commun.*: 11, 931-933, (1993).
57. T Sugiyama, H Yamakoshi, M Nojima. Ozonolysis of beta-(Alkoxycarbonyl).-Substituted and beta-Acyl-Substituted Vinyl Ethers - Cycloaddition Chemistry of the Derived alpha-Keto Ester O-Oxides and alpha-Diketone O-Oxides. *J.Org.Chem,* 58: 16, 4212-4218 (1993).
58. K Ishiguro, T Nojima, Y Sawaki. Novel aspects of carbonyl oxide chemistry. *J.Phys.Org.Chem.,* 10: 11, 787-796 (1997).

59. Nojima, T., Ishiguro, K., Sawaki, Y. Mechanistic study on the reaction of phenyldiazomethanes with singlet oxygen: Formation and cycloreversion of 1,2,3,4-dioxadiazole intermediates. *J Org Chem,* 62, 20, 6911-6917 (1997).
60. Ishiguro, K., Nojima, T., Sawaki, Y. Novel aspects of carbonyl oxide chemistry. *J Phys Org Chem,* 10, 11, 787-796 (1997).
61. R Fukagawa, M Nojima Synthesis and silica gel-catalysed decomposition of 3-, (1-arylcycloalkyl).-substituted 1,2,4-trioxolanes and 1,2,4-dioxazolidine. *Journal of the Chemical Society - Perkin Transactions 1* : 17, 2449-2454 (1994).
62. M Mori, H Yamakoshi, M Nojima, S Kusabayashi, KJ Mccullough, K Griesbaum, P Kriegerbeck, IC Jung Ozonolysis of Cyclopentadiene Derivatives - Competitive Participation of [3+2] and [3+4] Cycloadditions of Carbonyl Oxide Moieties to alpha, beta-Unsaturated Carbonyl Groups. Journal of the Chemical Society - *Perkin Transactions I* : 12, 1335-1343 (1993).
63. KJ Mccullough, K Teshima, M Nojima Unprecedented Formation of a Cyclic Tetramer from the Acidolysis of Indene Ozonide - Isolation and Characterisation of a Novel Dodecaoxacycloicosane Derivative. *J.Chem.Soc, - Chem.Commun.* : 11, 931-933 (1993).
64. T Sugiyama, H Yamakoshi, M Nojima Ozonolysis of beta-(Alkoxycarbonyl).-Substituted and beta-Acyl-Substituted Vinyl Ethers - Cycloaddition Chemistry of the Derived alpha-Keto Ester O-Oxides and alpha-Diketone O-Oxides. *J.Org.Chem,* 58: 16, 4212-4218 (1993).
65. KJ Mccullough, T Sugimoto, S Tanaka, S Kusabayashi, M Nojima. Cycloadditions Between Carbonyl Oxides and Dicarbonyl Compounds - Isolation and Characterisation of Novel Polycyclic 1,2,4,6-tetroxepane Derivatives. *Journal of the Chemical Society - Perkin Transactions 1* : 6, 643-651 (1994).
66. S Satake, Y Ushigoe, M Nojima, KJ Mccullough. The synthesis of novel 3,4-dihydro-1,2,5,7,4-tetroxazocine derivatives via extended [3+3+2] cycloaddition reactions between a carbonyl oxide, a nitrone and an aldehyde. *Journal of the Chemical Society - Chemical Communications,,* : 14, 1469-1470 (1995).
67. K Teshima, SI Kawamura, Y Ushigoe, M Nojima, KJ Mccullough. Ozonolyses of indene and of 1-alkyl- and 1,1-dialkyl- substituted indenes in protic solvents. Remarkable effects of the substituent steric bulk and the solvent nucleophilicity on the direction of cleavage of the primary ozonides and on the mode of capture of the carbonyl oxide intermediates by the solvents. *Journal of Organic Chemistry,* 60: 15, 4755-4763 (1995).
68. KJ Mccullough, S Tanaka, K Teshima, M Nojima. Polycyclic ozonides and peroxides derived from Hexamethyl(Dewar benzene). *Tetrahedron,* 50: 25, 7625-7634 (1994).
69. (69) Schobert, B.D. Ozonolysis and reduction in fine chemikals industry. *Chim.Oggi, 13(6), 21-4, (1995), /Pub.1995/.*
70. Appell, RB, IA Tomlinson, I Hill. New reagents for reductive quenching of ozonolysis reaction. *Syn. Comm., 25 (22), 3589- 3595, (1995)*
71. (71) Yang, YT, TS Li, YL Li. Ozone oxidation of oximes to carbonyl compounds. *Synthetic Comm., 23(8), 1121- 1124, (1993).*
72. Jachak, M., M.Mittelbach, U.Kriessmann, H.Junek. Ozonolysis of olefins. 6.Cyanoacetaldehyde by ozonolysis of (E)-1,4-Dicyano-2-Butene or allyl cyanide. *Synthesis- Stuthgart, 3, 275 - 276, (1992).*
73. Jellen, W, M Mittelbach, H Junek. Ozonolysis of olefins. VIII.Synthesis of phenoxyacet-aldehydes by ozonolysis of allyl phenyl ethers. *Monatsh.Chem., 127(2), 167-72, (1996).*
74. Matsumoto, H, H Kanda, Y Obada, H Ikeda, T Murakami. Preparation of quinolinaldehyde derivative as intermediate for quinoline type mevalonolactones. *Jpn.Kokai Tokkyo Koho JP 08 03,138 [96 03,138] (Cl. C07D215/14), 09 Jan 1996, JP Appl. 94/28,596, 25 Feb 1994, 10pp.*
75. Allen, DE, ChE Ticker, ChC Hobbs, R. Chidambaram. Process for preparing 6-chloro-2-hexanone from 1-methylcyclopentane. *U.S. US 5,491,265 (Cl. 568-347; C07C45/51), 13 Feb 1996, Appl. 395,266, 28 Feb 1995, 7pp.*
76. Lee, H.W., S.K.Ji, I.-Y.Ch.Lee, J.H.Lee. Convenient and practical synthesis of (R)-(+)-4-methyl-2-cyclohexen-1-one. *J.Org.Chem., 61(7), 2542-3, (1996).*
77. Mikaev, FZ, FZ Galin. Synthesis of (+)-4α-[1-(trimethylsiloxy)ethyl]-2-carene. *Izv. Akad. Nauk SSSR, ser. khim., 10, 1984-7,(1995).*

78. Mccullough K.J., M.Mori, T.Tabuchi, H.Yamakoshi, S.Kusabayashi, M.Nojima. [3+2] cycloaditions of carbonyl oxides to imines: An alternative approach to the synthesis of 1,2,4-dioxazolidines. *J.Chem.Soc.-Perkin Transactions, 1(1), 41-48, (1995).*

79. Mori, M., T.Tabuchi, M.Nojima, S.Kusabayashi. Synthesis of 3-Vinyl-1,2,4-trioxolanes by a [3+2] cycloaddition of carbonyl oxides with alpha, beta-unsaturated carbonyl compounds. *J.Org.Chem., 57(6), 1649-1652, (1992).*

80. (80) Casey,M., A.J.ulshaw. Cycloadditions of carbonyl oxides- a novel stereoselective approach to 1,3-diols. *SYNLETT, 3, 214-216, (1992).*

81. Kuczkowski, R.L. Ozonolysis of vinyl ethers. *Advances in Oxygenated Process, 3(Series: Advances in Oxygenated Process: A Research Annual 3(1991)).*

82. Reiser, R., R.-. Seeboth, C.Suling, G.Wagner, J.Wang, G.Schroder. Synthesis of 1,2-dioxolanes by ozonolysis of 1,1-disubstituted nonactivated olefins.*Chem.Ber., 125(1), 191-195, (1992).*

83. Hon, Y.-S., J.-L. Yan. Synthesis of bifunctional compounds from cycloalkenes via ozonide intermediates. *Tetrahedron, 53(14), 5217-5232, (1997).*

84. Marshall, JA, AW Garofalo, RC Sedrani. The direct conversion of olefins into esters through ozonolysis. *SYNLETT., 643-645, (1992).*

85. Odinokov. VN. Â.Í. Controlled ozonolysis of acyclic alkenes in organic synthesis. *Bashk. Khim. Zhurn., 1(3), 29-33, (1994).*

86. Odinokov. VN. Ozonolysis in specialized transformations of petrochemical olefins. *Bashk. Khim. Zhurn., 1, (1), 11-14, (1994).*

87. (87) Ishmuratov, GYu, RYa Kkharisov, VN Odinokov, GA tolstikov. Ozonolysis of unsaturated compounds in the synthesis of insect pheromones and juvenoids. *Uspehi Khimii*, 64, (6), 580-608 (1995).

88. Malhotra, R., S.Kumar, A.Satyam. Ozonolysis of [60] fulerene. *J.Chem.Soc., Chem. Communications, 11, 1339-40, (1994).*

89. Zwilichovsky, G., V.Gurvvich. Ozonolysis of (cyclohexa-1,4-dienyl)-L-alanine. An approach to the synthesis of new unnatural amino acids. X-ray molecular structure of 2-hydroxy-7-methyl-3-phenyl-pyrazolo[1,5-a] pyrimidine. *J.Chem.Soc., Perkin Trans.1., 19, 2509-15, (1995).*

90. Mueller, T., R.R.Schmidt. Thymidine diphospho-6-deoxy-a-D-ribo-3-hexulose- synthesis of a central intermediate in the biosynthesis of di- and trideoxy sugars. *Angew.Chem., Int.Ed.Engl., 34(12), 1328-9, (1995).*

91. Pitzer, K., T.Hudlicky. The synthesis of 4-amino-4-deoxy-D-mannose and glucosamine: a new methodology for the design of homohiral amino sugars from non-carbohydrate precursors. *Synlet., 8, 803-5, (1995).*

92. Kashima,Ch., T.Maruyama, H.Arao. Ozonolysis of azoles and its appliication to the organic synthesis. *Yuki Gosei Kagaku Kyokaishi, 54(2), 132-8, (1996).*

93. Kashima, C., K.Harada, A.Hosomi. Ozonolysis of 1-substituted imidazoles.*Heterocycles, 33(1), 385-90, (1992).*

94. Wu, H.-J., Ch.-Ch. Lin. Novel oxa-cage compounds: synthesis, strukture and the formation mechanism of tetraacetal oxa-cages and complex tetraquinane oxa-cages. *J.Org.Chem., 60(23), 7558-66, (1995).*

95. McVay,K.R., D.G.Gaige, W.S.Kain. Process and oxidation catalysts for forming two terminal carboxylic acid groups from an ozonide in the presence of fine oxygen bubbles.*U.S. US 5,543,565 (Cl. 562-523; C07C51/16) 06 Aug 1996, Appl. 376,173, 20 Jan 1995,7pp.*

96. Gesser, H.D., G.Zhu, N.R.Hunter. Conversion of ethane to methanol and ethanol by ozone sensitized partial oxidation at near atmospheric pressure.*Catalysis Today, 24, 321- 325, (1995).*

97. Gesser, H.D., N.R.Hunter, P.A.Das. The ozone sensitised oxidative conversion of methane to methanol and ethane to ethanol. *Ñatalysis Letters, 16(1-2), 217-221, (1992)*

98. Hitomi, S., T.Ishibashi, M.Takashi, T.Kenkichi. Ortho-selective nitration of acetanilides with nitrogen dioxide in the presence of ozone.*Tetrahedron Letters, 32(45), 6591-94, (1991)*

99. Suzuki, H., T.Murashima, K.Shimizu, K.Tsukamoto. The ozone mediated reaction of nonactivated arenes with nitrogen oxides. *Chemistry Letters, 817-18, (1991)*

100. Suzuki, H., T.Mori, K.Maeda. Ozone mediated reaction of polychloro benzenes and some related halogeno compounds with nitrogen dioxide: A novel non-acid methodology for the selective mono nitration of moderately deactivated aromatic systems. *SYNTHESIS- Stutgart, 8, 841-45, (1994).*

101.Suzuki, H., S.Yonezawa, T.Mori, K.Maeda. The ozone mediated reaction of aromatic acetals and acylal with nitrogen dioxide: A novel methodology for the nuclear nitration of acid sensitive aromatic compounds under neutral conditions. *J.Chem.Soc.-Perkin Trans., 1(11), 1367-69, (1994).*
102.Suzuki, H., T.Murashima, K.Tsukamoto. Mineral acid- free preparation of nitro compounds from pyridine or its derivatives. *Jpn.Kokai Tokkyo Koho JP 04,282,368 [92,282,368] (Cl. C07D213/61), 07 Oct 1992, Appl. 91/67,642, 08 Mar 1991, 3pp.*
103.Suzuki, H., T.Murashima. Nitrogen oxide (NO_x). Nitration of aromatic compounds in presence of ozone. *Kagaku to Kagyo (Tokyo), 46(1), 43-5, (1993).*
104.Lyakumovich, AG, YD Samuilov, YaM Goldberg, MN Kondrashova. Method of succinic acid preparation. *Russ. RU 2,036,895 (Cl. C07C51/34), 9Jun 1995, Appl. 93,015,890, 25 Mar 1993.*
105.Ebdon, JR, NJ Flint. Preparation of α-phenyl ketone, ω-carboxylate-ended telehelic methyl methacrylate oligomers by the ozonolysis of regioregular methyl methacrylate-phenylacetylene copolymers. *Macromolecules, 27, 6704, (1994).*
106.Sawaki, T, S Watanabe, K Shimada. Manufacture of silicon carbide ceramics. *Jpn.Kokai Tokkyo Koho JP 02 26,868 [90,26,868] (Cl. C04B35/56), 29 Jan 1990, Appl. 88/176,938, 18 Jul 1988, 7pp.*
107.Wasinger, E, D Hall. Method of decolorizing of fabrics. *U.S. US 5 261,925 (Cl. 8-111; D06L3/14), 16 Nov 1993, US Appl. 560,357, 31 Jul 1990, 7pp.*
108.Bradley, RH, IL Clackson, DE Sykes. UV/Ozone modification of wool fibre surfaces. *Applied Surface, 72(2), 143-47, (1993).*
109.Connor, HG, D Budgell. Process for producing cellulosic fibers and moldings. *PCT Int. Appl. WO 96 18,761 (Cl. D01F2/00), 20 Jun 1996, DE Appl. 4,444,700, 15 Dec 1994, 21pp.*
110.Miura, M., T Naito, T Hino, H Kuroda. Pitch-based carbon or graphite fibers with high tensile strenght and their manufacture. *Jpn.Kokai Tokkyo Koho JP 01,250,415 [89,250,415] (Cl. D01F9/14), 05 Oct 1989, Appl. 88/74,471, 30 MAR 1988, 9pp.*
111.Queslati, R, S Roudesli. EPR analysis of thermal decomposition of peroxides in ozonized polypropylene fiber for grafting. *European Polymer Journal, 27(12), 1383-90, (1991).*
112.Catoire, B, V Verney, R Hagege, A Michel. Study of polypropylene peroxidation by ozonization using electron spin resonance and transmission electron microscopy. *Polymer, 33(11), 2307-11, (1992).*
113.Kulik, EA, M Ivanchenko, K Kato, S Sano, Y Ikada. Peroxide generation and decomposition on polymer surface. *J.Polym.Sci., Part A- Polymer Chemistry, 33(2), 323-30, (1995).*
114.fujimoto, k., y.takebayashi, h.inoue, y.ikada. Ozone-induced graft polymerization onto polymer surface. *J.Polym.Sci., Part A- Polymer Chemistry, 31(4), 1035-43, (1993).*
115.Wang, Ch, A Wang, B Che, C Zhou, I Su, S Lin, B Wang. Ozone-induced graft copolymerization of polyethylene glycol monomethyl ether methacrylate onto poly(etherurethane). *Gaofenzi Xuebao, (1), 114-18, (1997).*
116.Jian, Y., Sh. Shiraishi. Modification of the surface properties of polymers by ozone treatment. I. Modification of the surface properties of polypropylene film by ozone treatment. *Hyomen Gijutsu, 40(11), 1251-5, (1989).*
117.Jian, Y., SH Shiraishi. Modification of the surface properties of polymers by ozone treatment. II. Low temperature nickel plating onto polypropylene with ozone treatment method. *Hyomen Gijutsu, 40(11), 1256-60, (1989).*
118.Jian, Y, SH Shiraishi. Surface improvement of polymers by ozone treatment. III. Improvement of the coating properties of polypropylene by ozone treatment. *Hyomen Gijutsu, 41(3), 273-7, (1990).*
119.Yamamoto, H. Manufacture of moistureproof packaging materials with sealing resin layers. *Jpn.Kokai Tokkyo Koho JP 02,198,819 [90,198,819] (Cl. B29C47/04), 07 Aug 1990, Appl. 89/16,184, 27 Jan 1989, 6pp.*
120.Yoshito, M. Manufacture of methylmethacrylate polymers with good thermal degradation resistance. *Jpn.Kokai Tokkyo Koho JP 03,45,604 [91,145,604] (Cl. C08F8/06), 27 Feb 1991, Appl. 89/181,558, 13 Jul 1989, 2pp.*
121.Shiraishi, SH, U Ken, K Suchiro, K Nitsuta. Ozone modification of propene polymer moldings. *Jpn.Kokai Tokkyo Koho JP 03,103,448 [91,103,448] (Cl. C08J7/12), 30 Apr 1991, Appl. 89/239/357, 14 Sep 1989, 5pp.*

122.Marshall, RA, DK Parker, JW Hershberger. Ozonization of PVC latex for removal of vinyl chloride residues. *U.S. US 5,498,693 (Cl. 528-483; C08F6/24), 12 Mar 1996, Appl. 313,504, 26 Sep 1994, 5pp.*

123.Iwasaki, T. Graft polymerization of vinyl monomers onto Nylon 6 fiber by ozone oxidation method. *Kobunshi Ronbunshu, 50(2), 115-20, (1993).*

124.Hosoda, Y, K Ikishima, A Yanai. Adhesion improvement of PVC-, fluoropolymer-, or polyester-coated steel panels. *Jpn.Kokai Tokkyo Koho JP 04,197,472 [92,197,472] (Cl. B05D3/10), 17 Jul 1992, Appl. 90/331,521, 29 Nov 1990, 4pp.*

125.Ebdon, JR, NJ Flint. Synthesis of water-soluble telehelic methyl-ketone-ended oligo- N,N-dimethylacrylamides by the ozonolysis of poly(N,N- dimethylacrylamide- stat- 2,3-dimetylbutadiene)s. *J.Polym.Sci., Part A- Polymer Chemistry, 33(3), 593-97, (1995).*

126.Rl, L, JR Ebdon, P Hodge. Synthesis of telehelic methyl-methacrylate and styrene oligomers witth fluorophenyl ketone end groups by the ozonolysis of copolymers containing 4-fluorophenyl butadiene units. *Polymer, 34(2), 406, (1993).*

127.Eftushenko, AM, IP Chikhachova, GV timofeeva, ST Stavrova, VP zubov. Graft polymerization of acrylic acid into ozonated siloxane matrixes. *Visoko molek. soed., ser. A i B, 35(6), Â312-15, (1993).*

128.Tashibana, SH, R Yokoyama. Manufacture of base disk for laser recording material. *Jpn.Kokai Tokkyo Koho JP 01,248,339 [89,248,339] (Cl. G11B7/26), 03 Octl 1989, Appl. 88/75,079, 28 Mar 1988, 4pp.*

129.Rimmer, S, JR Ebdon, JM Sherpherd. Polymer-supported reagents as aids in controlled degradation of polymers containing ozonides in the main chain to give telechelic oligomers. *Reactive and Functional Polymers, 26, 145-55, (1995).*

130.Chtourou, H, B Riedl, BV Kokta, A Adnot, S Kaliguine. Synthetic pulp fiber ozonation: an ESCA and FTIR study. *J.Appl.Polym.Sci., 49, 361-73, (1993).*

131.Vallet, A, M Delmas, TH Fargere, G Sacher. Processing fiber-reinforced plastics and ceramics for recycling. *PCT Int.Appl. WO 96 18,484 (Cl. B29B17/02), 20 Junl 1996, FR Appl. 94/15,174, 16 Dec 1994, 16pp.*

132.Kawabe, K, Y Hatakeyama. Heat-sealing method for laminates. *Jpn.Kokai Tokkyo Koho JP 09 01,665 [91 01,665] (Cl. B29C65/02), 07 Janl 1997, Appl. 95/172,699, 15 Jun 1995, 8pp.*

133.Czerwiec, W, A Kwaitek. Manufacture of pour point depressants from waste polyethylene and/or polypropyllene. *Eur. Pat.Appl. EP 383,969 (Cl. C08F8/50), 29 Aug 1990, Appl. 89/103,098, 22 Feb 1989, 3pp.*

134.Shimizu, SH, N Iwata. Surface treatment of stainless steel parts la. *Jpn.Kokai Tokkyo Koho JP 08 337,867 [96 337,867] (Cl. C23C8/18), 24 Dec 1996, Appl. 95/141,448, 08 Jun 1995, 6pp.*

135.Lagana, V. System for passivation of metal surfaces. *EUR Pat. Appl. EP 504,621 (Cl. C23C8/12), 23 Dec 1992, IT Appl. 91/MI715, 18 Mar 1991, 7pp.*

136.Berneron, R, P de Gelis. Increasing corrosion resistance of metals and alloys. *EUR Pat. Appl. EP 340,077 (Cl. C23C8/36), 02 Nov 1989, FR Appl. 88/5,091 18 Apr. 1988, 10pp.*

137.Lu, HH, DJ Diguette. Effect of dissolved ozone on the corrosion behaviour of Cu-30 Ni and 304L stainless steel in 0.5N sodium chloride solutions. *Report 1989, Order No.AD-A213804, 37pp.*

138.Havlik, T, M Skrobian. Acid oxidizing leaching of chalcopyrite with ozone. *Rudy, 37(10), 295-301, (1989).*

139.Havlik, T, M Skrobian. Acid leaching of chalcopyrite in the presence of ozone. *Can. Metall.Q., 29(2), 133-9, (1990).*

140.Clough, T, JW Siebert, AC Riese. Redox leaching of precious metals from manganese-containing ores. *Pat. Specif.(Aust.) AU 607,523 (Cl. C22B11/04), 07Mar 1991, US Appl. 213,884, 30 Jun 1988, 40pp.*

141.Ye, Y, WH Jang, MR Yalamanchili, JD Miller. Molybdenite flotation from copper/molybdenum concentrates by ozone conditioning. *Trans.Soc.Min., Metall., Explor., 288, 173-9, (1990), /Pub. 1991/.*

142.Chkoniya, TK. Manufacture of potassium permanganate. *U.S.S.R. SU 1,541,190 (Cl. C01G45/12), 07 Feb 1990, Appl. 4,246,700, 24 Mar 1987.*

143.Saski, T, H Sasaki, K Iida. Manufacture of ITO ceramics by firing in air atmosphere containing ozone. *Jpn.Kokai Tokkyo Koho JP 08 169,769 [96 169,769] (Cl. C04B35/64), 02 Jul 1996, Appl. 94/316,464, 20 Dec 1994, 4pp.*

144.Saski, T, H .Sasaki. Manufacture of ITO sinters. *Jpn.Kokai Tokkyo Koho JP 07 277,821 [95 277,821] (Cl. C04B35/495), 24 Oct 1995, Appl. 94/66,030, 04 Apr 1994, 4pp.*

145.Caropreso, FE, HM Castrantas, JM Byne. Method for oxidizing carbonaceous material, and especially for bleaching gray kaolin for subsequent use as coating or filler for paper. *U.S. US 4,935,391 (Cl. 501-146; C04B20/06), 19Jun 1990, Appl. 331,423, 31 Mar 1989, 6pp.*

146.Emel'yanova, OV, BV Emel'yanov. Manufacture of mercury (I)chloride. *U.S.S.R. SU 1,567,520 (Cl. C01G13/04), 30 May 1990, Appl. 4,342,566, 14 Dec 1987.*

147.Ruhlandtsende, AD, AD Bacher, U Muller. Formation of arsenic acid from $(As_2Cl_8)^{2-}$ -ions and ozone crystal structure of $PPh_4Cl.H_3AsO_4$. . *Zeitschrift fur Naturforschung, Section B, 47(12), 1677-80,*

148.Belfadhel, H, A Ratel, A Ouederni, RS Bes, JC Mora. Removal of color and organic matter in industrial phosphoric acid by ozone: effect on activated carbon treatment. *Ozone: Sci.Eng., 17(6), 637-45, (1995).*

149.Achutharaman, VS, KM Beauchamp, N Chandrasekhar, GC Spalding, BR Johnson, AM Goldman. Fabrication of high-Tc superconductors ysing ozone-assisted molecular beam epitaxy. *Thin Solid Films, 216(1), 14-20, (1992).*

150.Locquet, JP. Characterization of a ratio frequency plasma sourse for molecular beam epitaxial growth of high-Tc superconductor films. *J.Vacuum Sci.&Tech., A.Vacuum surfaces and films, 10(5), 3100-3103, (1992).*

151.Berkley, DD, AM Goldman, BR Johnson, J Morton, T Wang. Techniques for the drowth of superconducting oxide thin films using pure ozone vapor. *Rev.Sci.Instrum., 60(12), 3769-74, (1989).*

152.Usui, T, Y Osanai. Ozone oxidation in preparation of oxide superconductors. *Jpn.Kokai Tokkyo Koho JP 01 122,921 [89 122,921] (Cl. C01G3/00), 16 May 1989, Appl. 87/278,772, 04 Nov 1987, 5pp.*

153.Odan, K, H Miura, Y Bando. Manufacture of bismuth copper alkaline earth oxide superconductors by using ozone. *Jpn.Kokai Tokkyo Koho JP 01 224,258 [89 224,258] (Cl. C04B25/00), 07 Sep 1989, Appl. 88/45,774, 01 Mar 1988, 3pp*

154.Goldman, AM, DD Berkley, BR Johnson. Preparation of superconductive ceramic oxides using ozone. *PCT Int. Appl. WO 90 02,215 (Cl. C23C14/26), 08 Mar 1990, US Appl. 234,421, 19 Aug 1988, 29pp.*

155.Nakamori, T. Manufacture of rare earth barium copper oxide high-temperature supercondctor ceramics. *Jpn.Kokai Tokkyo Koho JP 01 179,752 [89 179,752] (Cl. C04B35/00), 17 Jul 1989, Appl. 88/648, 07 Jan 1988, 3pp*

156.Johnson, BR, KM Beauchamp, DD Berkley, JX Liu, T Wang, AM Goldman. Growth of co-evaporated superconducting yttrium barium copper oxide ($YBa_2Cu_3O_{7-x}$) thin films oxidized by pure ozone. *Proc. SPIE-Int.Soc.Opt.Eng., 1187(Process.Films High Tc Supercond.Electron), 27-36, (1990).*

157.Vig, JR. UV/ozone cleaning of surfaces. *Proc. Electrochem. Soc., 90-9 (Semicond.Clean.Tachnol. 1989), 105-13, (1990).*

158.Suemitsu, M, T Kaneko, N Nobuo. Low temperature silicon surface cleaning by fluoric acid (HF) etching/UV-ozone cleaning (HF/UVOC) method. (1).Optimization of the HF treatment. *Jpn.J.Appl.Phys., Part 1, 28(12), 2421-4, (1989).*

159.Kaneko, T, M Suemitsu, N Miyamoto. Low temperature silicon surface cleaning by hydrofluoric acid etching/UV-ozone cleaning (HF/UVOC) method. (II).*In situ* UVOC. *Jpn.J.Appl.Phys., Part 1, 28(12), 2425-9, (1989).*

160.Inada, A, K Kawasumi. Removal of organic contaminants from semiconductor substrates by ozone. *Jpn.Kokai Tokkyo Koho JP 01 135,023 [89 135,023] (Cl. H01L21/30), 26 May 1989, Appl. 87/292, 016 20 JNov 1987, 2pp.*

161.Ohmi, T. Total room temperature wet cleaning of silscon surface. *Semicond. Int.*, 19, (8), 323-4, 326, 328, 330, 332, 334, 336, 338, (1996).

162. Kamikawa, Y, K Matsumura. Removal of resist films using liquid ozone. *Jpn.Kokai Tokkyo Koho JP 02 106,040 [90 106,040] (Cl. H01L21/304), 18 Apr 1990, Appl. 88/259,204, 20 Oct 1988, 3pp.*
163. Maeda, K. Ozone applied in VLSI (very large scale integration) manufacturing. *Kurin Tekunoroji, 4(2), 62-9, (1994).*
164. Choi, H, K Kyoo, SD Jung, H Jeon. Effects of UV/O3 and SC-1 step in the HF last silicon wafer cleaning on the properties of gate oxide. *Han'guk Chaehyo Hakhoechi, 6(4), 395-400, (1996).*
165. Hansen, RWC, J Welske, D Wallace, M Bissen. Cleaning of optical surfaces with photogenerated reactants. *Nucl.Instrum.Methods Phys.Res., Sect.A, 347(1-3), 249-53, (1994).*
166. Ohmi, T, T Sagawa, M Kogure, T Imaoka. Native oxide growth and organic impurity removal on silicon surface with ozone-injected ultrapure water. *J.Electrochem.Soc., 140(3), 804-10, (1993).*
167. Takaaki, K, Y Akimasa, M Yasuji. Reaction mechanism of chemical vapor deposition using tetraethylorthosilicate and ozone at atmospheric pressure. *Jpn.J.Appl.Phys., 31, 2925-30, part 1, No.9A (1992).*
168. Hishamune, Y. Covering semiconductor devices with silica films. *Jpn.Kokai Tokkyo Koho JP 01 82,634 [89 82,634] (Cl. H01L21/316), 28 Mar 1989, Appl. 87/242,113, 25 Sep 1989, 6pp.*
169. Kurihara, H, M Saito, T Kozuki. Fabrication of nondoped silicate glass film with flat structure by O3-TEOS deposition. *Jpn.Kokai Tokkyo Koho JP 09 08,026 [97 08,026] (Cl. H01L21/316), 10 Jan 1997, JP Appl. 95/118,005, 19 Apr 1995, 8pp.*
170. Homma, T. Future trends for interlayer dielectric films and their formation technologies in ULSI multilevel interconnections. *Mater.Chem.Phys., 41(4), 234-9, (1995).*
171. Toris, K, S Saiton, Y Ohji. Preparation of lead zirconate thin films by reactive evaporation. *Jap.J.Appl.Physics, Part 1- nRegular Papers Short Notes & Review Papers; 33(9B), 5287-90, (1994).*
172. Haluska IA. Formation of ceramic coatings on substrates in the presence of ozone. *Fr. Demande FR 2,697,852 (Cl. C23C16/22), 13 May 1994, Appl. 91/565, 18 Jan 1991, 23pp.*
173. Nishida, SH. Method of forming zinc oxide light-shielding film. *Ger. offen DE 3,916,983 (Cl. C23C16/40), 30 Nov 1989, JP Appl. 88/125,792, 25 May 1988, 14pp.*
174. Hollinger, G, D Gallet, M Gendry, MP Besland, J Joseph. Evidence for a new passivating indium rich phosphate prepared by UV/O_3 oxidation of indium phosphide. *Appl.Phys.Lett., 59(13) 1617-19, (1991).*
175. Tawada, H, M Saitoh, CH Isobe. Method of heat-treating an oxide optical crystal and a heat treatment for carrying out the same. *Eur Pat. Appl. EP 407,945 (Cl. C30B29/28), 16 Jan 1991, JP Appl. 89/178,586, 11 Jul 1989, 23pp.*
176. Tamada, H, A Yamada, M Saitoh. Lithiumniobate thin-film optical waveguide grown by liquid phase epitaxy and its application to second harmonic generation. *J. Appl.Phys., 70(5), 2536-41, (1991).*
177. Sadamoto, M, Y Ashida, N Fukuda. Manufacture of solar cell modules. *Jpn.Kokai Tokkyo Koho JP 08 148,707 [96 148,707] (Cl. H01L31/04), 07 Jun 1996, Appl. 94/280,407, 15 Nov 1994, 14pp.*
178. Bianchetti, MF, HR Canepa, S Larrondo, NE Walsel de Reca. Surface activation of photoconductive PbS films. *An Asoc.Quim.Argent.*, 84,(1), 73-7, (1996).
179. Enomoto, R, M Asai. Adhesion-producing materials for electroless plating and printed circuits and methods for fabricating them using the materials. *Ger. offen DE 3,913,966 (Cl. C09J3/00), 09 Nov 1989, JP Appl. 88/104,044, 28 Apr 1988, 15pp.*
180. Binz. *Real Encyclopedi der gesamten Helkunder* Bd. 18, S. 189, Berlin und Wien (1898).
181. Lambert, BK. *Nature,* 181, 1202, (1958).
182. Shepetinovskii, Z, I Mikashinovich. Second All-Union Conference "Ozone obtaining and application". Theses of reports. p. 157, Moscow (1991).
183. *Eur. Pat.* Appl. EP 261 032 (1988).
184. Carpendiae, J Freeberg. *J.Clin. Gastroenterol.* 17, 2, 142 (1993).
185. Belianin, E, I Shmelev. *Ter. Arkh.*, 66, 2, 29 (1994).
186. Sato. *Exp. Anim.*, 39, 2, 223 (1990).
187. Grits, A Formichev. *Microbiologiya*, 59, 5, 831 (1990).
188. Heindl. *Zentralbl.Hyg.Unwelmed.*, 194, 5/6, 464 (1993).
189. Ancowski. *Complement Inflammation*, 7, 2, 57 (1990).

190.Kontorshtikova, KN, NN Andreeva. Second All-Union Conference "Ozone obtaining and application. Theses reports, p. 157, Moscow (1991).

191.Kitanova, S., G. Koynova, S. Rakovsky, V. Kagan. Xenobiotic Metabollism and Toxicity Workshop of Balkan Countries, P35, March 10-21st 1991, Novi Sad, Yugoslavia

192.Rakovsky, S. Pross. I-st Bulg.-Russ. Sympos. "Free Radicals and Biostabilzers" I-st Bulg. - Russ. Symp. p.113, 23-26 November 1987, Sofia.

193.Rakovsky, S. International Conference on "Regulation of Free Radical Reactions" - (Biomedical Aspects), Abstracts, No.140, 13-16 September, 1989.

194.Rakovsky, S., R. Fotty, V. Ershov. International Conference on "Regulation of Free Radical Reactions" - (Biomedical Aspects), Abstracts, No.141, 13-16 September, 1989.

195.Rakovsky, S., O Todorova, R Fotty, A. Terebenina, G. Borisov. International Conference on "Regulation of Free Radical Reactions" - (Biomedical Aspects), Abstracts, No.142, 13-16 September, 1989.

196.Haruf, M, E Serbinova, A Erin, L Ulhin, V Savov, S Rakovsky, V Kagan. International Conference on "Regulation of Free Radical Reactions" - (Biomedical Aspects), Abstracts, No.198, 13-16 September, 1989.

197.Aliev, MA, L. Ioffe. Second All-Union Conference "Ozone obtaining and application". Theses reports, p. 161 Moscow (1991).

198.Krivatkina, SD. Second All-Union Conference "Ozone obtaining and application". Theses reports, p. 156, Moscow (1991).

199.Gloor, M, B Lipphardt. *Z.Hautkr.*, 51, 3, 97 (1976).

200.Sweet, F. *Science(Washington-D.C.)*, 209 (4459), 931 (1980).

201.Richter, A. *J. Toxicol. Environ. Health*, 25, 3, 383 (1988).

202.Shinriki, N. *Eur. Pat.* Appl. EP 601,891, Cl A61K9/08 (1994).

203.Burleson, GR, TM Murray. *Appl. Microbiol.*, 29, 340 (1975).

204.Karzenelson, E, G Koerner, N Biederman. *Appl. Environ. Microbiol*, 37, 715 (1979).

205.J.Vaughn, J, YS Chen. *Appl. Environ. Microbiol.*, 53, 9, 2218 (1987).

206.Lee, RE. *Semin. Oncol.*, 1, 254 (1974).

207.Rokitansky. *Hospitalis*, 52, 643 (1982).

208.Stroczynski, Z Antoszewski. *Pol. Tyd. Lek.*, 47 (42-43), 964 (1992).

209.Turzynski, J Stroczynski. *Pol. Tyd. Lek.*,46 (37-39), 700-710 (1991).

210.Rilling, S. *"Ozone" Science Engineering Intern.*, 7, 4, 257 (1985).

211.Schulzs. *Lab. Anim.*, 20, 1, 41-48 (1986).

212.Pribuda, S. *Semana Med.*, 123, 26, 10-269 (1963).

213.Wolff. *Das medizinische Ozon. vfm Publications*, Heilderberg (1979).

214.Payr. *Munchner Medizinische Wochenschrift*, 82, 220 (1935).

215.Dorstewitz, H. *Kongres-bericht der artztlichen geselltschafen fur ozontherapie*, Baden-Baden (1981).

216.S.H.Stadlaender, SH. *Erfahrungsheilkunde*, 30, 4, 277 (1981).

217.Rokitansky. *Hospitalis*, 52, 711 (1982).

218.Rotter, M, H Mittermayer. *Wien Klin. Wochenschr.*, 86, 24, 776 (1974).

219.Washuttl, J. *Erfahrungsheilkunde*, 28, 811 (1979); 32, 815 (1983).

220.Pryor, W. *Chem. Biol. Interact.*, 79, 41 (1991).

221.Washuttl, J, R Viebahn. *"Ozone" Science Enginering Intern.*, 11, 4, 411 (1989).

222.Sweet, F. *Science*, 202, 931 (1980).

223.R.Viebahn, R, I Steiner. *"Ozone" Science Engineering Intern.*, 12, 65 (1989).

224.Brinkman, R, H Lamberts. *Nature*, 181, 1202 (1958).

225.Chow, C, M Mustafa. *Environ. Physiol. Biochem.*, 5, 142 (1975).

226.Brickley, R, J Hackney, K Clark. *Arch. Environ. Health*, 30, 40 (1975).

227.Goldstein, BD. *Arch. Environ. Health*, 26, 279 (1975).

228.Vender, R. *Toxicol. Ind. Health*, 10, 1/2, 53 (1994).

229.Lamberts, H, TS Veninga. *Lancet*, 1, 133 (1964).

230.Schulz, E, G Moore, E Calabrese. *Bull. Environ. Contam. Toxicol.*, 26, 273 (1981).

231.Chow, C, AL Tappel. *Arh. Environ. Health*, 26,205 (1973).

232.Sato, S, M Kawakami. *Am. Rev. Resp. Dis.*, 113, 809 (1976).

233.Chow, C, J Kanenko. *Environ. Res.*, 19, 49 (1979).
234.Chan, PC, RJ Kindya. *J.Biol.Chem.*, 252, 23, 8537 (1977).
235.A.Savino, A, ML Peterson. *Environ. Res.*, 15, 1, 65 (1978).
236.A.Canada, A, G Airriess. *Nutrit Res.*, 7, 797 (1987).
237.Murphy, S, H Davis. *Toxicol. Appl. Pharmacol.*, 6, 528 (1964).
238.Giri, S, M Hollinger. *Environ. Res.*, 21, 467 (1980).
239.Veninga, T, J Wagenaar. *Environ. Health Perspect.*, 39, 153 (1981).
240.Shimasaki, H, T Takatori. *Biochem. Biophys. Res. Comm.*, 68, 1256 (1976).
241.Jankovski, S, W Doroszkiewicz. *Complement Inflammation*, 7, 2, 57 (1990).
242.Jegier, Z. *J.Am. Ing. Hyg. Assoc.*, 137, 6, 329 (1974).
243.M.Kleinfeld, M, P Gielrnynb. *Am. J. Med. Sci.*, 231, 638 (1956).
244.Laperwerff, M. *Aerospace Med.*, 34, 479 (1963).
245.Trams, E, C Lauter. *Arch. Environ. Health.*, 24, 153 (1972).
246.Berney, WB, RS Dyer. *U.S. Environ. Prot. Agency OFF Res. Dev., EPA*, 1, PB-264,233,555 (1976).
247.Hackney, JD. From *Gov. Rep. Announce Index(U.S.)*, 77, 14, 98 (1977).
248.Takanashi, Y, T Miura. *J. Toxicol. Environ. Health*, 15, 855 (1985).
249.Kitanova, S., G. Koynova, S. Rakovsky, V. Kagan. *Xenobiotic Metabollism and Toxicity Workshop of Balkan Countries*, P34, March 10-21st 1991, Novi Sad, Yugoslavia
250.Kitanova, S, G Koynova, S Rakovsky, V. Kagan. Pross. II-nd USSR Conference "*Ozone - Obtainment and Application*". Moscow, p. 168, 30.01-1.02.1991.
251.Zelac, R, H Crompoy. *Environ. Res.*, 4, 325 (1971); 4, 262 (1971).
252.Vaughan, T, W Moorman. *Toxicol. Appl. Pharmacol.*, 20, 404 (1971).
253.Iwao, U. *Environ. Res.*, 48, 1, 78 (1989).
254.Fairchild, E, S Grahm. *Toxicol. Appl. Pharmacol.*, 6, 607 (1964).
255.Gordon, T, B Taylor, M Amdur. *Arch. Environ. Health*, 36,6, 284 (1981).
256.Eskerv, M, M Scheuchenzuber. *Environ. Res.*, 40, 274 (1976).
257.Criegee, R. *Rec. Chem. Prog.*, 18, 111 (1957).
258.Menzel, DB. *Adv. Mod. Environ. Toxicol.*, 5 , 183 (1982).
259.C.K.Chow, CK, MZ Hussain. *Exprl. Path.*, 25, 182 (1976).
260.J.A.Last, JA, DB Greenberg. *Toxicol. Appl. Pharm.*, 51, 247 (1979).
261.Hussain, ZM, CE Goss. *Life Sci.*, 18, 897, (1976).
262.Veninga, TS. *Environ. Health Perspect.*, 39, 153 (1981).
263.Chow, CK, CE Goss. *J.Toxicol. Environ. Health*, 3, 5/6, 877 (1977).
264.Saymor. *Arch. Environ. Health*, 29, 3, 164 (1974).
265.Mohammad, D Anthany. *J.Chest*, 66, 1, 165, suppl. pt 21 (1974).
266.Summer, W. *Process Biochem.*, 11, 7, 26 (1976).
267.Plopper, CG. *Toxicol. Appl. Pharmacol.*, 121, 1, 124 (1994).
268.Derbal, E Kimberly. *Environ. Health Perspect. Suppl.*, 102, 61 (1994).
269.Kelly, FJ. *Hum.Exp. Toxicol.*, 13, 6, 407 (1994).
270.Drechsler, P, M Deborah. *Exp. Geront.*, 30, 1, 65 (1995).
271.Miller, FJ, JH Oventon, RH Kaskot. *Tox. and Appl. Pharm.*, 79, 11 (1985).
272.Stock, TH, DJ Kotchmar. *J.Air Pollut. Contr. Assoc.*, 35, 1266 (1985).
273.Dillar, C, N Urribarri. *Arch. Environ. Health*, 25, 6, 426 (1972).
274.Mohammad, D Anthony. *J.Lab. Clin. Med.*, 82, 3, 357 (1973).
275.Peister, A. *Presse Term. Clim.*, 1, 58 (1977).
276.Godstein, B, O Belchum. *Toxicol. Appl. Pharmacol.*, 27, 2, 330 (1974).
277.Elaine, S, D Dziedzic. *Toxicol. Lett.*, 51, 125 (1990).
278.Schesinger, RB, KE Driscoll. *J. Toxicol. Environ. Health*, 20,125 (1987).
279.Abraham, WM, W Sielezak. *Eur. J. Respir. Dis.*, 68, 114 (1986).
280.W.M.Foster, WM, DL Costa, EG Langenbak, *J. Appl. Physiol.*, 63, 996 (1987).
281.Boorman, GA, LW Schwartz, DL Dungworth. *Lab. Invest.*, 43, 108 (1980).
282.K.Nikua, K, DW Wilson, SN Giri. *Am.J.Pathol.*, 131, 373 (1988).
283.Harkema, JR, CG Plopper, DM Hyde. *Am.J.Pathol.* ,128, 29, 44 (1987).
284.Kirkpatrik, JR, RF Henderson. *Exp. Lung. Res.* 15, 1 (1989).

285.Schlesinger, RB. *J.Toxicol. Environ. Health*, 21, 27 (1987).
286.Lum, H, DG Plopper. *Exp. Lung Res.*, 5, 61 (1983).
287.Michael, WF. *J. Appl. Physiol.*, 75, 5, 1938 (1993).
288.Driscoll KE, RB Schleinzenger. *Toxicol. Appl. Pharmacol.*, 93, 312 (1988).
289.Kimura, I, E Godstein. *J. Infect. Dic.*, 143, 247 (1981).
290.Sherwood, RL, E Goldstein. *Environ. Res.*, 41, 378 (1986).
291.Bleavins, MR, HJ Whith. *Toxicologist*, 7, 129 (1987).
292.Blevins, MR, D Dziedzic. *Toxicologist*, 8, 198 (1988).
293.Sell, S. *Immunology, Immunopathology and Immunity*. Elsevier. New York (1987).
294.Aranyi, C, SC Vana, PT Thomas. *J.Toxicol. Environ. Health*, 12, 55 (1983).
295.Eskew, ML, WJ Scheuchenzuber, RW Scholz. *Environ. Res.*, 40, 274 (1986).
296.Amoruso, MA, G With. *Life Sci.*, 28, 20, 2215 (1981).
297.Nikola, KJ, DW Wilson, DL Dungworth, CG Plopper. *Toxicol. Appl. Pharmacol.*, 93, 394 (1989).
298.Clark, T. *Toxicol. Letter*, 72,1-3, 279 (1994).
299.Stephens, RJ, MF Sloan, G Reeman. *Am. J. Pathol.*, 74, 31 (1973).
300.Evans, MJ, N.P.Dekker. *Exp. Mol. Phatol.*, 42, 366 (1985).
301.Plopper, CG, CK Chow, DL Dunworth. *Exp. Mol. Pathol.*, 29, 400 (1978).
302.Nikula, KJ, DW Wilson, DL Dungworth, CG Plopper. *Am. J. Pathol.*, 131,373 (1988).
303.Hussain, MZ, CE Cross. *Life Sci.*, 18, 897, 904 (1976).
304.Last, JA, DB Greenberg. *Toxicol. Appl. Pharmacol.*, 51, 257 (1979).
305.Fukase O, K Hashimoto. *Nippon Eiseigaku Zassi.*, 37, 694 (1982).
306.Kramer F. *Erfahrungsheilkunde*, 24, 120 (1975).
307.Kramer F. *Ozonachrichten*, 2, 65 (1983).
308.Turk, R. *Erfahrungsheilkunde*, 25, 177 (1976).
309.Salzer, N, M Metka, M Shonbauer. *Erhfahrungsheilkunde*, 32, 9731 (1983).
310.Werkmeister H. *Erfahrungsheilkunde*, 25, 180 (1976)
311.Geabelin K. *Seifen, Oile, Fette, Wasche*, 112, 1, 1718 (1986).
312.Eur. Pat. 222 309 (1987).
313.Ger.Offen 3,806,203 Cl A 61 L9/01.
314.US Pat. 5,266,275 Cl 422-116 A 61 L2/20.
315.German Pat. 66 489, Cl A611 (1969).
316.Gal G . *Intern. J. of Artifical Organs*, 15, 8,461 (1992).
317.Hiromichi. JP 05329, 196(93,329,196) Cl A 61 L 2/18 (1993).

INDEX

H

I

J

K

L

M

N

O

P

R